Interdisciplinary Applied Mathematics

Volume 43

More information about this series at http://www.springer.com/series/1390

Geneviève Dupont • Martin Falcke • Vivien Kirk
James Sneyd

Models of Calcium Signalling

Geneviève Dupont
Unité de Chronobiologie Théorique
Université Libre de Bruxelles
Brussels, Belgium

Martin Falcke
Max Delbrück Center
for Molecular Medicine
Berlin, Germany

Vivien Kirk
Department of Mathematics
University of Auckland
Auckland, New Zealand

James Sneyd
Department of Mathematics
University of Auckland
Auckland, New Zealand

ISSN 0939-6047 ISSN 2196-9973 (electronic)
Interdisciplinary Applied Mathematics
ISBN 978-3-319-29645-6 ISBN 978-3-319-29647-0 (eBook)
DOI 10.1007/978-3-319-29647-0

Library of Congress Control Number: 2016933449

Mathematics Subject Classification (2010): 34C15, 34C23, 34E13, 37H10, 92B25, 92C05, 92C30, 92C42, 92C45

Printed on acid-free paper

This Springer imprint is published by Springer Nature
The registered company is Springer International Publishing AG Switzerland

Michael J. Sanderson, one of my closest friends and colleagues, died suddenly and unexpectedly on the 24th of April, 2016, while this book was in production.

I worked with Mike for the past 25 years on a range of problems to do with the dynamics of calcium, and he contributed an enormous amount to this book, both directly and indirectly.

Mike was a wonderfully talented scientist and a tremendous inspiration in so many ways. His death is a great loss, not just to his family, his friends, and to me, but to the entire field of calcium signalling.

Indeed, all the authors of this book knew Mike, and respected him highly for his experimental work and his understanding of, and sympathy with, modelling.

Geneviève, Martin, Vivien and I therefore dedicate this book to the memory of Michael Sanderson. It is a source of great sadness that Mike was never able to hold a completed copy of this book (and make fun of me for all those equations), but a source of great pride that I was able to count Mike as my colleague and my very dear friend.

Auckland, New Zealand
28 April, 2016

James Sneyd

Preface

The interactions between mathematics and physiology have a long and illustrious history, of which the study of calcium signalling is but a small chapter. Nevertheless, for those involved, it is a fascinating chapter, in which a wide range of mathematical and computational techniques are used to study a problem of central physiological interest; the spatio-temporal dynamics of the concentration of free intracellular calcium ions.

Since this calcium concentration is central to a vast range of intracellular signalling pathways, and since its spatiotemporal properties are so complex, it is an ideal subject for the joint study of experimentalists and theoreticians. And so, perhaps unsurprisingly, for the past 30 years or more, the study of calcium signalling has been a popular and active area of investigation for scientists from a wide range of disciplines. Indeed, we, the authors, show this by example; one of us is a chemist, one a physicist, one a mathematician, and one a modeller. Our interests have converged, along quite different paths, to the same interdisciplinary region, one of great richness and variety.

However, what we are not is experimentalists. This is not by any means to suggest that experimental approaches are not important in the study of calcium signalling. We would not be so foolish. To the contrary, each one of us has worked very closely with our experimental colleagues, as we recognise that, without data, models are worthless. We also believe that the main aims of modelling are to help the interpretation of observations, to make testable predictions, and to guide future experimental investigation.

Nevertheless, in a single book, it is simply not possible to cover the entire field of calcium signalling, in both its experimental aspects and the associated modelling. We have thus chosen to take a somewhat narrow view of the field, focusing entirely on the modelling aspects, as we try to show how quantitative models of calcium signalling are constructed, how they can help advance our understanding, and how they can help build bridges between seemingly unrelated observations. We refer to the experimental literature often, of course, and we expect our readers to do so also. However, our references to the experimental literature are neither comprehensive nor complete.

Because of the interdisciplinary nature of calcium modelling, there are some unavoidable prerequisites for the reader. We present no advanced mathematical theory, but the reader will be assumed to be familiar with the basic properties of ordinary and partial differential equations. Conversely, some knowledge of elementary physiology will be assumed. Concepts such as cells, the cytoplasm, the plasma membrane, and elementary enzyme biochemistry will be needed and will not be explained in detail.

It is true that readers unfamiliar with the basic physiology will find it easier to read and learn what they need than will a reader who is unfamiliar with the basic theory of differential equations. Such is the nature, unfortunately, of physiological modelling. Nevertheless, any reader who has taken mathematics, physics, or engineering to the third year level at university will be well equipped to read this book.

To learn the necessary mathematics, we suggest referring to almost any basic text on ordinary and partial differential equations. Such texts are commonly used in engineering and applied math courses; we ourselves have used Kreyszig (1994) and Kaplan (1981). Similarly, the background biology can be most easily learned by referring to one of the many cell physiology textbooks, such as Levy et al (2006) or Alberts et al (2002). Finally, there are a number of books that describe the modelling approaches used in quantitative biology and physiology. In this book we refer often to Keener and Sneyd (2008), but Murray (2003), Fall et al (2002), Hoppensteadt and Peskin (2002), Beuter et al (2003), and Edelstein-Keshet (1988) are all excellent references.

This book is not designed as a textbook. It could well be used as additional reading material for a graduate, or advanced undergraduate, course, but is not designed to stand alone in such a context. It is aimed instead at any researcher, from any discipline, who wishes to learn more about how mathematical and computational modelling may be used in the study of intracellular calcium signalling.

Indeed, the methods we use here, and the interdisciplinary approach we present, are not confined merely to calcium signalling, but are applicable to any field of study where biological importance is combined with dynamic complexity. Thus, our wider goal is to show how the combination of modelling and experiment can lead to understanding that would be difficult, if not impossible, to gain using only a single viewpoint, and to inspire others to take a similarly broad view of interdisciplinary science.

Brussels, Belgium — Geneviève Dupont
Berlin, Germany — Martin Falcke
Auckland, New Zealand — Vivien Kirk
Auckland, New Zealand — James Sneyd
November 2015

Acknowledgements

This book could not have been written without the help, over many years, of many different people. We have relied on the expertise (and the patience!) of our experimental colleagues and have been inspired by our fellow modellers. In particular, we thank Michael Berridge, Don Bers, Frank Borgese, Mark Cannell, John Challiss, Philippe Champeil, David Clapham, Laurent Combettes, Edmund Crampin, Paul De Koninck, Jean-Francois Dufour, Rémi Dumollard, Christophe Erneux, David Gall, Albert Goldbeter, Didier Gonze, Allan Herbison, James Keener, Jean-Christophe Leloup, Luc Leybaert, Hinke Osinga, Ian Parker, Sandip Patel, Jose Puglisi, Michael Sanderson, Trevor Shuttleworth, Stéphane Swillens, Akihiko Tanimura, Colin Taylor, Rudiger Thul, Thierry Tordjmann, Guy Tran Van Nhieu, Je-Chiang Tsai, Larry Wagner, Martin Wechselberger, and David Yule.

We also thank 25 years (or thereabouts) of PhD students, Master's students, and post-doctoral researchers, without whose efforts much of this work would never have been done. They are too numerous to mention individually, but their efforts are not forgotten.

Contents

Acronyms

5-HT	5-hydroxytryptamine. Also called serotonin. 5-HT is a common neurotransmitter with a multitude of functions. It binds to G-protein-coupled receptors (GPCR) to inhibit adenylate cyclase or to activate phospholipase C (PLC), which leads to production of IP_3 and release of Ca^{2+} from internal stores.
ACh	Acetylcholine. A common neurotransmitter and the first to be identified as such. It can bind to ionotropic receptors (that form an ion channel directly) or to metabotropic receptors that lead to the formation of second messengers such as IP_3.
ADP	Adenosine diphosphate. See ATP.
AMP	Adenosine monophosphate. See ATP.
ARC	Arachidonate-regulated Ca^{2+} channels. These channels are situated in the plasma membrane and open in response to arachidonic acid, which is formed from diacylglycerol (DAG) by DAG lipase. Since DAG is produced by the cleavage of PIP_2 by phospholipase C (PLC), DAG is formed whenever IP_3 is made. ARC channels are thus one pathway for Ca^{2+} entry through receptor-operated Ca^{2+} channels (ROCC).
ASM	Airway smooth muscle. A type of smooth muscle that surrounds the airways in the lung.
ATP	Adenosine triphosphate. The ubiquitous molecule necessary for intracellular energy storage and transfer. It consists of an adenine molecule connected to a sugar ring (ribose), with three phosphate groups connected to the 5' carbon atom of the ribose. (The adenine and ribose together form adenosine). If only two phosphate groups are attached, the molecule is called adenosine diphosphate

	(ADP) and with one phosphate group attached is called adenosine monophosphate (AMP).
ATPase	An enzyme that catalyses the hydrolysis of ATP into ADP and a free phosphate, releasing energy in the process, energy that can be used to drive other reactions, such as the transport of Ca^{2+} up its concentration gradient.
BAPTA	1,2-bis(o-aminophenoxy)ethane-N,N,N',N'-tetraacetic acid. A molecule that binds Ca^{2+} and is commonly used experimentally as an exogenous Ca^{2+} buffer. In comparison to another commonly used exogenous buffer, EGTA, BAPTA binds Ca^{2+} quickly.
CaM	Calmodulin, an abbreviation for calcium-modulated protein. A Ca^{2+}-binding protein present in all eukaryotic cells. When bound to Ca^{2+}, it controls a wide variety of cellular processes via its action on target proteins. Many of these target proteins are unable to bind Ca^{2+} directly, and thus calmodulin is a crucial Ca^{2+} sensor and signal transducer.
CaMKII	Ca^{2+}/calmodulin-dependent protein kinase II. Often written as CaM kinase II. A protein kinase (i.e. a protein that phosphorylates a target protein) that is regulated by Ca^{2+}/CaM.
cAMP	Cyclic adenosine monophosphate. An important intracellular second messenger, cAMP is synthesised from ATP by adenylate cyclase.
CCh	Carbachol. An agonist that stimulates the ACh receptor.
CCK	Cholecystokinin. A hormone, secreted in the duodenum, and responsible for stimulating the digestion of fat and protein. The word is derived from a combination of Greek roots: *chole*, meaning "bile"; *cysto*, meaning "sac"; *kinin*, meaning "move". Hence, cholecystokinin means "move the bile sac", i.e. "move the gallbladder", which is not inappropriate.
cGMP	Cyclic guanosine monophosphate. Like cAMP, cGMP is an important intracellular second messenger. It is synthesised from GTP by guanylate cyclase.
CHO cell	Chinese Hamster Ovary cell. A cell line, first derived in 1957 and used extensively since then in experimental work.
CICR	Calcium-induced calcium release. The autocatalytic release of Ca^{2+} from the endoplasmic or sarcoplasmic reticulum through IP_3 receptors or ryanodine receptors. CICR causes the fast release of large amounts of Ca^{2+} from internal stores and is the basis for Ca^{2+} oscillations and waves in a wide variety of cell types.

COS cell — A cell line derived from monkey kidney tissue. The acronym "COS" comes from the cells being **C**V-1 (simian) in **O**rigin, and carrying the **S**V40 genetic material.

CPA — Cyclopiazonic acid. A SERCA channel blocker.

CRU — Calcium release unit. A mathematical construct that describes, in a simplified form, a combination of L-type Ca^{2+} channels, a diadic cleft, and RyR.

CSQ — Calsequestrin. A Ca^{2+} buffer in the sarcoplasmic reticulum, with a high Ca^{2+} binding capacity. Each calsequestrin protein can bind up to 50 Ca^{2+} ions.

CV — Coefficient of variation. The ratio of the standard deviation to the mean, and a measure of how spread out a distribution is.

DAG — Diacylglycerol. One of the products of the cleavage of PIP_2 by activated PLC. The other product is IP_3.

DHPR — Dihydropyridine receptor. Another name for the L-type Ca^{2+} channels in cardiac and skeletal muscle.

DT40-3KO cells — DT40 cells are a cell line derived from chicken lymphocytes. DT40-3KO cells are DT40 cells for which both copies of the gene for all three IPR subtypes have been disrupted. They thus provide a clean genetic background for expression of only a single subtype of IPR.

ECF — Extracellular fluid.

EGTA — Ethylene glycol tetraacetic acid. A Ca^{2+}-binding molecule that is commonly used experimentally as an exogenous buffer. EGTA binds Ca^{2+} more slowly than the other commonly used Ca^{2+} buffer, BAPTA.

EPSP — Excitatory post-synaptic potential.

ER — Endoplasmic reticulum. An internal cellular compartment in non-muscle cells that acts as an important Ca^{2+} store (as well as having a variety of other important functions, less relevant to this book). The analogous compartment in muscle cells is called the sarcoplasmic reticulum (SR).

GDP — Guanosine diphosphate. See GTP.

GMP — Guanosine monophosphate. See GTP.

GFP — Green fluorescent protein. A protein, originally derived from a jellyfish, that exhibits bright green fluorescence when exposed to blue or ultraviolet light. Since the GFP gene can now be incorporated relatively easily into a wide range of cells in living animals and can be targeted to specific cell types, GFP is now widely used as a biosensor. The 2008 Nobel Prize in Chemistry was awarded to Martin Chalfie, Osamu Shimomura, and Roger Y. Tsien for their discovery and development of the green fluorescent protein.

GHK	Goldman-Hodgkin-Katz. Usually used in the context of the GHK equation, which describes a common model for current flow through an ion channel.
GPCR	G-protein-coupled receptor. A receptor in the cell membrane that converts an extracellular signal into an intracellular signal via coupling to G proteins.
GSPT	Geometric singular perturbation theory. A mathematically rigorous technique that uses limiting cases of a model to make predictions about full model dynamics.
GTP	Guanosine triphosphate. Similar to ATP except that adenine is replaced by guanine. The combination of guanine and ribose is called guanosine. GTP is used for synthesis of RNA and DNA and is also a source of energy, like ATP.
HEK 293 cell	A cell line derived from human embryonic kidney cells.
HeLa cell	A cell line derived from cervical cancer cells taken from Henrietta Lacks, hence their acronym.
HSY cell	A cell line derived from human parotid epithelial cells.
IPI	Interpuff interval. The time interval between successive Ca^{2+} puffs (punctate release of Ca^{2+} from a cluster of IP_3 receptors).
IPR	Inositol trisphosphate receptors. Proteins in the membrane of the endoplasmic/sarcoplasmic reticulum that bind IP_3, leading to the release of Ca^{2+} from the ER/SR.
IP_3	Inositol 1,4,5-trisphosphate. A second messenger responsible for the release of intracellular Ca^{2+} from internal stores, through IP_3 receptors.
ISI	Interspike interval. The time interval between successive Ca^{2+} spikes (whole-cell Ca^{2+} responses).
JSR	Junctional sarcoplasmic reticulum in cardiac cells. That part of the sarcoplasmic reticulum that is close to the diadic cleft.
LCC	L-type Ca^{2+} channel. Also called a dihydropyridine receptor (DHPR).
LTD	Long-term depression. Activity-dependent reduction of synaptic strength that lasts for hours (or longer).
LTP	Long-term potentiation. Activity-dependent increase of synaptic strength that lasts for hours (or longer).
MCh	Methacholine. A muscarinic receptor agonist.
MCU	Mitochondrial Ca^{2+} uniporter. The protein that transports Ca^{2+} into the mitochondria.
MDCK cells	Madin-Darby canine kidney cells. A cell line derived in 1958 by S. H. Madin and N. B. Darby from the kidney tissue of an adult female cocker spaniel.
mGlu5	Metabotropic glutamate receptor 5. Metabotropic glutamate receptors are G-protein-coupled receptors and are

	encoded by 8 different genes, GRM1–GRM8. mGlu5 is encoded by GRM5.
MLCK	Myosin light chain kinase. A Ca^{2+}-dependent kinase that phosphorylates the myosin light chain in smooth muscle, allowing for the myosin to bind actin and create force in the crossbridge cycle.
MLCP	Myosin light chain phosphatase. A phosphatase (Ca^{2+}-dependent in some species) that dephosphorylates the myosin light chain in smooth muscle, preventing the myosin binding to actin and thus preventing the generation of force in the crossbridge cycle.
MMO	Mixed-mode oscillations. Small, subthreshold oscillations interspersed with large amplitude spikes.
NAADP	Nicotinic acid adenine dinucleotide phosphate. A second messenger that releases Ca^{2+} from the acidic organelles.
NCX	Na^+/Ca^{2+} exchanger. An antiport transporter that, under normal conditions, removes Ca^{2+} from the cytoplasm at the expense of Na^+ entry.
NCLX	Mitochondrial Na^+/Ca^{2+} exchanger. An isoform of the NCX and the principal pathway for Ca^{2+} efflux from the mitochondria.
NMDA receptor	N-methyl-D-aspartate receptor. A type of ionotropic glutamate receptor, found in neurons, that binds the agonist N-methyl-D-aspartate.
Orai	A class of membrane Ca^{2+} channels that are encoded by the orai genes (with three homologs; Orai1, Orai2, and Orai3). Orai channels interact with STIM proteins, which are themselves activated by depletion of Ca^{2+} in the ER. Thus, Orai channels are a principal component of store-operated Ca^{2+} entry.
PKA	Protein kinase A. Also known as cAMP-dependent protein kinase. A kinase with multiple cellular functions that is active only when cAMP is present.
PKC	Protein kinase C. A large family of protein kinase enzymes that play crucial roles in many signal transduction cascades.
PIP_2	Phosphatidylinositol 4,5-bisphosphate. Also known as PtdIns(4,5)P_2. A membrane-bound phospholipid that is split by PLC into IP_3 and DAG, in one of the early steps of the IP_3 signalling pathway.
PLA cells	Processed lipoaspirate cells. Fibroblast-like stem cells isolated from human adipose tissue.
PLC	Phospholipase C. The enzyme that splits PIP_2 into DAG and IP_3. There are 13 kinds of PLC, classified into 6 iso-

	types, and they are mostly activated by G-protein-coupled receptors.
PMCA	Plasma membrane Ca^{2+} ATPase. A membrane ATPase Ca^{2+} pump that removes Ca^{2+} from the cell.
PTH	Parathyroid hormone. A hormone that promotes uptake of Ca^{2+} from the bones and inhibits secretion of Ca^{2+} in urine.
RIN-m5F cell	A rat insulinoma cell line from pancreatic islets.
ROCC	Receptor-operated Ca^{2+} channels. Membrane Ca^{2+} channels that are opened, not by depletion of the internal Ca^{2+} stores, but indirectly by the activated G-protein-coupled receptor. The molecular identity of ROCC remains unclear, but they are often activated by arachidonic acid (thus, ARC are a type of ROCC).
RyR	Ryanodine receptor. A Ca^{2+} channel (with three isoforms) on the ER/SR membrane, that is blocked by the alkaloid ryanodine. RyR are similar in structure and function to IP_3 receptors, and are the principal mediators of Ca^{2+}-induced Ca^{2+} release in muscle cells.
SERCA	Sarcoplasmic/endoplasmic reticulum Ca^{2+} ATPase. A Ca^{2+} ATPase pump that transports Ca^{2+} up its concentration gradient from the cytoplasm to the ER/SR.
Sf9 cell	A cell line derived from *Spodoptera frugiperda*, the larval stage of the Fall armyworm moth.
SH-SY5Y cell	A cell line derived from human neuroblastoma cells.
SOCC	Store-operated Ca^{2+} channels. Membrane Ca^{2+} channels that are opened by depletion of the Ca^{2+} concentration in the ER/SR. In many cell types (but probably not all), SOCC are Orai channels that have been activated by STIM.
SPCA	Secretory pathway Ca^{2+} ATPase. The secretory pathway is the way in which proteins are secreted from a cell and involves the rough endoplasmic reticulum, the Golgi apparatus, and secretory vesicles.
SR	Sarcoplasmic reticulum. An internal cellular compartment in muscle cells that acts as an important Ca^{2+} store (as well as having a variety of other important functions, less relevant to this book). The analogous compartment in non-muscle cells is called the endoplasmic reticulum (ER).
STDP	Spike-timing-dependent plasticity. Changes in synaptic strength caused by nearly simultaneous spikes of the pre-synaptic and post-synaptic neurons.
STIM	Stromal interaction molecule. A protein, situated in the ER/SR membrane, that senses depletion of Ca^{2+} in the ER/SR.

TIRF	Total internal reflection fluorescence. A type of microscopy that can be used to visualise very thin regions of the cell close to the cell membrane.
TRPC channels	Transient receptor potential (canonical) channels. A large family of membrane Ca^{2+} channels (although not specific to Ca^{2+}) that are regulated by a variety of molecules, including arachidonic acid. TRPC channels are thought to form receptor-operated Ca^{2+} channels, but may also be involved in store-operated Ca^{2+} entry.
TPC	Two-pore channel. NAADP-regulated channels that release Ca^{2+} from acidic organelles.
VGCC	Voltage-gated Ca^{2+} channels. Membrane Ca^{2+} channels that open in response to depolarisation of the cell membrane.

Part I
Basic Theory

Chapter 1
Some Background Physiology

1.1 Introduction

Calcium physiology is a vast field, far too large for us to do it justice in this book. Here, instead, we shall be concerned almost entirely with the physiology and dynamical behaviour of ionised intracellular calcium (Ca^{2+}). In almost every cell type, the cytosolic concentration of Ca^{2+} exhibits complex spatiotemporal behaviour, ranging from stochastic spiking to regular oscillations, to periodic waves, and to spiral waves. Since these so-called Ca^{2+} oscillations and waves control a vast array of cellular functions, they have been studied intensively for at least the past 40 years. However, although nothing can be learned without a solid experimental foundation, phenomena of such complexity simply cannot be well understood without the parallel use of quantitative models. It is thus in the realm of Ca^{2+} dynamics that mathematical modelling and physiology have come to have a particularly close and mutually beneficial relationship. It is this relationship that we explore here.

The problem of whole-body Ca^{2+} homeostasis is a fascinating topic in itself, but one that we omit as being rather outside our scope. Nevertheless, a brief discussion of this topic is useful in order to put intracellular Ca^{2+} into context.

In total, the human body contains approximately 1300 g of Ca^{2+}, of which about 99% is in the bones, 1% is inside cells, and about 0.1% is in the extracellular fluid (ECF). The movement of Ca^{2+} between these three compartments, as well as its absorption and excretion, is under the control of three hormones; parathyroid hormone (PTH), calcitriol, and calcitonin (see Fig. 1.1). PTH is secreted by the parathyroid gland and stimulates resorption of Ca^{2+} from the bones (i.e., movement of Ca^{2+} from the bones into the ECF), increases Ca^{2+} reabsorption by the kidneys, and stimulates the production of calcitriol. Calcitriol, a metabolite of vitamin D_3 and synthesised mainly in the proximal tubule, increases absorption of Ca^{2+} by the gastrointestinal tract and,

G. Dupont et al., *Models of Calcium Signalling*, Interdisciplinary Applied Mathematics 43, DOI 10.1007/978-3-319-29647-0_1

like PTH, stimulates resorption from bone. Calcitonin, on the other hand, increases the movement of Ca^{2+} into bone. All three hormones inhibit the excretion of Ca^{2+} by the kidneys, by promoting Ca^{2+} reabsorption, with PTH having the greatest effect.

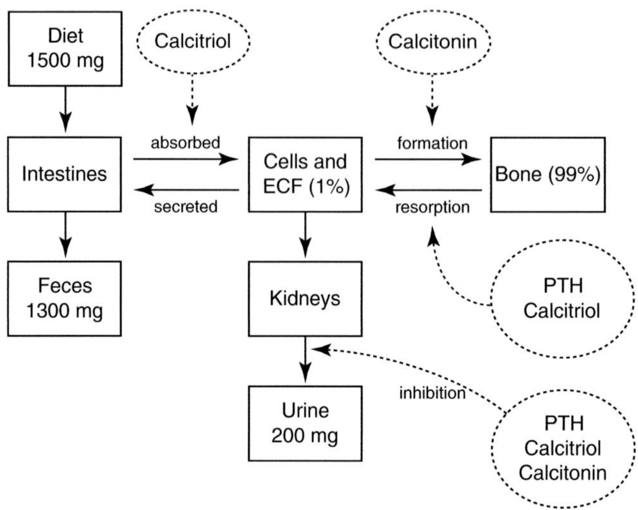

Fig. 1.1 Whole-body Ca^{2+} homeostasis. Most Ca^{2+} is stored in the bones, with only about 1% stored in the extracellular fluid (ECF) and cells. Movement of Ca^{2+} between the intestines, the ECF, and bone is controlled by three hormones: calcitriol, calcitonin, and parathyroid hormone (PTH).

Our concern here is primarily with intracellular Ca^{2+}. The ECF has a Ca^{2+} concentration of around 1 mM, while active (i.e., energy consuming) pumps and exchangers keep the cytoplasmic Ca^{2+} concentration about four orders of magnitude lower, at around 0.1 μM. Similarly, internal cellular compartments such as the endoplasmic/sarcoplasmic reticulum (ER/SR) and lysosomes have a high internal Ca^{2+} concentration. The cell cytoplasm is thus under a strong Ca^{2+} "pressure". Any opening of Ca^{2+} channels, either in the plasma membrane or in the membrane of the internal compartments, results in a large flux of Ca^{2+} into the cytoplasm. This arrangement has both advantages and disadvantages. One advantage is that a cell can quickly raise its internal $[Ca^{2+}]$ merely by opening Ca^{2+} channels in the plasma membrane or an internal store, and can equally quickly reduce the cytoplasmic $[Ca^{2+}]$ by closing these channels and using pumps and exchangers to remove Ca^{2+}. Thus, the cytoplasmic $[Ca^{2+}]$ is under very tight temporal control. One disadvantage is that such an arrangement requires the continual expenditure of energy to maintain the low cytoplasmic $[Ca^{2+}]$.

There is a very good reason for why cells expend considerable amounts of energy to keep their cytosolic $[Ca^{2+}]$ low. Prolonged high cytoplasmic $[Ca^{2+}]$ is highly toxic. For one thing, a maintained high $[Ca^{2+}]$ would result in the precipitation of Ca^{2+} phosphates, highly damaging to the cell.

For another, there are many Ca^{2+}-sensitive effector molecules in the cytoplasm (such as kinases and phosphatases), which control a wide array of cellular functions such as differentiation, secretion, gene transcription, and proliferation, and prolonged high $[Ca^{2+}]$ disrupts these vital control mechanisms. In addition, prolonged high $[Ca^{2+}]$ is implicated in both the initiation and execution phases of apoptosis, whereby cells destroy themselves with endogenous proteases and endonucleases (Orrenius et al, 2003).

Despite this inherent Ca^{2+} toxicity, almost every cell type uses Ca^{2+} as a second messenger, to control a wide array of cellular functions. Ca^{2+} is a necessary component of excitation-contraction coupling in muscle cells, in excitation-secretion coupling in synapses, for exocytosis, and for fluid transport in exocrine epithelia. It controls plasticity in both pre-synaptic and post-synaptic neurons, it controls gene regulation and differentiation in every cell type, and it is necessary for cell movement. It is one of the most important and ubiquitous cellular second messengers.

This raises the questions of why and how Ca^{2+} is used as a signal. For the "why", there is of course no definite answer, but some chemical considerations can throw some light on the subject (Carafoli, 1987; Jaiswal, 2001; Ochiai, 1991). The most abundant cations in biological systems, as in sea water, are Na^+, K^+, Ca^{2+}, and Mg^{2+}. Monovalent cations are less suitable as signalling molecules, since they normally form loose complexes with proteins. Hence, Na^+ and K^+ are widely used to control voltage in excitable cells, but not as chemical signals per se. This leaves the question of why evolution selected Ca^{2+} instead of Mg^{2+} as an intracellular messenger. One difference between the ions is that Ca^{2+} has a larger ionic radius. In addition, its electronic structure allows for a variable coordination number (i.e., a variable number of nearest neighbours in large molecular assemblies), which is not the case for Mg^{2+}. Both properties allow Ca^{2+} to bind to a larger variety of substrates, and to do so in a thermodynamically more favourable way. In addition, Ca^{2+} does so faster, as protein binding and dissociation are about 100 times more rapid for Ca^{2+} than for Mg^{2+}. Finally, it might also be relevant that Ca^{2+} has a greater affinity for carboxylate oxygens while Mg^{2+} prefers nitrogen ligands. As the frequency of expression of carboxylate-containing amino acids is slightly larger than that of nitrogen-containing ones, this also may have favoured the use of Ca^{2+} instead of Mg^{2+}.

In any event, in order to use Ca^{2+} as an intracellular signalling messenger, cells have evolved mechanisms by which they can tightly regulate their cytoplasmic $[Ca^{2+}]$, both in space and in time. Cells can raise $[Ca^{2+}]$ very quickly in small regions of the cytoplasm, while keeping $[Ca^{2+}]$ low in other areas, and can remove Ca^{2+} from these areas just as quickly. These tightly regulated increases and decreases in local $[Ca^{2+}]$ – which can, in some circumstances, be converted to global whole-cell responses – are the mechanisms by which cells use Ca^{2+} as a controlling signal.

1.2 Common Features of Calcium Dynamics: The Calcium Toolbox

Although there is enormous variety between cell types, when it comes to Ca^{2+} dynamics there are also many common features (Fig. 1.2). This has led to the concept of a Ca^{2+} "toolkit", or "toolbox" (Berridge et al, 2000), i.e.,

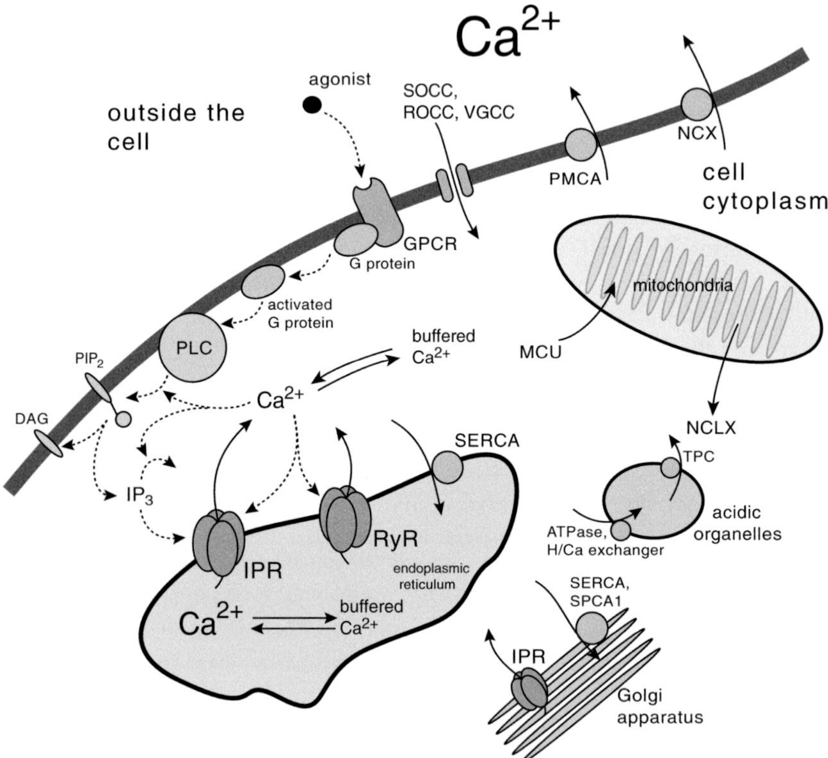

Fig. 1.2 Diagram of some of the major components involved in the control of cytoplasmic Ca^{2+} concentration. Calcium fluxes are denoted by solid lines. Binding of agonist to a G protein-coupled receptor (GPCR) leads to activation of a G protein, and subsequent activation of phospholipase C (PLC). This cleaves phosphatidylinositol bisphosphate (PIP_2) into diacylglycerol (DAG) and inositol 1,4,5-trisphosphate (IP_3), which is free to diffuse through the cell cytoplasm. When IP_3 binds to IP_3 receptors (IPR), which are mostly on the endoplasmic reticulum (ER) membrane, it causes the release of Ca^{2+} from the ER. Calcium can also be released from the ER through ryanodine receptors (RyR). Both IPR and RyR are modulated by Ca^{2+}, as are the rates of IP_3 production and degradation. Calcium enters the cytoplasm via the mitochondrial Na^+/Ca^{2+} exchanger (NCLX), and from the outside through store-operated Ca^{2+} channels (SOCC), receptor-operated Ca^{2+} channels (ROCC), and voltage-gated Ca^{2+} channels (VGCC). Calcium can be removed from the cytoplasm by SERCA pumps, plasma membrane Ca^{2+} ATPases (PMCA), Na^+/Ca^{2+} exchangers in the plasma membrane (NCX), or by the mitochondrial Ca^{2+} uniporter (MCU). Less is known about Ca^{2+} transport in and out of the Golgi apparatus and the acidic organelles, but some established mechanisms are indicated. TPC is the two-pore channel while SPCA is the secretory pathway Ca^{2+} ATPase.

a relatively limited number of components which are common to many cell types, but which can be expressed at different levels. Cells can then express whichever toolbox components are needed to carry out a specific function. Such expression can, and will, vary both in space and time, giving flexibility in the Ca^{2+} signalling pathways.

The toolbox components are described in greater quantitative detail in Chapter 2. Here, we give only a brief overview.

1.2.1 Agonists, Receptors, and Second Messengers

In almost all (eukaryotic) cell types, changes in intracellular $[Ca^{2+}]$ can be initiated by the binding of agonists (such as neurotransmitters or hormones) to receptors on the surface of the cell. These receptors are coupled to G proteins, forming a complex that transmits signals from the outside to the inside of the cell (Fig. 1.2). There are many types of G protein-coupled receptors (GPCR), and they connect to many intracellular signalling pathways (Luttrell, 2006). To underscore their importance, it is estimated that about 3% of the human genome codes for GPCR, and from 40–60% of all pharmaceuticals target GPCR (Howard et al, 2001). GPCR are transmembrane receptors; the binding of the agonist to the extracellular domain causes a change in conformation that activates the G protein on the intracellular side of the membrane. GPCR activate two principal signal transduction pathways, one associated with the formation of cyclic adenosine monophosphate (cAMP), the other, the phosphatidylinositol signalling pathway, associated with the formation of inositol 1,4,5-trisphosphate (IP_3) and diacylglycerol (DAG).

In the latter pathway (with which we shall mostly be concerned in this book) the GPCR is linked to a G_q protein that activates phospholipase C (PLC), an enzyme that, like G_q, is restricted to the plasma membrane. When activated, PLC cleaves another membrane-bound molecule, phosphatidylinositol 4,5-bisphosphate (PIP_2), into IP_3 and DAG. At this stage, the signal transduction pathway can leave the membrane. Although DAG is membrane-bound, IP_3 is soluble in the cytoplasm. The IP_3 that is released by the action of PLC can thus diffuse through the cell. In particular, it binds to IP_3 receptors (IPR), which are located mostly on the membrane of the endoplasmic/sarcoplasmic reticulum. Despite their name, IPR are simply intracellular Ca^{2+} channels that are activated by the binding of IP_3, and that release Ca^{2+} from the ER/SR.

To add additional complexity, DAG is a signalling molecule in its own right, being (among other things) an activator of protein kinase C (PKC), which is important in a variety of cellular signalling pathways.

The GPCR $\rightarrow$ PLC $\rightarrow$ IP_3 pathway is far more complex than is apparent from this brief overview. There are many different receptors, binding different agonists, these can be linked to a variety of different G proteins, and each G protein can activate one of four different PLC isoforms (PLCβ1–4). Other PLC isoforms can be activated by other mechanisms, including tyrosine-kinase-coupled receptors, or simply a rise in $[Ca^{2+}]$. Thus, different agonists, even when they all operate through the PLC $\rightarrow$ IP_3 pathway, can generate a huge variety of Ca^{2+} responses.

1.2.2 Internal Compartments

The most important internal compartment, at least from the point of view of Ca^{2+} dynamics, is the endoplasmic/sarcoplasmic reticulum. The endoplasmic reticulum (ER; in muscle cells it is called the sarcoplasmic reticulum, or SR) is a highly folded internal compartment that is mainly responsible for protein synthesis. The ER/SR is so convoluted that it is thought to contain about half of the total membrane in a cell, while forming a single, connected, compartment that contains approximately 10% of the total cellular volume.

For our purposes here, the most important feature of the ER/SR is its ability to sequester large amounts of Ca^{2+} and to release this Ca^{2+} quickly. This sequestration is carried out by energy-consuming Ca^{2+} pumps (Sarcoplasmic/Endoplasmic Reticulum Calcium ATPases, or SERCA pumps) that we describe in more detail below, while the release of Ca^{2+} from the ER/SR is through two major types of Ca^{2+} channels (the inositol trisphosphate receptor, or IPR, and the ryanodine receptor, or RyR), again described below.

Mitochondria are responsible for the production of ATP via oxidative phosphorylation, and function also as a major internal Ca^{2+} store. Ca^{2+} is taken up into the mitochondria by the action of the mitochondrial Ca^{2+} uniporter, which uses the energy of the mitochondrial membrane potential to transport Ca^{2+} (Rizzuto et al, 2012; Olson et al, 2012; Carafoli, 2003), and Ca^{2+} is released from the mitochondria by a sodium/calcium exchanger (NCLX). The mitochondrial membrane potential is itself maintained by proton extrusion as a result of respiration. The various ways in which mitochondria interact with cytosolic Ca^{2+}, and the effects on mitochondrial or cellular function, are not well understood. However, there seems to be a general consensus that the ER/SR is the primary driver of intracellular Ca^{2+} fluxes, while the mitochondria perhaps play more of a secondary, modulatory, role. There is also increasing evidence that mitochondria are often in close spatial proximity to the ER, forming microdomains of high [Ca^{2+}] that could serve as effective conduits for the transfer of Ca^{2+} between these organelles (Csordás and Hajnóczky, 2009; Csordás et al, 2001; Rizzuto et al, 1994, 1993, 1999).

The Golgi apparatus is another organelle that sequesters and releases Ca^{2+}. Sequestration is via two different types of ATPase pumps; a pump, the same as is found in the ER membrane, and a secretory pathway Ca^{2+} ATPase1, SPCA1. Release of Ca^{2+} from the Golgi apparatus is through IPR (Pizzo et al, 2011; Missiaen et al, 2004; Surroca and Wolff, 2000; Pinton et al, 1998), thus making this organelle another potential player in the dynamic modulation of Ca^{2+} signalling.

More recently, another family of internal Ca^{2+}-storing compartments have come into prominence, the *acidic organelles* (Zhu et al, 2010; Patel et al, 2011; Patel and Muallem, 2011). These are internal organelles that have a markedly acidic interior, and have a variety of names and functions: endosomes, lysosomes, acidocalcisomes, dense granules, and secretory vesicles are all examples of acidic organelles. Much less is known about Ca^{2+} transport to and from acidic organelles, and it is likely that each different type of acidic organelle has its own set of Ca^{2+} transport pathways. In some acidic organelles, Ca^{2+} is taken up by Ca^{2+} ATPase pumps, which are different from SERCA, or by a H^+/Ca^{2+} exchanger, which uses the existing proton gradient (remember that acidic organelles, by their nature, have a high internal proton concentration) to drive Ca^{2+} transport up its concentration gradient.

The best characterised Ca^{2+} release pathway from an acidic organelle is the two-pore channel (TPC), which is found in the endo-lysosomal system in animal cells. The TPC is opened by low concentrations of the intracellular second messenger nicotinic acid adenine dinucleotide phosphate (NAADP), thus forming what is now recognised to be another major Ca^{2+} release pathway from internal stores (Galione and Ruas, 2005; Galione, 2011; Patel et al, 2011). TPC are often in close physical proximity to IPR and RyR, and there is speculation that Ca^{2+} release through TPC acts as a trigger for additional Ca^{2+} release from the ER. However, the properties of NAADP-induced Ca^{2+} release through TPC, and its interaction with other Ca^{2+} release pathways, remain unclear.

1.2.3 Internal Calcium Channels: IPR and RyR

Release of Ca^{2+} from the ER/SR occurs principally through two channels. These channels have somewhat confusing names, as both are called "receptors", but they are complex and highly regulated Ca^{2+} channels nonetheless. One is the IP_3 receptor (IPR), so-called because it is activated, among other things, by IP_3; the other is the ryanodine receptor (RyR), so-called because it is inhibited (with very high affinity) by the poisonous alkaloid ryanodine. These two Ca^{2+} channels play central roles in almost all Ca^{2+} signalling, and will play a correspondingly important role in this book.

The RyR is the largest known ion channel, with a molecular weight of over 2 MDa (Van Petegem, 2015, 2012; Fill and Copello, 2002). It is a homotetrameric protein, constructed from four individual isoforms, all of the same type. There are three mammalian isoforms, RyR1-3, with two non-mammalian isoforms, RyRα and RyRβ. Each isoform is expressed in different amounts in different tissues. For example, RyR2 is the isoform found in cardiac cells, while RyR1 is most predominant in skeletal muscle.

It would be misleading to think of an RyR as a simple pore that allows the passage of Ca^{2+}. The RyR monomers themselves are associated with a wide array of regulatory proteins, forming a huge macromolecular signalling structure that can be modulated by a variety of effectors such as kinases and phosphatases. Indeed, the size and complexity of this signalling structure attest to the importance of RyR function and control.

One of the most important modulators of RyR behaviour is Ca^{2+} itself. When $[Ca^{2+}]$ is low (around 100 nM), an increase in $[Ca^{2+}]$ causes an increase in the open probability of the RyR. Thus, a small increase in $[Ca^{2+}]$ at the mouth of the RyR can initiate a positive feedback process that ends in the explosive release of a large amount of Ca^{2+} through the RyR. This process, first identified in skeletal muscle by Endo et al (1970), is called Ca^{2+}-induced Ca^{2+} release, or CICR.

CICR is the central feature of Ca^{2+} dynamics and excitation-contraction coupling in skeletal and cardiac myocytes (Ca^{2+} is equally important in smooth muscle, and CICR plays a role there also, but the process is rather different). In cardiac myocytes, depolarisation of the cell membrane causes entry of Ca^{2+} through L-type Ca^{2+} channels, and this entering Ca^{2+} initiates CICR through closely apposed RyR. The resultant large Ca^{2+} transient in the cytoplasm activates the actin/myosin contractile machinery to generate force and contract the cell. In skeletal muscle, the L-type Ca^{2+} channels and the RyR appear to be physically linked in such a way that a depolarisation-induced change in conformation of the L-type channels directly causes opening of the RyR, and subsequent CICR.

The second major intracellular Ca^{2+} channel is the IPR (Joseph 1996, Patel et al 1999; Taylor and Laude 2002; Foskett et al 2007; Mak and Foskett 2015), a tetrameric protein similar to the RyR, but somewhat smaller, with a molecular weight of around 1 MDa. The IPR channel is a tetrameric combination of four IPR molecules, each molecule having a molecular weight of about 260 kDa. In mammalian cells the IPR molecule exists in three isoforms, IPR1-3, each with slightly different properties, and it is likely that every cell type expresses the three IPR isoforms in different proportions (Taylor et al, 1999). There is, however, more variety than appears; each of the isoforms can exist in a number of splice variants, while heterotetrameric IPR channels can be constructed from combinations of different isoforms. There is thus a considerable diversity of IPR channels, a diversity which remains poorly understood, both in extent and in function (Alzayady et al, 2013; Yule et al, 2010).

Although when discussing the structure of the IPR it is necessary to make a careful distinction between the IPR molecule (which exists in different isoforms) and the IPR channel (which is a combination of four IPR molecules), in this book we shall usually neglect to do so. Instead, for convenience, we shall use IPR to denote IPR channels.

Ca^{2+} and IP_3 are the most important ligands of the IPR. The steady-state open probability (P_o) of the IPR is a biphasic function of $[Ca^{2+}]$ (as discussed in more detail in Section 2.7, and illustrated in Fig. 2.14). At low $[Ca^{2+}]$, an increase in $[Ca^{2+}]$ leads to an increase in P_o (as it does for the RyR also), while the reverse happens at higher $[Ca^{2+}]$. Although this general picture is true for practically all isoforms, there is still a great deal of variability in the exact shape of this biphasic curve.

The effect of IP_3 on the IPR appears to be via modulation of the inhibitory effect of high $[Ca^{2+}]$. For a given cell type, a change in $[IP_3]$ appears to affect mostly the falling portion of the steady-state P_o curve (i.e., the inhibition by high $[Ca^{2+}]$) while having a lesser effect on the rising portion of the curve. An increase in $[IP_3]$ also increases the maximal open probability of the IPR. However, we still lack a good understanding of the mechanistic basis of such behaviour.

Of equal importance to the steady-state P_o curve is the dynamic response of the IPR to an increase in $[Ca^{2+}]$ or $[IP_3]$. Given that, during normal cell behaviour, $[Ca^{2+}]$ is typically changing in complex ways, it might be that IPR are rarely allowed to settle down to their steady-state behaviour, but are instead responding in a dynamic fashion to changes in $[Ca^{2+}]$. Unfortunately, since dynamic behaviour is much more difficult to measure experimentally, there are fewer data on dynamic responses than there are on steady-state behaviour. The most detailed data on the dynamic responses of single IPR are those of Mak et al (2007), who measured IPR single-channel behaviour in an experiment that allowed for fast changes in the environment of the IPR. When $[Ca^{2+}]$ is increased from under 10 nM (resting state) to 2 μM (optimal), the IPR respond quickly, with a mean latency of around 40 ms. However, when $[Ca^{2+}]$ is increased from 2 μM (optimal) to 300 μM (inhibitory), the IPR respond more slowly, with a mean latency of around 300 ms. Recovery from the inhibited state is even slower, with a mean time to recovery of over 2 s. Such results are similar in spirit, although differing in detail, from the earlier work of Parker and Ivorra (1990), Finch et al (1991), Finch and Goldin (1994), Dufour et al (1997), and Parker et al (1996), who also studied time-dependent responses of the IPR to changes in Ca^{2+} and IP_3 concentration.

Thus, although Ca^{2+} exerts multiple effects on the IPR, it does so at different rates. Activation by Ca^{2+} is the fastest, followed by slower inhibition, followed by the slowest process of recovery from inhibition. This combination of fast activation and slow inhibition is characteristic of a wide range of physiological processes; in particular, such timescale separation lies at the heart of many physiological excitable systems, such as action potentials in neurons. It thus comes as no surprise to learn that Ca^{2+} dynamics (mediated by either the IPR or the RyR) forms an excitable system also, with a mathematical structure similar in many ways to other excitable systems. Excitable systems – and Ca^{2+} excitability is no exception – typically exhibit threshold behaviour; a small perturbation from rest merely returns (usually monotonically) to the steady state, but a large enough perturbation will cause a large transient response before returning to rest. In neurons, this large transient response is called an action potential. In Ca^{2+} dynamics, it is called a Ca^{2+} spike.

1.2.4 IP_3 Metabolism

Not only is the Ca^{2+} flux through the IPR modulated by IP_3 and Ca^{2+} in multiple and complex ways, the concentration of IP_3 is itself controlled by a complex set of reactions that are controlled by $[Ca^{2+}]$. As we saw above, IP_3 is made when PLC splits PIP_2 into DAG and IP_3. In humans there are thirteen isoforms of PLC, grouped into six subfamilies – β, γ, δ, ϵ, ζ, and η (interestingly, there is no α isoform, as the original identification of PLCα turned out to be a mistake) – and each of these isoforms is regulated in different ways (Vines, 2012). For example, PLCβ1-4 are activated by G proteins, but the other isoforms are activated by other signalling pathways, or simply by a rise in $[Ca^{2+}]$. This allows for considerable complexity in the control of IP_3 production.

Although the activity of PLC is critically dependent on the presence of Ca^{2+}, the dynamic effects of this are not always clear. For example, if the activity of PLC is an increasing function of $[Ca^{2+}]$ within the physiological range, there is a positive feedback loop involving $[Ca^{2+}]$ and $[IP_3]$. In some cell types this appears not to be important, as PLC is fully activated even at resting $[Ca^{2+}]$ (Renard et al, 1987; Bird et al, 1997). However, in other cell types an increase in $[Ca^{2+}]$ causes a dramatic increase in PLC activity (Horowitz et al, 2005). These differences might be due, at least in part, to the different properties of PLC isoforms.

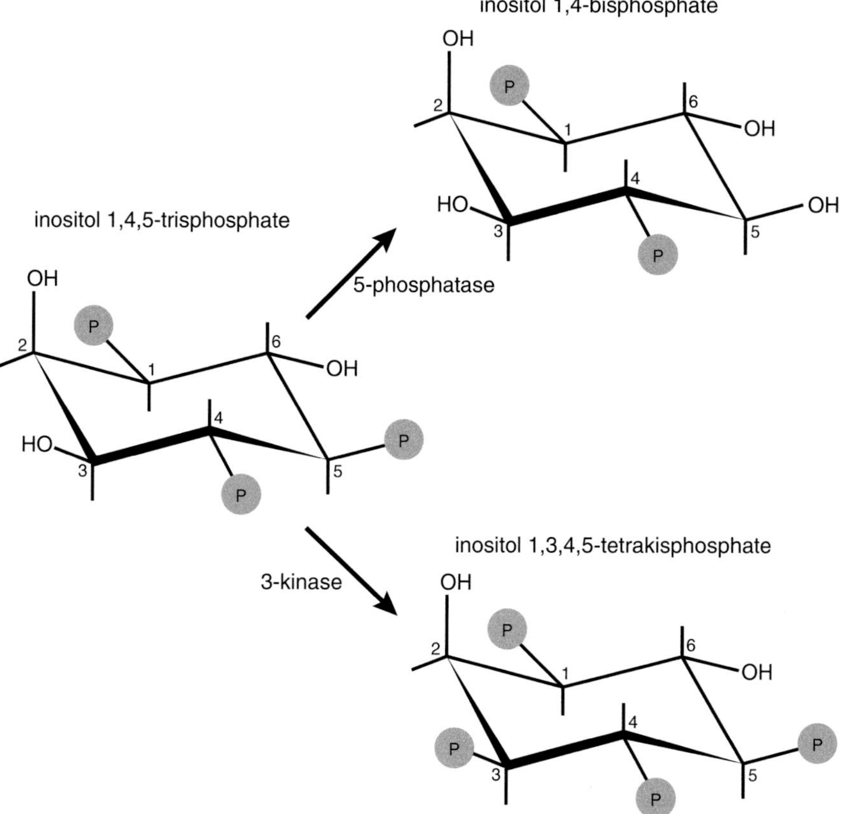

Fig. 1.3 The two ways that inositol 1,4,5-trisphosphate (IP_3) can be metabolised in cells. The P in the shaded circle denotes a phosphate group, PO_4^{3-}. A kinase can add an additional phosphate group at the third carbon of the inositol ring, to give inositol 1,3,4,5-tetrakisphosphate, or a phosphatase can remove the phosphate group from the fifth carbon, to give inositol 1,4-bisphosphate.

IP_3 can be metabolised in two different ways (Fig. 1.3). For the following it is helpful to keep in mind that IP_3 consists of an inositol ring with phosphate groups (PO_4^{3-}) attached to the carbon atoms at positions 1, 4, and 5. Thus, IP_3 can be metabolised either by removing one of the phosphate groups or by adding another. Of the six possibilities, only two occur in the cytosol. A phosphatase can remove the phosphate group at position 5, leaving inositol 1,4-bisphosphate ($Ins(1,4)P_2$), or, alternatively, a kinase can add a phosphate group at position 3, thus producing inositol 1,3,4,5-tetrakisphosphate, or $Ins(1,3,4,5)P_4$, which is itself a signalling molecule (Eva et al, 2012). An excellent review of inositol biochemistry and its relationship to Ca^{2+} signalling is Berridge and Irvine (1989).

The rate of the kinase depends on [Ca^{2+}] (via the formation of a Ca^{2+}/calmodulin complex), while, since Ins(1,3,4,5)P_4 is also a substrate for the phosphatase, it acts as a competitive inhibitor of that reaction. For the same reason, IP_3 also acts as a competitive inhibitor for the action of the phosphatase on Ins(1,3,4,5)P_4. There is thus a complex web of interactions involving Ca^{2+} and IP_3, some with positive feedback, others with negative.

1.2.5 Calcium Influx

The rate of entry of Ca^{2+} into the cell is an important modulator of intracellular Ca^{2+} dynamics in all cell types. There are four principal entry pathways for Ca^{2+}; through voltage-gated channels, through receptor-operated channels, through store-operated channels, and through ligand-gated channels.

Voltage-gated Ca^{2+} channels (VGCC) are of greatest importance in excitable cells and muscle cells (Catterall, 2011; Dolphin, 2006; Catterall, 2000; Tsien and Tsien, 1990). VGCC are closed at the resting membrane potential, but open when the membrane potential increases (i.e., when the cell is depolarised), to allow fast entry of large amounts of Ca^{2+}. In some cells these fast Ca^{2+} currents can replace the usual Na^+ current as the mediator of the action potential. There are a number of different types of VGCC; L-type channels are particularly important in cardiac and skeletal muscle, where they initiate excitation-contraction coupling. The other types of VGCC (N, P/Q, R, and T) occur in a variety of other cell types and are distinguished partly by their sensitivity to different toxins and Ca^{2+}, partly by their physiological roles, and partly by their sensitivity to voltage. VGCC are also modulated by Ca^{2+}, both positively and negatively, allowing for a rich array of possible feedbacks and behaviours (Christel and Lee, 2012). There is thus a variety of VGCC models, ranging from simple to highly complex.

There are two other major Ca^{2+} influx pathways, usually called, respectively, store-operated Ca^{2+} channels (SOCC) and receptor-operated Ca^{2+} channels (ROCC). SOCC open when the ER is depleted, while ROCC open in direct response to agonist stimulation, independently of store depletion or depolarisation.

The idea that Ca^{2+} entry is dependent on the amount of Ca^{2+} in the internal store was first proposed by Putney (1986); the paper was, somewhat confusingly, called *A model for receptor-regulated calcium entry*, which conflicts with the modern terminology. Since then, a number of store-operated currents have been identified in a range of cell types (Parekh and Putney, 2005); possibly the best known is the Ca^{2+}-release-activated current, I_{crac}, of mast cells and T cells (Hoth and Penner, 1992; Lewis, 2007), but there are many others, with differing electrophysiological properties.

In the last 10 years a great deal has been learned about the molecular mechanisms underlying SOCC (Smyth et al, 2010; Soboloff et al, 2012). The protein that senses the filling state of the ER is STIM (STromal Interaction Molecule) which occurs in two mammalian homologs, STIM1 and STIM2. STIM sits in the ER membrane, but has a portion projecting into the lumen of the ER. At rest, STIM is distributed evenly around the ER membrane, but, when the ER $[Ca^{2+}]$ decreases, STIM forms clusters in regions of the ER membrane that are close to the plasma membrane. At the same time, another protein, Orai (with three mammalian homologs, Orai1, Orai2, and Orai3), that spans the plasma membrane, also congregates in these regions and, upon interaction with STIM, forms a Ca^{2+} channel to allow Ca^{2+} influx from the extracellular medium into the cytosol. Thus the Orai protein forms the pore of the I_{crac} current, giving a definite molecular identity to at least one SOCC. However, despite these dramatic recent advances, the exact nature of the STIM/Orai interaction is still not well understood, and neither do we know whether all SOCC are built from Orai/STIM interactions.

Receptor-operated Ca^{2+} channels, or ROCC, open in response (either directly or indirectly) to agonist binding to a receptor, and are independent of the amount of Ca^{2+} in the ER. As discussed above, agonist binding to a GPCR initiates a series of membrane-bound reactions that results in the production of DAG. DAG in its turn can be cleaved by DAG lipase, to yield arachidonic acid, an activator of Ca^{2+} influx channels (Shuttleworth, 2012; Shuttleworth and Mignen, 2003) called arachidonate-regulated Ca^{2+} channels (ARC channels). Although one might naturally expect that a store-independent pathway such as this shares little in common with the store-dependent pathway mediated by the STIM/Orai interaction, such is not the case. Although STIM is best known as an ER Ca^{2+} sensor, between 10–20% of STIM1 is situated in the plasma membrane, and the presence of STIM1 is essential for the function of ARC channels (Shuttleworth, 2012). The mechanism behind this dependency is unknown, but it is thought to be independent of ER Ca^{2+} levels. Furthermore, ARC channels are constructed from a combination of Orai1 and Orai3 subunits, thus drawing an even closer analogy between SOCC and ROCC.

The story is complicated still further by the presence, but unknown function, of TRPC channels (Transient Receptor Potential channels - the C stands for canonical, rather than channel, and denotes the subtype of TRP channel. The somewhat strange name for these channels arises from their initial identification in photoreceptors of a Drosophila mutant, in which they caused a transient depolarisation in response to light). TRPC channels are situated in the plasma membrane, and are believed to act as Ca^{2+} channels. They seem to be regulated by STIM1, Orai1, and also by products of PLC activation (Nilius, 2003; Salido et al, 2009), but exactly how they interact with other ROCC and SOCC, or their function, remains unclear.

Thus, agonist binding to a receptor initiates a complex and interwoven set of pathways that regulate Ca^{2+} influx by a number of different mechanisms. Why this should be so is unclear, but guesses can be made. There is some evidence that, at low agonist levels when depletion of the ER is less, ROCC are the predominant Ca^{2+} influx pathway (at least initially), but at higher agonist concentrations, when ER depletion is more significant, SOCC take over. As one would expect, this is not the only possible explanation, as spatial effects and Ca^{2+} microdomains might also play important roles, even at low agonist concentrations. In addition, recent evidence suggests that STIM2 is more sensitive than STIM1 to small decreases in ER Ca^{2+} concentration, and thus might mediate store-operated Ca^{2+} entry at lower agonist concentrations (Thiel et al, 2013). What is certain, however, is that Ca^{2+} influx is vital for cellular function, and is regulated in multiple, interconnected ways that we do not yet fully understand.

Finally, Ca^{2+} entry through ligand-gated channels is of particular importance in neuronal signalling (Burnashev, 1998), as ionotropic glutamate receptors, ACh receptors and ATP-gated receptors are all permeable to Ca^{2+}. Although it is likely that Ca^{2+} entry through ligand-gated channels can modulate synaptic transmission and neuronal signalling, the exact role played by such Ca^{2+} entry remains unclear (Karlstad et al, 2012).

In some specific cell types, the role of Ca^{2+} entry through non-specific cation channels is clear. For example, in vertebrate photoreceptors, light-sensitive non-specific cation channels are kept open in the dark by cGMP, and about 15% of the current through these light-sensitive channels is carried by Ca^{2+}. Stimulation by light leads to a drop in [cGMP], closure of the light-sensitive channels, and a decrease in the cytosolic [Ca^{2+}]. This results in, among other things, the activation of guanylate cyclase, an increase in [cGMP], and the reopening of the light-sensitive channels, a feedback loop that lies at the heart of light adaptation (Hamer, 2000; Akopian and Witkovsky, 2002; Korenbrot, 2012; Koch and Dell'Orco, 2013).

1.2.6 Calcium Removal from the Cytoplasm

Ca^{2+} is removed from the cytoplasm in a number of ways. Active pumps (i.e., pumps that consume energy that comes from the hydrolysis of ATP) pump Ca^{2+} either into the ER/SR (SERCA pumps) or out of the cell (plasma membrane Ca^{2+} ATPase pumps, or PMCA), while the Na^+/Ca^{2+} exchanger (NCX) uses the energy in the Na^+ gradient to move Ca^{2+} out of the cell at the expense of Na^+ entry. Additional pathways for Ca^{2+} removal are the mitochondrial Ca^{2+} uniporter (MCU), or the mitochondrial H^+/Ca^{2+} exchanger, both of which take up Ca^{2+} into the mitochondria, although the H^+/Ca^{2+} exchanger can also release Ca^{2+} into the cytoplasm under some conditions.

PMCA and SERCA pumps have a high affinity for Ca^{2+} but relatively low maximal pumping rates. They thus respond quickly to small increases of $[Ca^{2+}]$ over baseline. The MCU tends to have the reverse properties, with low affinity and high maximal pumping rates. Thus, it can transport large amounts of Ca^{2+} but tends to operate only when $[Ca^{2+}]$ is in the μM range. For this reason, transport of Ca^{2+} into the mitochondria is a particularly fascinating topic, as it is closely linked with microdomains between the mitochondria and the ER (Spät et al, 2008; Csordás and Hajnóczky, 2009). However, much remains unclear about how exactly $[Ca^{2+}]$ regulates the threshold and gain of the MCU (Csordás et al, 2013).

Each of these transporters comes in several flavours with slightly different kinetics, and with different prevalence in different cell types. In chromaffin cells, for example, when $[Ca^{2+}] = 1000$ nM Ca^{2+} clearance is performed mostly by the MCU, with little being taken up by SERCA or PMCA, while the reverse is true in pancreatic β cells and superior cervical ganglion neurons (Duman et al, 2008). It is also important to keep in mind that SERCA, PMCA, and NCX can run in reverse. Thus, for example, given a large enough ER Ca^{2+} concentration, the SERCA pumps will transport Ca^{2+} out of the ER, generating ATP in the process.

1.2.7 Calcium-Binding Proteins and Fluorescent Dyes

Of all the Ca^{2+} ions that flow into a cell through one of the influx pathways described above, only very few remain as unbound Ca^{2+} in the cytoplasm. The remainder are bound by a wide variety of Ca^{2+}-binding proteins, called Ca^{2+} buffers. There is a bewildering array of Ca^{2+} buffers. In principle, any molecule that binds Ca^{2+} can act as a Ca^{2+} buffer, including such proteins as calmodulin, or even SERCA pumps. In addition, Ca^{2+} binds to many parts of the cellular structure, including the plasma membrane, or the membranes of internal organelles. There is thus a rather blurred line between proteins that act as simple Ca^{2+} buffers, and proteins that act as Ca^{2+} buffers by virtue of the fact that they bind Ca^{2+} in order to carry out some cellular function (Schwaller, 2010).

The most important endogenous cytosolic Ca^{2+} buffers are parvalbumin, calbindin, and calretinin, each of which has different concentrations, ranging from tens to thousands of μM, in different cell types. These cytosolic buffers have K_d values (i.e., the value at which they bind half their maximal capacity) well above the resting cytosolic Ca^{2+} concentration, and thus most of the buffer is unbound at rest.

Each internal compartment also has its own complement of Ca^{2+} buffers (Prins and Michalak, 2011). In the ER, calreticulin is responsible for approximately half the Ca^{2+} buffering, with a variety of other buffers (GRP78 and GRP94, for example) buffering the remainder. Buffering in the SR is done mostly by calsequestrin.

Additional Ca^{2+} buffers can be added to the cell as part of the experimental process. The most important such exogenous buffers are fluorescent Ca^{2+} dyes. The usual procedure for observing and measuring intracellular Ca^{2+} concentrations is to add to the cell molecules that emit light when they bind Ca^{2+} (Grynkiewicz et al, 1985; Paredes et al, 2008). The emitted light can then be observed by light microscopy. However, in order to emit light, the dyes must first bind Ca^{2+}, thus acting as Ca^{2+} buffers. It follows that, in order to measure the Ca^{2+} concentration inside cells, it is first necessary to change the buffering capacity of the cell, making it a difficult task to observe intracellular Ca^{2+} in an entirely undisturbed environment. Popular Ca^{2+} fluorescent dyes are fura-2, indo-1, and fluo-4. Of these, fura-2 and indo-1 are *ratiometric* dyes; by comparing the ratio of the emissions at different wavelengths, the experimentalist can determine the absolute, not just the relative, Ca^{2+} concentration. However, this requires a more complex experimental setup, and so non-ratiometric dyes are often used in experiments for which measurement of absolute concentrations is not critical. The measurement of $[Ca^{2+}]$ in intracellular organelles such as the ER or mitochondria is a task that is even more complex, and requires genetically encoded Ca^{2+} indicators targeted to organelles with the addition of appropriate tags (Suzuki et al, 2014).

In the last few years, genetically encoded Ca^{2+} dyes – based on Green Fluorescent Proteins, or GFPs (Tsien, 1998) – have become progressively more popular and important (Horikawa, 2015; Kotlikoff, 2007; Lock et al, 2015; Rose et al, 2014). Genes for the expression of these dyes can be transfected into cell lines or even whole animals (mice are a particularly common example of this), and specific cell types within an animal can be targeted. This technique has allowed for the observation of real-time $[Ca^{2+}]$ in cells such as gonadotropin-releasing hormone neurons (GnRH neurons), which, in the absence of the genetically encoded dye, are difficult to find by eye in a brain slice. However, in mice which express a Pericam genetically encoded Ca^{2+} indicator in their GnRH neurons, the neurons can be found much more easily (Herbison et al, 2001).

Other important exogenous buffers that are used extensively by experimentalists are BAPTA (1,2-Bis(o-AminoPhenoxy)ethane-N,N,N',N'-Tetraacetic Acid) and EGTA (Ethylene Glycol Tetraacetic Acid). Because they have different binding rates (BAPTA binds Ca^{2+} much more quickly than does EGTA), and because their concentration can be controlled experimentally, they are widely used to manipulate the buffering capacity of the cytoplasm in (partially) predictable ways.

1.2.8 Microdomains and Nanodomains

Because Ca^{2+} is so heavily buffered, it is unable to diffuse far through the cytoplasm before being bound to something. Hence the effective diffusion coefficient of Ca^{2+} is at least an order of magnitude lower than the diffusion coefficient of Ca^{2+} in aqueous solution. Because of this limited mobility, cells are able to raise $[Ca^{2+}]$ in small regions of the cell while leaving $[Ca^{2+}]$ unaffected in the remainder of the cell. Thus, control of cellular processes by Ca^{2+} is often a highly local affair, with $[Ca^{2+}]$ being raised in very small regions (often on the scale of tens of nanometres) to bind to effector molecules in that region alone. Such nanodomains are very difficult to observe experimentally, and computational modelling is often the only way to infer the behaviour of Ca^{2+} in such small regions. Although this increases the need for computational modelling, it makes such modelling more difficult; the construction and solution of models in nanodomains, requiring – as it usually does – partial differential equations in three dimensions and stochastic simulations, is a highly nontrivial task, and experimental verification of such models is not usually possible.

1.3 Spatiotemporal and Hierarchical Organisation

The above Ca^{2+} toolbox components (not to mention the many additional signalling pathways not even mentioned) can be put together in different ways, to give many different behaviours at the level of the whole cell, or groups of cells. At the whole-cell level, for example, the cytoplasmic $[Ca^{2+}]$ often increases and decreases in a periodic manner (although rarely with a period that remains unchanged for long periods of time), behaviour which we call a Ca^{2+} oscillation, or Ca^{2+} spiking. Similarly, if the release of Ca^{2+} is more spatially restricted, or if the cell is big enough, waves of increased cytosolic [Ca^{2+}] can propagate from one release site to another (as long as the release sites are not too far apart), to form global travelling waves of increased [Ca^{2+}].

When viewed at the level of an entire cell, or group of cells, Ca^{2+} oscillations and waves appear to be smoothly and regularly varying in space and time. However, closer investigation reveals that Ca^{2+} waves are built from multiple stochastic events at smaller spatial scales, and that these events themselves are built from yet smaller stochastic events. Thus, Ca^{2+} dynamics is organised in a hierarchical manner, with layers of interacting stochastic events.

At the lowest level, both IPR and RyR open in a stochastic manner (Fig. 1.4). Single-channel openings can be observed experimentally as small, spatially localised Ca^{2+} releases. If the opening of a single IPR or RyR then causes the opening of a cluster of receptors, a larger Ca^{2+} response is seen. A colourful, although not reliably consistent, terminology has arisen around these localised Ca^{2+} releases. For clarity, we shall use "blip" to denote Ca^{2+} release through a single IPR or RyR, and use "puff" and "spark" to denote Ca^{2+} release through groups of IPR and RyR, respectively. Puffs and sparks are the building blocks from which global events are built. If enough puffs combine, then a whole-cell wave can be formed. Hence, the time taken for a wave to form is essentially the time taken for enough blips to combine into a puff, and then enough puffs to combine into a wave. Each of these levels relies on the combination of stochastic events, and thus the time between Ca^{2+} waves follows a distribution, rather than being a fixed interval.

Calcium puffs have been extensively studied in *Xenopus* oocytes and HeLa cells (Marchant et al, 1999; Sun et al, 1998; Callamaras et al, 1998; Marchant and Parker, 2001; Thomas et al, 2000; Bootman et al, 1997a,b), while Ca^{2+} sparks were discovered by Cheng et al (1993) and studied by a multitude of authors since (Cheng et al, 1996; Lipp and Niggli, 1996; Cannell and Soeller, 1999; Jaggar et al, 2000).

Further detailed discussion of the stochastic aspects of Ca^{2+} dynamics is deferred until Chapter 4.

1.4 Examples of Calcium Signalling

In this section we describe briefly some examples of Ca^{2+} dynamics. Each of these examples will be discussed in much more depth later in this book, and is introduced here merely to give the reader some preliminary idea of the different kinds of Ca^{2+} responses.

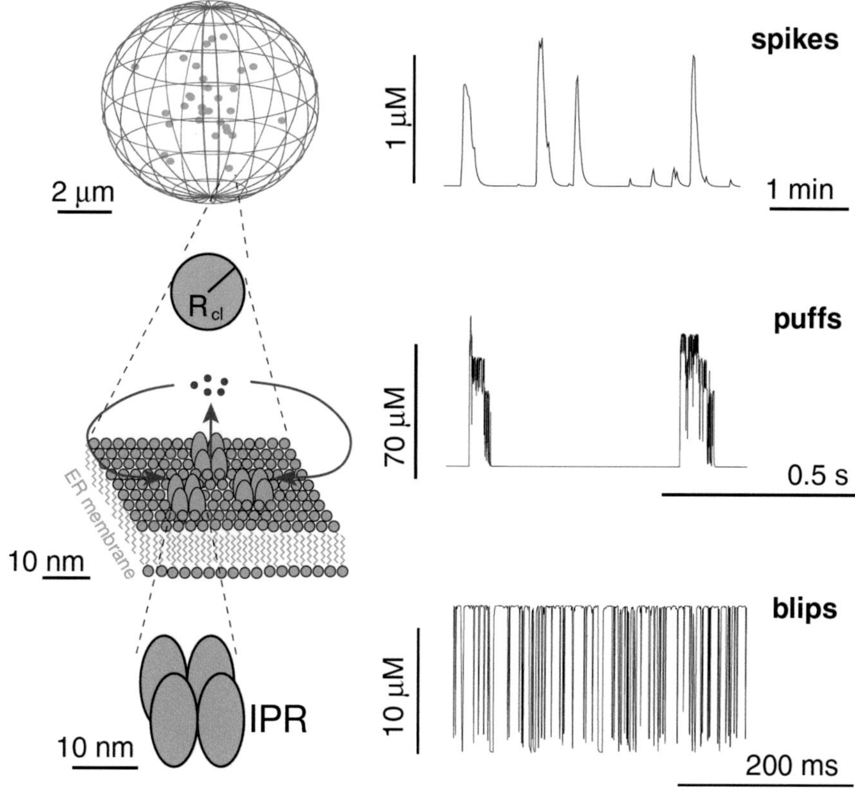

Fig. 1.4 Hierarchical organisation of IP_3-induced Ca^{2+} signalling. Stochastic Ca^{2+} signalling through RyR is identical in principle, although the specific details differ. Typical stochastic model simulations are shown on the right. Lowest panels: at the smallest spatial scale, stochastic binding of IP_3 and Ca^{2+} to IP_3 receptors (IPR) leads to stochastic opening and closing of each channel. These openings lead to the release of a small amount of Ca^{2+}, and are called "blips". At the mouth of the channel, the $[Ca^{2+}]$ reaches about $10\,\mu$M. Middle panels: when IPR are situated in clusters, the opening of one IPR can initiate the opening of neighbouring IPR, leading to the simultaneous opening of a group of IPR. This is called a Ca^{2+} "puff", or "spark", in the case of ryanodine receptors. R_{cl} is the radius of the cluster. Top panels: at the highest level of spatiotemporal organisation, one Ca^{2+} puff can excite neighbouring clusters of IPR, leading to large coordinated release of Ca^{2+} across an entire cell, called a Ca^{2+} spike. Note that the Ca^{2+} concentration of a spike is much lower than that of a puff; this is because spike concentrations are measured in the cytosol, while puff concentrations are measured inside a cluster of IPR, much closer to the site of release. Reproduced from Thurley et al (2012), Fig. 1, with permission from Elsevier.

1.4.1 Cardiac Myocytes

Certainly the most-studied example is that of the Ca^{2+} transient in cardiac myocytes (discussed in more detail in Section 7.2). In that cell type, there are multiple regions, called *diadic clefts*, where the SR and the plasma membrane are closely juxtaposed, separated only by around 15 nm. When the cardiac action potential reaches the diadic cleft, the resultant depolarisation of the membrane opens L-type Ca^{2+} channels, which allow the influx of Ca^{2+} into the diadic cleft. The raised $[Ca^{2+}]$ in the diadic cleft initiates further Ca^{2+} release through the RyR that are also located in the diadic cleft, and thus very close to the L-type channels. CICR through the RyR leads to a large Ca^{2+} transient in the cardiac cell, which activates the contractile machinery and causes contraction. A typical cardiac Ca^{2+} transient is shown in Fig. 1.5. The diadic cleft is the prototypical microdomain. Inside the cleft, Ca^{2+} concentrations reach very high levels very quickly, but these high concentrations are not seen in the bulk of the cell. Such close juxtaposition between the L-type channels and the RyR allows for rapid communication from the outside of the cell to the inside, a feature that is necessary for muscular contraction. For this reason, the diadic cleft is sometimes called a Ca^{2+} synapse.

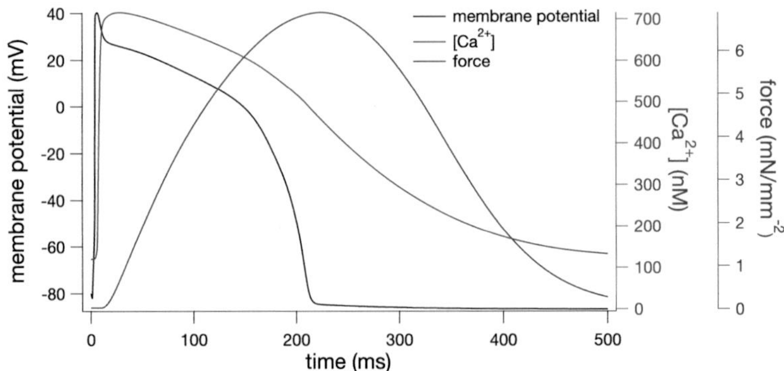

Fig. 1.5 A single action potential, the associated whole-cell Ca^{2+} transient, and the force, in a rabbit ventricular myocyte. Simulation data provided by Jose L. Puglisi (California Northstate University), using LabHEART (`www.labheart.org`) (Puglisi and Bers, 2001). Although the peak of the whole-cell Ca^{2+} transient is less than $1\,\mu$M, the $[Ca^{2+}]$ in the diadic cleft is larger by approximately two orders of magnitude.

1.4.2 Airway Smooth Muscle

In airway smooth muscle (i.e., the smooth muscle that surrounds the airways in the lung. It is necessary to specify *airway* smooth muscle as there is so much variety between the different smooth muscle types), the excitation-contraction process is quite different from that in cardiac or skeletal muscle. Agonist stimulation leads to the formation of IP_3, which releases Ca^{2+} from the ER. However, the Ca^{2+} response is far more complex than a simple transient increase. In response to intermediate concentrations of acetylcholine (ACh), waves of increased [Ca^{2+}] are initiated periodically from one end of the cell, and sweep across the entire cell. At a fixed position in the cell [Ca^{2+}] exhibits oscillations upon a raised baseline (Fig. 1.6 A), and the extent of contraction of the airway smooth muscle depends on the frequency of the oscillations (Perez and Sanderson, 2005b). The response to 5-HT (serotonin) is similar, but, interestingly, the response to an increase in extracellular K^+ is quite different, being baseline spikes of low frequency, much wider than the oscillations seen in response to the other two agonists (Fig. 1.6 B). These different types of oscillations rely on a complex interplay between the various kinds of Ca^{2+} channels (Section 7.4.1).

Fig. 1.6 illustrates a number of important features of Ca^{2+} dynamics. There is the immediate observation that Ca^{2+} dynamics is clearly complex. But, more than that, it illustrates that the same cell type can exhibit a wide range of complex Ca^{2+} responses depending on the stimulus applied. It is also important to note that, although the experimental results plot [Ca^{2+}] as a function of time only, one must be careful not to forget that the oscillations are actually periodic waves. The final point to be made about these Ca^{2+} oscillations in airway smooth muscle is that the oscillation per se has a function. Over the physiological range, as the oscillation frequency increases, so does the contraction of the muscle cell; a simple rise in [Ca^{2+}] is far less effective at causing contraction than is an oscillating [Ca^{2+}]. That the signal is carried by the frequency of the Ca^{2+} oscillation is a feature seen in many cell types (but not all, as we shall see later in this chapter), and is one of the central tenets of Ca^{2+} signalling.

1.4.3 Xenopus Oocytes

One of the most visually impressive examples of intracellular Ca^{2+} waves occurs in *Xenopus* oocytes. In 1991, Lechleiter and Clapham and their coworkers discovered that intracellular Ca^{2+} waves in immature *Xenopus* oocytes showed complex spatiotemporal organisation, forming (among other things) concentric circles and multiple spirals (Fig. 1.7). This is possible because of the large size of these cells; *Xenopus* oocytes can have a diameter larger than 600 μm, an order of magnitude greater than most other cells. In a smaller cell, a typical Ca^{2+} wave (often with a width of close to 100 μm) cannot be observed in its entirety, and there is not enough room for a spiral to form, in which case the Ca^{2+} waves take a form that is almost planar. However, in a large cell it may be possible to observe both the wave front and the wave back, as well as spiral waves, and this has made the *Xenopus* oocyte an important system for the study of Ca^{2+} waves (Section 6.1).

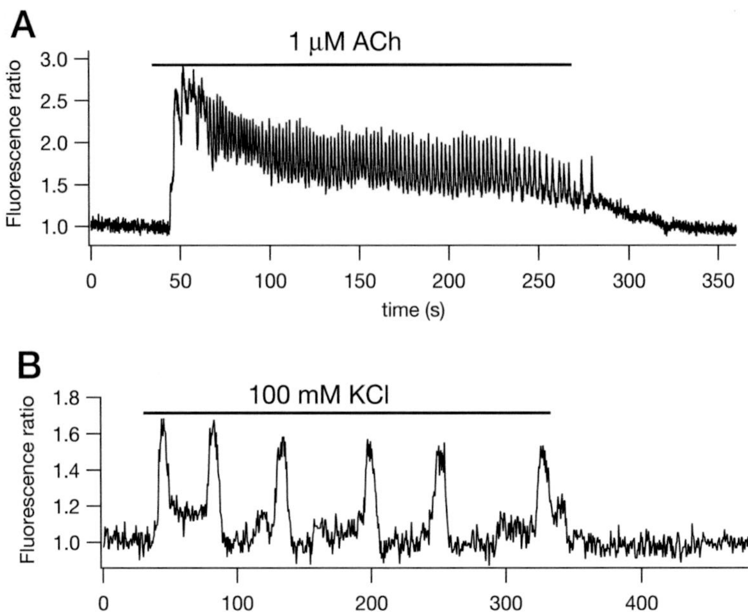

Fig. 1.6 Calcium oscillations in mouse airway smooth muscle (Perez and Sanderson, 2005b). **A:** In response to acetylcholine (ACh), $[Ca^{2+}]$ measured at a fixed position in the cell shows oscillations superimposed upon a raised baseline. **B:** In response to increased extracellular KCl, the waves and oscillations are quite different, being much broader spikes with a longer period and no raised baseline. Although the fluorescence ratio is not a precise measure of the $[Ca^{2+}]$, approximate values for $[Ca^{2+}]$ can be calculated. In particular, the fluorescence ratio provides an accurate indication of whether $[Ca^{2+}]$ is increasing or decreasing. Original data provided by Michael Sanderson (University of Massachusetts Medical School).

1.4.4 Pancreatic and Parotid Acinar Cells

Parotid acinar cells are exocrine epithelia in the parotid gland. They are grouped in clumps that look like bunches of grapes, or a lobular berry such as a raspberry. Hence their name, from the Latin *acinus*, meaning a berry. The cells in an acinus are grouped around a central duct, and their principal function is to secrete water and ions, mostly NaCl, into this duct. This primary saliva travels through an extensively branched system of ducts, which modifies its ionic and protein composition, before emerging in the mouth as secondary saliva.

Production of primary saliva is accomplished by an osmotic mechanism that relies on the regulation of ion channels at the basal and apical ends

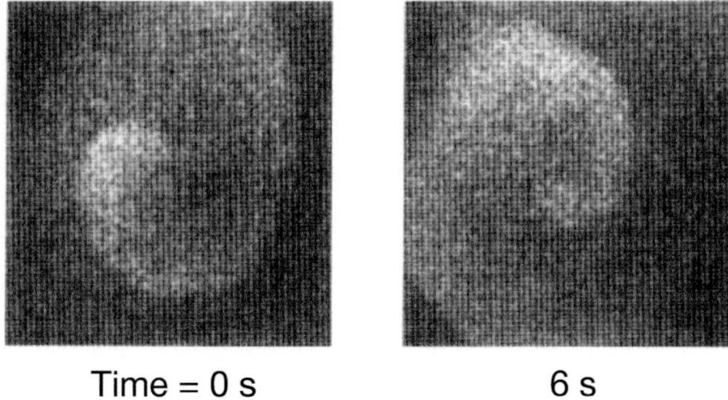

Fig. 1.7 Spiral Ca^{2+} wave in the *Xenopus* oocyte. The image size is $420 \times 420\,\mu m$. The spiral has a wavelength of about $150\,\mu m$ and a period of about 8 seconds. Lighter shades denote higher $[Ca^{2+}]$. Reproduced from Atri et al (1993), Fig. 11a, with permission from Elsevier.

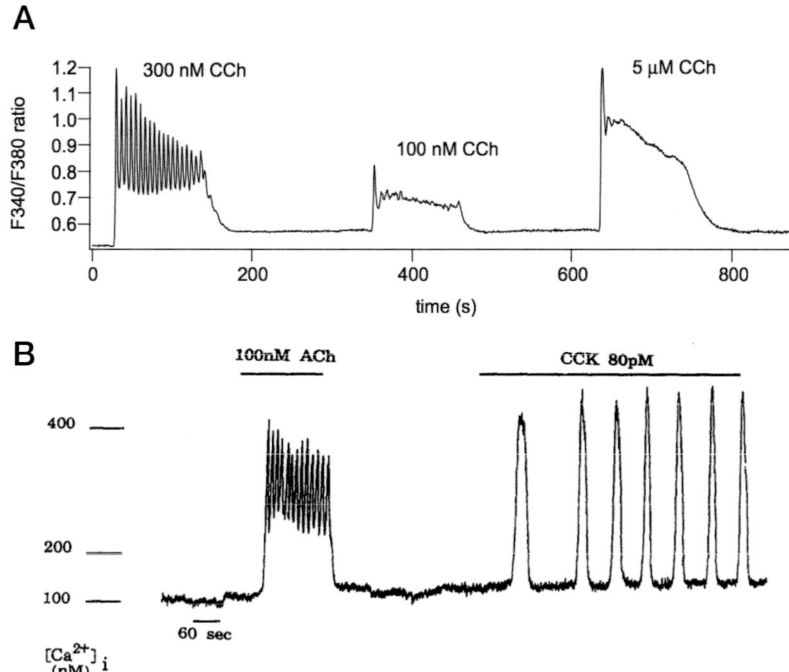

Fig. 1.8 **A:** Calcium oscillations induced by carbachol (CCh) in parotid acinar cells. Reproduced from Palk et al (2010), Fig. 5b, with permission from Elsevier. **B:** Ca^{2+} oscillations in pancreatic acinar cells induced by acetylcholine (ACh) and cholecystokinin (CCK). Reproduced from Yule et al (1991), Fig. 1, with permission from Elsevier.

of the cell (Section 6.3), and this regulation is carried out largely by Ca^{2+} oscillations. Typical experimental data are shown in Fig. 1.8 A. In response to carbachol (CCh; this binds to ACh receptors) parotid acinar cells exhibit Ca^{2+} oscillations on a raised baseline. As the CCh concentration varies, so do the oscillations. At low or high [CCh] there are no oscillations, while at intermediate [CCh] the oscillation frequency increases with [CCh]. Not unexpectedly, each of these "oscillations" is in fact a wave which starts in the apical region and travels quickly across the cell, but for convenience we shall usually continue to refer to them as oscillations. However, in contrast to airway smooth muscle, the frequency of the Ca^{2+} signal appears to play little role in determining the amount of water transported. Instead, the rate of secretion of primary saliva appears to be governed principally by the average $[Ca^{2+}]$ (Palk et al, 2012). The oscillations appear to be a way merely of raising the average $[Ca^{2+}]$. This is a timely reminder that, although the central dogma of Ca^{2+} oscillations is that the message is carried by the frequency of the oscillations, there are exceptions to this general rule.

Pancreatic acinar cells are also exocrine epithelia, similar in structure and physiology to parotid acinar cells, but specialised for the secretion of digestive enzymes. They are a well-known example of a cell type that displays widely different Ca^{2+} oscillations in response to different agonists, even when the agonists all operate via the release of Ca^{2+} through IPR. (Although we also saw two quite different oscillatory patterns in airway smooth muscle, that situation is quite different from that of pancreatic acinar cells, as the KCl-induced oscillations in airway smooth muscle are not based on Ca^{2+} release through IPR.)

Two oscillatory patterns in pancreatic acinar cells are shown in Fig. 1.8 B. Acetylcholine (ACh) induces oscillations very similar to those induced by CCh in parotid cells, with a higher frequency on a raised baseline. However, cholecystokinin (CCK) induces oscillations of quite a different sort, being baseline spiking of much longer period.

1.4.5 Airway Epithelial Cells

Mechanical stimulation of a single airway epithelial cell results in a wave of increased $[Ca^{2+}]$ that spreads across multiple cells, as shown in Fig. 1.9 (Sanderson et al, 1990; Boitano et al, 1992; Sanderson et al, 1994). The wave spreads across each cell at a speed of around $15\,\mu m\,s^{-1}$, and then has a short delay at the cell border before appearing in the neighbouring cell. Since the wave is accompanied by an increase in ciliary beat frequency, it results in an increased ciliary beat frequency over a much wider region than the initial mechanical stimulation.

Similar intercellular waves have been observed in intact livers, slices of hippocampal brain tissue, glial cell cultures, and many other preparations (Leybaert and Sanderson, 2012; Charles et al, 1991, 1992; Cornell-Bell et al,

1990; Kim et al, 1994; Robb-Gaspers and Thomas, 1995). Not all intercellular coordination is of such long range; synchronised oscillations are often observed in small groups of cells such as pancreatic or parotid acinar cells (Yule et al, 1996) or multiplets of hepatocytes (Tordjmann et al, 1997, 1998).

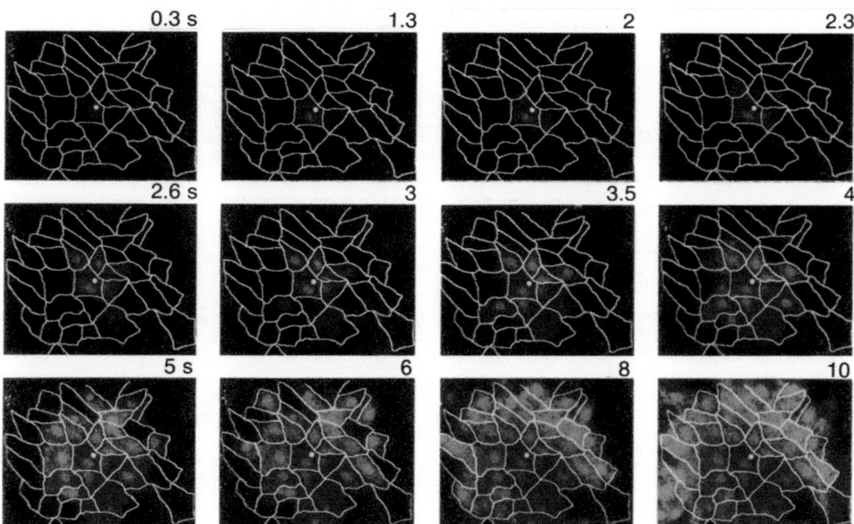

Fig. 1.9 Mechanically-stimulated intercellular wave in airway epithelial cells (Sanderson et al, 1990). The solid white lines denote the cell borders, and the white dot in the approximate centre of each frame denotes the place of mechanical stimulation. Warmer colours (red and orange) denote higher $[Ca^{2+}]$. After mechanical stimulation a wave of increased $[Ca^{2+}]$ can be seen spreading from the stimulated cell through other cells in the culture. The time after mechanical stimulation is given in seconds in the upper right corner of each panel.

Chapter 2
The Calcium Toolbox

The toolbox concept that was introduced in the previous chapter is a particularly useful way to approach Ca^{2+} modelling. The overall cell model can be constructed by picking and choosing which toolbox components to include, after which the component models then just fit together in a modular fashion. If the modeller wishes to change, for example, how they model the IPR, then one IPR model can be taken out and an alternative one inserted, without significant change (if any) in the remainder of the model.

Thus, our job as modellers is to understand, firstly, the many different toolbox components and their behaviour, and, secondly, what kinds of behaviour are to be expected from different combinations of components. The first of these tasks is difficult enough, and requires a deep knowledge of experimental data, but the second is far more complicated, requiring also an understanding of the emergent properties of nonlinear coupled systems.

In this chapter we focus on the first of these tasks, the quantitative description of the various toolbox components.

2.1 G Protein-Coupled Receptors

In many cell types, the first step in the Ca^{2+} signalling pathway is the binding of an agonist to a G protein-coupled receptor, or GPCR, on the cell membrane (Luttrell, 2006). GPCR are not merely large molecules sitting statically in the cell membrane, waiting for an agonist to come along and bind. They diffuse, form dimers, bind to G proteins, are internalised and recycled, are phosphorylated, bind to other molecules, change conformation, and initiate and interact with multiple signalling pathways, on time scales ranging from milliseconds to hours. Modelling of GPCR is correspondingly complex. Lauffenburger and Linderman (1993) give a comprehensive and detailed discussion of receptor models, including GPCR, and there has been a great deal of work

G. Dupont et al., *Models of Calcium Signalling*, Interdisciplinary Applied Mathematics 43, DOI 10.1007/978-3-319-29647-0_2

done since then (Linderman, 2008). A readable and interesting review of the history of general receptor modelling (although focused more on ion-channel receptors rather than GPCR) is Colquhoun (2006).

It must be admitted, however, that the world of GPCR modelling has not, as yet, intersected to any great extent with the world of Ca^{2+} modelling. Few Ca^{2+} models incorporate much detail at the level of the GPCR, instead making the simpler assumption that the concentration of activated PLC ([PLC*]) is just an increasing algebraic function (often a linear function) of the concentration of activated receptor. However, if other molecules further down the signalling pathway, such as DAG, IP_3, or Ca^{2+}, influence the rate of these initial reactions, there is potential for complex solutions. Such feedbacks are the subject of the seminal paper by Cuthbertson and Chay (1991) who present a series of models showing oscillatory behaviour; a more complex version of the same basic idea was published by Kang and Othmer (2007). Similarly, the model of Kummer et al (2000) assumes that α-GTP catalyses its own production, is degraded at a rate that is dependent on [PLC*], and is degraded at a separate rate that is dependent on $[Ca^{2+}]$, thus generating complex behaviour, including chaos.

However, with one exception, there is as yet little direct evidence that such feedbacks are the fundamental oscillatory mechanism that generates Ca^{2+} oscillations. This exception is the generation of Ca^{2+} oscillations in astrocytes by PKC-induced phosphorylation of the metabotropic glutamate receptor, $mGlu_5$, and we discuss this example in detail in Section 6.4.

One of the most intriguing questions in the study of Ca^{2+} dynamics is how different receptors, both working through IPR to release Ca^{2+} from the ER, can cause such drastically different Ca^{2+} responses. We saw this in Fig. 1.8 in pancreatic acinar cells, but it happens in many other cell types also. It is most likely that there are many reasons for these differences. Different receptors can activate quite different downstream pathways, which can affect the upstream reactions in multiple ways. These downstream pathways can also have widely diverging effects on things such as Ca^{2+} entry or the phosphorylation of transporters and exchangers. Nevertheless, it remains highly likely that some of these differences are to be explained by differing feedback mechanisms at the level of the receptor and GPCR. Despite this, in the search for mechanisms that underlie such widely differing Ca^{2+} responses, models have not yet used receptor-based feedback mechanisms to their full potential.

2.1.1 A Simple GPCR Model

Mahama and Linderman (1994) constructed a simple model of a GPCR that serves as a good starting point for considering more complex models. In their model, the agonist receptor is modelled in the simplest possible way, as

$$\mathrm{R} + \mathrm{A} \underset{k_{-1}}{\overset{k_1}{\rightleftharpoons}} \mathrm{R}^*, \tag{2.1}$$

where R denotes the receptor unbound to agonist, A denotes the agonist, and R* denotes the receptor bound to agonist. The corresponding differential equation (from mass action kinetics) is

$$\frac{d[\mathrm{R}^*]}{dt} = k_1[\mathrm{R}][\mathrm{A}] - k_{-1}[\mathrm{R}^*]. \tag{2.2}$$

If we assume further that the total amount of receptor is fixed (i.e., we ignore internalisation and recycling), we get

$$\frac{d[\mathrm{R}^*]}{dt} = k_1([\mathrm{R}]_{\mathrm{tot}} - [\mathrm{R}^*])[\mathrm{A}] - k_{-1}[\mathrm{R}^*], \tag{2.3}$$

where $[\mathrm{R}]_{\mathrm{tot}}$ is the total amount of the receptor, assumed to be constant.

Such a formulation is the simplest possible model of receptor activation, and ignores all the complexity of receptor diffusion, dimerisation, conformational changes, or binding to other molecules (including, for instance, G proteins).

Matters become more complex when interactions with G proteins are considered. An unactivated G protein has three subunits, α, β, and γ, with a guanosine diphosphate (GDP) attached to the α subunit. When an activated receptor binds to the $\beta\gamma\alpha$-GDP form of the G protein it forms a ternary complex. In this ternary complex the α subunit has a decreased affinity for GDP and an increased affinity for GTP. The GDP is exchanged for GTP, the G protein dissociates into the $\beta\gamma$ subunit and the active α-GTP subunit, and the ternary complex then itself dissociates. The active α-GTP subunit binds to PLC to form the activated PLC* complex. When the α-GTP in the PLC* complex is hydrolysed back to α-GDP, the PLC* complex rapidly dissociates to PLC and α-GDP.

This is illustrated in Fig. 2.1. The input to this diagram is activated receptor, R*, while the output is the amount of PLC*, which breaks down PIP_2 to form IP_3 and DAG. Another useful way to visualise this system of reactions is to write out each reaction explicitly. This gives

$$\mathrm{G} \xrightarrow{k_2} \alpha\text{-GTP} + \beta\gamma, \tag{2.4}$$

$$\mathrm{R}^* + \mathrm{G} \xrightarrow{k_1} \alpha\text{-GTP} + \beta\gamma, \tag{2.5}$$

$$\alpha\text{-GTP} + \mathrm{PLC} \xrightarrow{k_3} \mathrm{PLC}^*, \tag{2.6}$$

$$\mathrm{PLC}^* \xrightarrow{k_4} \alpha\text{-GDP} + \mathrm{PLC}, \tag{2.7}$$

$$\alpha\text{-GTP} \xrightarrow{k_4} \alpha\text{-GDP}, \tag{2.8}$$

$$\alpha\text{-GDP} + \beta\gamma \xrightarrow{k_5} \mathrm{G}. \tag{2.9}$$

We have written each of these equations as unidirectional, even though we know from thermodynamics that every reaction must be reversible. However,

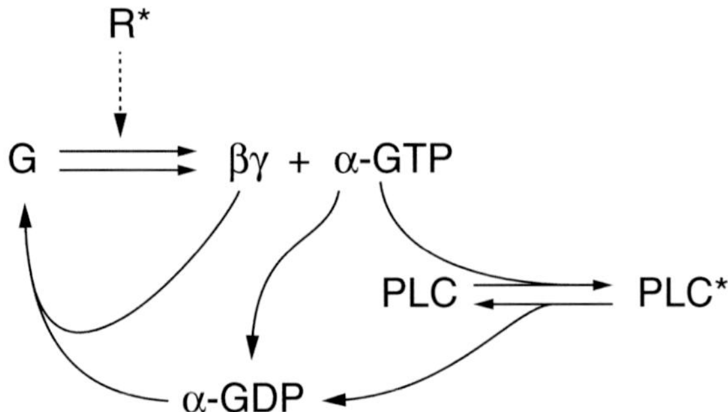

Fig. 2.1 Simplified diagram of the reactions involved in GPCR activation. Activated receptor (R^*) catalyses exchange of the α-GDP subunit for α-GTP, whereupon the G protein (G) dissociates into its $\beta\gamma$ and α-GTP subunits. The α-GTP subunit then forms a complex with PLC, forming activated PLC (PLC^*), which, upon hydrolysis of the α-GTP subunit, itself rapidly dissociates to PLC and α-GDP. The α-GTP subunit can be hydrolysed directly to α-GDP, which can then reassociate with the $\beta\gamma$ subunit to reform the inactive G protein. Note that the dissociation of G can also be catalysed by an inactivated receptor. However, dissociation catalysed by an activated receptor is much faster.

in many biological reaction networks unidirectional reactions are a useful approximation, as in the physiological regime the reverse reactions are often very slow. Also note that the first reaction (GPCR activation in the absence of activated receptors) is much slower than the second (GPCR activation as a result of receptor activation), which is the whole point of receptor activation. Indeed, since $k_2 \ll k_1$, the first reaction can be omitted without significant difference in the model behaviour. Finally, note that we are assuming that the rate of hydrolysis of α-GTP is the same whether or not α-GTP is attached to PLC, which is why both (2.7) and (2.8) have the same rate constant.

The corresponding differential equations are

$$\frac{d[\alpha\text{-GTP}]}{dt} = k_1[\mathrm{R}^*][\mathrm{G}] + k_2[\mathrm{G}] - k_3[\alpha\text{-GTP}][\mathrm{PLC}] - k_4[\alpha\text{-GTP}], \quad (2.10)$$

$$\frac{d[\alpha\text{-GDP}]}{dt} = k_4[\mathrm{PLC}^*] + k_4[\alpha\text{-GTP}] - k_5[\alpha\text{-GDP}][\beta\gamma], \quad (2.11)$$

$$\frac{d[\mathrm{PLC}^*]}{dt} = k_3[\alpha\text{-GTP}][\mathrm{PLC}] - k_4[\mathrm{PLC}^*], \quad (2.12)$$

together with the conservation equations

$$[\mathrm{PLC}] + [\mathrm{PLC}^*] = \text{constant} = [\mathrm{PLC}]_{\text{total}}, \quad (2.13)$$

$$[\beta\gamma] = [\alpha\text{-GTP}] + [\alpha\text{-GDP}] + [\mathrm{PLC}^*], \quad (2.14)$$

$$[\mathrm{G}] + [\beta\gamma] = \text{constant} = [\mathrm{G}]_{\text{total}}. \quad (2.15)$$

Since there is no feedback in this model, its behaviour is not particularly complicated. When $[\mathrm{R}^*]$ is raised, $[\mathrm{PLC}^*]$ increases also to a steady state, either monotonically or with one turning point (i.e., with [PLC*] first increasing then decreasing). Thus, for a maintained agonist concentration, this model by itself does not generate any complex long-term behaviour.

2.1.2 More Complex Receptor Models

There are a number of more complex GPCR models in the literature, based on assumptions that the receptor can exist in a number of different states as it binds agonist and G protein, or activates.

For example, the *ternary complex* model (De Lean et al, 1980) assumes that the receptor can exist in four forms: R (the base form), RG (bound to G protein), RA (bound to agonist), or RGA (bound to both agonist and G protein). Such a model is most easily described by the reaction diagram in Fig. 2.2 A.

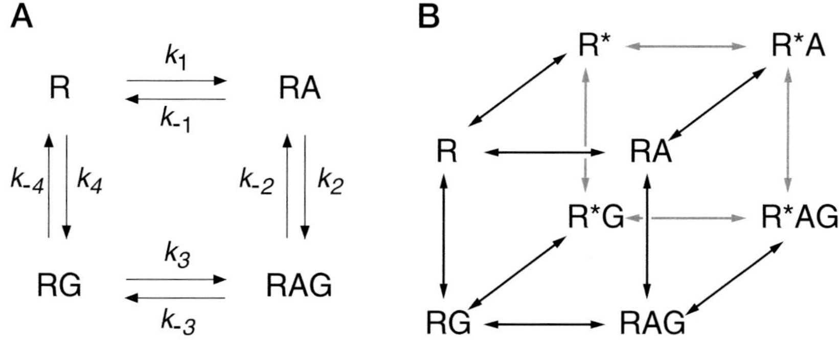

Fig. 2.2 A: The ternary complex model (De Lean et al, 1980). The receptor can exist in four forms: R (the base form), RG (bound to G protein), RA (bound to agonist), or RAG (bound to both agonist and G protein). **B:** The cubic ternary complex model (Weiss et al, 1996). The receptor can exist in eight forms, as each of the four states of the ternary complex model in panel A can be either activated or inactivated. The activated form of the receptor is denoted by *.

Differential equations for this model are written down using the principle of mass action, together with the assumption that the total amount of receptor is conserved, and thus

$$[\mathrm{R}] + [\mathrm{RA}] + [\mathrm{RG}] + [\mathrm{RAG}] = \mathrm{R}_{\mathrm{tot}}. \tag{2.16}$$

Note that, in order to satisfy detailed balance (sometimes called microscopic reversibility) we must have $K_1K_2 = K_3K_4$, where $K_i = k_{-i}/k_i$.

The ternary complex model was later extended by Weiss et al (1996), who incorporated the assumption – motivated by experimental data from a number of receptor types – that the receptor could also exist in activated and inactivated states, no matter whether agonist and G protein were bound or not. This extends the ternary complex model to the *cubic ternary complex* model, with the 8-state reaction diagram as shown in Fig. 2.2 B. Again, the differential equations follow from mass action and conservation of the total number of receptors.

2.1.3 A Kinetic Model of GPCR Signalling

Undoubtedly the most detailed modelling of GPCR is the work of Falkenburger et al (2010a,b), with related papers by Falkenburger et al (2013) and Dickson et al (2013). This series of papers began with the experimental work of Jensen et al (2009) who used fluorescence resonance energy transfer (FRET) microscopy to measure, as functions of time, the concentrations of various intermediate species in the GPCR signalling pathway, after activation of the M1 muscarinic receptor. These data were used to fit a kinetic model of GPCR signalling, which was then used as the basis for the interpretation of other experimental results. These papers are an impressive combination of theory and experiment, and are the most detailed joint theoretical/experimental study yet performed of GPCR signalling and its connection to Ca^{2+} dynamics.

The model of Falkenburger et al (2010a) is a combination of the models discussed above. A diagram is shown in Fig. 2.3. In complexity, the receptor model is between a ternary complex model and a cubic ternary complex model, having six states (instead of the four of the ternary complex model and the eight of the cubic ternary complex model). This decision was motivated by experimental data that did not allow for the determination of the additional two states.

This receptor model is then coupled to a G protein model very similar to that in Fig. 2.1, just with one additional step in the PLC reactions. The additional step is inclusion of the PLC-α-GDP intermediate complex, which exists after the GTP has been converted to GDP, but before the intermediate complex dissociates.

We do not give a complete listing of the model equations and parameters. These can be found in the original paper of Falkenburger et al. What is more important is to see how the model solutions compare to experimental data, and these curves are shown in Fig. 2.4. For these curves, a saturating concentration of the agonist was applied (10 μM Oxo-M).

- The rate of solution change was measured experimentally, and mimicked in the model by adding agonist at the same rate (yellow curve).
- When agonist is applied, the first quantity to increase is the amount of receptor bound to agonist (black dots). In the model this is the sum of all the species with R bound to A (i.e., RA, RAG, and RAGβ), and is shown by the black curve.

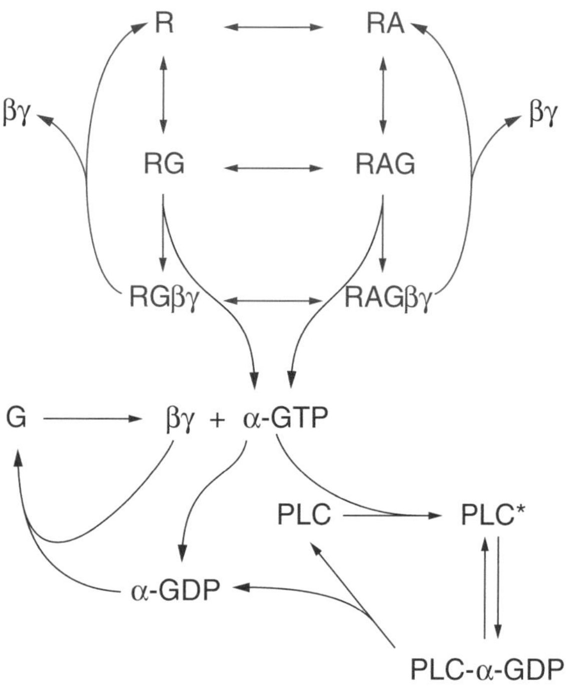

Fig. 2.3 Schematic diagram of the model of Falkenburger et al (2010a). The M1 muscarinic receptor states and reactions are given in red, while the reactions and species associated with the G protein are given in blue. The distinction is somewhat arbitrary, and is for clarity only. To avoid excessive complexity in the diagram, the binding of G and A to the receptor are not shown explicitly. However, those dependencies must be included in the kinetic equations.

- Next to rise is the concentrations of species with R bound to Gβ (red dots), which in the model is the sum of RG, RAG, RG$\beta\gamma$, and RAG$\beta\gamma$ (red curve). A comparison of the black and red curves shows that binding of G protein is a slower process than agonist binding, which is not unexpected.
- The green curve is the sum of PLC* and PLC-α-GDP. Although it might seem counterintuitive that PLC* increases faster than does the dissociation of G protein (the green curve increases faster than the blue curve does), this is partially due to the fact that the curves are both plotted relative to their maximum values. On an absolute scale both curves increase together. However, PLC saturates before the G protein has finished dissociating, which causes the observed lag when plotted on a relative scale.
- The final curve to rise, G$\alpha\beta\gamma$ (the blue curve), is the sum of all species with undissociated G proteins (i.e., RG, RAG, and G), but plotted as a fraction of the minimum value, not the maximum value. The G$\alpha\beta\gamma$ curve is plotted this way since the sum of all the undissociated G proteins is, necessarily, a decreasing curve (all G protein starts undissociated, and

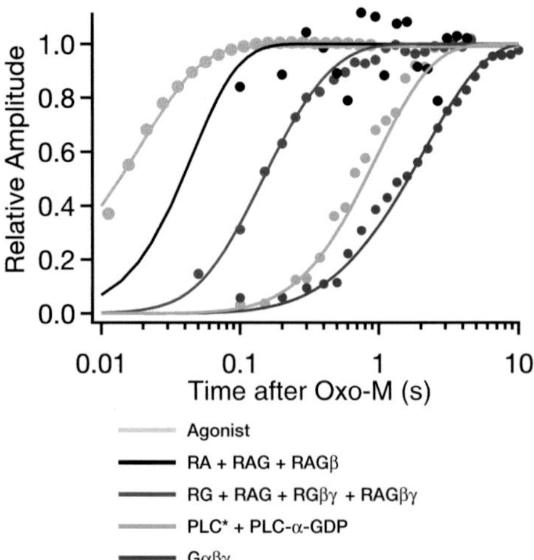

Fig. 2.4 Experimental data (solid circles) and model fits (smooth curves) from the model of Falkenburger et al (2010a), following the notation of Fig. 2.3. Oxo-M is the agonist. Original data provided by Bjorn Falkenburger (Aachen), Jill Jensen (University of Washington), and Bertil Hille (University of Washington).

addition of agonist causes dissociation into the α and $\beta\gamma$ subunits). Thus, the rate of rise of the G$\alpha\beta\gamma$ curve is a measure of the rate of dissociation of the G protein.

These curves tell us a great deal about the inner workings of the GPCR signalling cascade, and for further discussion and interpretation the interested reader is referred to the original article. However, it is worth pointing out that the monotonicity of these curves is a possible indication that the GPCR signalling cascade might be well approximated by simple increasing functions. For example, the amount of activated PLC appears to be a monotonically increasing function of time (after application of agonist), reaching steady state in approximately 5 seconds. For this receptor/agonist combination (and possibly for others also) these results suggest that simple models of PLC activation will be adequate for more detailed investigations of the downstream pathways, such as IPR and RyR activation.

One important feature that does not appear in the above model is desensitisation (by phosphorylation, for example) and recycling of the receptors. These aspects are modelled in more detail by Lemon et al (2003). Receptor phosphorylation also forms the basis for the oscillations observed in astrocytes in response to stimulation of the glutamate mGlu$_5$ receptor (Dupont et al, 2011b). However, we shall leave discussion of that model until Section 6.4, when we consider it in the context of a model of Ca^{2+} oscillations.

2.2 SERCA and PMCA

Although SERCA and PMCA are not identical ATPases, it is rare for models to make a detailed distinction between these two Ca^{2+} efflux pathways. Even if models were to do so, it is unlikely that such differences would make significant qualitative changes (or possibly even significant quantitative changes) to model results. Thus, here we present models for a generic Ca^{2+} ATPase. Such a model can be adapted for either SERCA or PMCA merely by the adjustment of parameters, an approach likely to be sufficiently accurate for most models.

SERCA models (as do most models in the Ca^{2+} toolbox) come in many different flavours, from exceedingly simple to considerably more complex. We shall discuss a selection of these models, starting with the simplest possible, to show how one can make models progressively more complex. Of course, we emphasise as strongly as possible that more complex does not necessarily mean better. The choice of SERCA model – or, indeed, the choice of any model – must be governed by the question under consideration.

2.2.1 Unidirectional Models

For each ATP consumed, a SERCA pump transfers two Ca^{2+} ions from the cytoplasm to the ER/SR (MacLennan et al, 1997). Thus, the simplest possible model, based on nothing but the law of mass action, would be

$$2\mathrm{Ca}^{2+}(\mathrm{cytoplasm}) + \mathrm{E} \xrightarrow{k} 2\mathrm{Ca}^{2+}(\mathrm{ER/SR}) + \mathrm{E}. \tag{2.17}$$

Here, the SERCA pump is denoted by E, to suggest that the pump is acting like an enzyme. The corresponding differential equation is

$$\frac{dc}{dt} = -2k[\mathrm{E}]c^2 = -\tilde{k}c^2, \tag{2.18}$$

where c denotes cytoplasmic $[Ca^{2+}]$. To make a modular model construction more convenient, we shall usually write this as

$$J_{\mathrm{serca}} = \tilde{k}c^2, \tag{2.19}$$

where J_{serca} denotes the Ca^{2+} flux due to the SERCA pump. (The use of J to denote a flux is standard.) Note that, by common usage, the flux is written as a function of Ca^{2+} that is always positive, and thus does not take the direction of the flux into account.

Although this model is startlingly simple, it is not necessarily a poor approximation to actual pump fluxes in a physiological regime. Nevertheless, it suffers from some major disadvantages. The first disadvantage is that the flux does not saturate as c increases, which is clearly unphysiological.

To correct this defect, we can model the pump using the Michaelis-Menten theory of enzyme reactions (Keener and Sneyd, 2008; Segel, 1975). Thus

$$2\text{Ca}^{2+}(\text{cytoplasm}) + \text{E} \underset{k_{-1}}{\overset{k_1}{\rightleftharpoons}} \text{C} \overset{k_2}{\longrightarrow} 2\text{Ca}^{2+}(\text{ER/SR}) + \text{E}. \tag{2.20}$$

C is some intermediate complex formed when the pump binds two cytoplasmic Ca^{2+}, and dissociates to release Ca^{2+} into the ER/SR. The corresponding differential equations are

$$\frac{dc}{dt} = 2k_{-1}\gamma - 2k_1 ec^2, \tag{2.21}$$

$$\frac{de}{dt} = k_{-1}\gamma - k_1 ec^2 + k_2\gamma, \tag{2.22}$$

$$\frac{dc_e}{dt} = 2k_2\gamma, \tag{2.23}$$

where $\gamma = [\text{C}]$ is the concentration of the intermediate complex, c_e is the concentration of Ca^{2+} in the ER/SR, and $e = [\text{E}]$.

Note that we do not need an additional equation for γ, since $e + \gamma = e_{\text{tot}}$, where e_{tot}, the total concentration of SERCA, is a constant.

These equations themselves are not particularly useful for modelling, as they contain rate constants and variables that are difficult to measure, and about which information is rarely, if ever, available. Instead, most analyses of enzyme kinetics are based on one of two approximations to these equations; the *equilibrium* approximation or the *quasi-steady-state* approximation. These approximations are discussed in detail in the literature (for example, Keener and Sneyd (2008)), and we shall not cover the basic theory again here. In this book we shall mostly just use the equilibrium approximation, as this is simpler to apply to complex models.

For the SERCA pump model, the equilibrium approximation assumes that the first of the reactions is in instantaneous equilibrium, and thus

$$k_{-1}\gamma = k_1 ec^2 = k_1(e_{\text{tot}} - \gamma)c^2, \tag{2.24}$$

from which it follows that

$$\gamma = \frac{k_1 e_{\text{tot}} c^2}{k_{-1} + k_1 c^2} = \frac{e_{\text{tot}} c^2}{K^2 + c^2}, \tag{2.25}$$

where

$$K^2 = \frac{k_{-1}}{k_1}. \tag{2.26}$$

Since the net rate of transport of Ca^{2+} is $k_2\gamma$, it follows that

$$J_{\text{serca}} = \frac{k_2 e_{\text{tot}} c^2}{K^2 + c^2}. \tag{2.27}$$

Finally, since $k_2 e_{\mathrm{tot}}$ is the maximal Ca^{2+} flux, attained when all the SERCA proteins have bound Ca^{2+}, we write the flux as

$$J_{\mathrm{serca}} = \frac{V_{\max} c^2}{K^2 + c^2}, \tag{2.28}$$

where $V_{\max}$ is the maximum velocity of the reaction. If c is small with respect to K then (2.19) becomes a good approximation.

In general, two Ca^{2+} ions do not bind simultaneously, but sequentially. We model this as

$$\mathrm{Ca}^{2+}(\text{cytoplasm}) + \mathrm{E} \underset{k_{-1}}{\overset{k_1}{\rightleftharpoons}} \mathrm{C}_1 \overset{k_2}{\longrightarrow} \mathrm{Ca}^{2+}(\text{ER/SR}) + \mathrm{E}, \tag{2.29}$$

$$\mathrm{Ca}^{2+}(\text{cytoplasm}) + \mathrm{C}_1 \underset{k_{-3}}{\overset{k_3}{\rightleftharpoons}} \mathrm{C}_2 \overset{k_4}{\longrightarrow} \mathrm{Ca}^{2+}(\text{ER/SR}) + \mathrm{C}_1. \tag{2.30}$$

Applying the equilibrium approximation to the reversible equations gives

$$ce = K_1 \gamma_1, \tag{2.31}$$

$$c\gamma_1 = K_3 \gamma_2, \tag{2.32}$$

where $K_i = \frac{k_{-i}}{k_i}$, and γ_i denotes the concentration of complex C_i.

Together with the conservation equation

$$e + \gamma_1 + \gamma_2 = e_{\mathrm{tot}}, \tag{2.33}$$

we can solve to find that

$$\gamma_1 = \frac{K_3 \gamma_2}{c}, \tag{2.34}$$

$$\gamma_2 = \frac{c^2 e_{\mathrm{tot}}}{c^2 + K_3 c + K_1 K_3}. \tag{2.35}$$

Hence, the velocity of the reaction is

$$J_{\mathrm{serca}} = k_2 \gamma_1 + k_4 \gamma_2 = \frac{e_{\mathrm{tot}}(k_4 c^2 + k_2 K_3 c)}{c^2 + K_3 c + K_1 K_3}. \tag{2.36}$$

Note how this is merely a general form of (2.28). The correspondence between (2.28) and (2.36) becomes even clearer in the limit as the binding of the second Ca^{2+} ion becomes very fast compared to the binding of the first ion (i.e., $K_3 \to 0$, $K_1 \to \infty$, $K_1 K_3 \to K^2$, for some constant K). Taking this limit gives

$$J_{\mathrm{serca}} = \frac{e_{\mathrm{tot}} k_4 c^2}{c^2 + K^2}. \tag{2.37}$$

This is called *positive cooperativity*, whereby the binding of the first ion makes the binding of the second ion much faster. Equation (2.37) is called a *Hill* function, with Hill coefficient 2, after the physiologist A.V. Hill, who won the 1922 Nobel Prize in Physiology and Medicine for his work on muscle.

As a general rule, enzyme reactions that are thought to be cooperative are modelled by Hill functions, with the coefficient determined, where possible, by comparison to experimental data. Indeed, Hill equations are used even in cases where the data indicate that the Hill coefficient is non-integer, which is clearly not physically possible according to the derivation of the equation. However, the more general form, (2.36), is rarely used, as experimental data is rarely accurate enough to allow for a clear distinction to be made between (2.36) and (2.37).

Parameters for SERCA pump models are often taken from Lytton et al (1992), who showed that the rate of pumping by a variety of SERCA pumps (types 1, 2A, 2B and 3) can be well described by Hill functions, with Hill coefficients of around 2, and K values ranging from 0.27 to 1.1 μM.

2.2.2 Bidirectional Models

One major flaw of the above models is that the Ca^{2+} flux is only from the cytoplasm to the ER/SR, or to outside the cell. In reality, ATPase pumps are bidirectional, and once the $[Ca^{2+}]$ in the ER/SR gets high enough, the net pump flux will approach zero. This plays a significant role in setting the steady-state concentration of Ca^{2+} in the ER/SR.

There are two major ways to construct bidirectional SERCA models. Firstly, one can make the above unidirectional models bidirectional by adding in a reverse reaction. Secondly, one can construct more elaborate Markov models of the pump. We shall consider both approaches briefly.

The above unidirectional models can be made bidirectional by the simple addition of a reverse reaction, to give

$$2\mathrm{Ca}^{2+}(\mathrm{cytoplasm}) + \mathrm{E} \underset{k_{-1}}{\overset{k_1}{\rightleftharpoons}} \mathrm{C} \underset{k_{-2}}{\overset{k_2}{\rightleftharpoons}} 2\mathrm{Ca}^{2+}(\mathrm{ER/SR}) + \mathrm{E}. \tag{2.38}$$

Applying the equilibrium approximation to this reaction (i.e., assuming that cytoplasmic Ca^{2+} is in instantaneous equilibrium with the ATPase), and following the same procedure as before gives

$$J_{\mathrm{serca}} = e_{\mathrm{tot}} \frac{k_1 k_2 c^2 - k_{-1} k_{-2} c_e^2}{k_1 c^2 + k_{-1}}. \tag{2.39}$$

Note that the steady state is reached when

$$c^2 = K_1 K_2 c_e^2. \tag{2.40}$$

Thus $K_1 K_2$ determines the concentrating power of the ATPase; the smaller are K_1 and K_2, the slower are the reverse reactions, and the greater is the concentrating power of the pump. Usually $K_1 K_2 \ll 1$, which explains why $c \ll c_e$ at steady state.

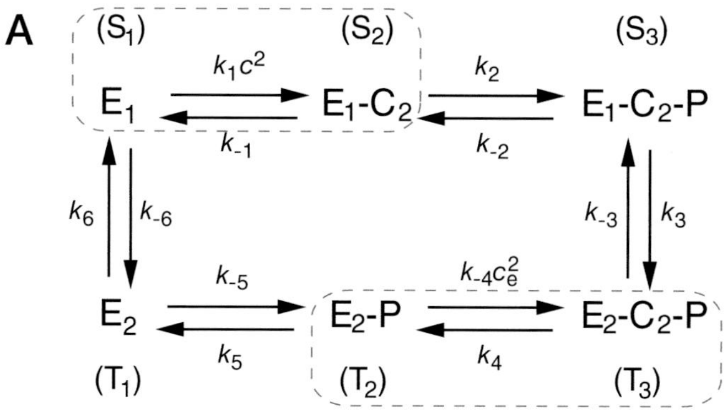

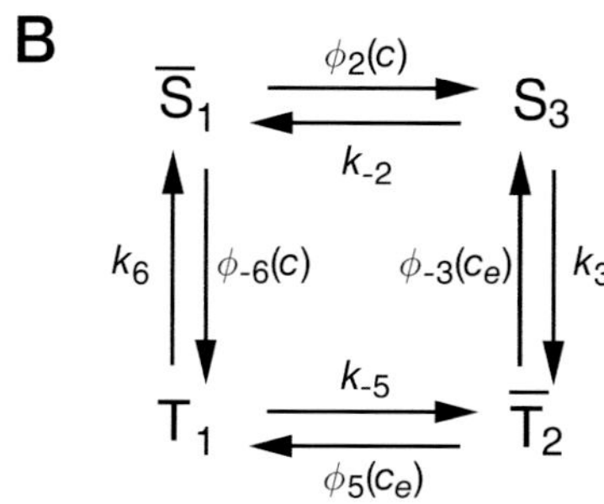

Fig. 2.5 **A:** Schematic diagram of the SERCA model of MacLennan et al (1997). E_1 is the conformation in which the Ca^{2+} binding sites are exposed to the cytoplasm, and E_2 is the conformation in which they are exposed to the ER/SR lumen and have a much lower affinity for Ca^{2+}. The P denotes that the pump has been phosphorylated; for simplicity ATP and ADP have been omitted from the diagram. The transport of H^+ has also been omitted. **B:** By assuming fast equilibrium between the pairs of states shown grouped in panel A (the dashed boxes), this diagram in which there are fewer states, but with Ca^{2+}-dependent transitions, can be derived. The functions in the transition rates are $\phi_2 = c^2 k_2/(c^2 + K_1)$, $\phi_{-3} = k_{-3} K_4 c_e^2/(1 + K_4 c_e^2)$, $\phi_5 = k_5/(1 + K_4 c_e^2)$, and $\phi_{-6} = K_1 k_{-6}/(K_1 + c^2)$.

A similar expression can be derived from a more complex Markov model of the pump (MacLennan et al, 1997). The schematic diagram of the model is shown in Fig. 2.5 A. The pump can be in one of two basic conformations, E_1 and E_2. In the E_1 conformation the pump binds two Ca^{2+} ions from the cytoplasm, whereupon it exposes a phosphorylation site. Once phosphorylated, the pump switches to the E_2 conformation in which the Ca^{2+} binding sites are exposed to the ER lumen and have a much lower affinity for Ca^{2+}. Thus, Ca^{2+} is released into the ER, the pump is dephosphorylated, and the cycle is completed when the pump switches back to the E_1 conformation. For each Ca^{2+} ion transported from the cytoplasm to the ER, two protons are transported in the reverse direction. However, for simplicity, we ignore the effects of H^+ concentration on the pump rate because we are not concerned here with the effects of cellular pH. The rate-limiting step of the transport cycle is the transition from E_1-C_2-P to E_2-C_2-P.

In the original description of the model, MacLennan et al assumed that the binding and release of Ca^{2+} occurs quickly. Thus, in Fig. 2.5 A, states S_1 and S_2 have been grouped together by a box with a dotted outline, as have states T_2 and T_3. The assumption of fast equilibrium gives

$$s_1 = \frac{K_1}{c^2} s_2, \tag{2.41}$$

with a similar expression for t_2 and t_3. Here, we denote the fraction of pumps in state S_1 by s_1, and similarly for the other states. We now define two new variables: $\bar{s}_1 = s_1 + s_2$ and $\bar{t}_2 = t_2 + t_3$. From (2.41) it follows that

$$\bar{s}_1 = s_1 \left(1 + \frac{c^2}{K_1}\right) = s_2 \left(1 + \frac{K_1}{c^2}\right). \tag{2.42}$$

Hence, the rate at which $\bar{S}_1$ is converted to S_3 is $k_2 s_2 = \frac{c^2 k_2 \bar{s}_1}{c^2 + K_1}$. Similarly, the rate at which $\bar{S}_1$ is converted to T_1 is $k_{-6} s_1 = \frac{K_1 k_{-6} \bar{s}_1}{K_1 + c^2}$.

Repetition of this process for each of the transitions results in the simplified model shown in Fig. 2.5 B. The form of the functions ϕ can be understood intuitively. For example, as c increases, the equilibrium between S_1 and S_2 is shifted further towards S_2. This increases the rate at which S_3 is formed, but decreases the rate at which S_1 is converted to T_1. Thus, the transition from $\bar{S}_1$ to S_3 is an increasing function of c, while the transition from $\bar{S}_1$ to T_1 is a decreasing function of c.

Such Markov models can be incorporated as is into a model for Ca^{2+} dynamics. For example, the simplified ATPase model consists of three differential equations

$$\frac{d\bar{s}_1}{dt} = k_{-2} s_3 + k_6 t_1 - (\phi_2 + \phi_{-6}) \bar{s}_1, \tag{2.43}$$

$$\frac{ds_3}{dt} = \phi_2 \bar{s}_1 + \phi_{-3} - (k_{-2} + k_3) s_3, \tag{2.44}$$

$$\frac{dt_1}{dt} = \phi_{-6} \bar{s}_1 + \phi_5 \bar{t}_2 - (k_6 + k_{-5}) t_1, \tag{2.45}$$

together with the conservation equation

$$\bar{s}_1 + s_3 + t_1 + \bar{t}_2 = 1. \tag{2.46}$$

Equations (2.43)–(2.46) can be solved numerically and then J_{serca} calculated from

$$J_{\text{serca}} = \phi_2(c)\bar{s}_1 - k_{-2}s_3, \tag{2.47}$$

which gives a complete description of the ATPase pump flux.

Mostly, however, the detailed parameters of the Markov model are not known, in which case the steady-state flux is used instead. The steady-state flux is found by setting (2.43)–(2.45) each to be equal to zero, combining with (2.46) to solve for the four variables, and then substituting the result into (2.47). This gives

$$J_{\text{serca}} = \frac{c^2 - K_1K_2K_3K_4K_5K_6c_e^2}{\beta_1 c^2 + \beta_2 c_e^2 + \beta_3 c^2 c_e^2 + \beta_4}, \tag{2.48}$$

for some $\beta_1 \ldots \beta_4$ which are polynomial functions of the rate constants, of so little interest that we do not bother to say exactly what they are.

2.2.3 Coupling to ATP and pH

Since the SERCA pump is an ATPase it is necessarily dependent on ATP, and thus its operation is tied closely to cellular metabolism. In addition, SERCA pumps are also dependent on pH, as for every Ca^{2+} ion transported, two H^+ ions are transported in the opposite direction (Levy et al, 1990; Yu et al, 1993). Incorporation of ATP and pH dependence into the models allows for the construction of models of enormous complexity (Haynes and Mandveno, 1987; Gould et al, 1986), running (in one case) to well over 100 states.

However, Tran et al (2009) have shown that a thermodynamically accurate, pH-dependent and ATP-dependent SERCA model can be constructed in a much simpler fashion. Starting with a 12-state model that included binding of two Ca^{2+} ions, binding of ATP, and competitive inhibition by H^+, they showed that quasi-equilibrium approximations (essentially the same procedures as used above in Fig. 2.5) could be used to simplify the model to three states, without losing any significant ability to reproduce the data of Ji et al (1999). Indeed, for many conditions, their three-state model could be simplified even further to a two-state model, without undue loss of accuracy.

Since the simplification procedures are identical in spirit to those already demonstrated in this section, just slightly more complex, we do not give them in detail. However, it is useful to make two observations. Firstly, although the complex 12-state model can be simplified to a two or three-state model, the original model is not without use, as its structure determines the form of the rate constants in the simplified model. That is, the structure of the full model determines how the concentrations of H^+ and ATP appear in the rate constants of the simplified model. Secondly, models should be constructed appropriately, based on the available data. If you have limited data there is little point in trying to fit a model with 100 states and hundreds of rate constants. For example, the data of Ji et al (1999) shows, in essence, that the pump rate is a sigmoidally increasing function (i.e., a Hill function) of $[Ca^{2+}]$, and that the K_m of the Hill function shifts with pH. It does not take a hugely complex model to reproduce this. Indeed, making the model too complex does nothing more than ensure that the parameters are undetermined by the data. The model of Tran et al (2009) is thus an excellent example of sensible modelling; starting from a relatively complex model, incorporating what are believed to be the key steps in the SERCA cycle, they showed how the model can be progressively simplified without losing the ability to reproduce the data. Their final models are both elegant and accurate.

2.3 The Sodium/Calcium Exchanger

Because it relies on two concentration differences, an antiport transporter such as the Na^+/Ca^{2+} exchanger can act as a pump (Blaustein and Lederer, 1999). Although this transporter is a passive pump (because it consumes no chemical energy directly) it is often described as a secondarily active pump; it uses the Na^+ concentration gradient to pump Ca^{2+} out of the cell against its concentration gradient, but energy is required to establish and maintain the Na^+ concentration gradient. Na^+/Ca^{2+} exchange is an important mechanism for Ca^{2+} removal in a number of cell types, such as some neurons, as well as cardiac ventricular cells, in which much of the Ca^{2+} that enters the cell during an action potential is removed from the cell by the Na^+/Ca^{2+} exchanger. It has therefore been studied extensively, and a number of highly detailed models have been constructed (Mullins, 1977; Hilgemann et al, 1991; Matsuoka, 2002; Kang and Hilgemann, 2004). Nearly all of these models are of the isoform found in cardiac cells (the NCX1 isoform). Nevertheless, there is still considerable controversy over such basic questions as the exact stoichiometry of the exchange (Hilgemann, 2004).

Just as with SERCA models, NCX models come in two different flavours. Firstly, one can consider the NCX as an unidirectional enzyme with two substrates, cytosolic Ca^{2+} and external Na^+, or, secondly, one can write a bidirectional Markov model similar to that in Fig. 2.5. We shall consider simple examples of both types of model. We shall also briefly consider a bidirectional enzyme model that results in an expression for the flux that is very similar to that of the bidirectional Markov model.

2.3.1 Unidirectional Enzyme Model

If we consider the NCX as an enzyme with two substrates, internal Ca^{2+} (with concentration c_i) and external Na^+ (with concentration n_e), then we can write the reaction of the transporter in the form shown in Fig. 2.6. Assuming that all the reactions are at equilibrium gives the equations

$$Ec_i = K_1 E_1, \tag{2.49}$$
$$En_e^3 = K_2 E_2, \tag{2.50}$$
$$E_2 c_i = K_3 E_3, \tag{2.51}$$
$$E_1 n_e^3 = K_4 E_3, \tag{2.52}$$

where, as usual, $K_i = k_{-i}/k_i$. We also have the conservation equation $E + E_1 + E_2 + E_3 = E_t$, where E_t is the total amount of exchanger.

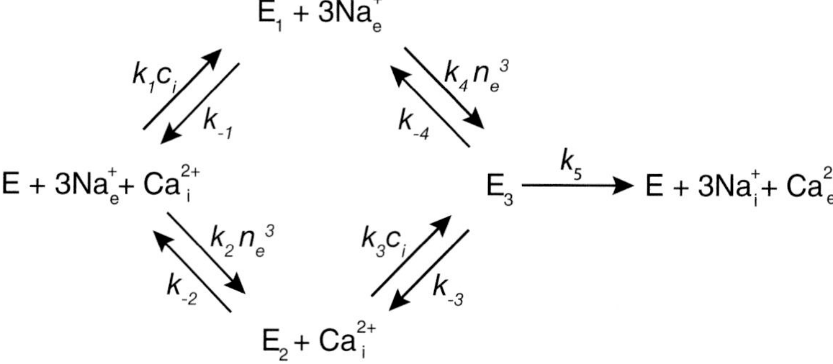

Fig. 2.6 Schematic diagram of the unidirectional enzyme model of the Na^+/Ca^{2+} exchanger. c_i is the concentration of cytosolic Ca^{2+}, n_e is the concentration of extracellular Na^+. E_1 denotes the complex where the exchanger has one Ca^{2+} bound, E_2 has three Na^+ bound, and E_3 has both one Ca^{2+} and three Na^+ bound.

Although this appears to be overdetermined, having five equations for only four unknowns (E, E_1, E_2, and E_3) this is in fact not so. In order for (2.49)–(2.52) to be consistent it must be that $K_4 K_1 = K_2 K_3$, which is simply an expression of detailed balance around the reaction loop. With this constraint, (2.49)–(2.52) are not independent and thus we have four equations to solve for the four unknowns. We shall also assume, again for convenience, that $K_1 = K_3$ and $K_2 = K_4$, and thus that binding of Ca^{2+} is independent of whether or not Na^+ is bound. Although this is not necessary, it gives an expression for the flux which is commonly used.

Solution of (2.49)–(2.51), together with the conservation constraint, gives

$$\frac{E_3}{E_t} = \left(\frac{c_i}{K_1 + c_i}\right)\left(\frac{n_e^3}{K_2 + n_e^3}\right). \tag{2.53}$$

The flux, J_{NCX}, through the exchanger is $k_5 E_3$, and since $k_5 E_t$ is the maximal velocity, $V_{\max}$, it follows that

$$J_{\mathrm{NCX}} = V_{\max}\left(\frac{c_i}{K_1 + c_i}\right)\left(\frac{n_e^3}{K_2 + n_e^3}\right). \tag{2.54}$$

For most studies of Ca^{2+} dynamics, a simple expression such as this is accurate enough.

2.3.2 Bidirectional Markov Model

Next we discuss the second flavour of model, where the exchanger is modelled as a bidirectional Markov model, one not specifically for the NCX1 isoform, but applicable more generally. In this model (see Fig. 2.7), E_i is the exchanger protein in the conformation for which the binding sites are exposed to the interior of the cell, and E_e is the conformation for which the binding sites are exposed to the exterior. Starting at state X_1 in the top left of the figure, the exchanger can bind Ca^{2+} inside the cell, simultaneously releasing three Na^+ ions to the interior. A change of conformation to E_e then allows the exchanger to release the Ca^{2+} to the outside and bind three external Na^+. A return to the E_i conformation completes the cycle. Of course, it is a crude approximation to assume that one Ca^{2+} and three Na^+ ions bind or unbind the exchanger simultaneously.

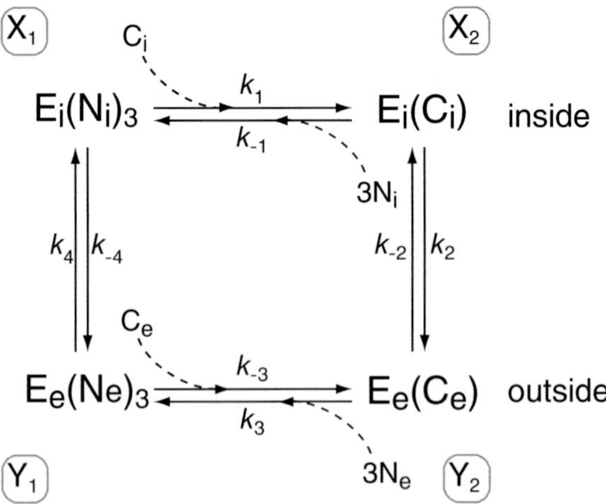

Fig. 2.7 Schematic diagram of the bidirectional Markov model of the $\mathrm{Na}^+/\mathrm{Ca}^{2+}$ exchanger.

To calculate the steady-state flux, we first solve for the steady-state values of x_1, x_2, y_1, and y_2, the fraction of exchangers in the state X_1, X_2, Y_1, and Y_2, respectively. There are four equations: three differential equations for exchanger states and one conservation equation. They are

$$\frac{dx_1}{dt} = k_{-1}n_i^3 x_2 + k_4 y_1 - (k_1 c_i + k_{-4})x_1, \tag{2.55}$$

$$\frac{dx_2}{dt} = k_{-2}y_2 + k_1 c_i x_1 - (k_2 + k_{-1}n_i^3)x_2, \tag{2.56}$$

$$\frac{dy_1}{dt} = k_{-4}x_1 + k_3 n_e^3 y_2 - (k_4 + k_{-3}c_e)y_1, \tag{2.57}$$

$$1 = x_1 + x_2 + y_1 + y_2. \tag{2.58}$$

Here, c and n denote, respectively, Ca^{2+} and Na^+ concentration, and the subscripts e or i represent external or internal concentrations.

Note that by taking $x_1 + x_2 + y_1 + y_2 = 1$, we have made a subtle but significant assumption, which can be interpreted in two equivalent ways. First, we can interpret the model as referring to a population of exchangers, in which case each variable refers to the fraction of exchangers in the relevant state, and the model describes the mean behaviour of the population. Alternatively, we can interpret this model as referring to the behaviour of a single exchanger, in which case the variables are probabilities of being in a given state, and the exchanger is modelled as a Markov process.

Solving for the steady state of these equations, the flux is found to be

$$\begin{aligned} J_{\mathrm{NCX}} &= k_4 y_1 - k_{-4}x_1 \\ &= \frac{c_i n_e^3 - K_1 K_2 K_3 K_4 c_e n_i^3}{\beta_1 c_i + (\beta_2 + \beta_3 c_i)c_e + (\beta_4 + \beta_5 c_e)n_i^3 + (\beta_6 + \beta_7 c_i + \beta_8 n_i^3)n_e^3}, \end{aligned} \tag{2.59}$$

where, as usual, $K_i = k_{-i}/k_i$. (Note that we could have defined J_{NCX} to be the difference between the forward and backward rates for any of the transitions, and we would have obtained the same answer, due to the cyclical nature of the model.) Each of the eight coefficients in the denominator is positive, and a function of the rate constants. The flux is a saturating function of the four concentrations, as it must be in order to make physical sense – since the number of exchangers is held fixed, the flux must necessarily have an upper limit.

Notice that here the units of J_{NCX} are 1/time, because the variables x_i and y_i are fractions of exchangers in a particular state (or probabilities) rather than concentrations of exchangers in a particular state. Hence, the flux in this model is a turnover rate, i.e., how many times the exchanger goes around the cycle per unit time. This can be converted to a concentration per time if the concentration of the exchangers is known.

Also note the similarity in form between (2.59) and (2.48). This is a general feature of Markov models of this type. The numerator is generally a relatively simple function of ratios of the rate constants, and of the concentrations on either side of the membrane. It is in the denominator that the majority of the complexity appears. Nevertheless, despite this seeming complexity, not every term in the denominator is always vital, and thus the expression for the flux can often be well approximated by simpler equations.

2.3.3 Modelling an Electrogenic Exchanger

Since each cycle of the Na^+/Ca^{2+} exchanger transports two positive charges out and three positive charges in, it generates an electric current. Such exchangers are said to be *electrogenic*. This implies that some of the rate constants must be functions of the membrane potential.

We begin with a simple example for illustration. Consider a ligand, L, with a charge z, and suppose that there is a process that moves L from the cell interior with potential V_i to the cell exterior with potential V_e, i.e.,

$$\mathrm{L}_i \underset{k_-}{\overset{k_+}{\rightleftharpoons}} \mathrm{L}_e. \tag{2.60}$$

The change in free energy (see Keener and Sneyd (2008) for more details) for this reaction is

$$\begin{aligned}\Delta G &= G^0_{\mathrm{L}_e} + RT\ln([\mathrm{L}_e]) + zFV_e - G^0_{\mathrm{L}_i} - RT\ln([\mathrm{L}_i]) - zFV_i \\ &= RT\ln\left(\frac{[\mathrm{L}_e]}{[\mathrm{L}_i]}\right) - zFV,\end{aligned} \tag{2.61}$$

where $V = V_i - V_e$ is the transmembrane potential, F is Faraday's constant, R is the gas constant, and T is temperature in Kelvin.

G^0_{L} is the standard free energy of L, which we assume to be the same on both sides of the membrane, and thus $G^0_{\mathrm{L}_e} = G^0_{\mathrm{L}_i}$. At equilibrium, we must have $\Delta G = 0$, so that

$$\frac{[\mathrm{L}_i]_{\mathrm{eq}}}{[\mathrm{L}_e]_{\mathrm{eq}}} = \exp\left(\frac{-zFV}{RT}\right). \tag{2.62}$$

However, at equilibrium we also have

$$\frac{[\mathrm{L}_i]_{\mathrm{eq}}}{[\mathrm{L}_e]_{\mathrm{eq}}} = \frac{k_-}{k_+} = K, \tag{2.63}$$

and thus

$$K = \exp\left(\frac{-zFV}{RT}\right). \tag{2.64}$$

This simple example shows that, for an electrogenic exchanger, the rate constants must be functions of the potential difference across the membrane, in such a way that the equilibrium constant satisfies (2.64).

Now we consider the more complex Na^+/Ca^{2+} exchanger. Here, the net reaction begins with three Na^+ outside the cell and one Ca^{2+} inside the cell, and ends with three Na^+ inside the cell and one Ca^{2+} outside. We can write this as

$$3\mathrm{Na}_e^+ + \mathrm{Ca}_i^{2+} \longrightarrow 3\mathrm{Na}_i^+ + \mathrm{Ca}_e^{2+}. \tag{2.65}$$

The change in free energy for this reaction is

$$\Delta G = RT \ln\left(\frac{n_i^3 c_e}{n_e^3 c_i}\right) + FV. \tag{2.66}$$

At equilibrium we must have $\Delta G = 0$, in which case

$$\frac{n_{i,\mathrm{eq}}^3 c_{e,\mathrm{eq}}}{n_{e,\mathrm{eq}}^3 c_{i,\mathrm{eq}}} = \exp\left(-\frac{FV}{RT}\right). \tag{2.67}$$

Next, note that the principle of detailed balance requires that around any closed reaction loop the product of the forward rates must be the same as the product of the reverse rates. For the model in Fig. 2.7 it follows that

$$k_1 c_{i,\mathrm{eq}} k_2 k_3 n_{e,\mathrm{eq}}^3 k_4 = n_{i,\mathrm{eq}}^3 k_{-1} k_{-4} c_{e,\mathrm{eq}} k_{-3} k_{-2}, \tag{2.68}$$

and thus

$$K_1 K_2 K_3 K_4 = \frac{c_{i,\mathrm{eq}}}{c_{e,\mathrm{eq}}} \frac{n_{e,\mathrm{eq}}^3}{n_{i,\mathrm{eq}}^3}. \tag{2.69}$$

Combining (2.68) and (2.69), we get

$$K_1 K_2 K_3 K_4 = \exp\left(\frac{FV}{RT}\right), \tag{2.70}$$

which, being independent of the concentrations, must hold in general.

Note that the NCX makes a net transfer of one positive charge into the cell, which is equivalent to one negative charge moving out of the cell. This explains the change in sign from (2.64) to (2.70).

It follows that

$$J_{\mathrm{NCX}} = \frac{c_i n_e^3 - \exp\left(\frac{FV}{RT}\right) c_e n_i^3}{\beta_1 c_i + (\beta_2 + \beta_3 c_i) c_e + (\beta_4 + \beta_5 c_e) n_i^3 + (\beta_6 + \beta_7 c_i + \beta_8 n_i^3) n_e^3}. \tag{2.71}$$

Since each β is a function of the rate constants, which, from (2.70), are themselves functions of V, the dependence on V can be substantially more complex than is immediately apparent.

In writing (2.70), no assumption was made about where the charge transfer takes place. From Fig. 2.7 it might appear that the charge transfer takes place during the transitions $Y_1 \rightarrow X_1$ and $X_2 \rightarrow Y_2$. However, this is not necessarily the case. If we assume that one Ca^{2+} ion is transferred from inside to outside during the $X_2 \to Y_2$ transition, and three Na^+ ions are transferred during the $Y_1 \to X_1$ transition, free energy arguments yield the additional constraints

$$\frac{k_{-2}}{k_2} = \tilde{K}_2 \exp\left(\frac{-2FV}{RT}\right), \qquad \frac{k_4}{k_{-4}} = \tilde{K}_4^{-1} \exp\left(\frac{-3FV}{RT}\right), \tag{2.72}$$

where $\tilde{K}_2$ and $\tilde{K}_4$ are independent of voltage, and where $K_1\tilde{K}_2K_3\tilde{K}_4 = 1$.

The most important observation is that, for given n_i and n_e (set by other mechanisms such as the Na^+–K^+ ATPase), a negative V enhances the rate at which the Na^+/Ca^{2+} exchanger removes Ca^{2+} from the cell. This makes sense; if V is negative, the potential inside the cell is negative compared to the outside and thus it is easier for the exchanger to move one positive charge into the cell. Since cells typically have a negative resting potential, the electrogenic nature of the exchanger increases its ability to remove Ca^{2+} in resting conditions.

2.3.4 Bidirectional Enzyme Model

A slightly different bidirectional model can be derived by extending the diagram of Fig. 2.6 to include both forward and reverse reactions. This kind of model is described in detail in Segel (1975) – where it is called a random bi-bi rapid equilibrium model – and is the basis of the more general NCX model of Pradhan et al (2010).

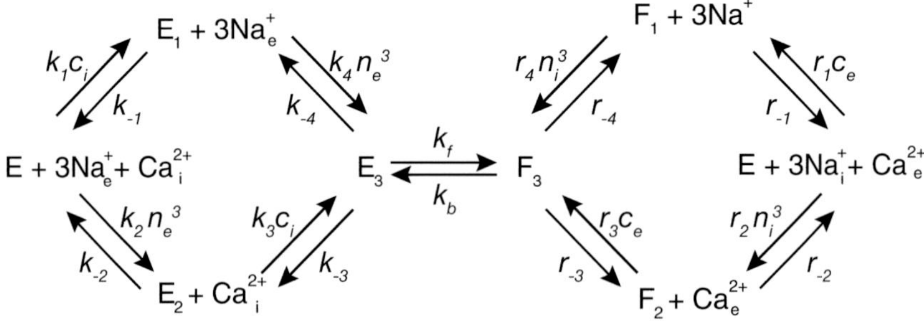

Fig. 2.8 Schematic diagram of the bidirectional enzyme model of the Na^+/Ca^{2+} exchanger. c_i and c_e are the concentrations of internal and external Ca^{2+}, respectively, with analogous definitions for n_e and n_i. E_1 denotes the complex where the exchanger has one internal Ca^{2+} bound, E_2 has three external Na^+ bound, and E_3 has both one internal Ca^{2+} and three external Na^+ bound. The definitions of F_1, F_2, and F_3 are analogous, but using external Ca^{2+} and internal Na^+.

The extended diagram is shown in Fig. 2.8. If we assume that all the reactions, with the exception of the $E_3 \rightleftharpoons F_3$ transition are at equilibrium, then we can write seven equations for the seven unknowns (each reaction loop gives only three equations, not four, because of the consistency requirement described in Section 2.3.1, and we also have the conservation equation that the total amount of enzyme remains constant, which we assume to be 1 for simplicity). For convenience, we shall also assume that $K_1 = K_3$ and $R_1 = R_3$, where, as usual, $K_i = k_{-i}/k_i$ and $R_i = r_{-i}/r_i$.

Use of a symbolic manipulation package such as Maple or Mathematica (the manipulations are too tedious to do by hand) gives

$$\begin{aligned} J &= k_f E_3 - k_b F_3 \\ &= \frac{k_f \tilde{c}_i \tilde{n}_e^3 - k_b \tilde{c}_e \tilde{n}_i^3}{1 + \tilde{n}_i^3 + \tilde{c}_e + \tilde{c}_e \tilde{n}_i^3 + \tilde{n}_e^3 + \tilde{c}_i + \tilde{c}_i \tilde{n}_e}, \end{aligned} \tag{2.73}$$

where $\tilde{c}_i = c_i/K_1$, $\tilde{c}_e = c_e/R_1$, $\tilde{n}_i^3 = n_i^3/R_2$, $\tilde{n}_e^3 = n_e^3/K_2$.

Note the similarity between (2.73) and (2.59). There is little practical difference between these expressions and it would be difficult to distinguish between them on the basis of experimental measurements.

2.3.5 A Model with Variable Stoichiometry

The most detailed sequence of models of the cardiac Na^+/Ca^{2+} exchanger (NCX1), in development for over 20 years, are those from Hilgemann, Matsuoka and their colleagues (Hilgemann et al, 1991, 1992; Matsuoka and Hilgemann, 1992; Matsuoka, 2002; Kang and Hilgemann, 2004), originally based on the earlier work of Mullins (1977), among others.

The most recent version of these models is based on data showing that the stoichiometry of the exchanger is not exactly 3:1, but is actually 3.2:1. To reproduce such non-integer stoichiometry the model assumes that the exchanger can go through multiple different reaction loops, each of which has a different stoichiometry. By balancing the strength of each reaction loop, the model can generate a variety of overall stoichiometries. Not only does the model reproduce the data well, but it was also used to make a number of predictions (about flux at a variety of Na^+ and Ca^{2+} concentrations) that were confirmed experimentally. Derivation of the steady-state flux through this model uses the same methods as before, just with many more terms because of the complexity of the model, so we do not give the details here.

2.4 Mitochondria

Mitochondria are a major intracellular Ca^{2+} store, taking up Ca^{2+} via the mitochondrial Ca^{2+} uniporter (MCU), and releasing Ca^{2+} into the cytoplasm through the mitochondrial Na^+/Ca^{2+} exchanger (NCLX), an isoform of the membrane NCX (Carafoli, 2003; Rizzuto et al, 2012). Ca^{2+} plays an important role in the control of mitochondrial function and thus the energetic balance of the cell is closely tied to Ca^{2+} homeostasis. Intuitively, it makes sense that ATP synthesis is promoted at the same time as a Ca^{2+}-induced physiological response, so that the cell can maintain its overall energy levels. Although these facts have been known for several decades, there is still debate over the extent to which mitochondria modulate Ca^{2+} signalling on the time scale of oscillations and waves. The current consensus seems to be that mitochondria are most likely to modulate Ca^{2+} signalling in microdomains between the ER and the mitochondria, although this is unlikely to be the only effect (Lukyanenko et al, 2009; Olson et al, 2012; Csordás and Hajnóczky, 2009; Csordás et al, 2001; Walter and Hajnóczky, 2005; Ishii et al, 2006).

In part because of this uncertainty, and in part because of the complexity involved in studying microdomains quantitatively, models of Ca^{2+} dynamics have sometimes paid less attention to mitochondrial transport than perhaps they ought to have done. This being said, it is worthwhile noting that one of the earliest observations of Ca^{2+} oscillations (albeit very heavily damped ones) was in fact that of Carafoli et al (1966) in mitochondrial suspensions, who pointed out the likely existence of feedback mechanisms to cause such behaviour.

One of the earliest models to consider mitochondrial Ca^{2+} in detail was that of Magnus and Keizer (1997), a model that was later presented in greater detail (Magnus and Keizer, 1998a,b). More recent studies of mitochondrial Ca^{2+} transport, with particular focus on cardiac cells, are those of Cortassa et al (2003) and Nguyen and Jafri (2005), while Patterson et al (2007) studied the relative contribution of Ca^{2+} fluxes from the ER, mitochondria and from the outside. Models of mitochondrial Ca^{2+} handling in cardiac myocytes, with references to models in other cell types also, are reviewed by Lukyanenko et al (2009).

The effects of mitochondrial Ca^{2+} transport on whole-cell Ca^{2+} dynamics have received relatively little attention. Falcke et al (1999), based on the experimental work of Jouaville et al (1995), showed how the inclusion of mitochondrial Ca^{2+} transport terms could affect the spatiotemporal properties of Ca^{2+} in the *Xenopus* oocyte (discussed in more detail in Section 6.1.2), while Marhl et al (1998) showed how mitochondria could help to maintain Ca^{2+} oscillations of constant amplitude. The possible involvement of mitochondria in complex oscillations of Ca^{2+} was studied by Marhl et al (2000), while Sneyd et al (2003), based on the experimental work of Tinel et al (1999) and Straub et al (2000), showed how the mitochondria can partially prevent the spread of Ca^{2+} waves from the apical to the basal region in parotid acinar cells, thus modulating the rate of fluid secretion (Section 6.3.2).

2.4.1 The Mitochondrial Uniporter

The major route for Ca^{2+} uptake into the mitochondria is via the mitochondrial Ca^{2+} uniporter (Olson et al, 2012; Santo-Domingo and Demaurex,

2010). The uniporter behaves like a Ca^{2+} channel that is activated by Ca^{2+} on the cytoplasmic side, and current through this channel is driven by the potential difference (usually denoted Ψ) across the inner mitochondrial membrane. Since $\Psi = V_{\text{in}} - V_{\text{out}} \approx -180$ mV, the Nernst equation says that Ca^{2+} will continue to flow into the mitochondria until the ratio of cytosolic to mitochondrial concentration is approximately 10^{-6}. For a resting cytosolic concentration of 100 nM this would imply the uniporter alone would result in a mitochondrial concentration of 100 mM.

Mitochondrial $[Ca^{2+}]$ never gets this high for two principal reasons. Firstly, the uniporter is activated only at high cytosolic $[Ca^{2+}]$, so that as soon as $[Ca^{2+}]$ falls the uniporter shuts off. Secondly, influx into the mitochondria is balanced by efflux through the mitochondrial Na^+/Ca^{2+} exchanger, which prevents mitochondrial $[Ca^{2+}]$ getting high.

Since the uniporter operates only when $[Ca^{2+}]$ is high on the cytosolic side of the channel (earlier studies reported a K_d of around 10–20 μM, although more recent studies now suggest that the value could be an order of magnitude lower (Csordás et al, 2013)), it is now believed that uniporter function often relies on spatially restricted microdomains of Ca^{2+} (Rizzuto et al, 1999; Pacher et al, 2000; Csordás and Hajnóczky, 2003). For instance, if IPR (or RyR) and the uniporter both open on to the same restricted spatial domain, release of Ca^{2+} through the IPR could cause a high $[Ca^{2+}]$ in this restricted domain, which would then activate the uniporter. In this way, Ca^{2+} could be transferred from the ER to the mitochondria, with only a short passage through the cytoplasm. However, there is evidence that the uniporter can operate at nM Ca^{2+} concentrations, with no absolute necessity for a microdomain, and this matter cannot be said to be entirely resolved (Santo-Domingo and Demaurex, 2010).

Since the uniporter is an ion channel that is opened, or activated, by binding of a ligand – in this case Ca^{2+}, with cooperativity of approximately two – the flux through the uniporter can be described as a product of two terms; the probability that the uniporter is open (or, equivalently, the fraction that are open in a population of uniporters) multiplied by the flux through the uniporter when it is open. Thus

$$J_{\text{uni}} = N g(c)\phi(\Psi, c), \tag{2.74}$$

where N is the total number of uniporters in the population, g denotes the fraction that are open, and ϕ is the current through a single open channel. In general, g will be a function of many things, including the potential difference across the membrane and the concentrations of various ligands; here, for simplicity, it is assumed to be a function only of $[Ca^{2+}]$, denoted, as usual, by c. For many ion channels g will also depend on time, as, for example, in the Hodgkin-Huxley model (Hodgkin and Huxley, 1952), where the time dependence of the Na^+ and K^+ channels plays such a vital role. However, neither time nor voltage dependence of g will play any role in our uniporter models.

There are many ways to model the current through a single open channel (see, for example, Hille (2001) or Keener and Sneyd (2008)). One of the most common is the Goldman-Hodgkin-Katz (GHK) equation, according to which

$$\phi(\Psi, c) = P_c \frac{2F}{RT}\Psi \left(\frac{c_m - c\exp\left(\frac{-2F\Psi}{RT}\right)}{1-\exp\left(\frac{-2F\Psi}{RT}\right)} \right), \tag{2.75}$$

where c_m is the mitochondrial Ca^{2+} concentration, and P_c is the permeability coefficient (with units of distance/time. Hence, the units of ϕ are concentration $\times$ distance/time, or, equivalently, moles per time per unit area). Often, the concentrations in (2.75) are multiplied by activity coefficients, which was the approach taken by Nguyen and Jafri (2005) in their model of the uniporter using the GHK equation.

Next, we need to model the activation of the uniporter by cytoplasmic Ca^{2+}. Data indicate that two Ca^{2+} act cooperatively to activate the uniporter, and thus we could simply write

$$g(c) = \frac{c^2}{K_d^2 + c^2}, \tag{2.76}$$

where K_d would be around 10–20 μM, giving a uniporter with low Ca^{2+} affinity. Combining (2.74)–(2.76) then gives a relatively simple model of the uniporter, similar to that constructed by Magnus and Keizer (1997). Magnus and Keizer used a more complicated allosteric model for Ca^{2+} cooperativity, based on a Monod-Wyman-Changeux approach, but the basic principle is the same as that described here.

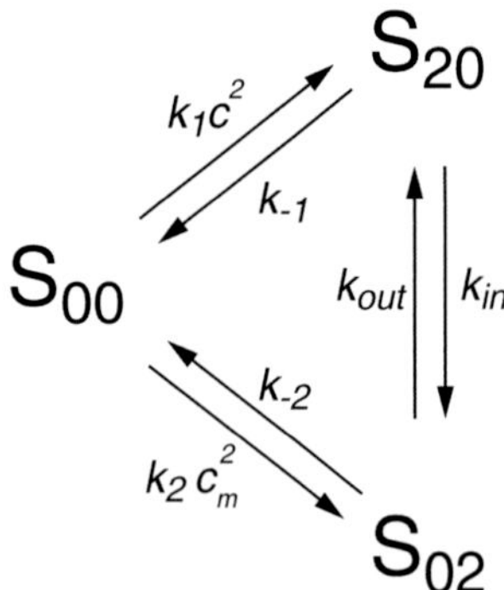

Fig. 2.9 Schematic diagram of one version of the uniporter model of Dash et al (2009). The uniporter can exist in three states: S_{00} (no Ca^{2+} bound), S_{20} (two Ca^{2+} bound to the cytoplasmic side), and S_{02} (two Ca^{2+} bound to the mitochondrial side). S_{00} is assumed to be in equilibrium with the other two states. The concentration of Ca^{2+} in the cytoplasm is c, and in the mitochondria is c_m.

A slightly different approach was taken by Dash et al (2009), who combined a barrier model of the ion channel with a Markov model of channel activation. In their model, the uniporter was assumed to exist in one of three states (Fig. 2.9), where we have simplified the Ca^{2+} binding by assuming that two Ca^{2+} ions bind simultaneously.

Equilibrium between S_{00} and the other two states gives

$$c^2 S_{00} = K_1 S_{20}, \tag{2.77}$$

$$c_m^2 S_{00} = K_2 S_{02}. \tag{2.78}$$

Since the total amount of uniporter is conserved, it follows that

$$\begin{aligned} S_{tot} &= S_{00} + S_{20} + S_{02} \\ &= S_{20}\left(1 + \frac{K_1}{c^2} + \frac{c_m^2}{K_2}\frac{K_1}{c^2}\right) \\ &= S_{02}\left(1 + \frac{K_2}{c_m^2} + \frac{K_2}{c_m^2}\frac{c^2}{K_1}\right). \end{aligned} \tag{2.79}$$

Hence

$$\begin{aligned} J_{uni} &= k_{in} S_{20} - k_{out} S_{02} \\ &= S_{tot}\frac{k_{in}\frac{c^2}{K_1} - k_{out}\frac{c_m^2}{K_2}}{1 + \frac{c^2}{K_1} + \frac{c_m^2}{K_2}}. \end{aligned} \tag{2.80}$$

Since the uniporter is moving two Ca^{2+} ions, i.e., four positive charges, across a potential difference of $-\Psi$, we know from the discussion in Section 2.3.3 that we must have

$$\frac{K_2}{K_1}\frac{k_{out}}{k_{in}} = \exp\left(\frac{4F\Psi}{RT}\right), \tag{2.81}$$

which gives an additional constraint on the rate constants.

It is worth noting that the existing models of the MCU were all constructed before the recent molecular identification of the MCU (Baughman et al, 2011; De Stefani et al, 2011), and the consequent expansion of experimental information about this protein (Csordás et al, 2013; Patron et al, 2014). Older models will have to be reexamined and, most likely, revised, to take these new data into account.

2.4.2 The Mitochondrial Sodium/Calcium Exchanger

The mitochondrial Na^+/Ca^{2+} exchanger (NCLX) is an isoform of the membrane NCX, with similar transport properties, and can thus be modelled in essentially the same way as described in Section 2.3. Such models give, in general, the flux through the NCLX as a rational function of the Ca^{2+} and Na^{+} concentrations in the cytoplasm and the mitochondria. More complex models give rational functions that look more complicated, but the underlying behaviour is essentially unchanged.

One of the earliest models of the NCLX is that of Wingrove and Gunter (1986) who fit their data to the unidirectional enzyme model (Section 2.3.1)

$$J_{\mathrm{NCLX}} = V_{\max} \left(\frac{c_m}{K_c + c_m} \right) \left(\frac{n_c^2}{K_n^2 + n_c^2} \right), \tag{2.82}$$

where n_c is the cytosolic Na^{+} concentration. This expression was also used by Magnus and Keizer (1997) and Cortassa et al (2003) who multiplied the flux equation by a voltage-dependent pre-factor to take into account how the flux changes with the inner membrane potential difference.

Interestingly, Colegrove et al (2000a,b) used the same unidirectional enzyme model in their study of depolarisation-induced Ca^{2+} fluxes in the bullfrog sympathetic neuron, but concluded that the dependence on Na^{+} concentration and voltage was negligible. Indeed, by assuming that the NCLX flux was a function only of c_m they were able to construct a highly accurate and predictive model. This suggests that in some situations a greatly simplified model of the NCLX might be all that is required, or, indeed, all that can be reliably constructed without falling into the trap of an underdetermined model, for which it is impossible to find the rate constants unambiguously.

More recent models (Cortassa et al, 2003; Nguyen and Jafri, 2005; Dash and Beard, 2008; Pradhan et al, 2010) have used the reversible NCX models discussed in Sections 2.3.2 and 2.3.4, with the model of Pradhan et al (2010) being the most general, and thus the most complex in form. However, just as with models of the NCX, the simpler models are accurate enough for most studies of whole-cell Ca^{2+} dynamics.

2.5 Voltage-Gated Calcium Channels

Influx of Ca^{2+} into the cell is a highly regulated process, occurring through a number of pathways, one of the most important of which is through voltage-gated Ca^{2+} channels (VGCC) (Catterall, 2011; Dolphin, 2006; Catterall, 2000; Tsien and Tsien, 1990). For example, in cardiac cells, Ca^{2+} influx through L-type voltage-gated Ca^{2+} channels is required to stimulate further release of Ca^{2+} from the SR, and thus contraction. Similarly, arrival of the action potential at a neuronal synapse causes depolarisation of the cell membrane, opening of VGCC, and thus entry of Ca^{2+}, which is required for the secretion of neurotransmitter.

There are ten different types of VGCC, with two different naming schemes. The older naming scheme, which distinguishes between L-type, N-type,

P/Q-type, R-type, and T-type channels was based on physiological properties such as their speed of inactivation, their single-channel conductance, and their sensitivity to various pharmacological agents. L-type channels, for example, are characterised by a high voltage of activation, a large single-channel conductance, and a slow voltage-dependent inactivation. They are the predominant VGCC in cardiac, smooth, and skeletal muscle cells. More recent structural information about VGCC has allowed for a more precise description of the various types of VGCC based on the type of α_1 subunit. Ca_v1.1 – Ca_v1.4 are all L-type channels, Ca_v2.1 – Ca_v2.3 are N, P/Q, and R-type channels, respectively, and Ca_v3.1 – Ca_v3.3 are all T-type channels. Ca_v1.1 are the L-type channels that mediate excitation-contraction coupling in skeletal muscle, while Ca_v1.2 mediate, among other things, excitation-contraction coupling in cardiac and smooth muscle (Catterall, 2011).

Although models of the different types of VGCC differ in their choice of parameters, and thus can exhibit quite different behaviours, there is less variability in the underlying model structure, as the same basic approaches are used for all the different channel types. For this reason we shall not present here a comprehensive list of all the various VGCC models that have been used, but shall concentrate instead on illustrating the types of model structures that have been used.

2.5.1 The Simplest Models

Many models of neuroendocrine cells, or other neurons, use relatively simple models of VGCC. Rather than using detailed Markov models of the ion channel (of the kind we shall see in the next section), they instead just assume that the current, I_{Ca}, is given by

$$I_{\mathrm{Ca}} = N_{\mathrm{Ca}} g_{\mathrm{Ca}}(V,t)\phi(V,c), \tag{2.83}$$

where V is the membrane potential, c is the intracellular $[\mathrm{Ca}^{2+}]$, N_{Ca} is the total number of channels, $g_{\mathrm{Ca}}(V,t)$ is the (possibly time-dependent) open probability of a single channel, and ϕ is the current through a single open channel.

Typical choices for ϕ are a linear I-V relationship or the GHK equation, but there are many other possibilities (see the next section). Similarly, $g_{\mathrm{Ca}}(V,t)$ can be chosen to satisfy any desired voltage and time dependencies. N_{Ca} is usually just considered constant, and amalgamated into g_{Ca}.

For example, the classic model by Chay and Keizer (1983) of electrical bursting in the pancreatic β cell simply modelled the Ca^{2+} current in exactly the same way as the Na^+ channel in the Hodgkin-Huxley model. Thus they set $g_{\mathrm{Ca}}(V,t) = \bar{g}_{\mathrm{Ca}} m^3 h$, where m and h satisfy the same equations as in the Hodgkin-Huxley model, and set $\phi = V - V_{\mathrm{Ca}}$, where V_{Ca} is the Nernst potential of Ca^{2+}.

This general approach, with minor modifications, has been widely used. To take another example, which is typical of many other neuroendocrine models, LeBeau et al (2000), in their study of bursting in GT1 neurons, included two Ca^{2+} currents; a T-type Ca^{2+} current modelled by

$$I_{\mathrm{CaT}} = g_{\mathrm{Ca,T}} m^2_{\mathrm{Ca,T}} h_{\mathrm{Ca,T}} (V - V_{\mathrm{Ca}}), \tag{2.84}$$

and an L-type current modelled by

$$I_{\mathrm{CaL}} = g_{\mathrm{Ca,L}} m^2_{\mathrm{Ca,L}} (V - V_{\mathrm{Ca}}). \tag{2.85}$$

Note that in this model the T-type VGCC inactivates (due to the term $h_{\mathrm{Ca,T}}$), while the L-type VGCC does not. However, as we shall see in the next section, more complex models of the L-type VGCC include voltage-induced inactivation.

Destexhe and Huguenard (2000), following on from the earlier work of Huguenard and Prince (1992), used a similar formulation for their model of the T-type VGCC , with the only difference being that they used a GHK expression for ϕ, rather than the linear expression of LeBeau et al.

2.5.2 Permeation Models of Calcium Channels

One important question, that we shall not consider in detail, is how to model the I-V curve, $\phi(V)$, of an open Ca^{2+} channel. There is a large and complex literature on the modelling of ion flow through channels, well beyond the scope of this book. The interested reader is referred to the classic book by Hille (2001), or a more introductory presentation by Keener and Sneyd (2008).

The major problem is to understand how a Ca^{2+} channel can be so selective for Ca^{2+}, while still allowing a large current. In addition, we would like to understand how the structure of the channel determines its function, and how experimentally observed I-V curves, over the entire range of measured voltages, can be explained by fundamental physical principles.

The commonly used linear and GHK I-V curves do not provide answers to such questions. They provide reasonable fits to I-V curves of a range of different channels (including Ca^{2+} channels), but provide only heuristic explanations at best. Slightly more complex models, the so-called barrier models, assume that the ions move through the channel by a process of jumping over energy barriers, and thus the rate of ion flow can be described by Eyring rate theory, or Kramer's rate theory (Eyring et al, 1949; Hänggi et al, 1990). An early model of this type was constructed by Hess and Tsien (1984), and provided an excellent fit to a wide range of data. However, despite the complexity of barrier models, they also are essentially heuristic, with little basis in the underlying physics of the channel.

More recently, another class of models was constructed by Robert Eisenberg and his colleagues (Chen et al, 1992; Chen and Eisenberg, 1993a,b; Nonner and Eisenberg, 1998; Nonner et al, 1999; Horng et al, 2012; Liu and Eisenberg, 2013, 2014b,a), based on a continuum model of the electric field within the channel, and using the Poisson-Nernst-Planck (PNP) equations. Both the linear and GHK I-V curves can be derived as limiting forms of PNP models. Although PNP models provide excellent fits to a wide range of data, they are not without controversy. Proponents of Brownian dynamics, or molecular dynamics simulations, are quick to point out perceived flaws in PNP theory, and propose their own models, based on the simulated movement of individual ions within a channel (Corry et al, 2000; Kuyucak and Chung, 2002).

We remain apart from this particular fray, merely observing that the proper way to model the I-V curve of a Ca^{2+} channel remains uncertain, but is certainly complex. We shall thus restrict ourselves to the use of linear or GHK I-V curves, while admitting that such simplicity may not be ideal.

2.5.3 Inactivation of Calcium Channels by Calcium

Many VGCC, including L-type VGCC, are inactivated by Ca^{2+}. This is an important control mechanism, particularly in cardiac cells, where L-type VGCC experience very high $[Ca^{2+}]$ on their cytoplasmic side. Attempts to model Ca^{2+}-induced inactivation of VGCC go back over 30 years, to the pioneering work of Chad et al (1984) and Chad and Eckert (1986), but one of the most interesting of the early models is due to Sherman et al (1990), as this model is based on an early concept of a Ca^{2+} microdomain, an idea that has since become ubiquitous.

According to Sherman et al (1990), Ca^{2+} entering through a Ca^{2+} channel immediately forms a region of high $[Ca^{2+}]$ at the mouth of the channel. The Ca^{2+} in this microdomain then binds to the channel, blocking it, but on a slower time scale than the formation of the Ca^{2+} microdomain at the channel mouth, or the voltage-induced opening of the channels. Although this Ca^{2+} microdomain hypothesis might seem standard today, for its time this model provided real insight into how Ca^{2+} channels might operate.

In the Sherman model, the channel can exist in three states: closed (C), open (O), or with Ca^{2+} bound (B). In state B the channel is closed. Movement between these states is governed by the transitions

$$\mathrm{C} \underset{k_{-1}(V)}{\overset{k_1(V)}{\rightleftharpoons}} \mathrm{O}, \tag{2.86}$$

$$\mathrm{O} + \mathrm{Ca}^{2+} \underset{k_{-2}}{\overset{k_2}{\rightleftharpoons}} \mathrm{B}, \tag{2.87}$$

with corresponding equations

$$\frac{dC}{dt} = k_{-1}(V)O - k_1(V)C, \tag{2.88}$$

$$\frac{dO}{dt} = k_1(V)C - k_{-1}(V)O + k_{-2}B - k_2[\mathrm{Ca}^{2+}]_m O, \tag{2.89}$$

$$C + O + B = 1. \tag{2.90}$$

Here we use the italic C to denote the fraction of channels in state C, and similarly for O and B. V is the membrane potential, and $[\mathrm{Ca}^{2+}]_m$ is the microdomain $[\mathrm{Ca}^{2+}]$, at the mouth of the channel.

The microdomain $[\mathrm{Ca}^{2+}]$ is not modelled explicitly (as, for example, by using partial differential equations, as described in Section 3.4). Instead it is assumed that $[\mathrm{Ca}^{2+}]_m$ is a linear function of the current through the channel, and thus

$$[\mathrm{Ca}^{2+}]_m = AI(V), \tag{2.91}$$

for some negative constant A, which depends on such things as the level of Ca^{2+} buffering in the cytoplasm.

$I(V)$ is the current-voltage relationship of the channel; typical examples would be a linear I-V curve or the GHK equation. We need to be a little careful about the sign of the current. By convention, positive ions moving into the cell is denoted by a negative current, and thus a negative current gives a positive $[\mathrm{Ca}^{2+}]_m$, as of course we require. This explains the requirement for $A < 0$ in (2.91). It is clear that this expression for $[\mathrm{Ca}^{2+}]_m$ will not work for positive Ca^{2+} currents, or even for zero Ca^{2+} currents, but we would not use this model in such circumstances.

A short calculation shows that the steady-state value for O, for a given fixed voltage, V, is

$$O_{\mathrm{ss}} = \frac{K_2}{K_2 + K_1K_2 + AI(V)}, \tag{2.92}$$

where, as usual, $K_i = k_{-i}/k_i$. As V increases, $I(V)$ increases (i.e., it becomes less negative), and thus $AI(V)$ decreases. It follows that, at steady state, O_{ss} is an increasing function of V, which makes sense, as we know that a VGCC opens in response to depolarisation.

However, in response to a step increase in voltage, O does not merely increase to its new value. Instead, O first increases and then decreases. We can see this by the following argument.

1. At the resting voltage, V_0 (typically $V_0 \approx -70$ mV), almost all the channels will be in the C state, as $K_1(V_0) \gg 1$. In particular, $B \approx 0$ at rest.
2. When V is suddenly increased in a step, and then held fixed at the new value, $\tilde{V}$, say, K_1 will immediately decrease, as k_1 increases and (possibly) k_{-1} decreases.

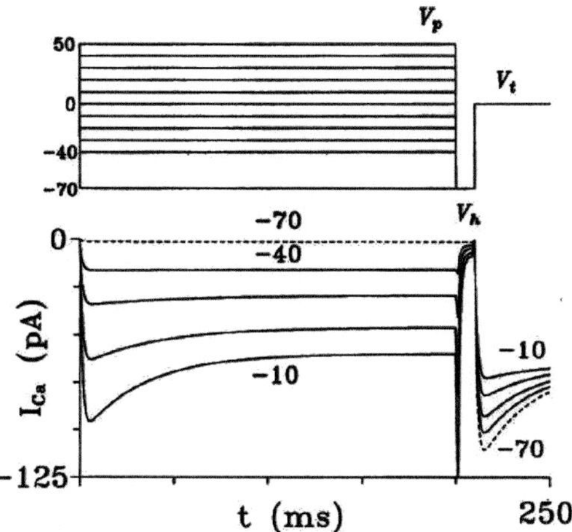

Fig. 2.10 Typical solutions of the model of Sherman et al (1990). The upper panel shows the voltage, V. The lower panel shows the Ca^{2+} current, $O(V,t)I(V)$. Voltage is stepped from the resting value of $V_h = -70$ mV to varying values (V_p ranging from -40 mV to -10 mV). At 200 ms V is stepped back to the resting value, V_h, and then increased again to $V_t = 0$ mV. These model simulations agree well with the experimental results of Plant (1988). Reproduced from Sherman et al (1990), Fig. 3, with permission from Elsevier.

3. However, if we assume that the C to O transition is much faster than the O to B transition, it follows that C will equilibrate with O, while B remains almost unchanged, at approximately zero. In this case we will have

$$C = K_1(\tilde{V})O, \tag{2.93}$$

while $B \approx 0$. Hence, $C + O \approx 1$ (by conservation of receptors), in which case we can solve for O to get the quasi-steady state

$$O_{\text{qss}} = \frac{1}{1 + K_1(\tilde{V})}. \tag{2.94}$$

4. Over a longer time scale, some of the channels in state O will move to state B, before O finally reaches its steady state of

$$O_{\text{ss}} = \frac{1}{1 + K_1(\tilde{V}) + AI(\tilde{V})/K_2} < \frac{1}{1 + K_1(\tilde{V})}. \tag{2.95}$$

Thus, as claimed, O first increases, then decreases. In other words, the channel inactivates in response to a Ca^{2+} current through the channel. Typical solutions are shown in Fig. 2.10.

2.5.4 A Two-Mode Model of Calcium-Induced Inactivation

The most detailed model of the L-type channel is that of Jafri et al (1998), based on an earlier model of Imredy and Yue (1994). It is a model of Monod-Wyman-Changeux type, a schematic diagram of which is shown in Fig. 2.11. There are two basic conformations of the channel, called, respectively, the normal (top row, denoted N) and Ca^{2+} (bottom row, denoted C) conformations. State O is the only open state. Depolarisation increases the rate at which the channel moves from left to right in either the normal or Ca^{2+} conformation. However, it is only from the normal conformation that the channel can change to the open state (transition N_4 to O). Binding of Ca^{2+} causes the channel to move from the normal conformation to the Ca^{2+} conformation (i.e., into a conformation from which it cannot access the open state) thus inactivating the receptor. The rate constants within each mode are voltage-dependent, but the final transition to the open channel state is not.

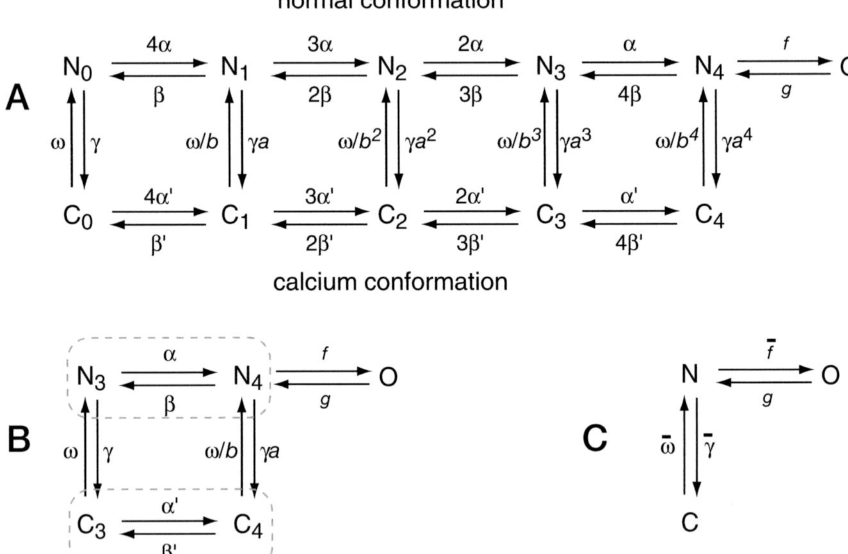

Fig. 2.11 **A:** The two-mode model of the L-type Ca^{2+} channel due to Jafri et al (1998). Activation by voltage occurs because α is an increasing function of V, while β is a decreasing function of V. Inactivation by Ca^{2+} occurs because γ is an increasing function of $[Ca^{2+}]$. To satisfy detailed balance, $\alpha' = a\alpha$ and $\beta' = \beta/b$. **B:** A simplified model obtained by truncating the leftmost three columns. **C:** The simplest version of the model, obtained by assuming a fast equilibrium between N_3 and N_4, and between C_3 and C_4, as shown by the dashed boxes in part B. From those fast equilibrium assumptions it follows that $\bar{\omega} = \omega\left(\frac{\beta'+\alpha'/b}{\alpha'+\beta'}\right)$, $\bar{\gamma} = \gamma\left(\frac{\beta+a\alpha}{\alpha+\beta}\right)$, and $\bar{f} = \frac{\alpha f}{\alpha+\beta}$.

The model assumes further that the channel consists of four independent and identical subunits. Each subunit can exist in an open or closed state (although, of course, a single open subunit does not allow Ca^{2+} current), and N_i denotes a channel that is in the normal conformation and has i subunits in the open state, with a similar definition for C_i. The channel can open only when it is in the normal conformation and all four subunits are in the open state.

If we know the $[Ca^{2+}]$ at the mouth of the channel, the schematic diagram in Fig. 2.11 allows us to compute the open probability of the channel by the reasonably simple process of writing down the differential equations of the Markov model, using the law of mass action, and then solving them numerically to find O as a function of time. Of course, it is not always a simple matter to calculate the $[Ca^{2+}]$ at the mouth of the channel, but that problem is discussed in more detail in Section 3.4.

Good agreement with experimental data can be obtained by simpler versions of this model (Hinch et al, 2006). First, six channel states are omitted, to get the simpler diagram shown in Fig. 2.11 B, and then the two remaining states in the normal conformation are amalgamated by assuming fast equilibrium, as are the remaining two states in the Ca^{2+} conformation (Fig. 2.11 C).

To complete the channel model, two things remain. Firstly, as well as Ca^{2+}-dependent inactivation, the channel also is inactivated by voltage, and this needs to be included in the model. Secondly, it remains to decide how to model the current through the open channel.

In the Jafri et al version of the two-mode model, time- and voltage-dependent inactivation of the channel is accomplished by a heuristic inactivation variable, of the same type as in the Hodgkin-Huxley equation. This approach is essentially identical to that discussed at the beginning of Section 2.5. Thus the current is proportional to a variable y, which satisfies the differential equation

$$\tau_y(V)\frac{dy}{dt} = y_\infty(V) - y, \tag{2.96}$$

for some specified functions of voltage, τ_y and y_∞, which we do not give here in detail. The function y_∞ is a decreasing function of V, and thus y acts to inactivate the channel when the membrane is depolarised. Finally, the I-V curve of an open channel was assumed to be a GHK equation.

If we put this all together, the current through the L-type Ca^{2+} channel is given by

$$I_{\rm Ca} = OP_{\rm Ca}y\frac{2F}{RT}V\frac{[{\rm Ca}^{2+}]_{\rm in} - [{\rm Ca}^{2+}]_{\rm out}\exp\left(\frac{-2FV}{RT}\right)}{1-\exp\left(\frac{-2FV}{RT}\right)}, \tag{2.97}$$

where $P_{\rm Ca}$ is the permeability of the channel to Ca^{2+}, $[Ca^{2+}]_{\rm in}$ is the Ca^{2+} concentration at the mouth of the channel, and $[Ca^{2+}]_{\rm out}$ is the concentration outside the cell. The variables y and O are determined by associated differential equations, as discussed above.

Of course, this expression for the current does not tell us how to calculate $[Ca^{2+}]_{in}$, which is a difficult problem itself. We discuss this in Section 3.4, when we discuss how to model Ca^{2+} microdomains.

More recent work from Winslow's group has used a simplified version of the above two-mode model (Greenstein et al, 2006; Gauthier et al, 2011). Firstly, most of the closed channel states, in both the normal and Ca^{2+} modes, were omitted. Secondly, voltage-dependent inactivation was not modelled as a separate independent variable, y, but was included directly in the Markov model, as a third mode. These simplifying assumptions lead to a 10-state model, but one that we shall not discuss in any further detail.

2.6 Receptor-Operated and Store-Operated Channels

Because they are so important in excitable cells such as neurons and striated muscle cells, voltage-gated Ca^{2+} channels have been studied in great detail, and models of them abound. However, when we turn to the study of other kinds of Ca^{2+} entry, we find far fewer models. This is due to a number of factors. Firstly, our knowledge of the proteins that mediate store-operated entry (i.e., STIM1, STIM2, and Orai1) is relatively recent, not to mention incomplete. Thus, there has been little opportunity for detailed mathematical models of SOCC, and few data. Secondly, it remains unclear how important are the quantitative details of ROCC and SOCC. Certainly we know that Ca^{2+} influx is highly important for the control of intracellular Ca^{2+} dynamics. What we do not know is whether the dynamic regulation of such entry is important. Models suggest that Ca^{2+} entry controls the intracellular Ca^{2+} dynamics over longer time scales than that of a single oscillation. For example, slowly increasing Ca^{2+} entry can increase the total amount of Ca^{2+} in the ER/SR, leading to activation of RyR and spontaneous Ca^{2+} release. As another example, changes in Ca^{2+} entry during Ca^{2+} oscillations can slowly modulate the average ER Ca^{2+} concentration, leading to modulation of the oscillations over a longer time scale, but without a significant dynamic effect on a single oscillation. However, these are just model hypotheses, and the exact situation in any particular cell type is unclear.

2.6.1 Receptor-Operated Channels

There is currently in the literature no detailed model of receptor-operated Ca^{2+} entry. In most models that use ROCC as the basis for Ca^{2+} entry, the

Ca^{2+} influx is merely assumed to be an increasing algebraic function, often even just a linear function, of agonist stimulation. Thus, as agonist stimulation increases, so does receptor-operated Ca^{2+} entry. A typical example is

$$J_{\text{in}} = \alpha_0 + \alpha_1 S, \tag{2.98}$$

where S denotes the level of agonist stimulation. In models that explicitly include GPCR activation, influx can be analogously described as a linear function of G protein activation (Kummer et al, 2000).

In cases where feedback on IP_3 production can be neglected, and $[IP_3]$ is a monotonic function of S, this can be simplified even further to

$$J_{\text{in}} = \beta_0 + \beta_1 p, \tag{2.99}$$

where p is the IP_3 concentration (Dupont and Goldbeter, 1993).

The few known (or suspected) details of how this occurs – for example the production of DAG, leading to the formation of arachidonic acid, and subsequent activation of ARC channels (Shuttleworth and Mignen, 2003) – are amalgamated into a single algebraic step, and any time dependencies are ignored. This is, of course, a drastic simplification, but in the absence of more detailed knowledge of the pathways involved there is usually little point in constructing more detailed models.

There is, as yet, no evidence that feedback modulation of ROCC (for example, via Ca^{2+} modulation of arachidonic acid production or of TRPC channels) plays an important role in the control of intracellular Ca^{2+} dynamics, although this is not unlikely. In the absence of any such direct evidence, a simple algebraic relationship suffices for most models.

2.6.2 Store-Operated Channels

The field of SOCC models is almost as thin. Although a number of models have used the basic principle that a decrease in ER Ca^{2+} concentration leads to increased influx across the plasma membrane (for a recent example, see Croisier et al (2013)), there are, as yet, few detailed models of the underlying mechanisms.

An early attempt to fit a SOCC model to experimental data was that of Ong et al (2006). Although this model is not based on data about STIM and Orai (these data mostly came later than the model), it is a useful study of the time course of SOCC activation. Experimentally this was done as follows:

- Thapsigargin was added in conditions of zero external Ca^{2+}. Thapsigargin blocks SERCA pumps, leading to a transient rise in cytosolic $[Ca^{2+}]$ and a permanent decrease in ER $[Ca^{2+}]$. SOCC might well be activated, but no Ca^{2+} current comes through them, as there is no Ca^{2+} outside the cell.

- At varying times after the addition of thapsigargin, Ca^{2+} is restored to the extracellular medium. This gives a large and persistent Ca^{2+} response. The size of the Ca^{2+} response is an indication of the extent to which SOCC have been activated.

To describe this behaviour requires an integrated model of whole-cell Ca^{2+} dynamics (of the type we discuss in Chapter 3), but we do not go into this level of detail here, studying only the SOCC model in isolation.

The model assumes that SOCC are activated by a decrease in ER $[Ca^{2+}]$, and inactivated by an increase in cytoplasmic $[Ca^{2+}]$ (denoted by c). However, it is not the $[Ca^{2+}]$ in the bulk ER that affects SOCC activation. Instead, the model splits the ER into two regions: one just under the ER membrane and the bulk ER. It is the $[Ca^{2+}]$ in the ER submembrane region (denoted by c_e) that controls SOCC activation.

A second important feature of the model is the incorporation of time dependency via a heuristic inactivation variable, h. There is no mechanistic basis to the form of h, which is determined solely by fitting to experimental data.

The SOCC model is thus

$$J_{\text{soc}} = f(c_e)h, \tag{2.100}$$

$$\tau_h \frac{dh}{dt} = h_\infty(c) - h, \tag{2.101}$$

where f and h_∞ are decreasing functions, whose exact form is of less immediate interest. Since there is no mechanistic interpretation of these functions, only their shape matters.

Maximal activation of SOCC occurs approximately 300 seconds after application of thapsigargin (Ong et al, 2006). In the model, this time scale is determined by the time scale of inactivation of SOCC by cytoplasmic Ca^{2+} (with time constant $\tau_h = 300$ s), not by the intrinsic dynamics of SOCC components such as STIM1 and Orai1. However, significant levels of Ca^{2+} entry through SOCC occur quickly; 20 seconds after application of thapsigargin, the level of SOCC entry has already reached approximately a third of the maximum value.

Steady-state properties of I_{CRAC} (in a Jurkat E6-1 cell line; the data are from Luik et al (2008)) are shown in Fig. 2.12. As $[Ca^{2+}]$ in the ER decreases, the SOCC current increases, with a functional form similar to a reverse Hill function, with half-maximal SOCC current when $[Ca^{2+}]_{\text{ER}} = 169\ \mu$M, and a Hill coefficient of 4.2. Thus it appears that the chain of events from STIM activation to the formation of the STIM/Orai complex has at least one step that is cooperative. Liu et al (2010) constructed a simple model of STIM and Orai that can reproduce these steady-state data, but their model incorporates no spatial aspects, and is highly preliminary at best.

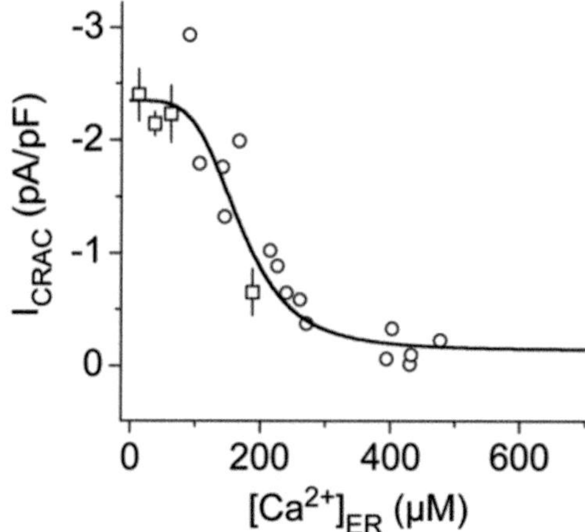

Fig. 2.12 Data from Jurkat T cells, showing how the SOCC current (in this case I_{CRAC}) depends on $[\mathrm{Ca}^{2+}]$ in the ER. The smooth curve is the Hill function fit. Reproduced from Luik et al (2008), Fig. 1c, with permission from Nature Publishing Group.

2.6.3 STIM–Orai Binding

Although the kinetics of how STIM diffuses to form punctate clusters with Orai has not been modelled in detail, a detailed steady-state model of STIM binding to Orai has been constructed (Hoover and Lewis, 2011). This model is partly a Monod-Wyman-Changeux model, partly a direct model of binding cooperativity, and is illustrated in Fig. 2.13 A.

Each Orai molecule is assumed to exist in an open configuration and a closed configuration. In either configuration, Orai is assumed to have four equivalent STIM binding sites, leading to a total of 10 states for the Orai molecule. When Orai is in the closed configuration, C, it can bind STIM at any of four equivalent binding sites. If each binding site has an equilibrium constant of K_a (defined here as the forward rate over the reverse rate, i.e., k_a/k_{-a}, the inverse of the usual definition), then the effective equilibrium constant of the reaction $\mathrm{C} \leftrightarrow \mathrm{C}_1$ is $4K_a$. Direct cooperativity of STIM binding is incorporated by assuming that binding of the second STIM (i.e., the $\mathrm{C}_1 \rightarrow \mathrm{C}_2$ transition) has an equilibrium constant, not of K_a, but of aK_a. At this stage a remains unspecified, and could be either greater than or less than 1, allowing for either positive or negative cooperativity. An identical argument works for STIM binding to Orai in the open configuration, with the same cooperativity factor a. However, the equilibrium constant of STIM binding is different for the open configuration, being modified by the factor f.

Applying these assumptions to the binding of each STIM, to either the open or closed configurations, leads to the binding diagram shown in Fig. 2.13. From experimental data we know that state O_4 has the greatest current; if O_4

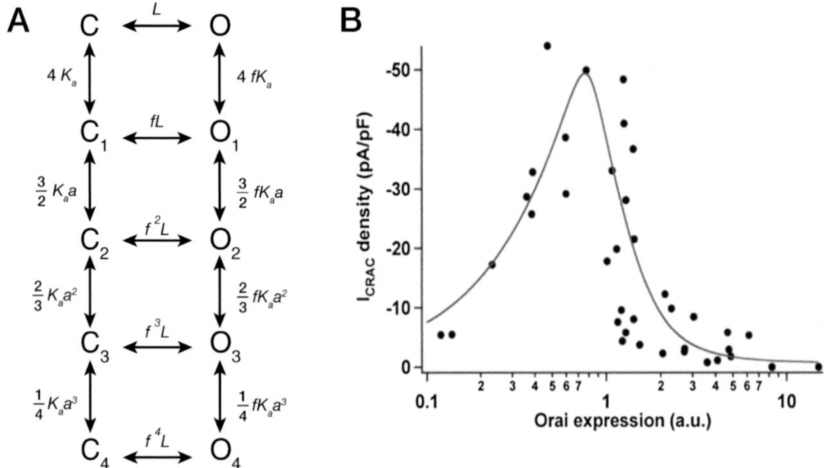

Fig. 2.13 **A:** Schematic diagram of the STIM/Orai equilibrium binding model of Hoover and Lewis (2011). C denotes the closed conformation of Orai, and O denotes the open conformation. Each conformation can bind up to four STIM; C_i denotes the conformation with i STIM bound, and similarly for O_i. Note that the vertical reactions all involve STIM binding or unbinding, although this is not shown explicitly. All the transitions are assumed to be at equilibrium and thus each can be described by a single equilibrium constant. **B:** experimental data (solid symbols) and model fit (smooth curve) showing I_{CRAC} current as a function of Orai expression. Reproduced from Hoover and Lewis (2011), Fig. 4B, with permission from PNAS.

is scaled to pass a current of 1, then O_3, O_2 and O_1 pass currents, respectively, of 0.275, 0.025 and 0.001.

The five parameters – L, K_a, f, a, and S_{total} (the total STIM concentration, in the same arbitrary units as the Orai concentration) – are now determined by fitting to experimental data. Hoover and Lewis (2011) begin by estimating L and f directly from some key experimental observations.

1. In the absence of STIM there is essentially no Orai conductance, and thus $L \ll 1$. For these data, $L = 10^{-4}$ is a good choice.
2. When the ER is heavily depleted, the open probability, P_o, of Orai is around 0.8. Since, in this case, STIM concentrations are high, it is reasonable to assume that all the Orai are in either the O_4 or the C_4 states, in which case $C_4 + O_4 = 1$, and $P_o = O_4$. Since, assuming equilibrium, we have $O_4/C_4 = Lf^4$, it follows that

$$Lf^4(1 - P_o) = P_o, \tag{2.102}$$

from which it follows that

$$f = \sqrt[4]{\frac{P_o}{L(1 - P_o)}} \approx \sqrt[4]{\frac{4}{L}} \approx 14.2. \tag{2.103}$$

Thus, we see immediately that the STIM binds preferentially to the open configuration. This leads to positive cooperativity of STIM binding, in the same way that a classical Monod-Wyman-Changeux model gives cooperativity. As STIM increases, more of the Orai is pulled into the open configuration, and thus STIM binding is increased.

We can now fit the model to the binding data shown in Fig. 2.13 B. This shows the SOCC current as a function of Orai density. As the Orai density increases, the SOCC current initially increases, as there is enough STIM to shift the Orai into the states of higher STIM occupancy, and thus of higher conductance. However, as Orai continues to increase, the STIM is overwhelmed, and the states of higher STIM occupancy become progressively less populated, leading to a decline in SOCC current. Thus the data shows a biphasic curve, with a steep decrease as Orai expression increases.

Fitting the model to these data gives the smooth curve shown in Fig. 2.13 B, which was drawn using L and f as given above, as well as $K_a = 100$, $a = 0.5$, and $S_{\text{total}} = 3.2$.

This model thus predicts that STIM binds to Orai with negative cooperativity ($a < 1$), but binds preferentially to the open configuration, leading to positive cooperativity via the induced preference for the open configuration. Although such details will be important for the detailed understanding of how Ca^{2+} influx is controlled, and how it in turn affects Ca^{2+} dynamics, detailed STIM/Orai binding models have yet to make their way into more macroscopic cell-level models.

2.7 Inositol Trisphosphate Receptors

The two major intracellular Ca^{2+} channels, the IP_3 receptor (IPR) and the ryanodine receptor (RyR), both have a long and complex modelling history. It is impossible here to do justice to the full range of models constructed for either channel, and thus, although we will give some historical overview, our discussion will focus on only a small number of models. Our choice of models to discuss is motivated partly by personal preference, and partly by our opinion about which models have been the most influential.

IPR are central to a wide variety of Ca^{2+} responses, including the oscillations and waves seen in many cell types, and thus play a major role in many models of Ca^{2+} dynamics. Their open probability is controlled by a variety of ligands, the most important of which are IP_3 and Ca^{2+}, although ATP is also an important modulator of IPR. Not only that, but there is significant time dependency in the response to ligand binding. Hence, in order to understand the behaviour of an IPR in situ it is necessary to measure both its steady-state open probability as a function of a variety of agonists, as well as its response to changes in the concentrations of these agonists. Recent reviews of the experimental literature on IPR are Mak and Foskett (2015),

Foskett et al (2007) and Parys and De Smedt (2012), while older reviews are Patel et al (1999) and Taylor and Laude (2002). IPR models have been reviewed by Sneyd and Falcke (2005) and Tang et al (1996).

An additional modelling complication is that the IPR molecule exists in three subtypes. The IPR channel is normally a homotetrameric combination of four subunits all of the same type, and models have been used to study how the different subtype properties are likely to affect oscillatory behaviour (Dupont and Combettes, 2006). However, this is not necessarily always the case, as the IPR can also exist in heterotetrameric combinations, about whose steady-state and kinetic properties very little is known. Apart from noting the existence of the difficulty, we shall otherwise entirely ignore the complication of heterotetrameric IPR.

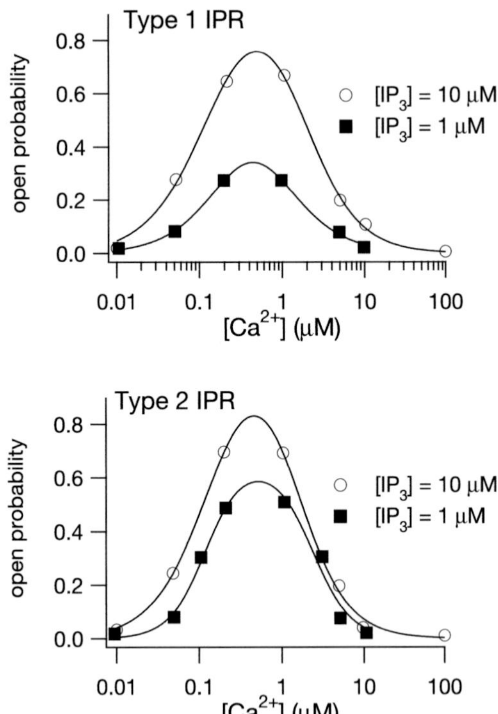

Fig. 2.14 Steady-state open probability curves measured in type 1 and type 2 IPR (Wagner and Yule, 2012). The symbols are the experimental data, measured by on-nucleus patch clamp in DT40-3KO cells. The smooth curves are heuristic biphasic curves, not based on any underlying model.

The steady-state properties of the IPR are now reasonably well understood. At any fixed [IP_3], the open probability, P_o, is a bell-shaped function of [Ca^{2+}], increasing at low [Ca^{2+}] and decreasing at high [Ca^{2+}] (Fig. 2.14), and, at any fixed [Ca^{2+}], is an increasing function of [IP_3]. The exact shape of the biphasic curve depends on the cell type and the type of IPR, but the basic shape is consistent across all IPR. Older studies of IPR in lipid bilayers (Kaftan et al, 1997; Moraru et al, 1999; Hagar et al, 1998; O'Neill et al, 2002) estimated the maximum P_o at less than 0.1, but more recent nuclear single-channel patch-clamp studies of IPR show that, in cells, the maximum P_o ranges from around 0.3 to around 0.8 (Foskett et al, 2007; Wagner and Yule, 2012). The interpretation of the bell-shaped curves shown in Fig. 2.14 is that the IPR has at least two Ca^{2+} binding sites as well as an IP_3 binding site. One of the Ca^{2+} binding sites is activating, the other is inhibitory.

This biphasic shape of the steady-state P_o curve suggests that, at low $[Ca^{2+}]$, there is a positive feedback loop such that the release of a small amount of Ca^{2+} through the IPR increases P_o, leading to an increase in Ca^{2+} release. This positive feedback loop is called Ca^{2+}-induced Ca^{2+} release, or CICR. However, it is important to emphasise that the steady-state properties of the IPR, although important, are not the full story. In many cases, an IPR will be in an environment in which $[Ca^{2+}]$ is changing continuously. If the changes in $[Ca^{2+}]$ are very slow, the steady-state properties of the IPR might be sufficient to understand behaviour. But if the changes in $[Ca^{2+}]$ are fast enough, the steady-state properties of the IPR are less important than its dynamic properties.

These dynamic properties remain far less clear. Data from Finch et al (1991), Dufour et al (1997), and Marchant and Taylor (1998) imply that, in response to a step increase in $[Ca^{2+}]$ or $[IP_3]$, P_o first increases to a peak before decreasing within a few seconds to a lower steady level, which can be an order of magnitude less than the peak of the response. Unfortunately, such large differences between the peak and the steady responses are not always consistent with the nuclear patch-clamp data, which show that the steady-state P_o gets as high as 0.8. These same patch-clamp data also indicate that IPR eventually inactivate at a constant $[IP_3]$, but the time scale is considerably longer, ranging from 10 to 100 seconds (Foskett et al, 2007). The time scales of binding of Ca^{2+} and IP_3 to their respective binding sites have been investigated by Mak et al (2007) using a rapid perfusion technique in combination with a single-channel patch clamp. These data can be reproduced by a model in which partial inactivation of the IPR occurs over a time scale of 1–2 seconds (Cao et al, 2013).

However, many uncertainties remain over exactly how an IPR responds to changes in ligand concentration. It is likely that CICR is a result of fast binding of Ca^{2+} to the activating IPR binding site, followed by slower binding of Ca^{2+} to the inactivating site. Such a process of fast activation followed by slower inactivation lies at the heart of excitability in other physiological systems such as the action potential. Nevertheless, many questions remain about the exact time scales involved.

Finally, modelling the response of an IPR in situ is complicated by the fact that the $[Ca^{2+}]$ at the mouth of the channel – which is, presumably, the Ca^{2+} concentration that controls Ca^{2+} binding to the IPR – is a complex function of time, the flux through the channel and the background cytosolic concentration. Indeed, as each IPR opens and closes stochastically, it is probable that the mouth of the receptor only ever experiences two different $[Ca^{2+}]$: a low $[Ca^{2+}]$ when the IPR is closed, and a high $[Ca^{2+}]$ when the IPR is open. This introduces additional complications for modelling. Since any particular IPR can only be open or closed, the whole-cell response has to be understood as the emergent behaviour of a large (or not so very large) number of IPR, opening and closing stochastically, and drastically modulating their local Ca^{2+} environment in a complex dynamic process. In addition, such complex dynamic stochastic behaviour cannot be dissociated from the larger environment of the IPR, as IPR in microdomains, or in clusters with other IPR, will also be significantly affected by their wider environment.

2.7.1 An Eight-State Markov Model

One of the earliest models of the IPR to incorporate sequential activation and inactivation by Ca^{2+} was that of De Young and Keizer (1992). In the twenty years since its publication, this model has become one of the most widely used IPR models, although it has now been superseded by models that are based on more recent data.

For this model, it is assumed that the IP_3 receptor consists of three equivalent and independent subunits, all of which must be in a conducting state for there to be Ca^{2+} flux. Each subunit has an IP_3 binding site, an activating Ca^{2+} binding site, and an inactivating Ca^{2+} binding site, each of which can be either occupied or unoccupied, and thus each subunit can be in one of eight states. Each state of the subunit is labelled S_{ijk}, where i, j, and k are equal to 0 or 1, with 0 indicating that the binding site is unoccupied and 1 indicating that it is occupied. The first index refers to the IP_3 binding site, the second to the Ca^{2+} activation site, and the third to the Ca^{2+} inactivation site. This is illustrated in Fig. 2.15. While the model has 24 rate constants, because of the requirement for detailed balance these are not all independent. In addition, two simplifying assumptions are made to reduce the number of independent constants. First, the rate constants are assumed to be independent of whether activating Ca^{2+} is bound or not. Second, the kinetics of Ca^{2+} activation are assumed to be independent of IP_3 binding and Ca^{2+} inactivation. This leaves only 10 rate constants, $k_1, \ldots, k_5$ and $k_{-1}, \ldots, k_{-5}$. Notice that with these simplifying assumptions, detailed balance is satisfied.

The fraction of subunits in the state S_{ijk} is denoted by x_{ijk}. The differential equations for these are based on mass-action kinetics, and thus, for example,

$$\frac{dx_{000}}{dt} = -(V_1 + V_2 + V_3), \tag{2.104}$$

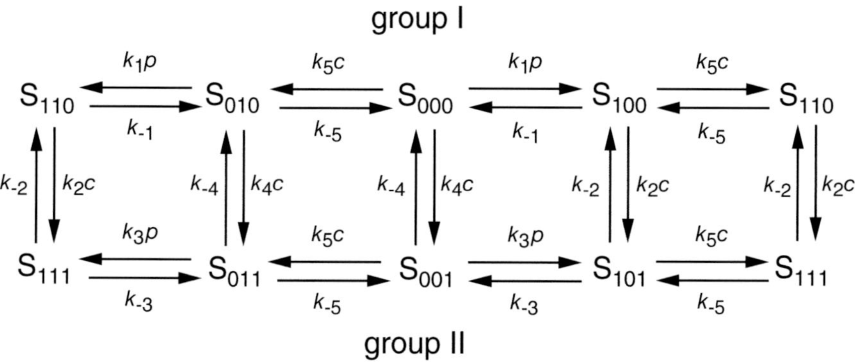

Fig. 2.15 The binding diagram for the IP_3 receptor model of De Young and Keizer (1992). Here, c denotes $[Ca^{2+}]$, and p denotes $[IP_3]$.

where

$$V_1 = k_1 p x_{000} - k_{-1} x_{100}, \tag{2.105}$$

$$V_2 = k_4 c x_{000} - k_{-4} x_{001}, \tag{2.106}$$

$$V_3 = k_5 c x_{000} - k_{-5} x_{010}, \tag{2.107}$$

where p denotes $[IP_3]$ and c denotes $[Ca^{2+}]$. V_1 describes the rate at which IP_3 binds to and leaves the IP_3 binding site, V_2 describes the rate at which Ca^{2+} binds to and leaves the inactivating site, and similarly for V_3. Since experimental data indicate that the receptor subunits act in a cooperative fashion, the model assumes that the IP_3 receptor passes Ca^{2+} current only when three subunits are in the state S_{110} (i.e., with one IP_3 and one activating Ca^{2+} bound), and thus the open probability of the receptor is x_{110}^3.

Parameters in the model were chosen to obtain agreement with the steady-state data of Bezprozvanny et al (1991). However, the kinetic properties of the IP_3 receptor are equally important; the receptor is activated quickly by Ca^{2+}, but inactivated by Ca^{2+} on a slower time scale. In the model, this is incorporated in the magnitude of the rate constants. We will see this in more detail in the next chapter (Table 3.1), where we see that k_1, k_3, and k_5 are substantially larger than k_2 and k_4, and k_{-1}, k_{-3}, and k_{-5} are also larger than k_{-2} and k_{-4}.

2.7.2 Reduction of the Eight-State Markov Model

Since IP_3 binds quickly to its binding site and Ca^{2+} binds quickly to the activating site, we can dispense with the transient details of these binding processes and assume instead that the receptor is in quasi-steady state with respect to IP_3 binding and Ca^{2+} activation (Li and Rinzel, 1994). This leads to the Li-Rinzel simplification of the De Young-Keizer IPR model.

As shown in Fig. 2.15, the receptor states are arranged into two groups: those without Ca^{2+} bound to the inactivating site ($S_{000}, S_{010}, S_{100}$, and S_{110}, shown in the upper line of Fig. 2.15; called group I states), and those with Ca^{2+} bound to the inactivating site ($S_{001}, S_{011}, S_{101}$, and S_{111}, shown in the lower line of Fig. 2.15; called group II states). Because the binding of IP_3 and the binding of Ca^{2+} to the activating site are assumed to be fast processes, within each group the binding states are assumed to be at quasi-steady state. However, the transitions between group I and group II (between top and bottom in Fig. 2.15), due to binding or unbinding of the inactivating site, are slow, and so the group I states are not in equilibrium with the group II states.

The quasi-steady-state equations for the states in group I are

$$x_{000}(k_5 c + k_1 p) = k_{-5} x_{010} + k_{-1} x_{100}, \tag{2.108}$$

$$x_{100}(k_5 c + k_{-1}) = k_{-5} x_{110} + k_1 p x_{000}, \tag{2.109}$$

$$x_{010}(k_{-5} + k_1 p) = k_5 c x_{000} + k_{-1} x_{110}. \tag{2.110}$$

The equation for the fourth receptor state, x_{110}, is superfluous, as we have the constraint

$$x_{000} + x_{010} + x_{100} + x_{110} = 1 - y, \tag{2.111}$$

where y is the fraction of receptors in the group II states, i.e.,

$$y = x_{001} + x_{011} + x_{101} + x_{111}. \tag{2.112}$$

We solve (2.108)–(2.111) to find the group I state probabilities in terms of y. This gives, for example,

$$x_{000} = \frac{K_1 K_5 (1 - y)}{(p + K_1)(c + K_5)}, \tag{2.113}$$

where $K_i = k_{-i}/k_i$. An identical procedure applied to the group II receptor states gives the quasi-steady-state equations for the group II states.

It now remains to derive a differential equation for y. Notice that y changes only on a slow time scale since any changes in y involve Ca^{2+} leaving or binding to the inactivating site, a process which is assumed to be slow. Thus we write the differential equations for the group II sites, taking care to include the transitions between the group I and group II sites, add the four equations, and substitute all the quasi-steady-state expressions to get, finally,

$$\frac{dy}{dt} = \left[\frac{(k_{-4} K_2 K_1 + k_{-2} p K_4) c}{K_4 K_2 (p + K_1)}\right](1 - y) - \left(\frac{k_{-2} p + k_{-4} K_3}{p + K_3}\right) y. \tag{2.114}$$

This can be written in the form

$$\tau_y(c,p)\frac{dy}{dt} = y_\infty(c,p) - y, \tag{2.115}$$

which is useful for comparison with other models.

The open probability, P_o, is obtained from the equation

$$P_o = x_{110}^3 = \left(\frac{pc(1-y)}{(p+K_1)(c+K_5)}\right)^3. \tag{2.116}$$

Note that $1-y$, which is the proportion of receptors that are not inactivated by Ca^{2+}, plays the role of an inactivation variable, similar in spirit to the variable h in the Hodgkin-Huxley equations. To emphasise this similarity, we can write $h = 1 - y$, in which case the reduced model can be written in the form

$$x_{110} = \frac{pc}{(p+K_1)(c+K_5)}h, \tag{2.117}$$

$$\tau_h(c,p)\frac{dh}{dt} = h_\infty(c,p) - h, \tag{2.118}$$

where τ_h and h_∞ are readily calculated from the corresponding differential equation for y.

2.7.3 Gating Models

Another common type of model is based on the idea of gating variables, in the manner of the Hodgkin-Huxley equations (Hodgkin and Huxley, 1952; Keener and Sneyd, 2008). In models of this type, the open probability, P_o, is written in the form

$$P_o \propto m_p(p,t)m_a(c,t)y(c,t), \tag{2.119}$$

where m_p is some IP_3-dependent activation variable, m_a is a Ca^{2+}-dependent activation variable, and y is a Ca^{2+}-dependent inactivation variable. In general, each of the gating variables will obey its own differential equation. However, in practice it is more common to assume that activation of the receptor (by IP_3 and Ca^{2+}) is fast enough to be at quasi-steady state, leaving the inactivation variable as the only one with its own differential equation.

For example, Dupont and Erneux (1997) assumed that

$$P_o \propto \frac{p}{K_p+p}\frac{c^3}{K_c^3+c^3}y, \tag{2.120}$$

$$\frac{dy}{dt} = k_-(1-y) - k_+\frac{c^4}{K_c^3+c^3}y. \tag{2.121}$$

At steady state, y is a decreasing function of c, and thus, at steady-state, P_o is a bell-shaped function of c, first increasing then decreasing.

In a similar model, Atri et al (1993) assumed that

$$P_o \propto \left(\mu_0 + \frac{\mu_1 p}{K_p + p}\right)\left(V_1 + \frac{V_2 c}{K_c + c}\right) y, \tag{2.122}$$

$$\tau_y \frac{dy}{dt} = \frac{K_2^2}{K_2^2 + c^2} - y. \tag{2.123}$$

For both these models the constants were chosen so as to reproduce a particular steady-state curve for P_o (for example, as measured by Parys et al (1992)), while the time constants in the equation for y give the delay between activation and inactivation. Early data from Finch et al (1991), Finch and Goldin (1994), Dufour et al (1997), and Marchant and Taylor (1998) suggest that inactivation occurs on a time scale of 1–2 s, and this is a typical value used in gating models. More recent single-channel data from IPR (see the next section) confirm the approximate accuracy of this estimate, but the single-channel data are usually insufficient to determine such longer time constants accurately.

Gating models are relatively easily adapted for the different IPR subtypes, and for IPR from a wide range of cell types (Dupont and Combettes, 2006). Since the steady-state biphasic curve is constructed simply as a product of two algebraic terms, one increasing and one decreasing, it is easily modified to take into account a wide range of experimental data.

2.7.4 Modal Models

The recent development of an experimental technique that allows for the measurement of time-dependent single-channel current through an IPR (Ionescu et al, 2007; Mak et al, 2007) has motivated the development of a new class of models that are fit directly to single-channel data from channels in the nuclear membrane, i.e., from channels that are presumably in a more natural environment than a lipid bilayer (Gin et al, 2009; Siekmann et al, 2011, 2012; Ullah et al, 2012).

These data and the associated models have shown clearly that the IPR exists in different modes, each with a different open probability, as shown in Fig. 2.16. The number of modes is variously claimed to be two (Siekmann et al, 2012) or three (Ionescu et al, 2007), but the principle is the same in either case. For convenience, our discussion here shall focus on the model with two modes.

By splitting the data into two modes (one called *park*, where the IPR is mostly closed, the other called *drive*, where the IPR is mostly open), Siekmann et al (2012) showed that the data could be reproduced by a model

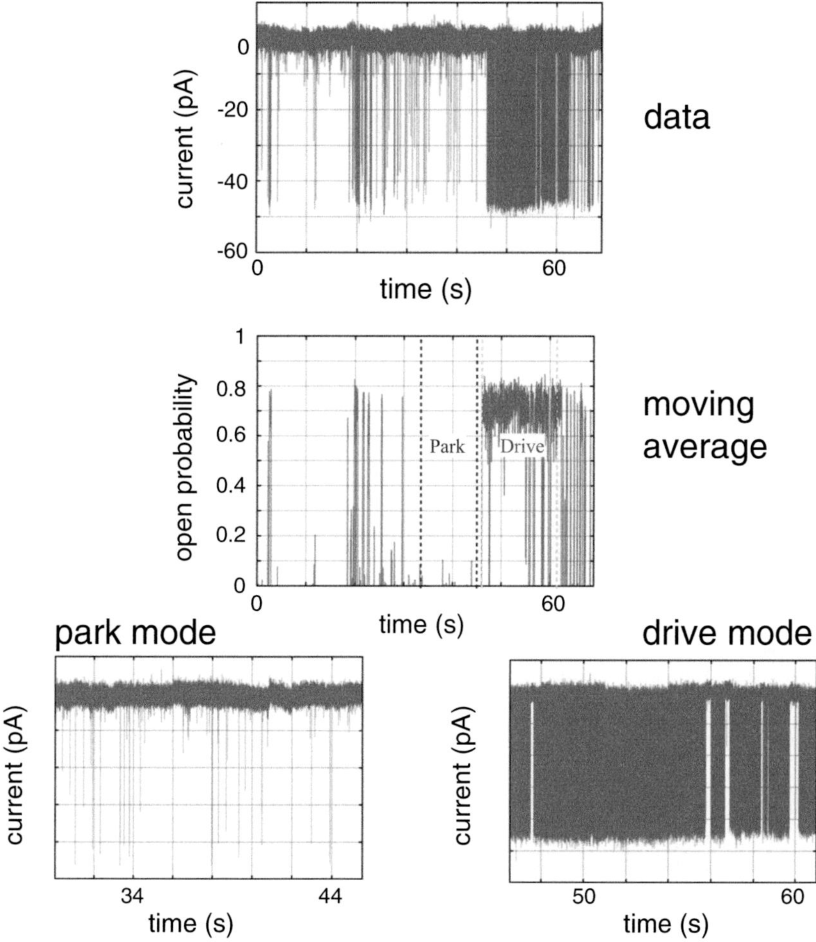

Fig. 2.16 Modal behaviour of the IPR. The top panel shows single-channel data collected from type 1 IPR in the nuclear membrane of DT40-3KO cells, at $[\mathrm{IP}_3] = 10\,\mu\mathrm{M}$, $[\mathrm{ATP}] = 5$ mM, and $[\mathrm{Ca}^{2+}] = 50$ nM (Wagner and Yule, 2012). The middle panel shows the moving average of the open probability of the data segment in the panel above, whence it can be seen that the open probability is usually either low or high. The mode in which the IPR has a low open probability is called the park mode, while the mode in which the IPR has a high open probability is called the drive mode. Expanded views of the data in the two modes are shown in the two lowest panels.

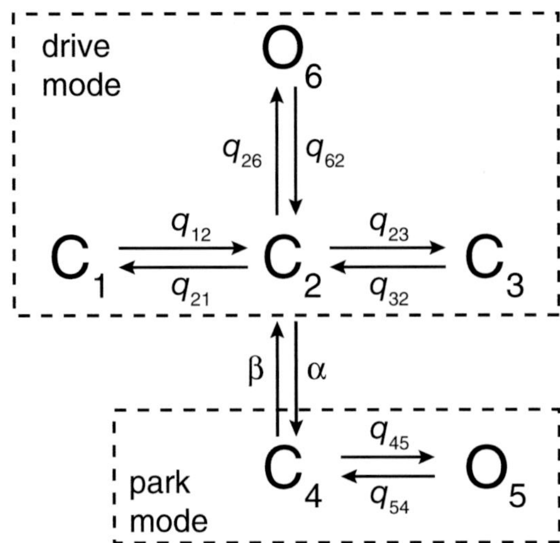

Fig. 2.17 Diagram of the park/drive model of Siekmann et al (2012). All the rates are constants, except α and β, which are functions of $[Ca^{2+}]$, $[IP_3]$, and time.

with four states in the drive mode, and two states in the park mode, and with two open states overall (Fig. 2.17). In addition, they showed that any more complex model would be underdetermined by the data they used. Thus, a six-state model is the most complex that can be unambiguously fitted using their data.

Interestingly, the transitions within each mode are independent of IP_3 and Ca^{2+}. Instead, those ligands exert their effects by changing the rate at which the IPR switches between modes. However, the transitions between the modes (i.e., β and α) are complex nonlinear functions of $[Ca^{2+}]$ and IP_3. Thus the mode transitions are not governed by simple binding of IP_3 or Ca^{2+}, with mass action kinetics. They are instead complex entities themselves, that must necessarily involve multiple additional states not present in the model.

The rates β and α are also functions of time. Cao et al (2013, 2014) have constructed a hybrid model, with characteristics of both modal and gating models, that has the stationary behaviour measured by Wagner and Yule (2012) as well as the time-dependent behaviour measured by Mak et al (2007). In this hybrid model, the transition rates between the modes are modelled by gating variables, while the remainder of the transitions are modelled by states in a modal model.

Thus, in Fig. 2.17 we have

$$\alpha = A_1(p) + A_2(p)(1 - m_\alpha h_\alpha), \tag{2.124}$$

$$\beta = B_1(p) + B_2(p) m_\beta h_\beta, \tag{2.125}$$

where m and h satisfy their own differential equations which are dependent on $[Ca^{2+}]$ and $[IP_3]$, while the A's and B's are functions only of $[IP_3]$. These functions are chosen by fitting to data, and have no underlying biophysical interpretation. More details are given in the next section, where we discuss a simplified version of the model.

This raises a thorny question about Markov models of the type shown in Figs. 2.17 or 2.15. Although it appears from the structure of the model that the IPR exists in a small number of discrete states, with (at least in the case of Fig. 2.15) transitions between the states governed by binding (with mass-action kinetics) of various ligands, one would be unwise to rely too heavily on such simplistic interpretations. The true story is likely to be very much more complicated, including, as it must, regulation by a host of other factors, and perhaps even a practical continuum of states. Certainly, IPR (and other receptors and channels) can be observed switching from clearly closed to clearly open states, and this indicates that Markov models are capturing some essential aspects of IPR behaviour. Nevertheless, the tidiness and elegance of a Markov model with a small number of states suggests a physical accuracy that is probably largely spurious. Thus, the hybrid nature of this model, combining a Markov model with a heuristic gating model, raises no serious difficulties. Neither aspect of the model is likely to be an accurate physical picture of an actual IPR in situ, but this does not prevent the combined model from being a useful predictive tool for modelling.

As of writing this, the only other modal Markov IPR model in the literature is that of Ullah et al (2012). This model is based on mass-action binding of Ca^{2+} and IP_3 to a 12-state Markov model, with the rate constants determined by fitting to single-channel data. This model assumes the existence of three IPR modes, rather than two as in the previous model, and is fit to data from insect Sf9 cells.

2.7.5 Simplifying the Modal Model

Although a Markov model is necessary for fitting single-channel data (and, indeed, can be completely determined by the data), for many other purposes, such as simulation of Ca^{2+} puffs and waves, a simplified model is sufficient. Indeed, it turns out that a model of the IPR, consisting of only a single differential equation, can do essentially as good a job as the full model, for any purpose except the fitting of single-channel data.

We construct such a simplified model by making a number of assumptions (Cao et al, 2014).

- States C_1, C_3, and O_5 all have low probabilities and thus have little effect on Ca^{2+} puffs and waves. They need to be in the model if one wishes to fit the exact details of the single-channel data, but are not necessary otherwise. So we simply throw these states away. The resulting simplified model is shown in Fig. 2.18.

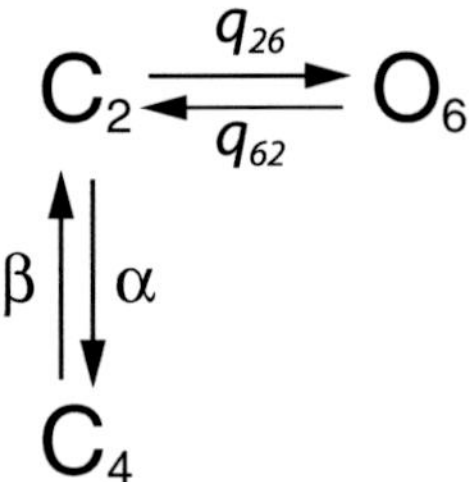

Fig. 2.18 Three-state simplified version of the modal model, obtained simply by omitting the states C_1, C_3, and O_5, all of which have very low probabilities. In addition, these three states are all assumed to be at equilibrium, as the transitions are much faster than the changes in α and β.

- Although all the four variables m_α, h_α, m_β, and h_β are functions of time, it turns out that m_α, m_β, and h_α are fast, and can be assumed to be at quasi-steady state, which we call $\bar{m}_\alpha$, $\bar{m}_\beta$, and $\bar{h}_\alpha$, respectively. (As an aside, we note that such assumptions of quasi-steady state are not without their dangers, a point which is discussed at greater length in Chapter 5.)
- Numerical simulations, using the parameter values determined from the data, show that transitions between the states occur much faster than do changes in the rate constants α and β. Hence, the transitions between all the states are assumed to be at equilibrium. It follows that

$$q_{26}C_2 = q_{62}O_6, \tag{2.126}$$

$$\alpha C_2 = \beta C_4, \tag{2.127}$$

$$O_6 + C_2 + C_4 = 1, \tag{2.128}$$

where, as usual, the italic O_6 denotes the fraction of IPR in the O_6 state, with analogous definitions for C_2 and C_4. It follows that the open probability of the IPR, P_o, is

$$P_o = O_6 = \frac{1}{1 + \frac{q_{62}}{q_{26}}\left(1 + \frac{\alpha}{\beta}\right)}. \tag{2.129}$$

With these simplifications, the IPR model condenses to a single differential equation

$$\frac{dh_\beta}{dt} = \tau_h(\bar{h}_\beta - h_\beta), \tag{2.130}$$

with the open probability given by (2.129).

This makes clear the analogy with previous two-variable models, such as the Li-Rinzel simplification of the De Young-Keizer model. Just as in the model of Li and Rinzel (1994), here the open probability is both an algebraic function of the concentrations and a function of an additional time-dependent variable that controls which state the IPR is in. Given a large step increase in $[\mathrm{Ca}^{2+}]$, P_o will change immediately to a different value, as α changes algebraically with $[\mathrm{Ca}^{2+}]$ (see (2.131)–(2.135) below). Furthermore, $\bar{m}_\beta$ changes algebraically with $[\mathrm{Ca}^{2+}]$ also, giving an immediate increase in β.

However, $\bar{h}_\beta$ has now decreased, and thus β will now decrease more slowly, with time constant τ_h, resulting in a slow decrease in P_o. If the initial change in $[\mathrm{Ca}^{2+}]$ caused an increase in P_o, we can interpret this as fast activation by Ca^{2+} followed by slower inactivation by Ca^{2+}, just as in the models of De Young and Keizer (1992), Li and Rinzel (1994) and Atri et al (1993). Although the underlying mechanism is quite different (in the modal model there is no Ca^{2+} binding to an inactivating binding site) and the details are more intricate, the overall effect is the same.

By fitting to data, both steady-state and time-dependent, Cao et al (2014) found that

$$A_1 = 1 + \frac{5}{p^2 + 0.25} \qquad A_2 = 62 + \frac{880}{p^2 + 4}, \tag{2.131}$$

$$B_1 = \frac{1.8p^2}{p^2 + 0.34} \qquad B_2 = \frac{110p^2}{p^2 + 0.1^2}, \tag{2.132}$$

$$\bar{m}_\alpha = \frac{\hat{c}^3}{\hat{c}^3 + 0.35^3}, \qquad \bar{h}_\alpha = \frac{80^2}{\hat{c}^2 + 80^2}, \tag{2.133}$$

$$\bar{m}_\beta = \frac{\hat{c}^3}{\hat{c}^3 + k_m^3}, \qquad \bar{h}_\beta = \frac{k_h^3}{\hat{c}^3 + k_h^3}, \tag{2.134}$$

$$k_m = 0.5 + \frac{0.5p^3}{p^3 + 64}, \qquad k_h = 0.4 + \frac{25p^3}{p^3 + 275}. \tag{2.135}$$

Note that we have written all Ca^{2+} concentrations in (2.133) and (2.134) as $\hat{c}$, rather than just as c. Throughout this book we are using c to denote the observable cytosolic Ca^{2+} concentration, but in this model the rate constants do not depend directly upon c. Instead, they depend on the Ca^{2+} concentration at the channel mouth or inside a cluster of IPR, both of which are higher than the cytosolic concentration. We discuss this further in Section 2.7.6.

To complete the description of the IPR, it remains to specify τ_h in (2.130). Although this is not well determined by single-channel data, puff data from Smith and Parker (2009) indicate that IPR are quickly inhibited by a high $[\mathrm{Ca}^{2+}]$, but are slow to reopen. This suggests that τ_h is large when $[\mathrm{Ca}^{2+}]$ is high, and smaller when $[\mathrm{Ca}^{2+}]$ is low. Although no precise estimates could be made, Cao et al (2014) found that, in the deterministic version of the model, it was sufficient to make τ_h a function of the proportion of open IPR. Hence we set

$$\tau_h = T_{\mathrm{high}} O_6 + T_{\mathrm{low}}(C_4 + C_2), \tag{2.136}$$

where $T_{\mathrm{high}} = 20\,\mathrm{s}^{-1}$ and $T_{\mathrm{low}} = 0.2\,\mathrm{s}^{-1}$. This relies on the assumption that a high $[\mathrm{Ca}^{2+}]$ at the mouth of the channel can affect, for some time, a transition rate between two closed states of the IPR. However, there is no proposed biophysical mechanism for such an effect, and so this aspect of the model must be considered purely heuristic.

2.7.6 The Question of Local Calcium Concentration

It is highly important to note that the $\hat{c}$ in (2.131)-(2.135) is *not* the background cytosolic Ca^{2+} concentration (Dupont and Swillens, 1996). It is, instead, either the Ca^{2+} concentration at the mouth of the channel or the Ca^{2+} concentration within a cluster of IPR.

- When an IPR is open, the Ca^{2+} concentration at the mouth of the channel will be very large. Indeed, it will be of similar magnitude to the Ca^{2+} concentration inside the ER (Rüdiger et al, 2007, 2010a,b). Thus, the rate at which the receptor closes will be dependent only upon this high Ca^{2+} concentration, which we shall call c_{open}. Hence, $\bar{m}_\alpha$ and $\bar{h}_\alpha$ depend on c_{open}, and so, in (2.133), $\hat{c} = c_{\text{open}}$. In practice, one option is to calculate the c_{open} using the approaches described in Sections 3.4.1 and 3.4.2, in which case c_{open} is a weighted average of the ER and cytosolic concentrations.
- When the IPR is closed, the Ca^{2+} concentration at the mouth of the channel will still not be c. This is because IPR are typically (possibly always) grouped in clusters, where the opening and closing of one IPR will strongly influence the opening and closing of its neighbours. Thus, we need to keep track of the Ca^{2+} concentration within the cluster, which we call c_{cluster}. The opening of the IPR will depend on c_{cluster}, and thus, in (2.134), $\hat{c} = c_{\text{cluster}}$. In practice, c_{cluster} needs to be computed as part of a larger model of Ca^{2+} dynamics, and will depend on the number of IPR in the cluster. An example can be seen in Section 3.2.3.

When fitting to single-channel data the distinctions between c, c_{open}, and c_{cluster} are not important. Single-channel data is collected at a fixed $[Ca^{2+}]$ on the cytosolic side of the IPR. This is possible since Ca^{2+} is not the charge carrier in these experiments (Wagner and Yule, 2012). If Ca^{2+} were used as the charge carrier, as soon as the IPR were to open, the channel mouth would experience an immediate large rise in $[Ca^{2+}]$, which would disrupt measurement of the open probability at the desired fixed $[Ca^{2+}]$. However, as soon as one wishes to use this IPR model in a whole-cell model of oscillations, these distinctions become crucial.

These complications will be addressed in more detail in Section 3.2.3, when we discuss how to incorporate this IPR model into whole-cell models of Ca^{2+} dynamics.

2.7.7 Open Probability and Flux

For each of the IPR models discussed above, the flux, J_{IPR}, through the IPR is proportional to the open probability of the IPR, and is often written as

$$J_{\mathrm{IPR}} = k_f P_o(c_e - \hat{c}), \tag{2.137}$$

where k_f is a scaling factor that controls the maximum total flux through the IPR. For example, k_f could be related to the density of IPR, or to the number of IPR in a cluster. In addition, the flux through the IPR is usually assumed to be proportional to the concentration difference across the IPR, which we denote here by $c_e - \hat{c}$. The specific choice for $\hat{c}$ will be dependent on which model is being used. For example, if the De Young-Keizer model is used, then $\hat{c} = c$, while if the modal model is used, $\hat{c} = c_{\mathrm{cluster}}$. It is a simplification to assume the dependence of the flux on the concentration difference is linear, but few data are available on this point, and none contradicts it.

It is important to note that, because of this dependence on $c_e - \hat{c}$, J_{IPR} will decrease as c_e decreases, whether or not Ca^{2+} in the ER plays any role in controlling P_o. Although control of IPR by ER Ca^{2+} (by means other than depletion) does not appear in any of the models discussed here, it may well be an important regulatory mechanism.

Use of (2.137) is a crude approach to the far more complex problem of calculating the flux through an open channel, which is discussed further in Section 3.4.2. However, this crude approach is likely to be sufficiently accurate for most purposes.

2.8 Ryanodine Receptors

Because of the importance of RyR for excitation-contraction coupling in skeletal and cardiac muscle, the history of RyR modelling is even longer and more varied than for IPR. For the same reason, modelling of RyR in skeletal and cardiac cells has tended to be closely linked to the spatial organisation of the diadic cleft (see Section 7.2). Although RyR play an important role in Ca^{2+} signalling in many other cell types, including neurons, oocytes, airway smooth muscle, and exocrine acinar cells, there are still few models of RyR that are constructed specifically for a non-muscle context.

RyR share some similarities with IPR. They are activated by a rise in $[Ca^{2+}]$, and thus mediate CICR. Indeed, RyR are the classic example of CICR, and the channel for which this phenomenon was first described (Endo et al, 1970). However, there is still controversy over whether or not RyR are inactivated by even higher $[Ca^{2+}]$, at least at physiological levels.

2.8.1 An Algebraic Model

One of the simplest models of the RyR was constructed by Friel (1995), who used it in a model of Ca^{2+} oscillations in a sympathetic neuron. In this model, Ca^{2+} flux through the RyR is an algebraic function of $[Ca^{2+}]$, with no time dependence or gating kinetics. Thus

$$J_{\mathrm{RyR}} = \left(\kappa_1 + \frac{V_r c^3}{K^3 + c^3} \right) (c_e - c), \tag{2.138}$$

where c and c_e are the cytoplasmic and ER Ca^{2+} concentrations, respectively. The first factor on the right-hand side of (2.138) can be interpreted as the open probability of the channel; it is an increasing function of c, with a sharp increase centred at $c = K$. Thus, around $c = K$, a small increase in c leads to a large increase in flux, which results in CICR, as expected. The second factor on the right, $(c_e - c)$, is a simple model of the driving force for Ca^{2+} flux through the channel, as discussed in the previous section.

The parameter κ_1 models a leak at zero $[Ca^{2+}]$. As we shall see in Chapter 3, where we describe how elements of the Ca^{2+} toolbox are combined into whole-cell models, such a leak is often necessary in order to ensure a physiological resting $[Ca^{2+}]$. It is not necessary for such a leak to come through the RyR, although it does so in Friel's model.

The simplicity of Friel's model makes it an attractive option for situations where data are lacking for the inclusion of more complex RyR models.

2.8.2 A Markov Model of RyR Inactivation

As we shall discuss in more detail in Section 7.2, RyR in skeletal and cardiac muscle are located in diadic clefts, regions only approximately 15 nm wide, into which open L-type Ca^{2+} channels as well as RyR. In cardiac cells, an initial influx of Ca^{2+} through the L-type channels causes a much larger release of Ca^{2+} through the RyR, the classic phenomenon of CICR. However, over a time scale of a few tens of milliseconds, this release spontaneously terminates. How this happens has been a puzzle for decades, and is still unresolved.

This puzzle motivated the development of RyR models that spontaneously inactivate. One early such model is that of Stern et al (1997), as shown in Fig. 2.19. In this model, binding of two Ca^{2+} opens the RyR, which then spontaneously binds a third Ca^{2+} to move to an inactivated state. Opening of the channel is faster than inactivation, and thus, in response to a step increase of $[Ca^{2+}]$, the RyR first opens then inactivates, in a biphasic response. Inactivation of the channel occurs within a few tens of milliseconds, thus causing termination of the RyR on the same time scale as seen experimentally.

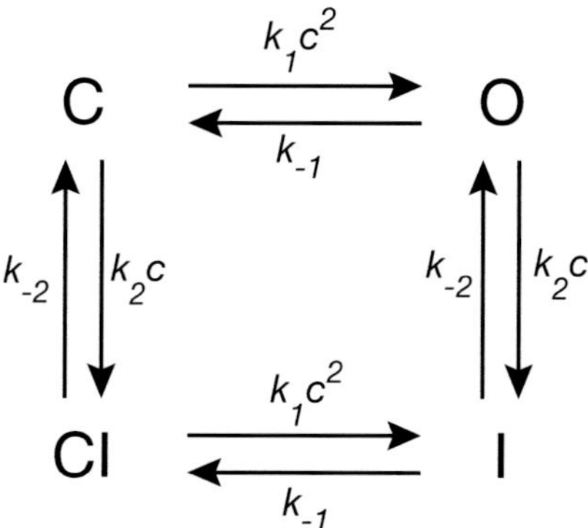

Fig. 2.19 The RyR model of Stern et al (1997). C denotes the closed state of the RyR, O the open state, I the inactivated state, and CI the intermediate state between C and I. The C to O transition is dependent on the binding of two Ca^{2+}, and thus the rate is proportional to c^2. Similarly, the transition from the open to the inactivated state depends on the binding of one Ca^{2+}.

2.8.3 Luminal Gating

There is compelling experimental evidence that gating of the RyR is also controlled by the luminal Ca^{2+} concentration. As $[Ca^{2+}]$ in the SR increases (at least until around 1 mM), so does the open probability of the RyR (Koizumi et al, 1999; Qin et al, 2008; Bassani et al, 1995; Shannon et al, 2000), probably via interaction with the luminal Ca^{2+} buffer, calsequestrin (Györke et al, 2004).

An early model of luminal RyR gating was constructed by Shannon et al (2004), who modified the Stern model (Fig. 2.19) by making k_1 an increasing function of c_e. Shannon et al (2004) also made the transition from O to I a decreasing function of c_e, thus increasing the effect of SR $[Ca^{2+}]$ on RyR opening.

A more mechanistic version of the same basic idea was constructed by Restrepo et al (2008), who incorporated simple calsequestrin dynamics. The core idea of their model is that calsequestrin exists as monomers and dimers, but that only the monomers can interact with RyR. When the calsequestrin monomer is bound to RyR, the channel has a lower open probability. Finally, the transition from the monomer to the dimer form of calsequestrin is dependent on luminal Ca^{2+}. Thus, as luminal $[Ca^{2+}]$ increases, the calsequestrin monomers convert to dimers, monomers unbind from the RyR, and the RyR open probability thus increases.

Letting M and D denote the calsequestrin monomers and dimers, respectively, we have

$$\mathrm{M} + \mathrm{M} \underset{k_{-1}}{\overset{k_1(c_e)}{\rightleftharpoons}} \mathrm{D}. \tag{2.139}$$

The rate k_1 is a steeply increasing function of c_e, the concentration of Ca^{2+} in the ER/SR. Restrepo et al (2008) go so far as to make k_1 almost a step function of c_e, with the step at $c_e = 850\ \mu$M.

Assuming a fast equilibrium between M and D, and assuming further that the total amount of calsequestrin is conserved (with a total concentration scaled to 1), we have

$$M^2 = K_1 D, \tag{2.140}$$

$$2M + D = 1, \tag{2.141}$$

where, as usual, the italic M and D denote the concentration of calsequestrin in the M and D states, respectively, and $K_1 = \frac{k_{-1}}{k_1}$. It follows that

$$M^2 - K_1(1 - 2M) = 0, \tag{2.142}$$

a quadratic equation that can be solved for M. Note that K_1 is a steeply decreasing function of c_e, and thus M is also a function of c_e.

The state diagram of the model is shown in Fig. 2.20. Binding of two cytosolic Ca^{2+} moves the RyR from left to right in the diagram, thus opening the RyR. Binding of the calsequestrin monomer moves the RyR from top to bottom. Since $k_{12} \gg k_{43}$, unbinding of the calsequestrin monomer results in an increase in RyR open probability.

Sato and Bers (2011) constructed a slightly modified version of this model, by making k_{12} and k_{43} saturating functions of c.

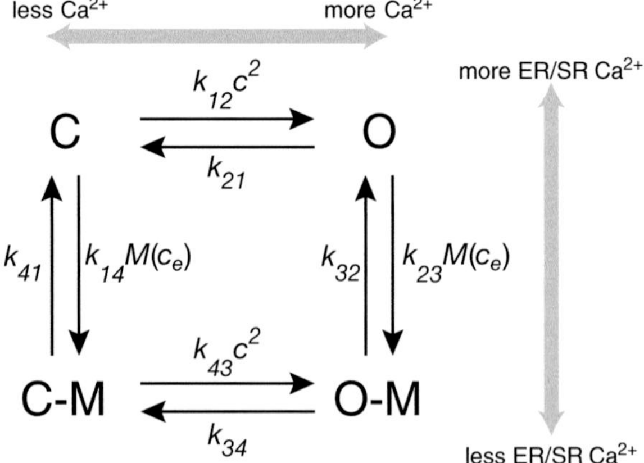

Fig. 2.20 The model of Restrepo et al (2008), which incorporates the effects of luminal Ca^{2+} on RyR gating, via interactions with the luminal Ca^{2+} buffer, calsequestrin, in monomer form (M). C denotes a closed state, O an open state. Both C and O states can exist with or without M bound, but O-M has a lower open probability than O.

2.8.4 Markov Models with Adaptation

In 1993 Györke and Fill showed how, in response to a step increase in $[Ca^{2+}]$, the open probability of the RyR first increases to a peak before declining to a lower plateau, a phenomenon they called adaptation. The time scale of this adaptation seems to depend on the details of the experimental method used to measure it, ranging from milliseconds to seconds.

Over the years there has been considerable debate about the relative importance of inactivation and adaptation of RyR (Fill and Copello, 2002). However, these terms are not precisely defined; what is called inactivation in the four-state model of Stern et al (1997) is not significantly different from the so-called adaptation in Györke and Fill (1993), as modelled by Keizer and Levine (1996). It is most likely that the fundamental property of RyR is that their open probability is dependent both on $[Ca^{2+}]$ and on time, possibly in a number of different ways, and thus inactivation and adaptation are merely different manifestations of the same underlying processes.

The first model of RyR adaptation was that of Keizer and Levine (1996), and was based on the data of Györke and Fill (1993). A modification of the original version was constructed by Keizer and Smith (1998) in their study of saltatory Ca^{2+} waves, it was modified further by Rice et al (1999), and then modified still further in a series of models by Greenstein and Winslow (Greenstein and Winslow, 2002; Greenstein et al, 2006; Greenstein and Winslow, 2011).

The model of Keizer and Smith (1998) is similar to that of Stern et al (1997), with the difference being that it contains six states rather than four (Fig. 2.21 A). Two states are open, and four are closed. At low $[Ca^{2+}]$ the RyR is mostly in state 1. As $[Ca^{2+}]$ increases, the RyR moves progressively towards states 6 and 4, thus increasing the open probability in a time-dependent fashion.

As $[Ca^{2+}]$ increases, the model can be progressively simplified by assuming fast equilibrium between some of the states; at intermediate $[Ca^{2+}]$, states 5 and 6 can be combined (Fig. 2.21 B), while at high $[Ca^{2+}]$ states 3 and 4 can also be combined, to give a four-state model (Fig. 2.21 C). The simplification is carried out by a procedure that we have already seen many times in this book. For example, if we assume a rapid equilibrium between states 5 and 6, with probabilities S_5 and S_6, respectively, then

$$k_{56}c^2S_5 = k_{65}S_6. \tag{2.143}$$

Letting $S_{56} = S_5 + S_6 = S_5(1 + \frac{k_{56}c^2}{k_{65}})$, it follows that the rate of transition from state 5 to state 2 (for example) is given by

$$k_{52}S_5 = \frac{k_{52}k_{65}}{k_{65} + k_{56}c^2}S_{56}, \tag{2.144}$$

which is the transition rate shown in Fig. 2.21 B. The other transition rates are similarly derived.

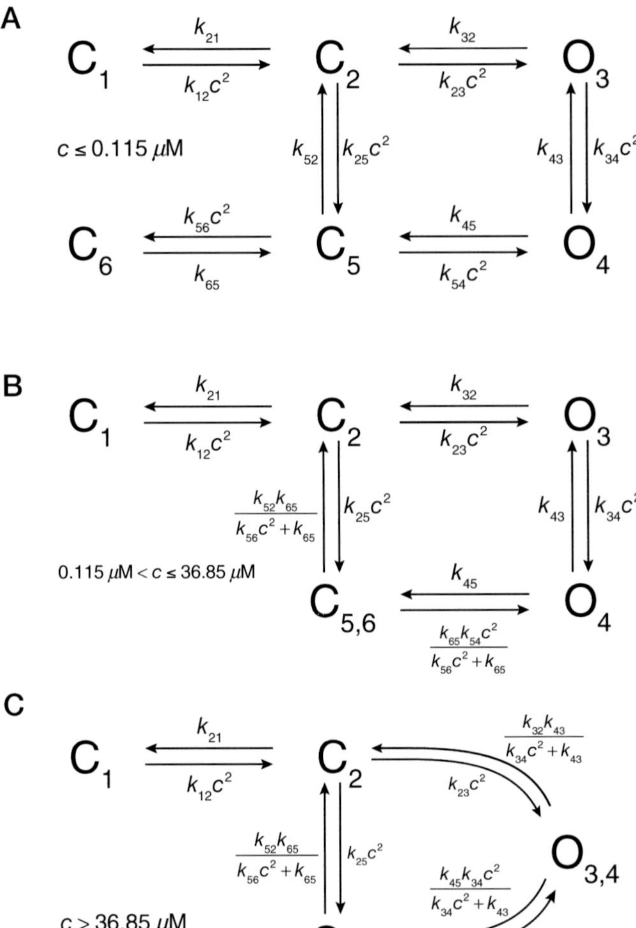

Fig. 2.21 The adapting RyR model of Greenstein and Winslow (2002). C_1 and C_2 are the resting closed states, O_3 and O_4 are the open states, and C_5 and C_6 are the refractory states. **A:** the original 6-state model. **B:** the model obtained by assuming that C_5 and C_6 are in instantaneous equilibrium. **C:** the model obtained by assuming that states O_3 and O_4 are also in instantaneous equilibrium. These different approximations are valid for different ranges of c, as indicated in the figure. Here, c is the Ca^{2+} concentration in the diadic cleft.

2.8.5 Two-State Models

Similar simplifications were also carried out by Keizer and Smith (1998), in their original presentation of the model. Indeed, they claimed (but did not show in detail) that their 6-state model could be condensed to a 2-state model, with one open state and one closed state, without changing its dynamic behaviour significantly.

The importance and relevance of simple models was recently reinforced by Cannell et al (2013), in their computational study of the diadic cleft. Based on data from RyR in lipid bilayers, they constructed a 2-state RyR model in which the rate constants were functions of the diadic cleft [Ca^{2+}], determined by fitting to open and closed time distributions. Their model, solved in an anatomically accurate three-dimensional domain (or, at least, reasonably anatomically accurate based on their synthesis of available data), was able to reproduce Ca^{2+} sparks accurately, both in time and space.

Most interestingly, the RyR model of Cannell et al (2013) contained no direct inactivating mechanism, such as adaptation, or inactivation by lower luminal [Ca^{2+}]. Nevertheless, the Ca^{2+} sparks terminated in a physiological realistic way. The reason for this termination is that the mean closed time of the RyR is a steeply decreasing function of [Ca^{2+}] in the diadic cleft. Thus, even a small decrease in diadic cleft [Ca^{2+}] leads to a large increase in the RyR mean closed time, making it more difficult for the RyR to reopen, and thus more difficult to sustain CICR. In such a case, the details of diadic cleft geometry are clearly important, influencing, as they do, the exact kinetics and amplitudes of the diadic cleft Ca^{2+} transients.

2.8.6 Modal Gating Model

Similarly to IPR, it is possible that RyR also exhibit modal behaviour, in that the open probability is not constant for fixed [Ca^{2+}] (Zahradníková and Zahradník, 1995). However, since the physiological importance of this modal behaviour is not yet clear (Fill and Copello, 2002), and because there are other data that suggest that RyR are not modal in nature (Wagner et al, 2014), we shall not discuss this model in any more detail.

2.9 Calcium Buffers

Calcium is heavily buffered in all cells, with at least 99% (and often more) of the available Ca^{2+} bound to large Ca^{2+}-binding proteins (Schwaller, 2010; Prins and Michalak, 2011). For example, calsequestrin and calreticulin are major Ca^{2+} buffers in the endoplasmic and sarcoplasmic reticulum, while in the cytoplasm Ca^{2+} is bound to calbindin, calretinin, and parvalbumin, among many others. Calcium pumps and exchangers and the plasma membrane itself are also major Ca^{2+} buffers. In essence, a free Ca^{2+} ion in solution in the cytoplasm cannot do much, or go far, before it is bound to something.

The basic chemical reaction for Ca^{2+} buffering can be represented by

$$\mathrm{P} + \mathrm{Ca}^{2+} \underset{k_-}{\overset{k_+}{\rightleftharpoons}} \mathrm{B}, \tag{2.145}$$

where P is the buffering molecule and B is buffered Ca^{2+}. Letting b denote the concentration of buffer with Ca^{2+} bound, and c the concentration of free Ca^{2+}, a simple model that includes Ca^{2+} buffering is

$$\frac{\partial c}{\partial t} = D_c \nabla^2 c + f(c) + k_- b - k_+ c(b_t - b), \tag{2.146}$$

$$\frac{\partial b}{\partial t} = D_b \nabla^2 b - k_- b + k_+ c(b_t - b), \tag{2.147}$$

where k_- is the rate of Ca^{2+} release from the buffer, k_+ is the rate of Ca^{2+} uptake by the buffer, b_t is the total buffer concentration, and $f(c)$ denotes all the other reactions involving free Ca^{2+} (for example, release from the IP_3 receptors, reuptake by pumps, etc). The terms $D_c \nabla^2 c$ and $D_b \nabla^2 b$ model the diffusion of Ca^{2+} and buffer, respectively, with diffusion coefficients D_c and D_b.

It is important to note that, by writing down (2.146), we are getting somewhat ahead of ourselves, as there are many vital assumptions implicit in such a simple formulation of Ca^{2+} diffusion. These are discussed in detail in Section 3.3. However, it is sufficient for now simply to accept that a reaction-diffusion equation such as (2.146) is a reasonable and useful model of spatially distributed intracellular Ca^{2+}.

2.9.1 Fast Buffers or Excess Buffers

If the reaction of Ca^{2+} with the buffer happens much more quickly than the other reactions involving Ca^{2+}, the buffer's effect on the intracellular Ca^{2+} dynamics can be analysed simply. In this case we take b to be in the quasi-steady state

$$k_- b - k_+ c(b_t - b) = 0, \tag{2.148}$$

and so

$$b = \frac{b_t c}{K + c}, \tag{2.149}$$

where $K = k_-/k_+$. Adding (2.146) and (2.147) we find the "slow" equation

$$\frac{\partial}{\partial t}(c+b) = D_c\nabla^2 c + D_b\nabla^2 b + f(c), \tag{2.150}$$

which, after using (2.149) to eliminate b becomes

$$\begin{aligned}\frac{\partial c}{\partial t} &= \frac{1}{1+\theta(c)}\left(\nabla^2\left(D_c c + D_b b_t \frac{c}{K+c}\right) + f(c)\right) \\ &= \frac{D_c + D_b\theta(c)}{1+\theta(c)}\nabla^2 c - \frac{2D_b\theta(c)}{(K+c)(1+\theta(c))}|\nabla c|^2 + \frac{f(c)}{1+\theta(c)},\end{aligned} \tag{2.151}$$

where

$$\theta(c) = \frac{b_t K}{(K+c)^2}. \tag{2.152}$$

Note that we assume that b_t is a constant, and does not vary in either space or time.

Thus, the assumption of fast buffering results in a significant change to the model. Whereas (2.146) is a reaction-diffusion equation, (2.151) is a reaction-diffusion-advection equation, where the advection is the result of Ca^{2+} transport by a mobile buffer (Wagner and Keizer, 1994). The new diffusion coefficient

$$D_{\text{eff}} = \frac{D_c + D_b\theta(c)}{1+\theta(c)} \tag{2.153}$$

is called the *effective diffusion coefficient*, and is a convex linear combination of the two diffusion coefficients D_c and D_b, so lies somewhere between the two. Since buffers are large molecules, $D_b < D_c$, and so $D_{\text{eff}} < D_c$. If the buffer is not mobile, i.e., $D_b = 0$, then (2.151) reverts to a reaction-diffusion equation. Also, when Ca^{2+} gradients are small, the nonlinear advective term can be ignored (Irving et al, 1990). Finally, the buffering also affects the qualitative nature of the nonlinear reaction term, $f(c)$, which is divided by $1+\theta(c)$. This may change many properties of the model, including oscillatory behaviour and the nature of wave propagation (Gin et al, 2006).

If the buffer is not only fast, but also of low affinity, so that $K \gg c$, (2.149) can be replaced by

$$b = \frac{b_t c}{K}, \tag{2.154}$$

in which case

$$\theta = \frac{b_t}{K} \tag{2.155}$$

is a constant. Thus, D_{eff} is constant also.

Commonly, it is assumed that the buffer has fast kinetics, is immobile, and has a low affinity. With these assumptions we get the simplest possible model of Ca^{2+} buffers (short of not including them at all) in which

$$\frac{\partial c}{\partial t} = \frac{K}{K+b_t}(D_c\nabla^2 c + f(c)), \tag{2.156}$$

wherein both the diffusion coefficient and the fluxes are scaled by the constant factor $K/(K+b_t)$; each flux in the model can then be interpreted as an *effective* flux, i.e., that fraction of the flux that contributes towards a change in free Ca^{2+} concentration.

It is important to note that, even if Ca^{2+} buffering does not appear explicitly in a model, it must be there in one form or another; Ca^{2+} buffering can never be simply ignored. Thus, any spatially distributed model that does not contain an explicit description of Ca^{2+} buffering must necessarily be using an effective diffusion coefficient and effective fluxes.

In 1995 Neher observed that if the buffer is present in large excess then $b_t - b \approx b_t$, in which case the buffering reaction becomes linear, the so-called excess buffering approximation:

$$\frac{\partial c}{\partial t} = D_c \nabla^2 c + f(c) + k_- b - k_+ c b_t, \tag{2.157}$$

$$\frac{\partial b}{\partial t} = D_b \nabla^2 b - k_- b + k_+ c b_t. \tag{2.158}$$

If we now assume the buffers are fast we recover (2.154) and thus (2.156). In other words, the simple approach to buffering given in (2.156) can be obtained in two ways; either by assuming a low affinity buffer or by assuming that the buffer is present in excess. It is intuitively clear why these two approximations lead to the same result — in either case the binding of Ca^{2+} does little to change the fraction of unbound buffer.

Typical parameter values for three different buffers are given in Table 2.1. BAPTA is a fast, high-affinity buffer, and EGTA is a slow high-affinity buffer, both of which are used as exogenous Ca^{2+} buffers in experimental work.

Buffer	D_b $\mu m^2\,s^{-1}$	k_+ $\mu M^{-1}\,s^{-1}$	k_- s^{-1}	K μM
BAPTA	95	600	100	0.17
EGTA	113	1.5	0.3	0.2
Endog	15	50	500	10

Table 2.1 Typical parameter values for three different buffers, taken from Smith et al (2001). BAPTA and EGTA are commonly used as exogenous buffers in experimental work, while Endog refers to a typical endogenous buffer. Typically $b_t = 100\,\mu M$ for an endogenous buffer. However, it is important to keep in mind that there is a huge variety of endogenous buffers, with widely varying properties (Schwaller, 2010; Prins and Michalak, 2011).

Despite the complexity of (2.151) it retains the advantage of being a single equation. However, if the buffer kinetics are not fast relative to the Ca^{2+} kinetics, the only way to proceed is with numerical simulations of the complete system, a procedure followed by a number of groups and discussed further in Section 3.8.2.

2.9.1.1 A Simplifying Transformation

In some situations it may be simpler to use a transformed variable, in which (2.151) takes a simpler form (Sneyd et al, 1998a). If we define

$$w = D_c c + D_b b_t \frac{c}{K+c}, \tag{2.159}$$

then w is a monotone increasing function of c, since

$$\frac{dw}{dc} = D_c + D_b \theta(c) \tag{2.160}$$

is positive. The unique inverse of this function is denoted by

$$c = \phi(w). \tag{2.161}$$

Although we can write an explicit equation for $\phi(w)$ as the solution of a quadratic equation, there is rarely any benefit in doing so.

In terms of w, (2.151) becomes

$$\frac{\partial w}{\partial t} = \frac{D_c + D_b \Theta}{1+\Theta} \left(\nabla^2 w + f(\phi(w))\right), \tag{2.162}$$

where $\Theta = \frac{b_t K}{(K+\phi(w))^2}$. Thus, we can remove the advection term at the price of working with a transformed variable.

2.10 Inositol Trisphosphate Metabolism

As illustrated in Fig. 2.22, and discussed in Section 1.2.4, both the production and degradation of IP_3 are functions of $[Ca^{2+}]$, leading to a complex web of positive and negative feedback loops.

Firstly, the activity of PLC is, in general, an increasing function of $[Ca^{2+}]$, leading to a positive feedback loop. Secondly, IP_3 can be metabolised either by removing one of the phosphate groups, or by adding another. The enzyme Ins(1,4,5)P_3-5-phosphatase will remove the phosphate group at position 5, leaving inositol 1,4-bisphosphate (Ins(1,4)P_2), or, alternatively, Ins(1,4,5)P_3-3-kinase will add a phosphate group at position 3, thus producing inositol 1,3,4,5-tetrakisphosphate, or Ins(1,3,4,5)P_4. The rate of the kinase depends on $[Ca^{2+}]$ (via the formation of a Ca^{2+}/calmodulin complex), while, since Ins(1,3,4,5)P_4 is also a substrate for the phosphatase, it acts as a competitive inhibitor of that reaction. For the same reason, IP_3 also acts as a competitive inhibitor for the action of the phosphatase on Ins(1,3,4,5)P_4.

Although the extent to which each reaction is important in any given cell type is not always clear, there is evidence that, in some cell types, one or more of these feedback loops play a vital role in the control of Ca^{2+} oscillations (Politi et al, 2006; Sneyd et al, 2006; Gaspers et al, 2014).

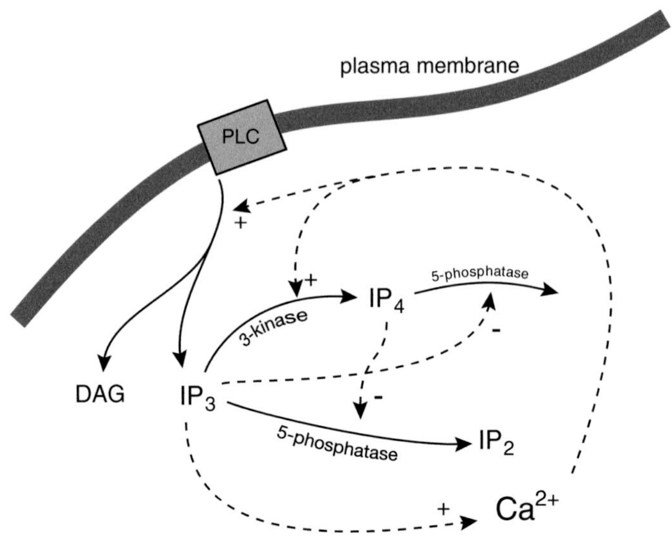

Fig. 2.22 Some of the reactions directly involved in the production and metabolism of IP_3. IP_3 is made by the membrane-bound protein phospholipase C (PLC), and is thus produced at the cell membrane. PLC can be activated by Ca^{2+}, leading to a positive feedback loop. In addition, IP_3 can be metabolised via two different pathways, one of which is Ca^{2+}-dependent, which forms a negative feedback loop. IP_3 denotes Ins(1,4,5)P_3, IP_4 denotes Ins(1,3,4,5)P_4, and IP_2 denotes Ins(1,4)P_2. DAG is diacylglycerol, an important second messenger in its own right.

2.10.1 IP_3 Production

Depending on the cell type and the context, the rate of production of IP_3 is either a function of PLC concentration alone, V_p say, or a function of both PLC and Ca^{2+} concentration. Thus, typically,

$$\frac{dp}{dt} = V_p \frac{c^n}{K_p^n + c^n}, \tag{2.163}$$

where $n = 0$, 1, or 2. The choice $n = 0$ was used by Dupont and Erneux (1997), based on the assumption that PLC is fully activated at resting $[Ca^{2+}]$ (Renard et al, 1987; Bird et al, 1997). The alternative approach of $n = 1$ or 2 was used by Meyer and Stryer (1988), Keizer and De Young (1992) and Politi et al (2006), who used the data of Blank et al (1991).

Note that the maximal rate of PLC activity, V_p, is determined from a model of the GPCR pathways (Section 2.1), and could thus be highly complex in its own right, including, for example, the concentration of the PLC substrate, PIP_2 (Brown et al, 2008).

2.10.2 IP_3 Removal

To model the conversion of IP_3 to other messengers, from Fig. 2.22 we write the enzyme reactions for the inositol phosphates:

$$[\mathrm{Ins}(1,4,5)\mathrm{P}_3] + 5\mathrm{P} \underset{k_{-1}}{\overset{k_1}{\rightleftharpoons}} \mathrm{C}_1 \xrightarrow{r_1} [\mathrm{Ins}(1,4)\mathrm{P}_2] + 5\mathrm{P}, \tag{2.164}$$

$$[\mathrm{Ins}(1,4,5)\mathrm{P}_3] + 3\mathrm{K} \underset{k_{-2}}{\overset{k_2}{\rightleftharpoons}} \mathrm{C}_2 \xrightarrow{r_2} [\mathrm{Ins}(1,3,4,5)\mathrm{P}_4] + 3\mathrm{K}, \tag{2.165}$$

$$[\mathrm{Ins}(1,3,4,5)\mathrm{P}_4] + 5\mathrm{P} \underset{k_{-3}}{\overset{k_3}{\rightleftharpoons}} \mathrm{C}_3 \xrightarrow{r_3} [\mathrm{Ins}(1,3,4)\mathrm{P}_3] + 5\mathrm{P}. \tag{2.166}$$

For clarity (at the expense of notational simplicity) we have expanded our way of writing the inositol phosphates, so that we can specify exactly where the phosphate groups are bound. Note that Ins(1,4,5)P_3 is our old friend, usually denoted more simply by IP_3, while Ins(1,3,4)P_3 is a different species, that does not bind to the IPR. The enzyme complexes are denoted by C_1, C_2, and C_3, 5P denotes the 5-phosphatase, while 3K denotes the 3-kinase.

Since both Ins(1,4,5)P_3 and Ins(1,3,4,5)P_4 are substrates for the 5-phosphatase they act as mutually competitive inhibitors. Although the theory for such competitive enzyme reactions is standard, for the convenience of the reader we briefly derive the rate equations here.

We assume that the first reactions in each of (2.164) and (2.166) are in instantaneous equilibrium to get

$$C_1 = \frac{p[5\mathrm{P}]}{K_1}, \tag{2.167}$$

$$C_3 = \frac{Z[5\mathrm{P}]}{K_3}, \tag{2.168}$$

where $K_i = k_{-i}/k_i$ and $Z = [\mathrm{Ins}(1,3,4,5)\mathrm{P}_4]$. As usual, p denotes [IP_3], or, more specifically, [Ins(1,4,5)P_3], while the italic C_i denotes [C_i]. From conservation of enzyme we get

$$C_1 + C_3 + [5\mathrm{P}] = C_t, \tag{2.169}$$

for some constant total amount of enzyme, C_t, and thus, solving for C_1, we get

$$C_1 = C_t \frac{p}{K_1\left(1 + \frac{Z}{K_3}\right) + p}. \tag{2.170}$$

It follows that the rate of degradation, V_{5P}, of Ins(1,4,5)P_3 by the 5-phosphatase is

$$V_{5P} = r_1 C_1 = V_1 \frac{p}{K_1\left(1 + \frac{Z}{K_3}\right) + p}, \tag{2.171}$$

where $V_1 = r_1 C_t$ is the maximum velocity of the reaction. Similarly, the rate of degradation, V_{ZP}, of Ins(1,3,4,5)P_4 by the 5-phosphatase is

$$V_{ZP} = r_3 C_3 = V_3 \frac{Z}{K_3\left(1 + \frac{p}{K_1}\right) + Z}, \tag{2.172}$$

where $V_3 = r_3 C_t$. Thus a competitive inhibitor merely shifts the K_d of the reaction (i.e., the value where the reaction attains half its maximum velocity), without changing the maximum velocity.

Finally, the rate of degradation, V_{3K}, of IP_3 by the 3-kinase is given by the usual Michaelis-Menten equation

$$V_{3K} = V_2 \frac{p}{K_2 + p}. \tag{2.173}$$

Here, however, the maximum velocity is a function of [Ca^{2+}]. Dupont and Erneux (1997) chose

$$V_2 = V_k \frac{c^2}{K_k^2 + c^2}. \tag{2.174}$$

Combining both production and degradation gives the model for IP_3 and IP_4:

$$\frac{dp}{dt} = \frac{V_p c^2}{K_p^2 + c^2} - V_{5P} - V_{3K} \tag{2.175}$$

$$= \frac{V_p c^2}{K_p^2 + c^2} - \frac{V_1 p}{K_1\left(1 + \frac{Z}{K_3}\right) + p} - \frac{V_k c^2}{K_k^2 + c^2}\frac{p}{K_2 + p}, \tag{2.176}$$

$$\frac{dZ}{dt} = V_{3K} - V_{ZP}$$

$$= \frac{V_k c^2}{K_k^2 + c^2}\frac{p}{K_2 + p} - \frac{V_3 Z}{K_3\left(1 + \frac{p}{K_1}\right) + Z}. \tag{2.177}$$

Chapter 3
Basic Modelling Principles: Deterministic Models

The task of combining toolbox components into an overall model is neither simple nor easy. Firstly, it cannot be emphasised too strongly that the type of model to be constructed depends critically on the question which one wishes to study with that model. For some questions, a simple well-mixed cell model – a set of ordinary differential equations – is sufficient. For other questions, it might be necessary to construct a stochastic partial differential equation in three dimensions, using a finite element model based on anatomical measurements. In most cases, it is simply not true to say that one type of model is "better" than another. A model may be more complicated, or less complicated, but more complication does not mean that the model is more accurate, or more realistic; such judgements are highly dependent on the context.

As a corollary to this, we must also always keep in mind that the purpose of a model is not to include, without discrimination, every single possible physiological complication, and thus reproduce, in silico, a miniature cell in all its complexity. Even if we wanted to do this, the task would be impossible. Something must always be left out. It is our job as modellers to decide, given the question under consideration, what must be included in our model, and what can be safely omitted.

Neither is the sole purpose of a model to reproduce, with impressive accuracy, a particular experimental curve. As a general rule, the purpose of a model is to predict the results of experiments that have not yet been performed, rather than reproduce the results of those that have been. Obviously, models need to be validated by comparison to existing results, but such validation by itself is of limited use. It is not until a model can tell the experimentalist something they did not already know that a model has been truly useful.

In this chapter we discuss some of the basic principles of how deterministic models of Ca^{2+} signalling are constructed, and the different types of models that are commonly used. Stochastic models are treated similarly in the next chapter. It is not our purpose here to present detailed models of particular cell types (this task is left for subsequent chapters) and thus the

G. Dupont et al., *Models of Calcium Signalling*, Interdisciplinary Applied Mathematics 43, DOI 10.1007/978-3-319-29647-0_3

discussion here has a more theoretical bent, with less reference to experimental data. Nevertheless, the reader should keep firmly in mind the necessity for comparison to, and prediction of, experimental data.

3.1 Types of Models

Models of Ca^{2+} signalling – as with most models of the real world – are divided into four major groups. The first major division is between deterministic and stochastic models; the second major division is between spatially homogeneous and spatially distributed models.

At the most fundamental level, Ca^{2+} signalling is driven by stochastic events. IPR, for example, exist in a number of discrete states, some open, some closed, and move from state to state in random fluctuations driven by thermal noise. When IPR open they release a small amount of Ca^{2+} into the cytoplasm, which can then open neighbouring IPR to form larger release events, and so on. If enough IPR happen to open, a global Ca^{2+} spike or wave can occur.

Because of this underlying stochasticity, there has been a lot of work done on stochastic models of Ca^{2+} spiking, or on smaller events such as blips, puffs, and sparks. Much of this work is discussed in more detail in Chapter 4.

However, despite the fact that Ca^{2+} signalling is inherently stochastic, deterministic models have played, and continue to play, important roles. Deterministic models can be useful for making predictions that can (and have) been verified experimentally; since they are easier to solve numerically than analogous stochastic models, they remain a highly important tool.

There is a similar relationship between spatially homogeneous models (modelled by ordinary differential equations, or ODEs) and spatially distributed models (modelled by partial differential equations, or PDEs). Although it is self-evident that cells are spatially distributed objects, and that the Ca^{2+} concentration will not, in general, be the same at each point inside the cell, nevertheless ODE models can still provide useful predictions that can help us understand how Ca^{2+} signalling works.

In short, one must avoid the temptation to conclude that, since cells are inherently stochastic and spatially distributed, deterministic or spatially homogeneous models must be "wrong". Models can be judged only on the usefulness of the predictions they make, and how much they have helped us to learn.

3.2 Spatially Homogeneous Models

The most elementary models assume that the cell is spatially homogeneous, i.e., that every part of the cell has the same Ca^{2+} and IP_3 concentrations, and that the cytoplasm and the ER coexist at each point of the cell (for a more detailed discussion of this point, see Section 3.3.2).

Fig. 3.1 shows a simple model of Ca^{2+} signalling. There are five Ca^{2+} fluxes: into the cell (J_{in}, through ROCC, for instance), out of the cell (J_{out}, through the PMCA pumps), from the ER into the cytoplasm (J_{IPR}, through the IPR, and J_{leak}, an unspecified leak), and from the cytoplasm into the ER (J_{serca}, the SERCA pumps).

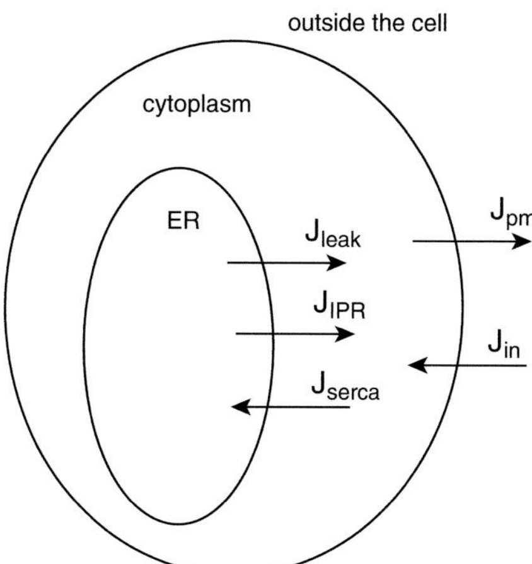

Fig. 3.1 Schematic diagram of a simple spatially homogeneous model of Ca^{2+} signalling. Although the diagram appears to imply that the ER is a separate compartment from the cytoplasm, in models of this type this is in fact not so. Rather, the cytoplasm and the ER are assumed to coexist at each point of the cell.

Note that Ca^{2+} buffering does not appear explicitly in this model. This is because we are making the implicit assumption that the buffers are fast and linear, and thus simply add a scaling factor to all the fluxes, as discussed in detail in Section 2.9. If we let c and c_e denote the Ca^{2+} concentrations in the cytoplasm and ER, respectively, with respective volumes V and V_e, then conservation of Ca^{2+} gives

$$\frac{d}{dt}(cV) = J_{\text{IPR}} + J_{\text{leak}} - J_{\text{serca}} + J_{\text{in}} - J_{\text{pm}}, \tag{3.1}$$

$$\frac{d}{dt}(c_e V_e) = -J_{\text{IPR}} - J_{\text{leak}} + J_{\text{serca}}. \tag{3.2}$$

Here, the units of each flux, J, is moles/time.

As long as the volumes of the cytoplasm and ER remain constant, we can divide out the volumes to get

$$\frac{dc}{dt} = \frac{1}{V}(J_{\text{IPR}} + J_{\text{leak}} - J_{\text{serca}} + J_{\text{in}} - J_{\text{pm}}), \tag{3.3}$$

$$\frac{dc_e}{dt} = \frac{1}{V_e}(-J_{\text{IPR}} - J_{\text{leak}} + J_{\text{serca}}). \tag{3.4}$$

In models like this it is usual to rescale all the fluxes, so that they have units of moles per time per cytoplasmic volume. Thus, we define, for example, a new $\tilde{J}_{\rm in} = J_{\rm in}/V$, with analogous expressions for the other fluxes, and rewrite both equations in these new units.

If we do this, and then (for notational convenience) drop the tildes, we get

$$\frac{dc}{dt} = J_{\rm IPR} + J_{\rm leak} - J_{\rm serca} + J_{\rm in} - J_{\rm pm}, \tag{3.5}$$

$$\frac{dc_e}{dt} = \gamma(-J_{\rm IPR} - J_{\rm leak} + J_{\rm serca}), \tag{3.6}$$

where $\gamma = \frac{V}{V_e}$. The factor of γ appears since a flux (of a given number of moles per time) from the cytoplasm to the ER causes a different change in concentration in each compartment, due to their different volumes.

At this stage the toolbox principle comes into play. Each of the fluxes can be modelled in ways described in the previous chapter, and then combined by the use of (3.5) and (3.6). To change, for example, the SERCA pump model, one pulls out the old $J_{\rm serca}$ model, and puts in the new one, without the necessity for any other structural changes to the model. Similarly, one can include any number of additional fluxes in and out of the cytoplasm and ER, or additional compartments with their own associated fluxes.

3.2.1 A Model Based on IPR Dynamics

For our first example, we use the simplified version of the De Young-Keizer model of the IPR ((2.115) and (2.116)), Hill functions for the PMCA and SERCA pumps (2.28), and a simple linear ROCC model for the influx (2.99). This gives

$$\frac{dc}{dt} = J_{\rm IPR} + J_{\rm leak} - J_{\rm serca} + \delta(J_{\rm in} - J_{\rm pm}), \tag{3.7}$$

$$\frac{dc_e}{dt} = \gamma(J_{\rm serca} - J_{\rm IPR} - J_{\rm leak}), \tag{3.8}$$

$$\frac{dy}{dt} = \phi_1(c)(1-y) - \phi_2(c)y, \tag{3.9}$$

where

$$J_{\rm IPR} = k_f P_o(c_e - c), \tag{3.10}$$

$$J_{\rm leak} = k_{\rm leak}(c_e - c), \tag{3.11}$$

$$J_{\rm serca} = \frac{V_s c^2}{K_s^2 + c^2}, \tag{3.12}$$

$$J_{\rm in} = \alpha_0 + \alpha_1 p, \tag{3.13}$$

$$J_{\rm pm} = \frac{V_p c^2}{K_p^2 + c^2}. \tag{3.14}$$

Here, y denotes the fraction of IPR that have been inactivated by Ca^{2+}, p denotes the IP_3 concentration, and P_o is the open probability of the IPR. All the parameters are given in Table 3.1.

The expressions for P_o, ϕ_1, and ϕ_2 are taken from (2.114) and (2.116). For convenience, we repeat them here:

$$P_o = \left(\frac{pc(1-y)}{(p+K_1)(c+K_5)} \right)^3, \tag{3.15}$$

$$\phi_1 = \frac{(k_{-4}K_1K_2 + k_{-2}pK_4)c}{K_4K_2(p+K_1)}, \tag{3.16}$$

$$\phi_2 = \frac{k_{-2}p + k_{-4}K_3}{p+K_3}. \tag{3.17}$$

$k_1 = 400\,\mu\text{M}^{-1}\,\text{s}^{-1}$	$k_{-1} = 52\,\text{s}^{-1}$
$k_2 = 0.2\,\mu\text{M}^{-1}\,\text{s}^{-1}$	$k_{-2} = 0.21\,\text{s}^{-1}$
$k_3 = 400\,\mu\text{M}^{-1}\,\text{s}^{-1}$	$k_{-3} = 377.2\,\text{s}^{-1}$
$k_4 = 0.2\,\mu\text{M}^{-1}\,\text{s}^{-1}$	$k_{-4} = 0.029\,\text{s}^{-1}$
$k_5 = 20\,\mu\text{M}^{-1}\,\text{s}^{-1}$	$k_{-5} = 1.64\,\text{s}^{-1}$
$\gamma = 5.5$	$\delta = 1$
$V_s = 0.9\,\mu\text{M}\,\text{s}^{-1}$	$K_s = 0.1\,\mu\text{M}$
$V_p = 0.1\,\mu\text{M}\,\text{s}^{-1}$	$K_p = 0.3\,\mu\text{M}$
$\alpha_0 = 0.01\,\mu\text{M}\,\text{s}^{-1}$	$\alpha_1 = 0.07\,\text{s}^{-1}$
$k_f = 1.11\,\text{s}^{-1}$	$k_{\text{leak}} = 0.02\,\text{s}^{-1}$

Table 3.1 Parameters of the model of Ca^{2+} oscillations (Section 3.2.1), based on the De Young-Keizer IPR model (De Young and Keizer, 1992). As usual, $K_i = k_{-i}/k_i$, for $i = 1, \ldots, 5$.

Before we analyse the model, it is worth looking more closely at the model construction.

- J_{leak} is an IPR-independent leak out of the ER. The pathway for this leak is unspecified. This leak is necessary in order for there to be a steady state when $p = 0$, i.e., when the IPR are all closed. For a steady state to exist we require that

$$J_{\text{IPR}} + J_{\text{leak}} = J_{\text{serca}}. \tag{3.18}$$

When $p = 0$, $P_o = 0$, and thus, to satisfy (3.18), either there is a background leak out of the ER, or the SERCA pump flux is zero. The latter is possible when we use a bidirectional SERCA pump model, such as (2.39). However, in this example our SERCA model is unidirectional, and thus a background flux from the ER is required in order to have a valid model.

- Both J_{IPR} and J_{leak} are assumed to be proportional to $c_e - c$. This is a relatively crude assumption; certainly the flux will be dependent on the concentration difference between the ER and the cytoplasm, and, since the transport of Ca^{2+} through the IPR is not an active process, it will be zero when $c = c_e$. However, there is no good reason to assume it is linear across the whole range of concentrations. Indeed, it is likely not to be. Nevertheless, the assumption of linearity is the simplest possible, and good enough for our present purposes.
- The constant k_f is a scaling factor that controls the maximum amount of Ca^{2+} flux through the IPR. It can be interpreted as representing the IPR density.
- The influx, J_{in}, is a linearly increasing function of p, and is thus clearly unphysiological once p gets high. This makes little difference to us as we shall not be considering the model at such high values of p. However, keep in mind that, in this model of J_{in}, p is being used as a proxy for agonist concentration. In reality, influx will be an increasing function of agonist concentration, not directly of p. In this model it is possible for us to use p in this manner, as there is no feedback from c to p, and thus p is a monotonic function of agonist concentration.
- The constant δ is not necessary from the physiological point of view, but is useful nonetheless. It can be used to adjust the ratio of the rate of Ca^{2+} transport across the plasma membrane to the rate of Ca^{2+} transport across the ER membrane. As δ gets small, transport across the plasma membrane becomes slower with respect to transport in and out of the ER, but the steady-state concentrations remain unchanged.

3.2.1.1 Steady States and Oscillations

The steady state of the model is determined by solving the three equations

$$J_{\text{in}} = J_{\text{pm}}, \tag{3.19}$$

$$J_{\text{IPR}} + J_{\text{leak}} = J_{\text{serca}}, \tag{3.20}$$

$$y = \frac{\phi_1}{\phi_1 + \phi_2}. \tag{3.21}$$

For convenience, we denote the steady state of c by $\bar{c}$, and similarly for c_e and y.

In the first of these equations, (3.19), there is a single variable, c, and thus $\bar{c}$ is determined entirely by the balance of the influx and efflux of Ca^{2+} across the plasma membrane. In other words, the resting cytoplasmic Ca^{2+} concentration is unaffected by Ca^{2+} transport across the ER membrane, or indeed across the membrane of any other internal compartment. For example, increasing the density of SERCA pumps (a reasonably common experimental technique), will cause a temporary decline in cytoplasmic $[Ca^{2+}]$, and a permanent increase in ER [Ca^{2+}], but no change in the final resting cytosolic [Ca^{2+}]. Once $\bar{c}$ is determined from (3.19), $\bar{c}_e$ is determined from (3.20), while $\bar{y}$ follows trivially.

It has been known experimentally for decades that the resting cytosolic [Ca^{2+}] increases with agonist concentration, and, in any model, this can only result from an increase in Ca^{2+} influx into the cell, not from an increase in the release of Ca^{2+} from the ER or any other internal compartment. The model incorporates this fact by making J_{in} an increasing function of p, from which it follows that $\bar{c}$ must also be an increasing function of p, and thus, indirectly, of agonist concentration.

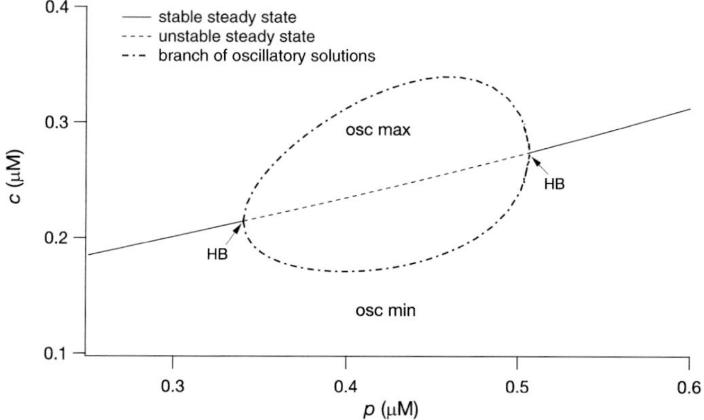

Fig. 3.2 Steady states and oscillatory solutions in the De Young-Keizer model (Section 3.2.1). As p increases, the steady state loses stability at a Hopf bifurcation (HB), giving rise to a branch of periodic solutions. Osc max and osc min denote, respectively, the maximum and minimum values of c over one oscillation.

The steady state, as a function of p, is shown in Fig. 3.2, where it can be seen that, as expected, $\bar{c}$ increases with p in an almost linear manner. At low and high values of p the steady state is stable, but at intermediate values it is unstable. Stability is lost and regained via two supercritical Hopf bifurcations. In Fig. 3.2 we plot, for each value of p, the maximum and minimum values of c on the limit cycle.

There is thus an intermediate range of values of p for which oscillations exist, and the period of these oscillations, although varying with p, is around 10 seconds. The oscillations have a sinusoidal form superimposed upon a slightly raised baseline, and arise principally through the inherent delay in the variable y. An increase in c tends first to open the IPR, leading to a further increase in c, but then, after a delay, y increases, leading to a decrease in P_o and the falling phase of the oscillations. If this delay in the dynamics of y is removed, the oscillations also disappear, and thus the oscillations arise from the intrinsic dynamics of the IPR. As we shall see in Section 3.2.7, such an oscillatory mechanism is called a Class I mechanism.

Although this model has some features that agree qualitatively with experimental data, one must be very careful not to push this analogy too far. Firstly, as we shall see in Chapter 4, we know that Ca^{2+} oscillations do not arise via Hopf bifurcations as p increases. They arise, rather, from intrinsic stochastic mechanisms, and are not limit cycles at all. This does not negate the utility and interest of deterministic models, but makes their interpretation a matter for great care. Secondly, the model is not in good agreement with oscillations seen in many cell types, which take the form of baseline spikes, not sinusoidal oscillations. Finally, although the range of periods and amplitudes is not entirely unreasonable, neither is it an accurate guide to most cell types.

Nevertheless, despite these caveats, such simple deterministic models can serve to illustrate important dynamical principles of Ca^{2+} oscillations, principles that are conserved over large families of models.

3.2.2 A Model Based on ER Refilling

One of the earliest models for IP_3-dependent Ca^{2+} release assumed the existence of two distinct internal Ca^{2+} stores, one sensitive to IP_3, the other sensitive to Ca^{2+} (Kuba and Takeshita, 1981; Goldbeter et al, 1990). Agonist stimulation leads to the production of IP_3 which releases Ca^{2+} from the IP_3-sensitive store through IP_3 receptors. The Ca^{2+} that is thereby released stimulates the release of further Ca^{2+} from the Ca^{2+}-sensitive store, possibly via ryanodine receptors. A crucial assumption of the model is that the concentration of Ca^{2+} in the IP_3-sensitive store remains constant, as the store is quickly refilled from the extracellular medium.

However, realisation that the IP_3-sensitive and Ca^{2+}-sensitive pool were almost certainly both the ER led to the development of a one-pool version of this original model (Dupont and Goldbeter, 1993), and it is this version that we present here.

The model equations are

$$\frac{dc}{dt} = J_{\text{in}} - J_{\text{pm}} - J_{\text{serca}} + J_{\text{IPR}} + J_{\text{leak}}, \tag{3.22}$$

$$\frac{dc_e}{dt} = \gamma(J_{\text{serca}} - J_{\text{IPR}} - J_{\text{leak}}), \tag{3.23}$$

where

$$J_{\text{in}} = \nu_0 + \nu_1 p, \tag{3.24}$$

$$J_{\text{serca}} = \frac{V_1 c^2}{K_1^2 + c^2}, \tag{3.25}$$

$$J_{\text{IPR}} = p\left(\frac{V_2 c_e^2}{K_2^2 + c_e^2}\right)\left(\frac{c^4}{K_3^4 + c^4}\right), \tag{3.26}$$

$$J_{\text{leak}} = k_f c_e. \tag{3.27}$$

As c increases, so does J_{IPR}, and so Ca^{2+} stimulates its own release through positive feedback, a feature that is central to the model's behaviour. However, this model of the Ca^{2+} flux through an IPR was developed some years before the properties of the IPR were known, and thus it includes no Ca^{2+}-dependent inactivation of the IPR. Instead, Ca^{2+} current through the IPR is terminated, not by Ca^{2+}-induced inhibition of the IPR, but by depletion of the ER, via the term $\frac{V_2 c_e^2}{K_2^2+c_e^2}$. Indeed, the period of oscillations in this model is determined principally by the time taken to refill the ER enough to allow another Ca^{2+} spike.

Although it is now recognised that Ca^{2+}-induced inhibition of the IPR is a crucial feature, with the result that this model is no longer widely used, it is worth keeping in mind that ER depletion might well, in many circumstances, be equally important.

3.2.3 A Model That Incorporates Microdomains Around IPR

To illustrate the complexities that can arise, we construct a different model by replacing the De Young-Keizer IPR model with the modal model given by (2.129)–(2.136).

Here, the most important thing to realise is that the parameters of the modal model are determined under conditions where the Ca^{2+} concentration at the mouth of the IPR remains constant. However, this will not be the case for a whole-cell model, as Ca^{2+} flowing through the IPR will cause a localised domain of high $[Ca^{2+}]$ at the channel mouth. This complication was first addressed by Dupont and Swillens (1996), but here we use the modal model as an illustration. (Of course, the exact same question arises when one incorporates the De Young-Keizer model into a whole-cell model. However, in that case the question was essentially ignored; the model parameters were simply chosen so that the background cytosolic concentrations could be used.)

To handle the Ca^{2+} microdomains, we introduce two additional Ca^{2+} concentrations:

- c_{open} is the Ca^{2+} concentration right at the mouth of an open IPR. This is the Ca^{2+} concentration that will control the rate of closing of an open IPR (Dupont and Swillens, 1996). Rüdiger et al (2007, 2010b) have shown that a good approximation is simply to assume that c_{open} is approximately equal to the concentration in the ER. Thus, the rate of IPR inactivation, α, will be dependent only on c_e, not c. This is incorporated in the model by replacing $\hat{c}$ in (2.133) with c_e. Thus,

$$\bar{m}_\alpha = \frac{c_e^3}{c_e^3 + 0.35^3}, \tag{3.28}$$

$$\bar{h}_\alpha = \frac{80^2}{c_e^2 + 80^2}. \tag{3.29}$$

- c_{cluster} is the Ca^{2+} concentration experienced by a closed IPR. This is not the background, observed Ca^{2+} concentration (denoted by c), but is instead the Ca^{2+} concentration within the cluster of IPR. It will thus depend on the number of IPR in the cluster.

Our whole-cell oscillation model is now based around the schematic diagram shown in Fig. 3.3. There are four variables in the model: the cytosolic concentration (c), the ER concentration (c_e), the microdomain cluster concentration (c_{cluster}), and the IPR slow variable (h_β). Conservation of Ca^{2+} gives the usual equations

$$\frac{dc}{dt} = J_{\text{diff}} + J_{\text{leak}} - J_{\text{serca}} - \delta(J_{\text{pm}} - J_{\text{in}}), \tag{3.30}$$

$$\frac{dc_e}{dt} = \gamma_e(-J_{\text{IPR}} + J_{\text{serca}} - J_{\text{leak}}), \tag{3.31}$$

$$\frac{dc_{\text{cluster}}}{dt} = \gamma_c(J_{\text{IPR}} - J_{\text{diff}}), \tag{3.32}$$

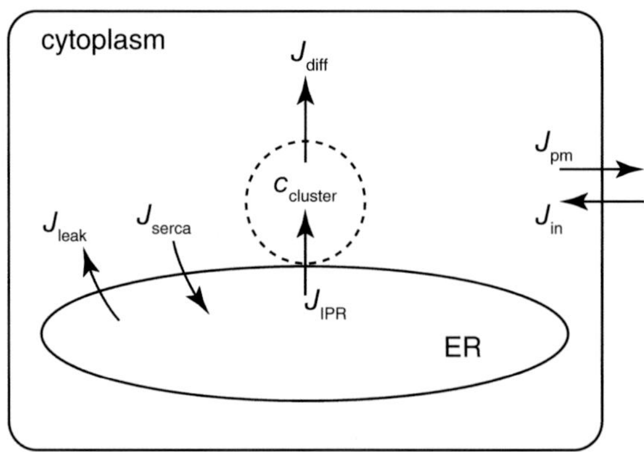

Fig. 3.3 Schematic diagram of the whole-cell model based around the modal IPR model. IPR are assumed to exist in clusters, with a local elevated Ca^{2+} concentration denoted by c_{cluster}. All the Ca^{2+} flowing through the IPR first enters this microdomain before then diffusing to the cytosol. However, spatial aspects of the model are not dealt with rigorously. Instead, it is simply assumed that the cluster microdomain and the cytosol form separate compartments (see Section 3.4 for a more detailed presentation of different ways of modelling microdomains).

which are coupled to (2.130) which governs h_β. The constants γ_e and γ_c are the usual volume ratios needed to compensate for the different volumes of the three regions.

The flux through the IPR is modelled as

$$J_{\mathrm{IPR}} = k_f N P_o (c_e - c_{\mathrm{cluster}}), \tag{3.33}$$

where N is the number of IPR in the cluster, k_f (with units of 1/time) is a constant of proportionality, and P_o, the open probability of a single IPR, is given by (2.129). Note that this equation for J_{IPR} is essentially the same as (3.10) in the previous model, with the only differences being the exact expression for P_o and the use of the cluster concentration rather than the cytosolic concentration.

The diffusion of Ca^{2+} from the cluster microdomain to the cytosol is treated as a linear process, and thus

$$J_{\mathrm{diff}} = k_{\mathrm{diff}}(c_{\mathrm{cluster}} - c). \tag{3.34}$$

The fluxes J_{leak}, J_{serca}, and J_{pm} remain the same as in the previous model, and are given by (3.11), (3.12), and (3.14), respectively, while (purely for convenience) J_{in} is modelled as a receptor-operated Ca^{2+} current that saturates as p gets large. Thus

$$J_{\mathrm{in}} = 0.2 + \frac{V_{\mathrm{rocc}} p^2}{p^2 + K_{\mathrm{rocc}}^2}. \tag{3.35}$$

All the parameter values are given in Table 3.2, and a typical solution (Fig. 3.4) exhibits baseline spiking, with large amplitude.

γ_e	$= 10$	δ	$= 0.3$
γ_c	$= 100$	k_{diff}	$= 3\,\mathrm{s}^{-1}$
V_s	$= 40\,\mu\mathrm{M\,s}^{-1}$	K_s	$= 0.26\,\mu\mathrm{M}$
V_p	$= 8\,\mu\mathrm{M\,s}^{-1}$	K_p	$= 0.5\,\mu\mathrm{M}$
V_{rocc}	$= 2\,\mu\mathrm{M\,s}^{-1}$	K_{rocc}	$= 0.2\,\mu\mathrm{M}$
$k_f N$	$= 4\,\mathrm{s}^{-1}$	k_{leak}	$= 0.04\,\mathrm{s}^{-1}$
q_{26}	$= 10500\,\mathrm{s}^{-1}$	q_{62}	$= 4010\,\mathrm{s}^{-1}$
T_{high}	$= 20\,\mathrm{s}^{-1}$	T_{low}	$= 0.2\,\mathrm{s}^{-1}$

Table 3.2 Parameters of the model of Ca^{2+} oscillations discussed in Section 3.2.3.

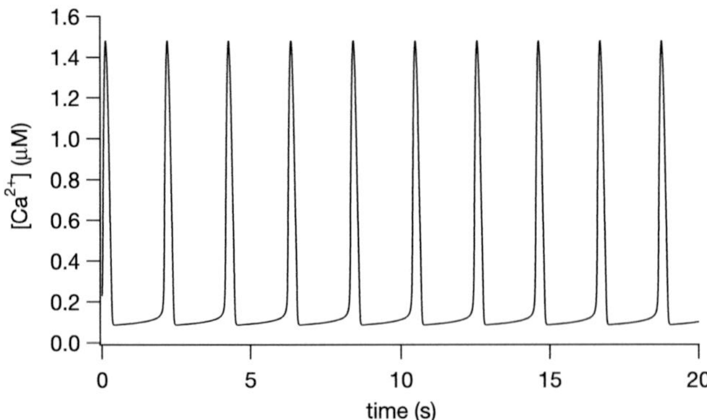

Fig. 3.4 A typical solution of the model of Section 3.2.3, computed with $p = 0.2$.

3.2.4 Calcium Excitability: Calcium-Induced Calcium Release

One important feature of Ca^{2+} dynamics, seen in a wide range of cell types, is Ca^{2+} *excitability*. This phrase, modelled on the analogous concept of electrical excitability, means that the addition of a small amount of Ca^{2+} to the cytoplasm can initiate the release of a much greater amount from the SR/ER, in a process of Ca^{2+}-induced Ca^{2+} release (CICR).

CICR was first observed in skeletal muscle (Endo et al, 1970), followed by the same observation in cardiac cells (Fabiato and Fabiato, 1975, 1978), and came thus to be closely associated with RyR in muscle. The canonical version of CICR is seen in cardiac cells, where a small amount of Ca^{2+} coming in through voltage-gated Ca^{2+} channels stimulates the release of a much larger amount of Ca^{2+} from the SR, through the RyR. However, Li et al (1995a) pointed out how essentially the same behaviour is exhibited by IPR also, and thus how CICR is important well beyond striated muscle. It is now known that CICR is the fundamental mechanism underlying Ca^{2+} oscillations in many cell types, and the propagation of many types of Ca^{2+} waves.

The theoretical understanding of CICR is based on general theories of excitable systems, which are based largely on the work of FitzHugh (FitzHugh, 1960, 1961, 1969a,b) and Nagumo (Nagumo et al, 1964; McKean, 1970). The FitzHugh-Nagumo model is the canonical excitable system, and has led to a wide array of theoretical investigations (Hastings, 1976; Troy, 1976; Rinzel, 1977; Rauch and Smoller, 1978; Hastings, 1982; Keener, 1980; Tyson and Keener, 1988). An introduction to excitable systems in a physiological context can be found in Keener and Sneyd (2008).

Basic concepts of excitability are fundamental to the theoretical study of Ca^{2+} oscillations and waves. In essence, the FitzHugh-Nagumo model can be used as a simple model of many Ca^{2+} oscillations. There are important differences, however, particularly when diffusion is introduced

(Sneyd et al, 1993; Tsai et al, 2012), as discussed in Section 5.8, and thus the general theory cannot be applied without care.

In Fig. 3.5 we illustrate CICR using the modal model of Section 3.2.3, although any IPR or RyR model would behave similarly. For these simulations, we first calculated the steady state of the model when $p = 0.07$. This value for p cannot be chosen arbitrarily. If p is too small or too large, the model will not exhibit CICR. Then, setting all other variables at their respective steady states, we set the initial condition for c to be just above the steady state. From the figure we see that if $c(0) = 0.22$ we get no significant response, but an increase of $c(0)$ to 0.23 causes a large transient response before the solution returns to steady state.

Note that the initial Ca^{2+} concentration must be large enough – must be above the *excitability threshold* – in order for a large transient to occur. This threshold is crucial for stability; one would not want a cardiac cell, for instance, to respond with a large Ca^{2+} transient to any perturbation, no matter how small. The threshold ensures that the cell is stable to small fluctuations, while remaining responsive to perturbations that are sufficiently large.

CICR plays a major role in Ca^{2+} wave propagation. When a small amount of Ca^{2+} is released from, say, a cluster of IPR, this Ca^{2+} will diffuse to neighbouring IPR clusters, and, if the concentration there goes above the threshold for excitable release, will initiate additional Ca^{2+} release from the neighbouring cluster, leading to the sequential activation of IPR clusters and thus propagation of a wave. Conversely, if release from the initial cluster

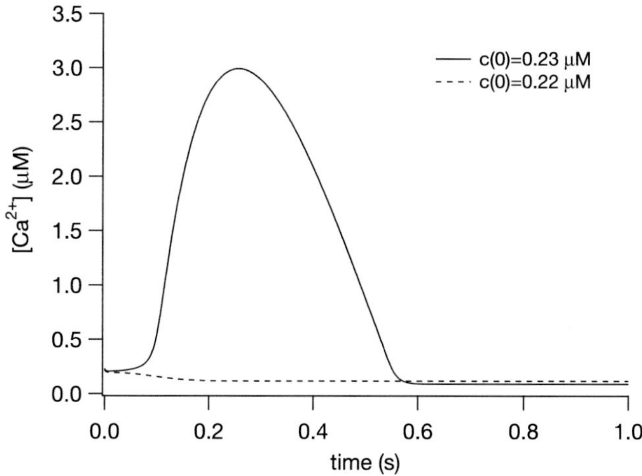

Fig. 3.5 Excitability in the modal model of Section 3.2.3. When $p = 0.07\,\mu$M the steady state is $c = 0.117\,\mu$M. When the initial condition for c is set at $0.22\,\mu$M, the solution just returns directly to steady state. However, if the initial condition for c is increased slightly to $0.23\,\mu$M, the solution undergoes a large excursion before returning to steady state.

is not large enough to drive the Ca^{2+} concentration above the excitability threshold at neighbouring IPR, no wave will form. This is discussed in more detail in Section 3.5.1.

3.2.5 Open-Cell and Closed-Cell Models

It is necessarily true that, in all cells, the resting cytosolic Ca^{2+} concentration is set by the balance between influx and efflux across the plasma membrane. However, in many cell types, Ca^{2+} transport across the plasma membrane has little direct effect on Ca^{2+} oscillations, which are caused, almost entirely, by Ca^{2+} transport in and out of the ER, with only small amounts of Ca^{2+} entering or leaving the cell over an oscillation cycle. In such cases, when Ca^{2+} is removed from outside the cell the oscillations persist for a considerable period of time, before the cell eventually gets depleted of so much Ca^{2+} that it can no longer sustain oscillations. An example of this, from airway smooth muscle cells, is shown in Fig. 3.6.

Thus, any model of Ca^{2+} oscillations in any cell type that exhibits such behaviour should exhibit oscillations when there is no influx of Ca^{2+} from outside the cell. However, since there can be no long-term oscillations in this case (as Ca^{2+} will necessarily decrease to zero eventually), it is more useful to look for oscillations when there is no Ca^{2+} transport either in or out of the cell. Such a model is called a *closed-cell* model. Similarly, a model in which there is transport to and from the outside is called an *open-cell* model.

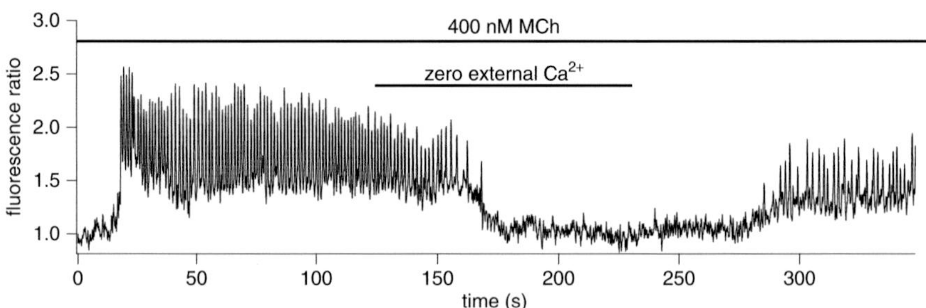

Fig. 3.6 Ca^{2+} oscillations (in response to MCh) in the presence and absence of external Ca^{2+}, measured in airway smooth muscle cells in a mouse lung slice. When external Ca^{2+} is removed the oscillations persist for some time but eventually disappear, while when external Ca^{2+} is restored the oscillations gradually reappear. Original data provided by Michael Sanderson and Jun Chen (University of Massachusetts Medical School).

We can convert the model of Section 3.2.1 to a closed-cell model simply by setting $\delta = 0$, in which case we get

$$\frac{dc}{dt} = (k_f P_o + J_{\text{leak}})(c - c_e) - J_{\text{serca}}, \tag{3.36}$$

$$\frac{dc_e}{dt} = \gamma(J_{\text{serca}} - (k_f P_o + J_{\text{leak}})(c - c_e)), \tag{3.37}$$

$$\frac{dy}{dt} = \phi_1(c)(1 - y) - \phi_2(c)y, \tag{3.38}$$

We see immediately that, in the closed-cell version of the model, $\gamma c + c_e =$ constant $= c_t$, say. Thus we have only two independent differential equations, (3.36) and (3.38), where $c - c_e$ is replaced by $c - (c_t - \gamma c)$.

The constant c_t is the total number of moles of Ca^{2+} in the cell – including both Ca^{2+} in the ER and in the cytoplasm – divided by the ER volume, and is an important factor controlling the properties of Ca^{2+} oscillations. To illustrate this, we treat c_t as a parameter, and plot the bifurcation diagram of the model as a function of c_t for fixed $p = 0.4$ (Fig. 3.7). We see that when c_t is too low or too high no oscillations exist for this value of p. Indeed, for any value of p, oscillations can exist only if c_t lies in the correct range.

We now return our attention to the open-cell version of the simplified De Young-Keizer model of Section 3.2.1. In Fig. 3.8 we show the response of the open-cell model when p is increased from 0 to 0.4 at time $t = 0$. The solution starts from the steady state corresponding to $p = 0$. For these parameters this steady state is $c = 0.1$, $c_t = 4.2$. When p is increased, c_t begins to decrease slowly, as Ca^{2+} is pumped out of the cell in response to the raised average cytosolic $[Ca^{2+}]$. The smaller δ is, the slower this decrease will be. However,

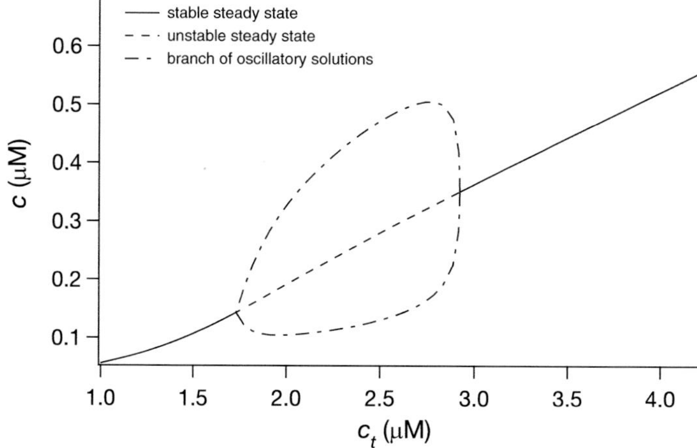

Fig. 3.7 Steady states and oscillatory solutions in the closed-cell version of the simplified De Young-Keizer model (Section 3.2.1), computed as functions of c_t for $p = 0.4$. Oscillations appear and disappear via two Hopf bifurcations, and occur only for intermediate values of c_t.

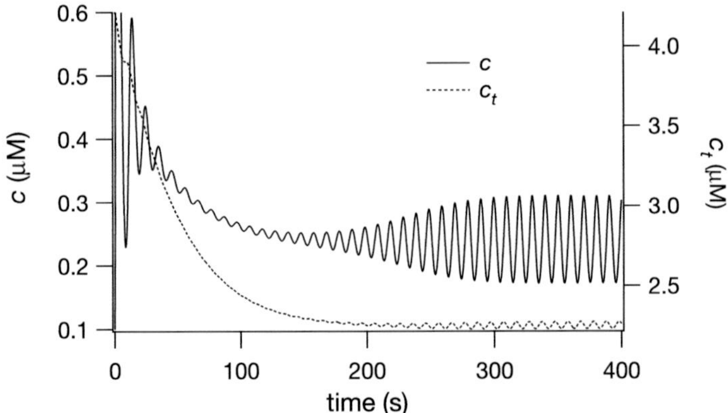

Fig. 3.8 Solution of the open-cell version of the simplified De Young-Keizer model (Section 3.2.1) showing the transient behaviour that occurs upon a step increase in p. At time $t = 0$, p was increased from 0 to 0.4, while c and c_t were started from their respective steady states when $p = 0$ (i.e., $c = 0.1$, $c_t = 4.2$).

we see from Fig. 3.7 that, when $p = 0.4$ and $c_t \approx 4$, the steady state is stable. Thus, the initial oscillations in c are damped out. But as c_t continues to decrease, the steady state becomes unstable again (as shown in Fig. 3.7) and the oscillations reappear, eventually stabilising with c_t oscillating close to 2.3. The generic shape of these oscillations – with an amplitude first decreasing then increasing – is commonly seen in a range of cell types.

Note that, when p is increased c_t decreases, even though the influx of Ca^{2+} into the cell is increased via the term J_{in}, which is an increasing function of p as given by (2.99). Because of the increased $[Ca^{2+}]$ in the cytoplasm, the plasma membrane pumps remove Ca^{2+} faster than it comes in through ROCC.

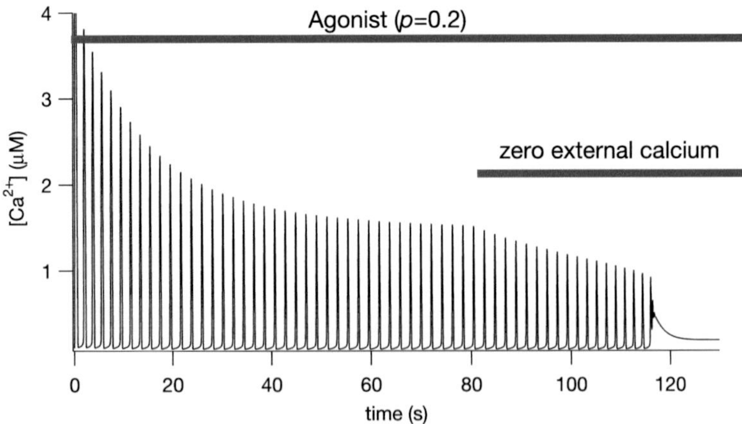

Fig. 3.9 The response of the modal model of Section 3.2.3 to the application of agonist, followed by the removal of extracellular Ca^{2+} (modelled by setting all influx terms to zero).

Although the shape of the Ca^{2+} oscillations is quite different, similar behaviour is seen in the modal model of Section 3.2.3, and is shown in Fig. 3.9. When p is increased to 0.2 μM at time $t = 0$, the model exhibits oscillations with decreasing amplitude, as c_t gradually decreases, before stabilising with an amplitude of just under 2 μM. When all Ca^{2+} influx is set to zero at time $t = 80$ s the oscillations gradually decrease in amplitude, before disappearing entirely at around 115 s. This is qualitatively similar to the experimental data shown in Fig. 3.6, although there are some clear quantitative discrepancies, particularly in the oscillation period in the absence of external Ca^{2+}.

Hence, modelling suggests that, by controlling c_t – for example, by regulating Ca^{2+} influx, or the rate of the plasma membrane pumps – a cell can regulate how it responds to agonists. An example of how this can be done experimentally is given by Sneyd et al (2004). In that paper c_t was manipulated experimentally, in order to show how, for example, oscillations can be manipulated by changes in c_t, at a constant p.

The approach used here – that of trying to explain model behaviour by treating one of the model variables as a bifurcation parameter – lacks greatly in mathematical rigour. Although it can be useful in simple cases, heuristic arguments like this should always be applied with caution. This is discussed in more detail in Section 5.4.

3.2.6 The Importance of Calcium Influx

Since c_t is such an important controlling variable, it comes as no surprise to learn that Ca^{2+} oscillations and waves can be modified by changing Ca^{2+} influx into the cell. This was shown experimentally by Girard and Clapham (1993) and Yao and Parker (1994), and has since become established as one of the central mechanisms by which a cell can control its Ca^{2+} signalling. For this reason, it is important to understand how Ca^{2+} influx is controlled, which explains the recent surge of interest in the mechanisms of receptor-operated and store-operated channels (Section 2.6).

The importance of Ca^{2+} influx is demonstrated in Fig. 3.10, using, for convenience, the De Young-Keizer model of Section 3.2.1, with the one change that J_{in} is now treated as a constant, not as a function of p. However, all other Ca^{2+} oscillation models have the same qualitative behaviour. When J_{in} is too low, oscillations cannot be stimulated by agonists, as there is no value of p for which oscillations occur. However, for a range of fixed values of p, oscillations can be caused simply by an increase in Ca^{2+} influx, without any change in agonist stimulation. In general, agonist stimulation will increase both p and J_{in}.

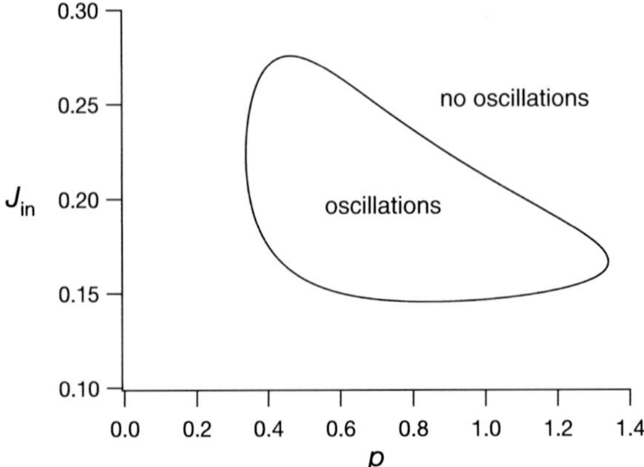

Fig. 3.10 Two-parameter bifurcation set of the simplified De Young-Keizer model (Section 3.2.1). Oscillations in the model occur for values of p and J_{in} inside the closed loop.

3.2.7 IP_3 Dynamics: Class I and Class II Models

In both the models discussed above, Ca^{2+} oscillations occur for a constant $[IP_3]$, as a result of the intrinsic dynamics of the IPR. Any oscillatory mechanism in which oscillations *can* occur for a constant $[IP_3]$ (even if $[IP_3]$ is not actually constant over the oscillations) is called a *Class I* mechanism.

However, in some cell types it is likely that other feedback mechanisms are equally, if not more, important than the intrinsic receptor dynamics. Stochastic mechanisms will be addressed in detail in Chapter 4, but there are other possible deterministic mechanisms that can help to set the oscillation period. One mechanism that has been widely studied is the interaction between Ca^{2+} and the production (and/or degradation) of IP_3 (Section 2.10), and models that use this mechanism to generate oscillations are called *Class II* models.

In Class II models, oscillations of Ca^{2+} are necessarily accompanied by oscillations in $[IP_3]$, and therefore it might be imagined that it is a relatively simple matter to distinguish these from Class I mechanisms by measuring $[IP_3]$ as a function of time. However, simultaneous measurement of $[Ca^{2+}]$ and $[IP_3]$ is not a simple experiment to do, and there are still relatively few such measurements. Some of the earliest results are those of Hirose et al (1999), who measured $[Ca^{2+}]$ and $[IP_3]$ simultaneously in Madin-Darby canine kidney epithelial cells (MDCK cells), and Nash et al (2001), who did so in Chinese Hamster Ovary cells (CHO cells). It is particularly fascinating to note that, at least in CHO cells, the type of oscillation depends on which receptor is activated. Activation of the metabotropic glutamate receptor mGluR5a gives synchronous oscillations in both $[Ca^{2+}]$ and $[IP_3]$, while activation of the M3 muscarinic receptor gives Ca^{2+} oscillations for a constant $[IP_3]$. Clearly there is enormous complexity even within a single cell type. Simultaneous measurements of $[Ca^{2+}]$ and $[IP_3]$ have also been performed in HSY cells

(Tanimura et al, 2009), where the peak in $[IP_3]$ occurs slightly after the peak in $[Ca^{2+}]$.

However, observing IP_3 oscillations that are closely synchronised with Ca^{2+} oscillations is no guarantee that both must occur together. The IP_3 oscillations may well modulate the exact nature of the Ca^{2+} oscillations, but might not be required.

The earliest Class II model was that of Meyer and Stryer (1988), who assumed that the activity of PLC was an increasing function of $[Ca^{2+}]$. Their model had only two variables, c and p, where

$$\frac{dp}{dt} = k_1 \frac{c}{c + K_c} - \beta p. \tag{3.39}$$

Since there was no consideration of IPR dynamics (i.e., the IPR was implicitly assumed to be always at steady state, although this is not the language used in the original paper), oscillations in this model were necessarily a result of the interaction between c and p.

The same expression for dp/dt was also used by Keizer and De Young (1992), in a paper that included IPR dynamics, using an early version of the model that subsequently became so well known (Section 2.7.1). Their conclusion was that the oscillations in their model were crucially dependent on the dynamics of the IPR, but not dependent on positive feedback from Ca^{2+} to IP_3 production. Thus, the model remained essentially a Class I model, where the oscillations were modulated by, but not dependent on, the accompanying IP_3 oscillations.

Dupont and Erneux (1997) constructed a more detailed model of IP_3 degradation, as discussed in Section 2.10, and used this as the basis of their model of Ca^{2+} oscillations. This model was used in a detailed study of Ca^{2+} oscillations in hepatocytes (Dupont et al, 2003), which is discussed in detail in Section 6.2, so we defer further discussion until then.

3.2.7.1 Hybrid Models

To investigate the differences between Class I and Class II models, it is useful to have a model that can be easily converted from one type to the other. This motivated the construction of hybrid models that could be either Class I or Class II (Politi et al, 2006; Domijan et al, 2006).

The model of Politi et al (2006) focused on studying the effects of positive and negative feedback from Ca^{2+} to IP_3 metabolism (Section 2.10). They used the Li-Rinzel model of the IPR (Section 2.7.2), standard Hill functions for the ATPases, and a simple model of Ca^{2+} influx. However, for their IP_3 dynamics they assumed that

$$\tau_p \frac{dp}{dt} = V_{\mathrm{PLC}} \frac{c^2}{K_{\mathrm{PLC}}^2 + c^2} - \left(\frac{\eta c^2}{K_{3K}^2 + c^2} + (1 - \eta) \right) p. \tag{3.40}$$

The parameters $K_{\rm PLC}$ and η tune the strength of the positive and negative feedback from Ca^{2+} to IP_3 metabolism. When $\eta = 0$ and $K_{\rm PLC}$ is not too small the feedback is all positive, as c increases the rate of p production, while when $\eta = 1$ and $K_{\rm PLC} \approx 0$ the feedback is all negative, as c does not affect the rate of p production, but just modulates the rate of p degradation. Finally, when $\eta = K_{\rm PLC} = 0$ the model is entirely a Class I model, as there is no feedback from c to p metabolism.

Typical results from the model of Politi et al (2006) are shown in Fig. 3.11. Here, only the positive feedback version is presented (i.e., $\eta = 0$), although both cases were studied in the original paper. When there is no feedback from Ca^{2+} to PLC (the dashed line), in which case the model is purely Class I, oscillations occur for values of $V_{\rm PLC}$ ranging from 0.7 to 1.2 μM/s. The period of these oscillations remains close to 10 seconds over the entire range (which is most likely a result of the simplicity of the IPR model). However, as feedback from Ca^{2+} to PLC is increased, i.e., as $K_{\rm PLC}$ increases, oscillations occur over a wider range of values of $V_{\rm PLC}$, and take on a much wider range of periods, from 10 s up to 200 s. Thus, the feedback from Ca^{2+} to PLC has a significant effect, slowing the oscillations and stabilising them for a wider range of agonist concentrations.

In addition, Politi et al (2006) used their model to make two important predictions:

- in the case of positive feedback, addition of an exogenous buffer will decrease the rise time of any Ca^{2+} oscillations, and can even eliminate oscillations entirely;
- in the case of negative feedback, an exogenous buffer will have little effect on the rise time of the oscillations, and cannot eliminate oscillations entirely.

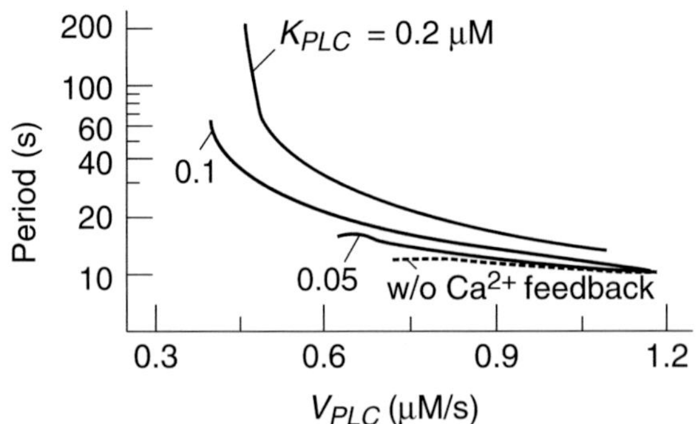

Fig. 3.11 The period of Ca^{2+} oscillations as a function of $V_{\rm PLC}$, the maximal rate of IP_3 production, for a range of values of $K_{\rm PLC}$ (see (3.40)). When $K_{\rm PLC} = 0$ PLC is insensitive to $[Ca^{2+}]$ (dotted line). Including Ca^{2+}-dependence of PLC activity significantly extends the range of observable periods. Reproduced from Politi et al (2006), Fig. 3C, with permission from Elsevier.

Experimental tests of these predictions showed that addition of the buffer slowed the rise time, and could eliminate oscillations if enough buffer was added. Their conclusion was that, in CHO cells, the Ca^{2+} oscillations were governed by a Class II mechanism involving positive feedback from Ca^{2+} to PLC. In a similar, and equally elegant, combination of theoretical and experimental approaches, Gaspers et al (2014) came to the same conclusion for COS cells.

A similar hybrid model was constructed by Domijan et al (2006), a model that has been used extensively in more theoretical investigations (Harvey et al, 2010). This hybrid model is based on the gating IPR model of Atri et al (1993) (Section 2.7.3), although similar qualitative results are obtained from a range of other models (Harvey et al, 2011). Instead of focusing on the differences between positive and negative feedback in a hybrid model, Domijan et al studied the difference in behaviour between a pure Class I model and a pure Class II model, converting one to the other via a hybrid model.

The model equations are

$$\frac{dc}{dt} = J_{\text{IPR}} - J_{\text{serca}} + \delta(J_{\text{in}} - J_{\text{pm}}), \tag{3.41}$$

$$\frac{dc_t}{dt} = \delta(J_{\text{in}} - J_{\text{pm}}), \tag{3.42}$$

$$\frac{dp}{dt} = \nu\left(1 - \frac{\alpha k_4}{c + k_4}\right) - \beta p, \tag{3.43}$$

where $J_{\text{IPR}} = k_f P_0(c_e - c)y$ and where P_0 and y are defined by (2.122) and (2.123). Note that ν is an increasing function of agonist concentration; for this reason, J_{in} will be, in general, an increasing function of ν.

Clearly, only positive feedback from Ca^{2+} to PLC is included; when $\alpha = 0$ the model is of pure Class I type, while when $\alpha = 1$ the model is a hybrid model, having both Class I and Class II mechanisms. However, by letting the speed of IPR inactivation tend to infinity (i.e., letting $\tau_y \to 0$) the model can be converted to a pure Class II model, in which the IPR is no longer time-dependent. Hence, by varying α and τ_y appropriately this model can be made pure Class I, pure Class II, hybrid, or even non-oscillatory. It is thus a useful tool for investigating the differences between these types of models. The model of Politi et al (2006) is similarly useful, and was also studied by Harvey et al (2011).

Further discussion of this model is left to Chapter 5.

3.2.7.2 Pulse Experiments

One possible way to distinguish between Class I and Class II mechanisms is by probing the cell with exogenous pulses of IP_3 and Ca^{2+}, as illustrated in Fig. 3.12 (Harootunian et al, 1988; Sneyd et al, 2006). In a Class I mechanism the oscillation frequency is a function, generally monotonically increasing, of $[IP_3]$. Thus, a temporary increase in $[IP_3]$ caused by the photorelease of a bolus of IP_3 might be expected to cause a temporary increase in oscillation frequency.

Conversely, in a Class II mechanism, $[IP_3]$ is one of the variables in the model. A sudden increase in $[IP_3]$ will cause a Ca^{2+} spike, and there will, in general, be a delay before the unperturbed oscillation is restored, with a corresponding phase change. This is a common way to investigate oscillators, and there are many studies of such phase resetting, or phase response, curves (Winfree, 1980; Guevara et al, 1986; Glass and Mackey, 1988; Ermentrout, 1996). Because of the experimental difficulties involved it has so far not been possible to construct a reliable phase resetting curve for a Ca^{2+} oscillator. Nevertheless, it is possible to obtain information of a more qualitative sort.

For example, in airway smooth muscle application of an exogenous pulse of IP_3 causes a transient increase in oscillation frequency (Fig. 3.13 A), while in pancreatic acinar cells, a similar application gives an immediate spike in $[Ca^{2+}]$, a delay to the next peak and a phase change (Fig. 3.13 B). Thus,

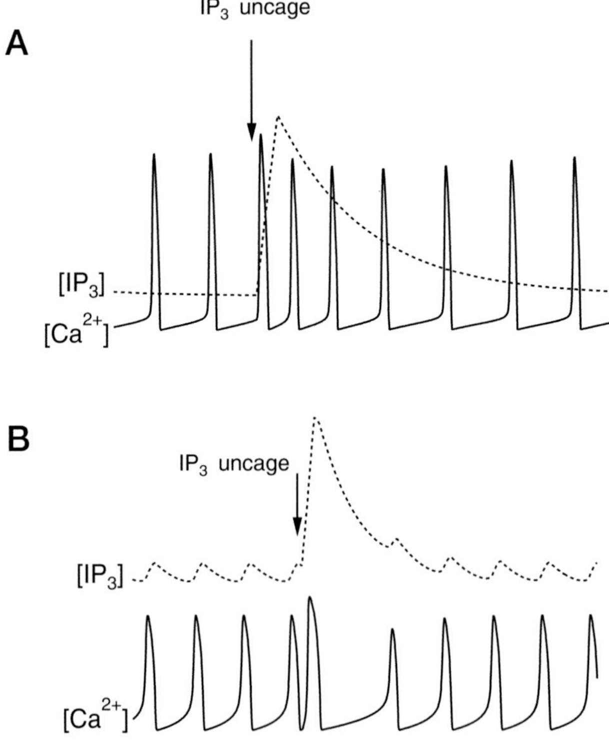

Fig. 3.12 Responses to exogenous pulses of IP_3; model predictions (Sneyd et al, 2006). **A:** If the oscillations are generated by a Class I mechanism, photorelease of IP_3 causes a transient increase in the frequency of Ca^{2+} oscillations. **B:** If the oscillations are generated by a Class II mechanism, an exogenous pulse of IP_3 causes an immediate response, and then a delay to the next peak.

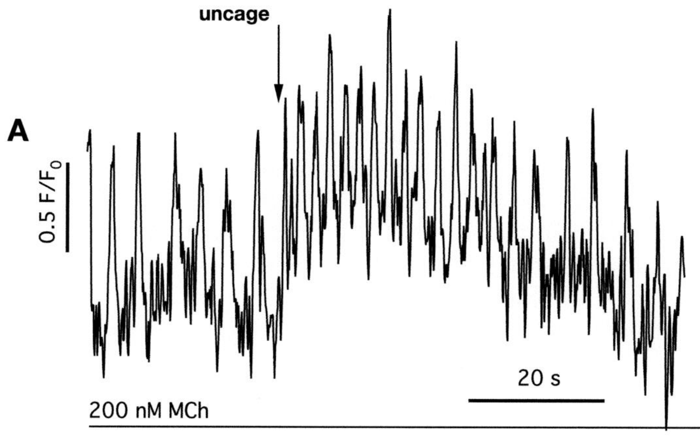

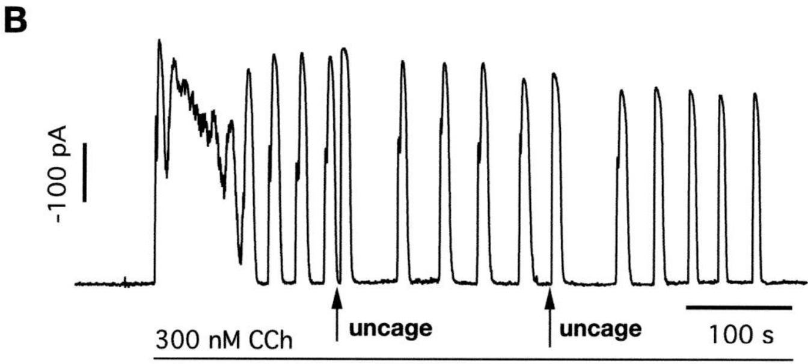

Fig. 3.13 Responses to exogenous pulses of IP_3; experimental tests of the model predictions (Sneyd et al, 2006). **A:** In airway smooth muscle cells, photorelease of IP_3 causes a transient increase in the frequency of MCh-induced Ca^{2+} oscillations. **B:** In pancreatic acinar cells, photorelease of IP_3 causes an immediate spike, and then a delay to the next peak of the CCh-induced Ca^{2+} oscillations (which in this case are measured via the conductance of a Ca^{2+}-activated Cl^- channel).

without monitoring $[IP_3]$, these experiments show that Ca^{2+} oscillations in smooth muscle cells are of Class I, while they are of Class II in pancreatic acinar cells. Although this might seem to be a clear-cut result, there are a number of hidden difficulties with this approach. Firstly, the response of Class I and Class II models to pulses of IP_3 is more complicated than the simple version described here. In fact, in some circumstances it is possible for a Class I model to respond in the manner of a Class II model. Such complications will be discussed in more detail in Chapter 5. In addition, the response to a pulse of IP_3 depends critically on the time scale of IP_3 degradation, something that is not usually known experimentally to any great accuracy. However, it is known that photoreleasable IP_3 degrades more slowly than native IP_3 (Dakin and Li, 2007), which allows for the appearance of a long delay after the photorelease of an IP_3 pulse.

3.3 Spatially Distributed Models

In some cell types, Ca^{2+} oscillations occur practically uniformly across the cell. In such a situation, measurement of the Ca^{2+} concentration at any point of the cell gives the same time course, and a well-mixed model is appropriate. Often, however, each oscillation takes the form of a wave moving across the cell; these intracellular "oscillations" are actually periodic intracellular waves. To model and understand such spatially distributed behaviour, inclusion of Ca^{2+} diffusion is necessary.

When attempting to model spatially distributed Ca^{2+} dynamics, it is tempting to generalise the whole-cell model simply by adding a diffusion term to (3.5). However, with a little reflection, one realises that this cannot be correct for several reasons. First, most of the flux terms represent movement of Ca^{2+} across a boundary, whether the plasma membrane, the ER, or mitochondria; the only exceptions are the fluxes on and off the buffers. Second, since the cytoplasmic space is highly inhomogeneous, it is not clear that movement of Ca^{2+} in this space is governed by a standard diffusive process.

A more proper formulation of the model would be to assume that the cell is a three-dimensional structure comprised of two connected, non-intersecting (but possibly highly interwoven) domains, the cytoplasm and the ER (ignoring all other subdomains, such as mitochondria, Golgi apparati, etc, to make the problem simpler). In the cytoplasm and the ER, Ca^{2+} is assumed to move by normal diffusion and to react with buffers, while all the other fluxes are across a domain boundary.

A highly simplified schematic diagram (in two dimensions) is shown in Fig. 3.14, where the ER domain is labelled Ω_e, the cytoplasmic domain is labelled Ω_c, the cell membrane is Γ_m, and the ER membrane is Γ_e. Thus,

$$\frac{\partial c}{\partial t} = \nabla \cdot (D_c \nabla c) - J_{\text{on}} + J_{\text{off}}, \qquad \text{in } \Omega_c, \tag{3.44}$$

and

$$\frac{\partial c_e}{\partial t} = \nabla \cdot (D_e \nabla c) - J_{\text{on,e}} + J_{\text{off,e}}, \qquad \text{in } \Omega_e. \tag{3.45}$$

Fluxes across the plasma membrane yield the boundary condition

$$D_c \nabla c \cdot \mathbf{n} = J_{\text{in}} - J_{\text{pm}}, \qquad \text{on } \Gamma_m, \tag{3.46}$$

while fluxes across the ER membrane yield

$$D_c \nabla c \cdot \mathbf{n} = D_e \nabla c_e \cdot \mathbf{n} = J_{\text{IPR}} - J_{\text{serca}}, \qquad \text{on } \Gamma_e, \tag{3.47}$$

where $\mathbf{n}$ is the unit outward normal vector to the cytoplasm. Notice that here we have assumed that there is no direct communication between the ER and extracellular space.

In models of this type it is not necessary to assume that the Ca^{2+} transporting molecules (IPR, for example, or SERCA pumps) are distributed

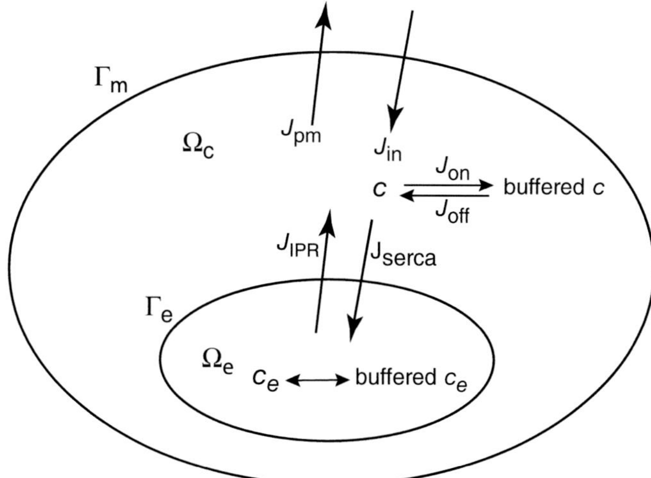

Fig. 3.14 Schematic diagram of a highly simplified model of a spatially distributed cell, showing a single internal compartment. Within each compartment the only Ca^{2+} fluxes are due to buffering (other reactions are ignored, for simplicity). All the other Ca^{2+} fluxes appear as boundary fluxes as Ca^{2+} flows from one compartment to another, or from outside the cell to inside.

evenly or continuously on the ER or cell membranes. They can even be modelled as individual molecules, with behaviour governed by statistical processes, an approach that is often taken in the modelling of microdomains (Section 3.4).

However, because of the intricate geometry of the ER, models of this type, although useful for modelling microdomains, are too complicated to be useful for modelling an entire cell, so some simplifying assumptions are needed. The fundamental idea is that since diffusion is rapid over short distances, local variations are smoothed out quickly and it is only necessary to know the average, or mean field, behaviour. Thus, one derives mean field equations that describe the behaviour over larger space scales. This is accomplished by a technique known as homogenisation.

3.3.1 A Brief Note on Terminology

Confusion can arise because the Ca^{2+} diffusion coefficient is modified by two different things. Firstly, it is decreased by the tortuosity of the ER and the cytoplasm; this is a purely geometrical effect. The resultant diffusion coefficient is called the *homogenised diffusion coefficient*, as discussed next, in Section 3.3.2. Secondly, it is decreased still further by binding to Ca^{2+} buffers, which is a biochemical effect and described in Section 2.9. In this case, the resultant diffusion coefficient is called the *effective diffusion coefficient*. Note that the effective diffusion coefficient includes both the effects of ER geometry and the effects of the buffers. In cells it is not practically possible to have a diffusion coefficient that includes the latter but does not include the former. Furthermore, the homogenised diffusion coefficient is not amenable to experimental manipulation, while the effective diffusion coefficient can be changed by changing the buffering capacity of the cell.

3.3.2 Homogenisation

We begin by considering a simple case. Suppose that a substance is reacting and diffusing in a one-dimensional region, and that the diffusion coefficient is rapidly varying in space. To be specific, suppose

$$\frac{\partial u}{\partial t} = \frac{\partial}{\partial x}\left(D\left(\frac{x}{\epsilon}\right)\frac{\partial u}{\partial x}\right) + f(u). \tag{3.48}$$

Here x is dimensionless, $D(x)$ is a periodic function of period one and of order one, and ϵ is small. Hence $D(\frac{x}{\epsilon})$ is periodic with period ϵ, and is thus rapidly varying on the length scale of order one. We expect that u should have some average or mean field behaviour with a characteristic length scale of order one, with small variations from this mean field that are of order ϵ.

Keener and Sneyd (2008) show that the mean field solution, u_0, is given by

$$\frac{\partial u_0}{\partial t} = \tilde{D}\frac{\partial^2 u_0}{\partial x^2} + f(u_0), \tag{3.49}$$

where

$$\frac{1}{\tilde{D}} = \int_0^1 \frac{1}{D}\,dx. \tag{3.50}$$

$\tilde{D}$ is the homogenised diffusion coefficient for this simple one-dimensional case.

It is helpful to illustrate this general procedure for a specific case. Suppose that we have an infinite line of cells in one dimension, each of length L, with $D = D_c$ inside each cell. However, at the boundary between each cell, D decreases to $D_i \ll D_c$ in a region of width ϵ, where $\epsilon \ll L$. This is a crude way of modelling restricted diffusion through gap junctions. Then,

$$\frac{1}{\tilde{D}} = \frac{1}{L}\int_0^{L-\epsilon} \frac{1}{D_c}\,dx + \frac{1}{L}\int_{L-\epsilon}^{L} \frac{1}{D_i}\,dx \tag{3.51}$$

$$= \frac{L-\epsilon}{L}\frac{1}{D_c} + \frac{\epsilon}{D_i L}. \tag{3.52}$$

Now take the limit as $\epsilon \to 0$, while also letting $D_i \to 0$ in such a way that $D_i/\epsilon \to \mathcal{F}$, where $\mathcal{F}$ is a constant, called the *permeability coefficient*, with units of distance/time. This gives

$$\frac{1}{\tilde{D}} = \frac{1}{D_c} + \frac{1}{\mathcal{F}L}. \tag{3.53}$$

This same result is derived in Section 3.6.2 in a slightly different way.

Homogenisation for the full three-dimensional problem is similar, although, once again, most of the details are omitted. We suppose that a substance is reacting and diffusing in a three-dimensional region that is subdivided into small periodic subunits Ω, called the microstructure, which are each contained in a small rectangular box. The subregion Ω is further subdivided into a cytoplasmic region, Ω_c, and an ER region, Ω_e, as shown in Fig. 3.15 A. Γ_e is the membrane boundary between the ER and the cytoplasm, and Γ_b is the intersection of the ER with the walls of the box. Because we assume that the microdomain is repeated periodically, the ER of one box connects, through Γ_b, to the ER of the neighbouring boxes.

We assume that the rectangular box is of length l (although the box need not be a cube), where l is much less than the natural length scale of the problem, in this case $\Lambda = \sqrt{D}$, the diffusion length. Thus, there are two natural length scales, l and Λ, and so $\epsilon = l/\Lambda \ll 1$ is a natural small parameter. Since l is small compared to the diffusion length scale, it is reasonable to assume that the solution is nearly homogeneous in each microstructural unit, and that the microstructure causes only small perturbations to the background solution.

Now we suppose that, in Ω_c, u obeys the reaction-diffusion equation

$$\frac{\partial u}{\partial t} = D\nabla^2 u + f, \tag{3.54}$$

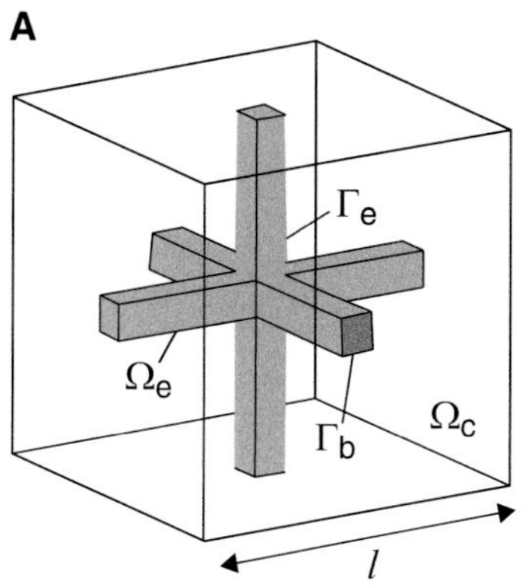

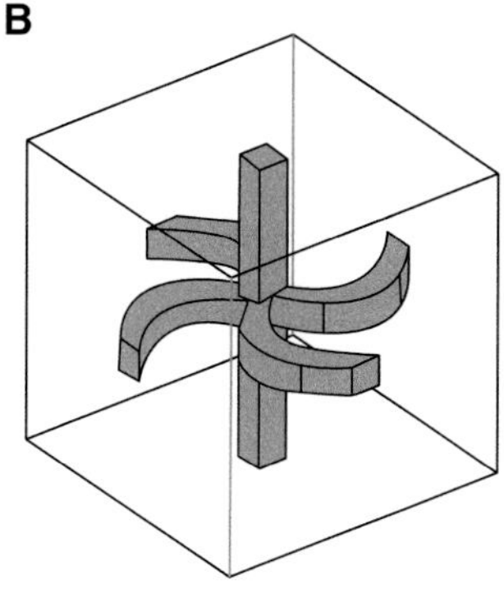

Fig. 3.15 Two possible ER microscale structures (Goel et al, 2006). The more complicated structure in panel B results in a nonisotropic homogenised diffusion tensor.

and that the flux across Γ_e is $\mathbf{n} \cdot D\nabla u = J$, where $\mathbf{n}$ is the outward normal. Another variable, v say, will be reacting and diffusing inside the region Ω_e, but it is unnecessary to carry the analysis through for both variables. Again, it is convenient to assume that these equations are in units of dimensionless time, so that D has units of length^2. We further assume that J is proportional to l. Notice that the total flux across Γ_e is proportional to the surface area, so that the total flux per unit volume is proportional to J/l. In order that this remain bounded in the limit that $\epsilon \to 0$, we assume that $J/l = j$, where j is bounded and non-zero.

After some calculations, the details of which are given in Keener and Sneyd (2008), one can show that, to a good approximation,

$$\frac{\partial u}{\partial t} = \nabla \cdot (\tilde{D}\nabla u) + \frac{S_c}{V_c} j(u) + f(u), \tag{3.55}$$

where V_c and S_c denote the (dimensionless) volume and surface area of Ω_c, respectively. $\tilde{D}$, the (dimensionless) homogenised diffusion tensor, is given by

$$\tilde{D} = \frac{1}{V_c} \int_{\Omega_c} (I + \nabla_\xi W)\, dV_\xi, \tag{3.56}$$

where W is the solution of

$$\nabla_\xi^2 W = 0 \qquad \text{in } \Omega_c, \tag{3.57}$$

$$\mathbf{n} \cdot (\nabla_\xi W + I) = 0 \qquad \text{on } \Gamma_e. \tag{3.58}$$

Hence, for any given periodic geometry, the homogenised diffusion tensor (3.56) can be calculated by solving (3.57) and (3.58). Goel et al (2006) calculated homogenised diffusion coefficients for a variety of possible ER microstructures. For the regular structure of Fig. 3.15 A, the homogenised diffusion coefficient of Ca^{2+} in the cytoplasm decreases by 60% as the volume fraction of the ER increases from 0 to 0.9. The more complicated structure of Fig. 3.15 B gives nonisotropic diffusion, with the diffusion coefficient in the vertical direction about 50% larger than the diffusion coefficient in the other two directions.

From the preceding argument, we see that, when constructing a spatially distributed model of Ca^{2+} dynamics, in many cases it is sufficient to assume that concentrations c and c_e coexist at every point in space, and

$$\frac{\partial c}{\partial t} = \nabla \cdot (\tilde{D}_c \nabla c) + \chi_c(J_{\text{IPR}} + J_{\text{RyR}} - J_{\text{serca}}) - J_{\text{on}} + J_{\text{off}}, \tag{3.59}$$

$$\frac{\partial c_e}{\partial t} = \nabla \cdot (\tilde{D}_e \nabla c_e) - \chi_e(J_{\text{IPR}} + J_{\text{RyR}} - J_{\text{serca}}) - J_{\text{on,e}} + J_{\text{off,e}}, \tag{3.60}$$

where $\tilde{D}_c$ and $\tilde{D}_e$ are homogenised diffusion coefficients for the cytoplasmic space and the ER, respectively, and χ_c and χ_e are the surface-to-volume ratios of these two commingled spaces.

It is usually assumed that the cellular cytoplasm is isotropic and homogeneous. Exceptions to the assumption of homogeneity are not uncommon, but exceptions to the isotropic assumption are rare. It is not known, however, how Ca^{2+} diffuses in the ER, or the extent to which the tortuosity of the ER plays a role in determining the homogenised diffusion coefficient of ER Ca^{2+}. It is typical to assume either that Ca^{2+} does not diffuse in the ER, or that it does so with a restricted diffusion coefficient, $\tilde{D}_e \ll \tilde{D}_c$.

The surface-to-volume ratios are necessary to get the units correct. In fact, one should notice that a flux term, for example J_{IPR}, in a whole-cell ODE model is not the same as in a spatially distributed model; in a whole-cell model the flux must be in units of concentration per unit time, while in a spatially distributed model the flux must be in units of moles per unit time per surface area. Multiplication by a surface-to-volume ratio changes the units to number of moles per volume per unit time. However, it is typical to scale all fluxes to be in units of moles per unit time per cytoplasmic volume, in which case (3.59) and (3.60) become (after rescaling)

$$\frac{\partial c}{\partial t} = \nabla \cdot (\tilde{D}_c \nabla c) + J_{\mathrm{IPR}} + J_{\mathrm{RyR}} - J_{\mathrm{serca}} - J_{\mathrm{on}} + J_{\mathrm{off}}, \tag{3.61}$$

$$\frac{\partial c_e}{\partial t} = \nabla \cdot (\tilde{D}_e \nabla c_e) - \gamma(J_{\mathrm{IPR}} + J_{\mathrm{RyR}} - J_{\mathrm{serca}}) - J_{\mathrm{on,e}} + J_{\mathrm{off,e}}, \tag{3.62}$$

where γ is the ratio of cytoplasmic to ER volumes.

In other words, after all the complicated derivation, the final equations are relatively simple. In order to construct a spatially distributed model, one can simply add diffusion terms to the whole-cell model equations, where the diffusion coefficients are calculated as integrals, and thus depend on the microgeometry. Of course, in practice these integrals are rarely performed, and the homogenised diffusion coefficients are either assumed to be close to the values in aqueous solution, or are set to zero.

For notational convenience, we usually omit the tilde denoting a homogenised diffusion coefficient; in such cases, the precise definition of D_c should be clear from the context.

3.3.3 Membrane Fluxes

The derivation in the previous section shows how the ER and the cytoplasm may often be assumed to coexist at all points inside the cell. However, treatment of fluxes across the plasma membrane is slightly more complex, and depends upon the exact cellular geometry.

In general, (3.61) and (3.62) are coupled to terms that describe the boundary fluxes across the cell membrane. Thus

$$D_c \nabla \cdot \mathbf{n} = J_{\mathrm{in}} - J_{\mathrm{pm}}, \qquad \text{on } \Gamma_m, \tag{3.63}$$

where, as in Fig. 3.14, Γ_m is the cell membrane and $\mathbf{n}$ is the outward normal. This is the only possible approach to take in three dimensions, when the cells are not thin.

However, if the cells are thin, the full three-dimensional model simplifies to a two-dimensional model, where the membrane fluxes appear as additional reaction terms. To show this, consider the domain bordered by two flat two-dimensional cell membranes situated at $z = 0$ and $z = L$. In this domain we have

$$\frac{\partial c}{\partial t} = D\nabla^2 c + f(c), \tag{3.64}$$

where $f(c)$ denotes all the intracellular reaction terms, such as IPR, RyR, SERCA pumps, and buffers. On the membrane we have the boundary conditions

$$D\frac{\partial c}{\partial z} = -J_0(c), \qquad \text{on } z = 0, \tag{3.65}$$

$$D\frac{\partial c}{\partial z} = J_1(c), \qquad \text{on } z = L. \tag{3.66}$$

The fluxes across the two boundaries will not necessarily be the same, and so $J_0 \neq J_1$ in general.

Averaging (3.64) in the z direction gives

$$\frac{1}{L}\int_0^L \frac{\partial c}{\partial t}\,dz = \frac{1}{L}\int_0^L D\nabla^2_{xy} c\,dz + \frac{1}{L}\int_0^L D\frac{\partial^2 c}{\partial z^2}\,dz + \frac{1}{L}\int_0^L f\,dz, \tag{3.67}$$

and thus, defining $\bar{c} = \frac{1}{L}\int_0^L c dz$, and $\bar{f} = \frac{1}{L}\int_0^L f dz$, we have

$$\frac{\partial \bar{c}}{\partial t} = D\nabla^2_{xy}\bar{c} + \bar{f} + \frac{1}{L}\left(J_1(c(x,y,L)) + J_0(c(x,y,0))\right). \tag{3.68}$$

So far the calculations are exact, and we have not used the fact that the area between the membranes is thin compared to the rate of diffusion of Ca^{2+}. However, if we now assume further that $L/\sqrt{D} \ll 1$ it can be shown (Keener and Sneyd, 2008) that c can be replaced by $\bar{c}$ throughout, to get

$$\frac{\partial \bar{c}}{\partial t} = D\nabla^2_{xy}\bar{c} + f(\bar{c}) + \frac{1}{L}\left(J_1(\bar{c}) + J_0(\bar{c})\right) + O\left(\frac{L}{\sqrt{D}}\right). \tag{3.69}$$

It follows that, for thin cells, the boundary fluxes can be incorporated as reaction terms, an approach that is commonly used to perform numerical simulations of Ca^{2+} waves in two spatial dimensions (see, for example, Fig. 6.1 or Fig. 3.23). For such simulations, it is also necessary to impose boundary conditions on the edges of the spatial domain, as, without these, no numerical solution is possible. Unfortunately, there is no physiological correlate to such boundary conditions for two-dimensional simulations. For simplicity, they are usually taken to be no-flux, but this condition cannot be interpreted literally as a physiological constraint, only as a required mathematical fiction. Care must then be taken when interpreting the solution close to the edges of the simulation domain.

3.3.4 Closed-Cell Spatial Models

We saw in Section 3.2.5 that, in the absence of membrane fluxes, the constant total Ca^{2+}, c_t, can be used to eliminate the equation for c_e. However, when diffusion is included, things are slightly more complicated. A typical closed-cell spatial model (with fast, linear buffering) has the form

$$\frac{\partial c}{\partial t} = D_c \nabla^2 c + J_{\mathrm{IPR}} - J_{\mathrm{serca}}, \tag{3.70}$$

$$\frac{\partial c_e}{\partial t} = D_e \nabla^2 c_e - \gamma (J_{\mathrm{IPR}} - J_{\mathrm{serca}}), \tag{3.71}$$

with no-flux boundary conditions at the cell membrane. Multiplying the first equation by γ, and adding to the second equation gives

$$\frac{\partial}{\partial t}(\gamma c + c_e) = \nabla^2 (D_c \gamma c + D_e c_e). \tag{3.72}$$

This cannot, in general, be written as a function of $c_t = \gamma c + c_e$, and thus the equation for c_e cannot, in general, be eliminated.

However, in the special case where $D_c = D_e$, we obtain

$$\frac{\partial c_t}{\partial t} = D_c \nabla^2 c_t, \tag{3.73}$$

again with no-flux conditions on the cell membrane. In this case, if c_t is initially spatially homogeneous, it remains so, in which case c_t is simply a constant across the entire domain, and so the equation for c_e can be eliminated.

3.4 Microdomains

As discussed previously, the rate of diffusion of Ca^{2+} is affected by two different and independent things. Firstly, geometric effects, as described in the previous section, can reduce the diffusion coefficient due to the tortuosity of the domain. Secondly, the effects of buffers, as described in Section 2.9, reduce the diffusion coefficient still further by at least another order of magnitude. Thus the movement of Ca^{2+} around the cytoplasm is highly restricted, far more so than the diffusion of ions such as Na^+ or K^+. This allows for a highly inhomogeneous cytosolic $[Ca^{2+}]$. Small regions where $[Ca^{2+}]$ is much higher than the rest of the cell are known to play critical roles in many cell types – cardiac and skeletal muscle, for example, and neurons – and are believed to be important in most cell types, although the details remain uncertain.

Localisation of effector molecules (i.e., molecules that respond in some way to an increase in $[Ca^{2+}]$) inside these microdomains means that the effectors will experience a much higher $[Ca^{2+}]$ than they would if they were sitting out in the cytoplasm, and this can confuse the interpretation of experimental results. Since cytoplasmic $[Ca^{2+}]$ can now be relatively easily measured it is tempting to conclude that all the other molecules affected by Ca^{2+} are being affected by the measured concentration. But the possible, indeed probable, existence of microdomains means that the actual situation can be much more complex.

3.4.1 Calcium at the Mouth of an Open Channel

One particularly important such microdomain is around the mouth of a Ca^{2+} channel. When Ca^{2+} comes through an open channel it diffuses away only slowly, thus potentially forming a region of high $[Ca^{2+}]$ at the mouth of the channel. Such localised high $[Ca^{2+}]$ is crucial in neuronal synapses, for example, as well as in the control of IP_3 receptors.

The question of buffered diffusion away from a point source has been studied extensively. Initial results by Stern (1992a) were extended by Naraghi and Neher (1997), Neher (1998a), Smith (1996), Smith et al (1996), and Bauer (2001). An excellent and clear summary of this work, and a redevelopment of the results in the language of asymptotics, is given by Smith et al (2001). A more detailed numerical study, including the full time-dependent problem and specialised for the study of Ca^{2+} release through IP_3 receptors, was presented by Thul and Falcke (2004a).

We begin by discussing the steady-state problem. We set up the problem by assuming that the channel is a point source situated in an infinite plane, $z = 0$ say, and releasing Ca^{2+} only into the region $z > 0$. The diffusion of Ca^{2+} is then restricted to the region $z > 0$. The solutions will be assumed to be spherically symmetric, we shall consider only the case of a single buffer, and transport to and from the ER (or any other internal compartment) will be ignored. Then, following the notation of Section 2.9, the differential equations for the steady-state profiles of c and b are

$$0 = D_c \nabla^2 c + k_- b - k_+ c(b_t - b), \tag{3.74}$$

$$0 = D_b \nabla^2 b - k_- b + k_+ c(b_t - b), \tag{3.75}$$

where, due to the spherical symmetry,

$$\nabla^2 = \frac{1}{r^2} \frac{d}{dr} \left[r^2 \frac{d}{dr} \right]. \tag{3.76}$$

Here, r is the distance to the point source.

At any given radius, r, the total Ca^{2+} flux (positive in the outward direction) across the hemisphere centred at $r = 0$ is given by

$$-2\pi r^2 D_c \frac{dc}{dr}, \tag{3.77}$$

i.e., the surface area ($2\pi r^2$) of the hemisphere, multiplied by the flux through each unit area of the hemisphere. Note that the minus sign is necessary as $\frac{dc}{dr} < 0$. From this it follows that

$$\lim_{r\to 0}\left(-2\pi r^2 D_c \frac{dc}{dr}\right) = \sigma, \tag{3.78}$$

where σ is the strength of the point source, i.e., the number of moles per second of Ca^{2+} that are coming out the source. Similarly,

$$\lim_{r\to 0}\left(-2\pi r^2 D_b \frac{db}{dr}\right) = 0. \tag{3.79}$$

As r gets large, c and b just tend to their concentrations in the cytoplasm, which are assumed to be in equilibrium. Hence,

$$\lim_{r\to\infty} c(r) = c_\infty, \tag{3.80}$$

$$\lim_{r\to\infty} b(r) = b_\infty = \frac{b_t c_\infty}{K + c_\infty}, \tag{3.81}$$

where $K = \frac{k_-}{k_+}$.

3.4.1.1 The Excess Buffer Approximation

The simplest case is when the source is weak, the buffer diffuses quickly relative to the diffusion of Ca^{2+}, and the buffer is present in excess (i.e., $b_t \gg K$). In this case the resting buffer profile is essentially undisturbed by the presence of the source, and thus $b = b_\infty$ (a rigorous justification of this approximation as the lowest order solution in an asymptotic expansion is given by Smith et al (2001)). Then the model reduces to the linear equation

$$0 = D_c \nabla^2 c + k_- b_\infty - k_+ c(b_t - b_\infty), \tag{3.82}$$

together with the boundary conditions (3.78) and (3.80). This can be solved explicitly to give

$$c(r) = \frac{\sigma}{2\pi D_c r} e^{-r/\lambda} + c_\infty, \tag{3.83}$$

where $\lambda = \sqrt{\frac{D_c}{k_+(b_t - b_\infty)}}$.

Here, c at the channel mouth (i.e., at $r = 0$) is bounded, which makes explicit the need for a weak source. A strong source will quickly saturate the buffer at the channel mouth, leading to a singularity in c at $r = 0$, as we shall see in the next section. Because of such potential singularities in c, it is not a simple matter to derive approximations to the solution that are valid both far away and close to the channel mouth. If there is a singularity in c at $r = 0$, then any such approximation must be valid both for finite values of c and when $c \to \infty$. This is discussed in more detail in Smith et al (2001), where various linearised approximations (Naraghi and Neher, 1997; Stern, 1992a) are shown to be valid in the limit as $r \to 0$.

3.4.1.2 The Rapid Buffer Approximation

As explained in Section 2.9.1, if we assume the buffers reach equilibrium quickly with respect to the other reactions, we get a simplified equation for Ca^{2+}, i.e.,

$$\frac{\partial c}{\partial t} = \frac{D_c + D_b\theta(c)}{1+\theta(c)}\nabla^2 c - \frac{2D_b\theta(c)}{(K+c)(1+\theta(c))}|\nabla c|^2, \tag{3.84}$$

where

$$\theta(c) = \frac{b_t K}{(K+c)^2}. \tag{3.85}$$

Smith (1996) used this equation to derive an explicit solution for c around a channel mouth,

$$c = \frac{1}{2D_c}\Bigg[-D_cK + \frac{\sigma}{2\pi r} + D_c c_\infty - D_b b_\infty + \sqrt{\left(D_cK + \frac{\sigma}{2\pi r} + D_c c_\infty - D_b b_\infty\right)^2 + 4D_cD_bb_tK}\,\Bigg]. \tag{3.86}$$

In this solution, $c \to \infty$ as $r \to 0$. Hence, since c and b are assumed to be in instantaneous equilibrium everywhere, we must have that $b \to b_t$ as $r \to 0$ (recall that b is the concentration of bound buffer). Thus, the rapid buffering approximation is only accurate when the source is strong enough to completely fill the buffer close to the channel mouth. In this sense, the rapid buffer approximation is the converse of the excess buffer approximation, which assumes that the buffer is essentially unperturbed by the source.

3.4.2 Incorporating ER Depletion

In the above discussion the strength of the Ca^{2+} source, σ, was assumed to be constant. However, this is not true in general, as depletion of the ER will decrease the driving force for Ca^{2+} efflux, leading to a decreased source strength. Thus we need to derive a more accurate relationship in which the flux through the channel decreases as the ER concentration decreases.

This property of the channel flux was discussed in Section 3.2.1, where we used the simple expression $J_{\text{IPR}} = k_f P_o(c_e - c)$. The term $c_e - c$ is a crude model of the fact that the flux through a channel will be dependent on the concentration difference across that channel.

It is not immediately obvious how to incorporate this property into the derivation of the Ca^{2+} gradients around the mouth of an open channel. This is because, as we saw above, if we use the rapid buffer approximation (which we wish to do in general, as the excess buffer approximation requires a weak source, which is often unrealistic, and using no approximation at all makes it more difficult to derive an analytic solution) $c \to \infty$ as $r \to 0$. In other words, at the channel mouth there is a mathematical singularity in the Ca^{2+} concentration. Therefore, right at the channel mouth (i.e., at $r = 0$), we cannot simply use the expression $c_e - c$, as c is undefined there.

To circumvent this difficulty we put a ball of radius r_0 around the channel, and solve the model equations (for both ER and cytosolic concentrations) only outside this ball. The Ca^{2+} flux from the ER into the ball must then be equal to the flux from the ball into the cytoplasm, and, in turn, this flux depends on the concentration difference between the ER and the cytoplasm.

To show the details, it is easier to begin with a simple diffusion equation with no buffering. Once the solution is worked out for that case, we can then use the transformed variable of (2.159), transform the buffered diffusion equation to the diffusion equation (2.162), and make some minor adjustments to get an implicit expression for the buffered case.

To simplify notation, in what follows we shall denote the ER $[Ca^{2+}]$ by e, rather than c_e. Assuming no buffering (and thus we are solving only the diffusion equation) at steady state we have

$$\frac{d}{dr}\left(r^2 \frac{dc}{dr}\right) = 0, \qquad r > r_0, \tag{3.87}$$

$$c(r_0) = c_0, \tag{3.88}$$

$$c(r \to \infty) = c_\infty, \tag{3.89}$$

with the analogous equations for e,

$$\frac{d}{dr}\left(r^2 \frac{de}{dr}\right) = 0, \qquad r > r_0, \tag{3.90}$$

$$e(r_0) = e_0, \tag{3.91}$$

$$e(r \to \infty) = e_\infty, \tag{3.92}$$

Note that we are solving for c and e only on the region outside the ball of radius $r = r_0$, but also that c and e are only defined on their respective hemispheres, as sketched in Fig. 3.16. The bulk cytosolic and ER concentrations are, respectively, c_∞ and e_∞, and are assumed to be known parameters. However, c_0 and e_0 are not known, and will need to be determined as part

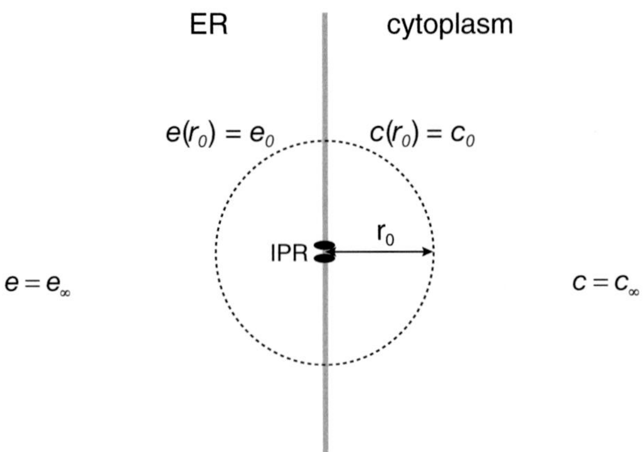

Fig. 3.16 Schematic diagram of the Ca^{2+} flux through a channel. Cytosolic $[Ca^{2+}]$ (c) and ER $[Ca^{2+}]$ (e) are defined on different sides of the membrane, and are assumed to be spherically symmetric. The solutions are computed only outside the ball of radius r_0.

of the solution. Note that, for a physiologically acceptable model, we must have $e_\infty > e_0 > c_0 > c_\infty$ so that Ca^{2+} flows out of the ER. The diffusion coefficients do not appear as we are calculating only the steady state, and we are assuming that the diffusion coefficients are constant.

Solving (3.87) and (3.90) gives

$$c = \frac{r_0}{r}(c_0 - c_\infty) + c_\infty, \tag{3.93}$$

$$e = \frac{r_0}{r}(e_0 - e_\infty) + e_\infty. \tag{3.94}$$

We now have to match the fluxes across the surface of the ball with radius r_0, keeping in mind that the flux out of the ER, for example, is only across half the surface of the ball. The flux, J, across the surface of the half ball into the cytoplasm is given by

$$J = -\frac{1}{2}4\pi r_0^2 D_c \frac{dc}{dr}\bigg|_{r=r_0} = 2\pi r_0 D_c(c_0 - c_\infty), \tag{3.95}$$

where D_c is the diffusion coefficient in the cytoplasm. Similarly, the flux from the ER into the half ball is given by

$$J = 2\pi r_0 D_e(e_\infty - e_0), \tag{3.96}$$

where D_e is the diffusion coefficient in the ER. Since these fluxes must be equal, it follows that

$$D_c(c_0 - c_\infty) = D_e(e_\infty - e_0). \tag{3.97}$$

The problem is not yet fully characterised, as we have four unknowns (c, e, c_0, and e_0) but only three equations (3.93), (3.94), and (3.97). To get a fourth equation we need to make some assumption about how the channel flux depends on the concentration difference.

The simplest assumption is

$$J = P(e_0 - c_0), \tag{3.98}$$

for some constant, P, denoting the permeability of the channel. Obviously this is a highly simplified model, but it suffices to show the principles involved.

Since we now have four equations for four unknowns we can solve, to get

$$c_0 = \frac{c_\infty\left(1 + \frac{2\pi r_0 D_e}{P}\right) + \frac{D_e}{D_c} e_\infty}{1 + \frac{2\pi r_0 D_e}{P} + \frac{D_e}{D_c}}. \tag{3.99}$$

Thus, c_0 is a weighted average of the bulk cytosolic and ER concentrations. A similar equation holds for e_0, which is also a weighted average of c_∞ and e_∞. In the limit as $P \to 0$, we have that $c_0 \to c_\infty$ (and $e_0 \to e_\infty$), as indeed they should, as in that case there is no flux through the channel.

Since c, e, and J can now all be calculated, this completes the specification of the case with no buffers.

When fast buffering is included, the procedure is essentially the same, except that now one must work with transformed variables, as discussed in Section 2.9.1. For simplicity assume that the ER and the cytoplasm have the same, single, buffer. The argument is easily extended to multiple different buffers. Following (2.159), define

$$w = D_c c + D_b b_t \frac{c}{K + c}, \tag{3.100}$$

$$v = D_e e + D_b b_t \frac{e}{K + e}, \tag{3.101}$$

with well-defined inverses

$$c = \phi(w), \tag{3.102}$$

$$e = \zeta(v). \tag{3.103}$$

We showed in Section 2.9.1 how w and v will each satisfy a diffusion equation, and thus, at steady state,

$$w = \frac{r_0}{r}(w_0 - w_\infty) + w_\infty, \tag{3.104}$$

$$v = \frac{r_0}{r}(v_0 - v_\infty) + v_\infty, \tag{3.105}$$

where $\phi(w_\infty) = c_\infty$, and similarly for v_∞. Note that, as before, w_0 and v_0 are unknown, but c_∞ and e_∞ are still given constants and so w_∞ and v_∞ are known also.

The additional two equations now appear implicitly as

$$D_c(\phi(w_0) - \phi(w_\infty)) = D_e(\zeta(v_\infty) - \zeta(v_0)) = P(\zeta(v_0) - \phi(w_0)). \quad (3.106)$$

One can no longer write down the analytic solution as a weighted average, but this is no bar to a numerical solution.

3.4.3 The Channel as a Disk

As we saw above, treating the Ca^{2+} channel as a point source creates mathematical difficulties, as the solution has a singularity at the channel. This problem was avoided by Bentele and Falcke (2007) who used a similar procedure but on a slightly different domain. They assumed that the channel extended all the way out to $r = r_0$, and thus on the region $r < r_0$ there is a flux from ER to cytoplasm that is proportional to $e - c$. Based on the numerical observations of Thul and Falcke (2004a), who showed that, with realistic parameters, buffering had only minor effects on the flux through a channel, Bentele and Falcke (2007) also omitted buffering.

Hence, their model equations are

$$\left.\begin{aligned} 0 &= D_c\nabla^2 c + P(e - c), \\ 0 &= D_e\nabla^2 e - \gamma P(e - c), \end{aligned}\right\} \qquad r < r_0, \quad (3.107)$$

and

$$0 = \nabla^2 c = \nabla^2 e, \qquad r > r_0, \quad (3.108)$$

with boundary conditions

$$\left.\begin{aligned} c &\to c_\infty, \\ e &\to e_\infty, \end{aligned}\right\} \qquad \text{as } r \to \infty, \quad (3.109)$$

and

$$\left.\frac{dc}{dr}\right|_{r=0} = \left.\frac{de}{dr}\right|_{r=0} = 0. \quad (3.110)$$

This system of equations and boundary conditions can be solved exactly; the details are given in Bentele and Falcke (2007). Here we note only that the solution is similar to that obtained previously by assuming the channel is a point source. The major difference is that the solution has no singularity as $r \to 0$. Nevertheless, Ca^{2+} around the channel still gets very high, and steep gradients develop quickly.

3.4.4 *Calcium Concentration Changes Quickly in Microdomains*

The rapid buffer approximation predicts that, for a constant source of sufficient strength, c within 10 nm of the channel mouth increases to hundreds of μM within a few milliseconds of channel opening, and decreases to the resting level just as quickly once the channel closes. If the ER can deplete, and thus the source strength declines as the channel remains open, Thul and Falcke (2004a) and Bentele and Falcke (2007) have shown that similar results occur; c at the channel mouth increases and decreases almost instantaneously as the channel opens and closes. Furthermore, c falls by over two orders of magnitude within approximately 1 μm of the channel mouth, and, after 15 ms, the [Ca^{2+}] 10 nm from the channel mouth is only half the value at the channel mouth (Thul and Falcke, 2004a). Similarly large gradients occur in the ER also.

Since channel opening and closing occurs on a time scale of a few ms – typical mean open times are between 4 and 12 ms for an IPR (Mak et al, 2001) and around 4 ms for an RyR (Wagner et al, 2014) – the channel mouth will effectively experience fast switching from a low [Ca^{2+}] environment to a high [Ca^{2+}] environment. The importance of these conclusions becomes apparent when one considers that IPR and RyR are continually opening and closing in a stochastic manner, and will therefore be controlled by a highly variable local [Ca^{2+}], quite different from observed concentrations in the cytoplasm.

3.4.5 *Microdomains Between Organelles*

In most cases, Ca^{2+} microdomains are considerably more complicated than just a high Ca^{2+} concentration at the mouth of an open channel. IPR and RyR can, and do, sit in regions of the ER/SR membrane that are closely apposed to the cell membrane or other organelles, such as the mitochondria or lysosomes. Release of Ca^{2+} into these restricted regions can cause local Ca^{2+} concentrations that cannot be simply described by the equations in the previous section. In cases like this, practically the only approach is to use numerical simulations. This has been done extensively for cardiac cells (Section 7.2) where the RyR release Ca^{2+} into the diadic cleft, a small region between the SR and the plasma membrane. Calcium microdomains in the pre-synaptic terminal have also been studied extensively (Section 8.2). Similar geometrically restricted microdomains have been studied numerically by Fameli et al (2007), van Breemen et al (2013), Means et al (2006), and Means and Sneyd (2010). However, because of the associated computational difficulties, detailed spatial studies of Ca^{2+} microdomains outside cardiac cells remain relatively rare.

3.4.6 Connecting a Microdomain to the Cell

Microdomains are interesting precisely because they can control cellular behaviour. However, although the microdomain itself might be reasonably amenable to computation, when one tries to incorporate the microdomain into the rest of the cell difficulties arise because of the different spatial scales involved. For example, suppose one wanted to model a microdomain between the ER and a mitochondria to see how this affects whole-cell Ca^{2+} responses. The microdomain itself can be modelled in a relatively straightforward manner, with an ER region, a mitochondrial region, and a cytoplasmic region, connected by boundaries with prescribed Ca^{2+} fluxes. However, this multidomain structure cannot be replicated throughout the entire cell as the computational cost would be prohibitive, and such extensive structural data are rarely, if ever, available.

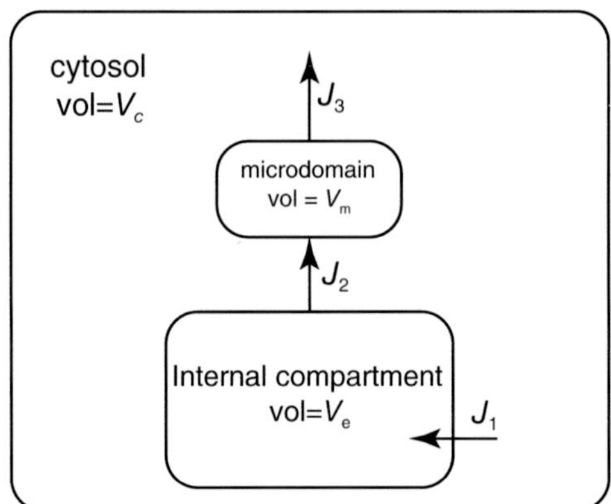

Fig. 3.17 Schematic diagram of the compartmental approach to modelling microdomains. The internal compartment has concentration c_e, the microdomain has concentration c_m, and the cytosol has concentration c.

There are essentially two solutions to this problem. Firstly, one can construct a compartmental model, wherein the microdomain, the cytoplasm, and the ER are all assumed to be homogeneous compartments. This has been a common approach in the modelling of cardiac cells, for example (Shannon et al, 2004; Jafri et al, 1998; Puglisi and Bers, 2001; Greenstein et al, 2006), and shall be discussed in more detail in Section 7.2. The same approach was also used to model the microdomain between a lysosome and the ER (Penny et al, 2014).

An example of such a compartmental microdomain model was discussed in the model of Section 3.2.3, and a slightly simpler version is illustrated in Fig. 3.17. To write down the equations for the concentrations in each compartment (c_e, c_m, and c) it is necessary to take the different compartmental volumes into account. Thus, if every flux is expressed in units of moles per litre cytoplasm per second, we have

$$\frac{dc}{dt} = J_3 - J_1, \tag{3.111}$$

$$\frac{dc_e}{dt} = \gamma_e(J_1 - J_2), \tag{3.112}$$

$$\frac{dc_m}{dt} = \gamma_m(J_2 - J_3), \tag{3.113}$$

where $\gamma_e = V_c/V_e$ and $\gamma_m = V_c/V_m$. In general, $\gamma_m \gg \gamma_e > 1$. A given flux out of the compartment into the microdomain, say, will change the microdomain concentration very quickly, but the compartment concentration much more slowly. The microdomain and compartmental volumes are thus important variables that control the dynamics of the response. Often, J_3 is modelled as a linear diffusive flux, such as $J_3 = k(c_m - c)$, but this is merely a convenience, not a requirement. The fluxes J_1 and J_2 typically have more complex descriptions, incorporating such things as IPR, RyR, or SERCA pumps.

Secondly, it is possible to couple a spatially inhomogeneous cytoplasm to a spatially inhomogeneous microdomain by utilising the techniques of homogenisation (see Section 3.3.2), as shown by Higgins et al (2007). In this approach, the microdomain is modelled using spatially separated cytosolic and ER/SR compartments, while the remainder of the cell is modelled using the homogenised equations.

This is illustrated in Fig. 3.18, which shows a microdomain (with Ca^{2+} concentration c_m) sitting between two organelles (concentrations c_{1m} and c_{2m}), the whole enclosed in a region of homogenised cytoplasm and organelles (with Ca^{2+} concentrations c, c_1, and c_2 for the cytoplasm, organelle 1, and organelle 2, respectively). Note that each organelle (for example, the ER) is assumed to extend throughout the entire cell, and is not restricted to the microdomain. Also note that there are six diffusion coefficients that need to be taken into account. Within the microdomain, the relevant diffusion coefficients are not the homogenised diffusion coefficients, but are just the diffusion coefficients corresponding to the relevant aqueous medium. We denote these diffusion coefficients D_c, D_1, and D_2. However, in the homogenised region we need to use homogenised diffusion coefficients, as shown in (3.56); these are denoted by $\tilde{D}_c$, $\tilde{D}_1$, and $\tilde{D}_2$.

Within the homogenised region the equations for c, c_1, and c_2 are just the usual equations, where we assume that the cytoplasm and both organelles coexist at each point in space (Section 3.3.2). Similarly, within the microdomain, c_m, c_{1m}, and c_{2m} satisfy a diffusion equation with buffering, (3.44) or (3.45), within their respective regions. The boundary B_{1m}, within the microdomain, connects the microdomain region of organelle 1 to the microdomain cytoplasm, and the boundary condition there takes the form of

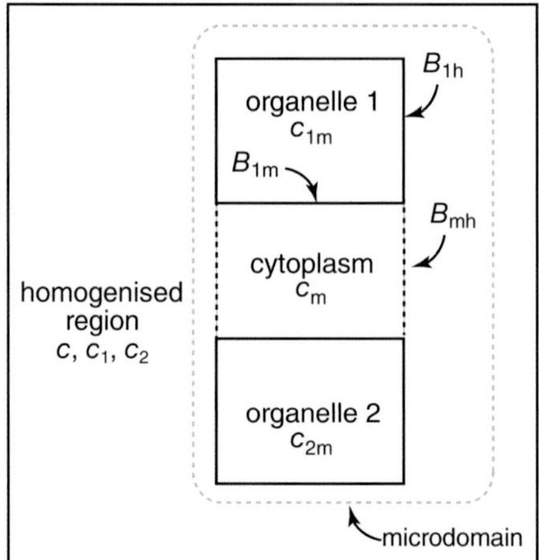

Fig. 3.18 Schematic diagram (not to scale) of a spatially distributed microdomain between two organelles sitting inside a region of cytoplasm.

(3.47). For example, if organelle 1 is the ER, then the fluxes on boundary B_{1m} could consist of an IPR flux and a SERCA pump flux, in which case the boundary condition is

$$D_c \nabla c_m \cdot \mathbf{n} = D_1 \nabla c_{m1} \cdot \mathbf{n} = J_{\mathrm{IPR}} - J_{\mathrm{serca}}, \qquad \text{on } B_{1\mathrm{m}}, \tag{3.114}$$

where $\mathbf{n}$ is the unit outward normal to the microdomain.

The remaining boundaries, B_{1h} and B_{mh} are more difficult to model, as they are boundaries between homogenised and non-homogenised regions. For example, B_{mh}, the boundary between the microdomain and the cytoplasm connects two regions, one of which contains only a cytoplasmic variable (c_m), the other of which contains three variables (c, c_1, and c_2). Higgins et al (2007) showed that, in this case, the correct boundary condition is derived naturally by requiring that Ca^{2+} be conserved across the boundary, i.e., that any flux out of the microdomain must equal any flux into the homogenised region. For example, on B_{mh} this gives

$$D_c \nabla c_m \cdot \mathbf{n} = \tilde{D}_c \nabla c \cdot \mathbf{n} \qquad \text{on } B_{\mathrm{mh}}. \tag{3.115}$$

Since the homogenised organelles are not directly connected to the microdomain cytoplasm, we also have, for example,

$$0 = \nabla c_1 \cdot \mathbf{n} = \nabla c_2 \cdot \mathbf{n} \qquad \text{on } B_{\mathrm{mh}}. \tag{3.116}$$

Similarly, the boundary conditions on B_{1h} are

$$D_1 \nabla c_{1m} \cdot \mathbf{n} = \tilde{D}_1 \nabla c_1 \cdot \mathbf{n}, \tag{3.117}$$

$$0 = \nabla c_2 \cdot \mathbf{n}, \tag{3.118}$$

$$0 = \nabla c \cdot \mathbf{n}. \tag{3.119}$$

The boundary conditions on B_{2h} are analogous.

With this method, the spatial structure of the microdomain can be resolved using a fine mesh, and then the microdomain can be connected to a large-scale homogenised model with a coarser mesh, allowed for a computation that retains spatial accuracy in both domains without undue computational demands. Multiple microdomains, with a periodic spatial distribution, can be computed using symmetry arguments and periodic boundary conditions. However, it remains a difficult problem to compute a whole-cell response when there are multiple microdomains distributed randomly. Before attempting such ambitious computations one would need to be confident that the random distribution of the microdomains is a crucial feature of the model.

3.4.7 Can Microdomains Be Modelled Deterministically?

Given that microdomains are regions of small volume, and given that the resting cytosolic [Ca^{2+}] is typically around a few tens of nM, it follows that, at rest, there are very few Ca^{2+} ions in a microdomain. For example, the diadic cleft of a cardiac myocyte (Section 7.2) is a roughly cylindrical region between the junctional SR and the T-tubule, with a width of about 15 nm and a radius of around 200 nm. It therefore has a volume of around 2×10^{-18} L. For a resting [Ca^{2+}] of 100 nM, there will be, on average, fewer than a single Ca^{2+} ion in any diadic cleft. Although it is possible that some microdomains have a much higher resting [Ca^{2+}], a situation that is not inconsistent with a lower resting cytosolic [Ca^{2+}], there is no reason to suppose this is the case in general.

Does it makes sense, therefore, to write down a deterministic, continuous model of the [Ca^{2+}] in microdomains? Fortunately, Hake and Lines (2008) have shown that, at least in the case of a diadic cleft, a deterministic model still provides a very good approximation to a detailed random walk simulation, as long as the diffusion coefficient of Ca^{2+} is large enough (which it is). This diadic cleft example is discussed in more detail in Section 7.2.4.

As yet there have been no analogous computations for other microdomains. However, there is nothing to suggest that the qualitative result will be any different. The simulations of Hake and Lines (2008) used only a simple two-state model of the RyR, and did not rely on complex models of the L-type Ca^{2+} channels or of buffers. Thus it is reasonable to assume that other microdomains of similar size can be described well by deterministic and continuous models. Using continuous deterministic models to describe much larger molecules such as buffers in microdomains is also valid; although the buffers do not diffuse quickly, we are interested only in their reaction with Ca^{2+}, which does. This approximation would break down if we needed to model reactions between small numbers of slow moving molecules, but this is not generally the case in Ca^{2+} dynamics.

3.5 Calcium Waves

The previous approaches have been used to construct a variety of deterministic models of wave propagation. These models share many similarities with the general theory of wave propagation in excitable systems. When Ca^{2+} is released from one release site, such as a cluster of IPR or RyR, it diffuses to neighbouring release sites where it initiates the release of additional Ca^{2+} via CICR. Thus a wave is propagated by passive diffusion of Ca^{2+} between active release sites. When the release sites are far enough apart, the wave is propagated in a punctate manner, and does not have a fixed profile for all times. However, one can still construct a series of snapshots of the wave profile such that it remains fixed as the wave propagates. In such a model, the wave is propagating through a spatially heterogeneous domain, in which the release sites are distributed discretely. If, on the other hand, the release sites are close enough together, they can be modelled as spatially homogeneous, in which case the wave front has a fixed profile for all time and travels with a constant speed. In both the discrete and the continuous cases, the wave can propagate indefinitely.

Such a mechanism is called *active* wave propagation, as diffusion alone is insufficient for the wave front to exist. An excitable release mechanism is also required, but such excitability depends on the existence of active, energy-consuming processes. In the case of Ca^{2+} waves, it is the active sequestration of Ca^{2+} into the ER or SR by SERCA pumps that ultimately provides the energy needed to propagate a wave.

3.5.1 The Fire–Diffuse–Fire Model

One particularly simple model of a Ca^{2+} wave is the *fire–diffuse–fire* model (Pearson and Ponce-Dawson, 1998; Keizer et al, 1998; Ponce-Dawson et al, 1999; Coombes, 2001; Coombes and Timofeeva, 2003; Coombes and Bressloff, 2003; Coombes et al, 2004), which is based on the assumption that the wave is driven by passive diffusion of Ca^{2+} between distinct, active, release sites. The model of excitability at each release site is highly simplified, leading to a model that is amenable to analytic approaches.

In this model, once [Ca^{2+}] reaches a threshold value, c^*, at a release site, that site fires, instantaneously releasing a fixed amount, σ, of Ca^{2+}. Thus, a Ca^{2+} wave is propagated by the sequential firing of release sites, each responding to the Ca^{2+} diffusing from neighbouring release sites. Hence the name fire–diffuse–fire.

We assume that Ca^{2+} obeys the reaction-diffusion equation

$$\frac{\partial c}{\partial t} = D_c \frac{\partial^2 c}{\partial x^2} + \sigma \sum_n \delta(x - nL)\delta(t - t_n), \tag{3.120}$$

where L is the spacing between release sites, δ is the Dirac delta function, and where, for simplicity, we are assuming a single spatial dimension. Although this equation looks linear, appearances are deceptive. Here, t_n is the time at which c first reaches the threshold value c^* at the nth release site. When this happens, the nth release site releases the amount σ. Thus, t_n depends in a complicated way on c.

The Ca^{2+} profile resulting from the firing of a single site, site i, say, is

$$c_i(x,t) = \sigma \frac{H(t - t_i)}{\sqrt{4\pi D_c(t - t_i)}} \exp\left(-\frac{(x - iL)^2}{4D_c(t - t_i)}\right), \tag{3.121}$$

where H is the Heaviside function. This is the fundamental solution of the diffusion equation with a delta function input at $x = iL$, $t = t_i$, and can be found in any standard book on analytical solutions to partial differential equations (see, for example, Keener (2000) or Kevorkian (2000)). If we superimpose the solutions from each site, we get

$$c(x,t) = \sum_i c_i(x,t) = \sigma \sum_i \frac{H(t - t_i)}{\sqrt{4\pi D_c(t - t_i)}} \exp\left(-\frac{(x - iL)^2}{4D_c(t - t_i)}\right). \tag{3.122}$$

Note that because of the instantaneous release, $c(x,t)$ is not a continuous function of time at any release site.

Now suppose that sites $i = N, N-1, \ldots$ have fired at known times $t_N > t_{N-1} > \ldots$. The next firing time t_{N+1} is determined by when c at x_{N+1} first reaches the threshold, c^*, that is,

$$c((N+1)L, t_{N+1}) = c^*, \qquad \frac{\partial}{\partial t} c((N+1)L, t_{N+1}) > 0. \tag{3.123}$$

Thus, t_{N+1} must satisfy

$$c^* = \sigma \sum_{i \le N} \frac{1}{\sqrt{4\pi D_c(t_{N+1} - t_i)}} \exp\left(-\frac{L^2(N+1-i)^2}{4D_c(t_{N+1} - t_i)}\right). \tag{3.124}$$

A steadily propagating wave corresponds to having $t_i - t_{i-1} = \text{constant} = \tau$ for all i, i.e., each site fires a fixed time after its leftward neighbour fires. Note that the resulting wave does not propagate with a constant profile, but has a well-defined wave speed L/τ. If such a τ exists, then $t_{N+1} - t_i = \tau(N+1-i)$ and τ is a solution of the equation

$$\frac{c^* L}{\sigma} = \sum_{n=1}^{\infty} \frac{1}{\sqrt{4\pi n\eta}} \exp\left(-\frac{n}{4\eta}\right) \equiv g(\eta), \tag{3.125}$$

where $\eta = \frac{D_c \tau}{L^2}$ is the dimensionless delay.

To find η we need to invert this equation. A plot of $g(\eta)$ is shown in Fig. 3.19. It can be shown that $0 \leq g(\eta) \leq 1$ and that g is monotonic with $g \to 0$ as $\eta \to 0$, and $g \to 1$ as $\eta \to \infty$. It follows that a solution to (3.125) exists only if $\frac{c^* L}{\sigma} < 1$. Thus, when the intercluster distance or the threshold is too large, or the amount of release is too small, there is no propagation. However, when $\frac{c^* L}{\sigma} < 1$, a unique solution of (3.125) is guaranteed, and thus there is a propagating wave.

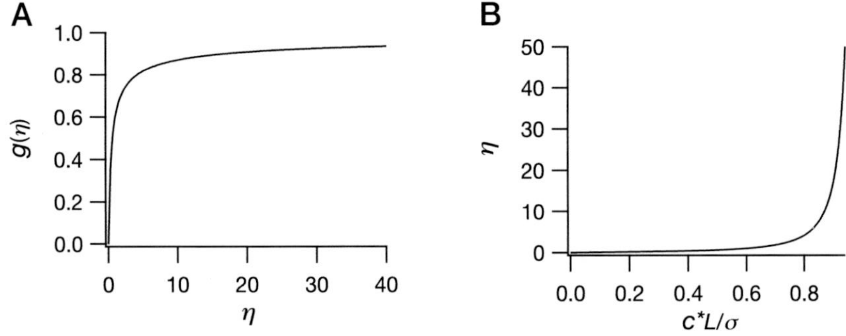

Fig. 3.19 **A:** Plot of $g(\eta)$, where $g(\eta)$ is defined by (3.125) (Keener and Sneyd, 2008). **B:** Plot of the delay η using (3.125).

It is an easy matter to plot the delay as a function of $\frac{c^* L}{\sigma}$ by appropriately exchanging the axes of Fig. 3.19 A. Thus, in Fig. 3.19 B we plot the dimensionless delay η as a function of $\frac{c^* L}{\sigma}$. It is equally easy to plot the dimensionless velocity $\frac{1}{\eta}$ as a function of $\frac{c^* L}{\sigma}$ (not shown). The result is that the velocity becomes infinitely large as $\frac{c^* L}{\sigma} \to 0$, and is zero for $\frac{c^* L}{\sigma} \geq 1$.

The fire–diffuse–fire model described above allows only for release of Ca^{2+}, but no uptake, and thus the Ca^{2+} transient is unrealistically monotone increasing. This deficiency is easily remedied by adding a linear uptake term that is homogeneous in space (Coombes, 2001), so that the model becomes

$$\frac{\partial c}{\partial t} = D_c \frac{\partial^2 c}{\partial x^2} - k_s c + \sigma \sum_n \delta(x - nL)\delta(t - t_n). \tag{3.126}$$

The analysis of this modified model is almost identical to the case with $k_s = 0$. The fundamental solution is modified slightly by k_s, with the Ca^{2+} profile resulting from the firing at site i given by

$$c_i(x,t) = \sigma \frac{H(t-t_i)}{\sqrt{4\pi D_c(t-t_i)}} \exp\left(-\frac{(x-iL)^2}{4D_c(t-t_i)} - k_s(t-t_i)\right). \tag{3.127}$$

Following the previous arguments, we learn that a propagating solution exists if there is a solution of the equation

$$\frac{c^* L}{\sigma} = \sum_{n=1}^{\infty} \frac{1}{\sqrt{4\pi n\eta}} \exp\left(-\frac{n}{4\eta} - \beta^2 n\right) \equiv g_\beta(\eta), \tag{3.128}$$

where $\eta = \frac{D_c \tau}{L^2}$ is the dimensionless delay, and $\beta^2 = \frac{k_s L^2}{D_c}$.

Plots of $g_\beta(\eta)$ are shown for several values of β, the dimensionless pumping rate, in Fig. 3.20. If $\beta \neq 0$, the function $g_\beta(\eta)$ is not monotone increasing, but has a maximal value, say $g_{\max}(\beta)$, which is a decreasing function of β. Furthermore, $g_\beta(\eta) \to 0$ as $\eta \to 0$ and as $\eta \to \infty$. If $\frac{c^* L}{\sigma} > g_{\max}(\beta)$, then no solution of (3.128) exists; i.e., there is propagation failure. On the other hand, if $\frac{c^* L}{\sigma} < g_{\max}(\beta)$, then there are two solutions of (3.128); the physically meaningful solution is the smaller of the two, corresponding to the first time that $c(x,t)$ reaches c^*.

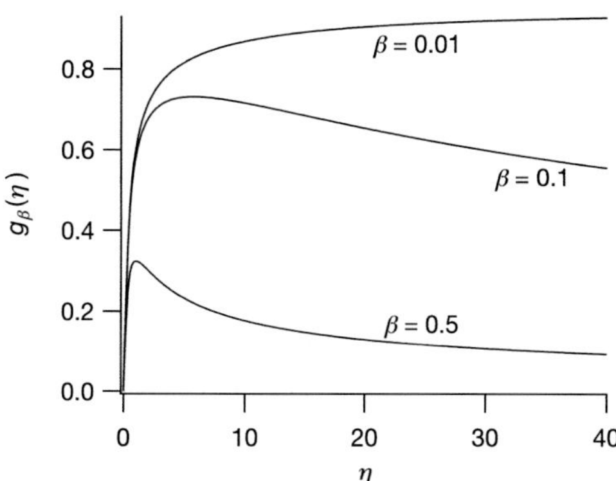

Fig. 3.20 Plots of $g_\beta(\eta)$ for three different values of β (Keener and Sneyd, 2008). Note that g_β is a decreasing function of β for all positive values of η.

3.5.2 Continuous Release Sites

If the release sites are continuously distributed, then we can generalise (3.120) to

$$\frac{\partial c}{\partial t} = D_c \nabla^2 c + \sum J, \tag{3.129}$$

where we are summing over all the relevant Ca^{2+} fluxes. If the cells are thin, then the sum includes the membrane fluxes due to Ca^{2+} influx and the plasma membrane pumps, as discussed in Section 3.3.3. In this case the PDE is two-dimensional. An example of waves in such a model is discussed in Section 6.1. If the cell is not thin, then three-dimensional computations are needed, with explicit computations of the fluxes across the boundary. Such computations are rarely done, as they do not necessarily give more insight than simpler models.

The canonical excitable systems for the study of wave propagation are the bistable equation and the FitzHugh-Nagumo equations. The mathematical theory of wave propagation in these systems is well developed; introductions can be found in Keener and Sneyd (2008), Murray (2003), Britton (1986), Grindrod (1991), or Fife (1979). In many cases, the structure of (3.129) turns out to be similar to that of the FitzHugh-Nagumo model, and thus waves can be partially (but not wholly) understood by analogy with wave propagation in excitable systems. However, the underlying physiology imposes certain constraints which make the theory of Ca^{2+} waves slightly different from either of the canonical models (Sneyd et al, 1993; Tsai et al, 2012), as shall be discussed further in Chapter 5.

If the equations in the absence of diffusion have a stable limit cycle, then the model can exhibit periodic waves and spatial structures of extreme complexity (Maginu, 1985; Kopell and Howard, 1973a, 1981; Grindrod, 1991; Sherratt, 1993, 1994). Such theory is well beyond the scope of this book.

In one dimension, Ca^{2+} waves can be studied by considering the bifurcation structure of the travelling wave equations. This in turn can give considerable insight into wave propagation in higher dimensions.

If we convert to a moving reference frame by introducing the travelling wave variable $\xi = x + st$, where s is the wave speed, we can write (3.129) as the pair of equations

$$c' = d, \tag{3.130}$$

$$D_c d' = sd - \sum J, \tag{3.131}$$

where a prime denotes differentiation with respect to ξ. Thus, a single reaction-diffusion equation is converted to two ordinary differential equations. In general, these two equations are coupled to the other equations for c_e, p, and the states of the various receptors.

Travelling pulses, travelling fronts, and periodic waves correspond to, respectively, homoclinic orbits, heteroclinic orbits, and limit cycles of the travelling wave equations. Thus, by studying the bifurcations of the travelling wave equations we can gain considerable insight into what kinds of waves exist in the model, and for which parameter values. However, such an approach does not give information about the stability of the wave solutions of the original reaction-diffusion equations, something that is much more difficult to determine.

Such approaches to the study of travelling waves are discussed in more detail in Chapter 5.

3.5.3 Waves in Multiple Dimensions

In more than one spatial dimension travelling waves of Ca^{2+} can form patterns that are more complex than one-dimensional travelling waves. In cells that are large enough – for example *Xenopus* oocytes, as described in Fig. 1.7 and Section 6.1 – Ca^{2+} waves can form spiral waves and target patterns (Lechleiter et al, 1991; Lechleiter and Clapham, 1992). However, such spatial patterns are not commonly seen. Most cells are far too small to allow any appreciable curvature of the wave front, which takes the form rather of a plane wave.

3.5.4 Phase Waves

The waves discussed above are dependent on the diffusion of Ca^{2+} from one release site to the next, followed by the excitable release of Ca^{2+}. However, there is another quite different type of wave that can occur. If the medium is oscillatory (i.e., if the reaction kinetics have a stable limit cycle), and the phase of the oscillation has a gradient across the domain, then there will be an apparent wave, occurring in the absence of diffusion, with a wave speed that depends on the phase gradient. Such waves are called *phase waves* or *kinematic waves* (Kopell and Howard, 1973b).

A periodic wave, with wavelength λ and period τ, moves at speed

$$s = \frac{\lambda}{\tau}. \tag{3.132}$$

Furthermore, if the oscillator phase is a linear function of distance, with slope α say, then we also have

$$\lambda = \frac{2\pi}{\alpha}. \tag{3.133}$$

Hence

$$s = \frac{2\pi}{\alpha\tau}, \tag{3.134}$$

which relates the slope of the phase curve to the speed of the phase wave. Clearly, as the gradient of the phase curve decreases, the wave speed increases, as expected intuitively.

This has an interesting and important conclusion for the study of Ca^{2+} waves (Jafri and Keizer, 1994). Suppose that a Ca^{2+} wave is initiated by the release of a bolus of IP_3 somewhere in the domain. The bolus of IP_3 will diffuse out, initiating Ca^{2+} oscillations at each point. However, at each point the oscillation will be initiated at different times, depending on exactly when the IP_3 concentration reaches the threshold for initiating oscillations. Eventually, IP_3 will equilibrate across the domain, and thus oscillations will have the same period (or approximately so) over the whole domain. But since the oscillations at each point were initiated at different times, they will all have a different phase. These phases will be distributed in approximately a monotonic fashion, with the more advanced phases being closer to the site of IP_3 release.

In a case like this one would expect to see phase waves propagating across the domain, and this was confirmed by Jafri and Keizer (1994). In their simulations, Ca^{2+} waves were initiated by the release of a bolus of IP_3, and these waves propagated across the domain practically unchanged when the diffusion of Ca^{2+} was removed. In addition, by using different amounts of IP_3 they showed that the waves obeyed the relationship (3.132), which is expected of phase waves, but which is not obeyed by waves which are propagated by an active mechanism.

There are a number of cell types where Ca^{2+} waves are initiated by the production of IP_3 at one specific part of the cell. Pancreatic and parotid acinar cells, for example, produce IP_3 only in their basal regions. This IP_3 diffuses across the cell and generates a Ca^{2+} wave that travels, somewhat paradoxically, from the apical region to the basal region. This cell type is discussed in more detail in Section 6.3.

In general it is not easy to tell whether an observed wave is mostly due to a kinematic mechanism, or whether it is propagated by an active mechanism. Jafri and Keizer (1994) suggest a number of experimental tests that could be used to distinguish between these two possibilities. These mostly rely on the fact that the speed of the phase wave will depend on the method used to initiate the wave (as this will set the phase gradient). Hence, by initiating waves with, say, ramps of IP_3 of differing slopes, by varying the size of the IP_3 bolus, or by changing the buffering properties of the cytoplasm, it should be possible to discover how important kinematic mechanisms are in the propagation of the wave.

3.6 Intercellular Waves

Calcium waves travel between cells, as well as through cells (see, for example, Fig. 1.9). Intercellular Ca^{2+} waves from the experimentalist's point of view are discussed in a comprehensive review by Leybaert and Sanderson (2012), and from a modelling point of view by Dupont et al (2007). They occur in a wide variety of cell types, including glial cells, neurons, various kinds of epithelial cells, smooth muscle, hepatocytes, chondrocytes, osteocytes, and cardiac myocytes.

There have been fewer models of intercellular Ca^{2+} waves than there have been of intracellular waves and oscillations. One example, in airway epithelial cells, is discussed below, as is a model for intercellular waves in astrocytes.

Intercellular waves in hepatocytes are another well-studied system, and are discussed in detail in Section 6.2. A comprehensive review of models of intercellular waves, and the different mechanisms therein, can be found in Dupont et al (2007).

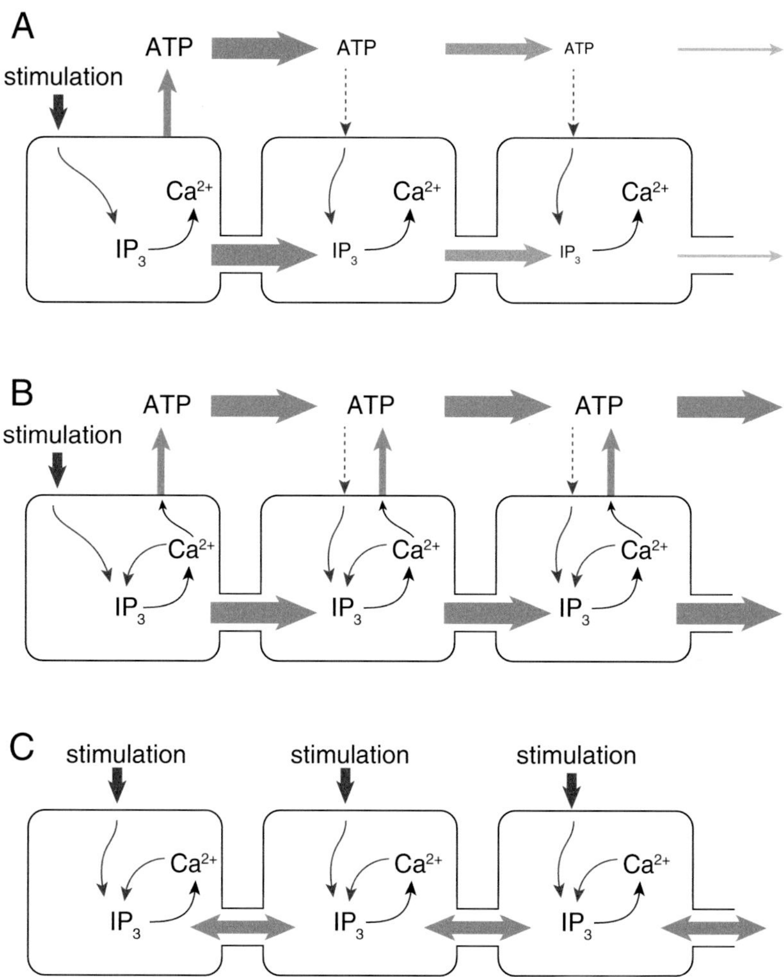

Fig. 3.21 The three mechanisms of intracellular Ca^{2+} wave propagation. **A:** stimulation of a cell leads to production of second messengers which diffuse, either intracellularly through gap junction, or extracellularly. Here, these messengers are assumed to be IP_3 and ATP, respectively. As they diffuse they release Ca^{2+} to form a spreading wave, but eventually the wave stops, as the messengers degrade. **B:** similar to case A, except that now the messengers are actively regenerated, and thus the wave can propagate indefinitely. **C:** every cell is now directly stimulated, leading to a field of coupled oscillators, which can exhibit a wide range of wave phenomena. In reality, most intercellular Ca^{2+} waves are propagated by a combination of these three mechanisms.

3.6.1 Mechanisms of Intercellular Wave Propagation

Intercellular Ca^{2+} waves travel by a combination of three principal mechanisms, illustrated in Fig. 3.21. In any particular case, it is a challenge to determine which is the most important mechanism controlling the wave.

A Passive diffusion (Fig. 3.21 A). Focal stimulation of a single cell (for example, by mechanical stimulation) causes the production of a second messenger, which then diffuses passively from cell to cell. Such diffusion can occur intracellularly or extracellularly.

1. Intracellular second messengers such as IP_3 diffuse from cell to cell through gap junctions. Although Ca^{2+} can also diffuse through gap junctions, it is so heavily buffered that its effective intercellular diffusion is very small, and can often be ignored in models (although see Höfer et al (2001)).
2. Some messengers (such as ATP or glutamate) diffuse extracellularly to propagate the wave by a paracrine mechanism. In this case, the extracellular messenger stimulates Ca^{2+} release in neighbouring cells, resulting in a wave of increased $[Ca^{2+}]$ that spreads from cell to cell.

However, in both cases, the diffusion of the messenger occurs via a passive process, and thus the spread of the wave is limited. There is also a clear site of wave initiation.

B Active propagation (Fig. 3.21 B). If the diffusing second messenger is regenerated, then the wave can travel by an active process. As in the previous case, there are two possibilities.

1. The wave is propagated actively across each cell in the way described in Section 3.3.3 (or, for example, by a fire–diffuse–fire mechanism). When it reaches the border of the cell the wave pauses, an intercellular messenger diffuses through the gap junction, leading to the initiation of another actively propagated wave in the next cell. This is called an intracellular active mechanism.
2. A similar active mechanism can be the result of a diffusing extracellular messenger also. For instance, if the extracellular messenger leads to the production of IP_3 and the release of Ca^{2+}, but this released Ca^{2+} leads to the release of additional extracellular messenger, the Ca^{2+} wave could be propagated by an extracellular active mechanism.

C Coupled oscillators (Fig. 3.21 C). If all the cells are stimulated by agonist, then each cell can be oscillatory, although mostly likely with slightly different periods.

1. If the cells are uncoupled, apparent waves can still appear, driven by the difference in oscillation phase across the field of cells, as discussed in Section 3.5.4. This results in intercellular phase waves which are unaffected by gap junction blockers, and persist in the absence of diffusion.

2. If the cells are coupled by diffusion of IP_3 or Ca^{2+} through gap junctions, the oscillations can be synchronised over groups of cells to form intercellular waves.
   a. If the coupling is strong each cell can synchronise in phase with its neighbours, leading to synchronous oscillations over the entire field of cells (although this is not the only possible behaviour).
   b. If the cells are weakly coupled, the oscillations can synchronise at different phases, leading to the appearance of intercellular waves. However, as opposed to the pure phase wave case discussed above in Section 3.5.4, weakly-coupled waves will de-synchronise in the absence of intercellular coupling. The theory of weakly-coupled oscillators is extensive (see, for example, Ermentrout and Kopell (1984), Kopell and Ermentrout (1986), or Kopell and Ermentrout (1990)).

3.6.2 Propagation by Gap Junctions

If each cell is modelled as well-mixed, then the flux through gap junctions is most easily modelled as linearly proportional to the concentration difference. For example, if u_i is the concentration of a species in cell i then the flux, $J_{i\to j}$, from cell i to cell j can be modelled as

$$J_{i\to j} = \alpha(u_i - u_j). \tag{3.135}$$

In this case, $J_{i\to j}$ merely appears as an additional term in the equations for u in cells i and j. Here, α is a rate constant, with units of 1/time. It is, in general, unrelated to the diffusion coefficient within the cells. It can be a function of other biochemical species; for example, high $[Ca^{2+}]$ closes gap junctions (Spray and Bennett, 1985) and thus α will tend to zero as $[Ca^{2+}]$ in either cell 1 or cell 2 gets large. In many nonexcitable cell models the diffusing variables are often Ca^{2+} or IP_3 (Wei and Shuai, 2011; Lallouette et al, 2014), while in many models of excitable cells such as neurons the diffusing variables usually include voltage. Models of coupled endocrine cells commonly include diffusion of both voltage and Ca^{2+}; good examples of that approach, using point cells, are Sherman et al (1988), Sherman and Rinzel (1991), and Tsaneva-Atanasova et al (2006). However, these models focus on electrical synchronisation rather than Ca^{2+} wave propagation.

More generally, u will satisfy a reaction-diffusion equation within each cell, with jump conditions at the cell boundaries to model intercellular transport. For simplicity we ignore the reaction terms, and study the problem in one spatial dimension. The generalisation to multiple spatial dimensions, and the inclusion of reaction terms, is a straightforward process.

Suppose that u is diffusing along one spatial dimension, through a line of cells each of length L, and with an intracellular diffusion coefficient D. Within each cell we thus have

$$\frac{\partial u}{\partial t} = D\frac{\partial^2 u}{\partial x^2}. \tag{3.136}$$

Although u is continuous within each cell, u need not be continuous across the intercellular boundary, and in general will not be. If there is a cell boundary at $x = x_b$, the flux through the boundary is assumed to be proportional to the concentration difference across the boundary. Thus,

$$-D\frac{\partial u(x_b^-, t)}{\partial x} = -D\frac{\partial u(x_b^+, t)}{\partial x} = \mathcal{F}[u(x_b^-, t) - u(x_b^+, t)], \tag{3.137}$$

for some permeability coefficient $\mathcal{F}$, with units of distance/time. The $+$ and $-$ superscripts indicate that the function values are calculated as limits from the right and left, respectively. Equation (3.137) is often called a *jump condition*.

When the cells through which u diffuses are short compared to the total distance that u moves, the movement of u can be described by an effective diffusion coefficient. The effective diffusion coefficient is defined by calculating the flux as a linear function of the total concentration difference from one end of a row of cells to the other. Thus, if we have N cells, each of length L, with $u = U_0$ at $x = 0$ and $u = U_1$ at $x = NL$, the effective diffusion coefficient D_e is defined by

$$J = \frac{D_e}{NL}(U_0 - U_1), \tag{3.138}$$

where J is the steady-state flux of u.

For convenience we assume that the cell boundaries occur at $L/2$, $3L/2, \ldots$, and thus the boundary conditions at $x = 0$ and $x = NL$ occur halfway through a cell. To calculate D_e, we look for a function $u(x)$ that satisfies $u_{xx} = 0$ when $x \neq (j + \frac{1}{2})L$ and satisfies (3.137) at $x = (j + \frac{1}{2})L$, $j = 0, \ldots, N-1$. Further, we require that $u(0) = U_0$, and $u(NL) = U_1$. A typical solution u that satisfies these conditions must be linear within each cell, with discontinuous jumps at the cell boundaries. Suppose that the slope of u within each cell is $-\lambda$, and that the jump in u between cells is $u(x_b^+) - u(x_b^-) = -\Delta$. Then, since there are $N-1$ whole cells, two half cells (at the boundaries), and N interior cell boundaries, we have

$$(N-1)\lambda L + 2\lambda\left(\frac{L}{2}\right) + N\Delta = U_0 - U_1, \tag{3.139}$$

and thus

$$U_0 - U_1 = N(\lambda L + \Delta). \tag{3.140}$$

Furthermore, it follows from (3.137) that

$$D\lambda = \mathcal{F}\Delta. \tag{3.141}$$

Finally, since the steady-state flux at each point of the domain must be same as the total flux through the domain, we know that

$$\begin{aligned} D\lambda &= \frac{D_e}{NL}(U_0 - U_1) \\ &= \frac{D_e}{L}(L\lambda + \Delta) \\ &= \lambda D_e\left(1 + \frac{D}{\mathcal{F}L}\right), \end{aligned} \tag{3.142}$$

from which it follows that

$$\frac{1}{D_e} = \frac{1}{D} + \frac{1}{L\mathcal{F}}. \tag{3.143}$$

This is the same result that was derived from homogenisation in Section 3.3.2.

3.6.2.1 An Example: Mechanically-Stimulated Waves in Airway Epithelial Cells

One of the earliest examples of such an intercellular Ca^{2+} wave was discovered by Sanderson et al (1990, 1994), who showed that, in epithelial cell cultures, a mechanical stimulus (for example, poking a single cell with a micropipette) can initiate a wave of increased intracellular $[Ca^{2+}]$ that spreads from cell to cell. Typical experimental results from airway epithelial cells are shown in Fig. 1.9. The epithelial cell culture forms a thin layer of cells, connected by gap junctions. When a cell in the middle of the culture is mechanically-stimulated, the $[Ca^{2+}]$ in the stimulated cell increases quickly. After a time delay of a second or so, the neighbours of the stimulated cell also show an increase in $[Ca^{2+}]$, and this increase spreads sequentially through the culture. An intracellular wave moves across each cell, is delayed at the cell boundary, and then initiates an intracellular wave in the neighbouring cell. Thus the intercellular wave moves via the sequential propagation of intracellular waves. Of particular interest is the fact that, in the absence of extracellular Ca^{2+}, the stimulated cell shows no response, but an intercellular wave still spreads to other cells in the culture. It thus appears that a rise in $[Ca^{2+}]$ in the stimulated cell is not necessary for wave propagation. Neither is a rise in $[Ca^{2+}]$ sufficient to initiate an intercellular wave. For example, epithelial cells in culture sometimes exhibit spontaneous intracellular Ca^{2+} oscillations, and these oscillations do not spread from cell to cell. Nevertheless, a mechanically stimulated intercellular wave does spread through cells that are spontaneously oscillating.

Sanderson and his colleagues (Boitano et al, 1992; Sanderson et al, 1994; Sneyd et al, 1994, 1995b,a) proposed a model of mechanically-stimulated intercellular Ca^{2+} waves in epithelial cells (Fig. 3.22). They proposed that mechanical stimulation causes the production of large amounts of IP_3 in the stimulated cell (by some unknown mechanism), and this IP_3 moves through the culture by passive diffusion, moving from cell to cell through gap junctions. Since IP_3 releases Ca^{2+} from the ER, the diffusion of IP_3 from cell to cell results in a corresponding intercellular Ca^{2+} wave. Experimental results indicate that the movement of Ca^{2+} between cells does not play a major role in wave propagation as it is so extensively buffered, and thus the model assumes that intercellular movement of Ca^{2+} is negligible. Relaxation of this assumption makes little difference to the model behaviour, as it is the movement of IP_3 through gap junctions that determines the intercellular wave properties.

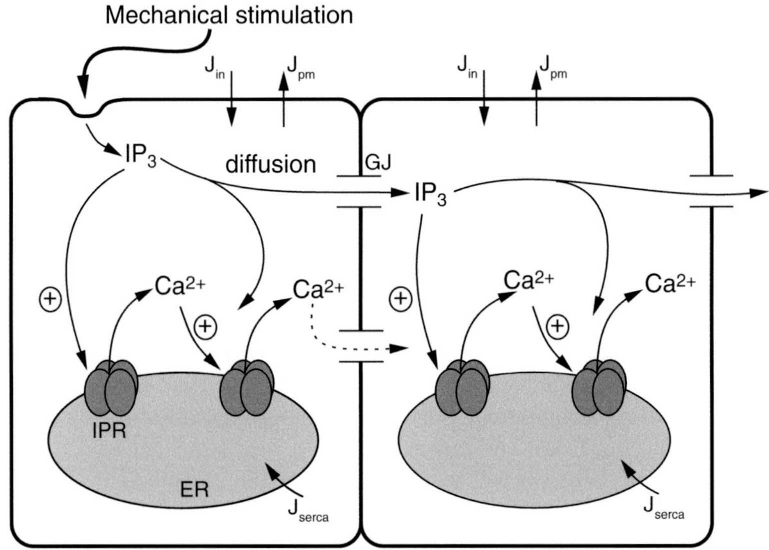

Fig. 3.22 Schematic diagram of the model of mechanically-stimulated intercellular Ca^{2+} waves in airway epithelial cells. GJ denotes a gap junction.

The epithelial cell culture is modelled as a grid of two-dimensional square cells. It is assumed that IP_3 moves by passive diffusion and is degraded with saturable kinetics. Thus

$$\frac{\partial p}{\partial t} = D_p \nabla^2 p - \frac{V_p p k_p}{k_p + p}. \tag{3.144}$$

When $p \ll k_p$, p decays with time constant $1/V_p$. Ca^{2+} is also assumed to move by passive diffusion, but it is released from the ER by IP_3 and pumped back into the ER by Ca^{2+} ATPases. Thus

$$\frac{\partial c}{\partial t} = D_c \nabla^2 c + J_{\rm IPR} - J_{\rm serca} + J_{\rm in} - J_{\rm pm}. \tag{3.145}$$

In the original model of Sneyd et al (1994), the model of Atri et al (1993) was used to model the IPR (Section 2.7.3), but the precise choice of IPR is not important for the qualitative behaviour. Similarly, the other fluxes were modelled in standard ways, as discussed previously, but the details are essentially unimportant for the present discussion. Note how the fluxes across the membrane of the cell are incorporated as reaction terms, and thus this model implicitly assumes that the cells are thin enough for this to be a reasonable approximation (Section 3.3.3).

The internal boundary conditions, or jump conditions, are given in terms of the flux of IP_3 from cell to cell. Thus, at the intercellular boundary between cell i and cell j

$$D_p \nabla p_i \cdot \mathbf{n} = D_p \nabla p_j \cdot \mathbf{n} = \mathcal{F}(p_i^+ - p_j^-), \tag{3.146}$$

where $\mathbf{n}$ is the unit normal directed from cell i to cell j. Our notation here is a little sloppy; p_i^+ and p_j^- are the concentrations in cells i and j, respectively, in the limit as the boundary is approached. This limit will depend in a complex way upon exactly where the boundary is being approached, and this must be taken into account. Thus, $p_i^+ - p_j^-$ is a function of the spatial variables, in a way that cannot be predicted without a numerical simulation.

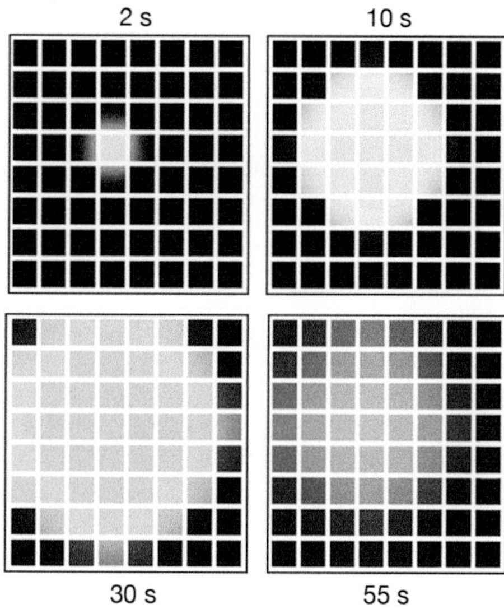

Fig. 3.23 Density plot of a two-dimensional intercellular Ca^{2+} wave, computed numerically from the intercellular Ca^{2+} wave model of Sneyd et al (1995b). Lighter shades denote higher $[Ca^{2+}]$, while the white lines denote the cell borders.

Initially, a single cell was injected with IP_3, which was then allowed to diffuse from cell to cell, thereby generating an intercellular Ca^{2+} wave. Figure 3.23 shows a numerical simulation of the model in two dimensions.

An intercellular wave can be seen expanding across the grid of cells and then retreating as the IP_3 degrades. The wave of IP_3 that spreads across the field of cells initiates Ca^{2+} oscillations, which are asynchronous in neighbouring cells, and these oscillations slowly disappear as IP_3 is degraded. Furthermore, as expected for a process based on passive diffusion, the intracellular wave speed (i.e., the speed at which the intercellular wave moves across an individual cell) decreases with distance from the stimulated cell, and the arrival time and the intercellular delay increase exponentially with distance from the stimulated cell (Sneyd et al, 1998b).

3.6.3 Regenerative and Partially Regenerative Waves

One of the major questions raised about the previous intercellular wave model was whether an IP_3 molecule is able to diffuse through multiple cells without being degraded. It is unlikely that an IP_3 molecule could survive long enough to cause intercellular waves propagating distances of up to several hundred micrometers. If the production of IP_3 is regenerative – for example, if PLC is activated by increased $[Ca^{2+}]$ – then the waves could potentially propagate indefinitely. However, if the waves eventually stop, this would seem, at first sight, to preclude a regenerative mechanism for IP_3 production.

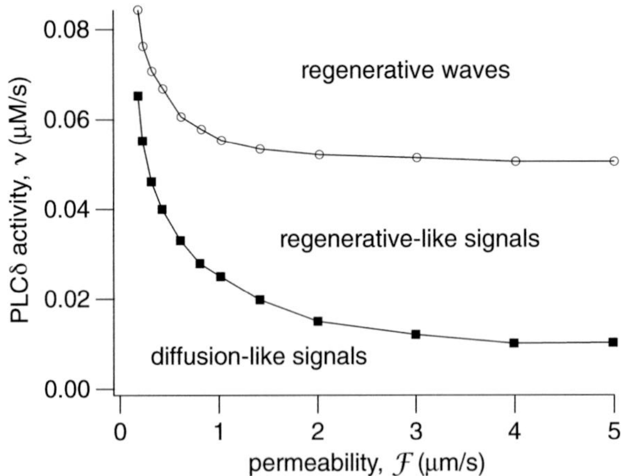

Fig. 3.24 Types of intercellular waves in the partially regenerative model as a function of the gap junction permeability of IP_3 and the maximal activity of PLCδ. These data are taken from Höfer et al (2002) and replotted.

This question was resolved by Höfer et al (2002) who showed that a regenerative mechanism could propagate intercellular Ca^{2+} waves much farther than would be possible by simple diffusion, but would not necessarily propagate them indefinitely; the waves can still eventually terminate. Their model was used to study long-range intercellular Ca^{2+} waves in astrocytes cultured from rat striatum, a preparation in which it is known that the intercellular waves propagate principally via diffusion of an intracellular messenger through gap junctions.

The model equations are similar to those in the previous section, with one major difference, in that they assume that the rate of production of IP_3 by one of the isoforms of PLC, PLCδ, is an increasing function of $[Ca^{2+}]$. Thus,

$$\frac{\partial p}{\partial t} = D_p \nabla^2 p + \frac{\nu c^2}{K^2 + c^2} - V_p p. \tag{3.147}$$

Results are shown in Fig. 3.24. When the positive feedback from Ca^{2+} to IP_3 production is small (i.e., when the maximal rate of PLCδ activity, ν, is small), the intercellular Ca^{2+} wave behaves as if it resulted from passive diffusion of IP_3 from the stimulated cell, as in the model of mechanically-stimulated intercellular Ca^{2+} waves. Conversely, when ν is large the waves become fully regenerative, propagating indefinitely. However, numerical computations show an intermediate range of ν values that give partially regenerative waves; waves that propagate further than they would have, had they been governed solely by passive diffusion of IP_3, but that eventually terminate and thus are not fully regenerative.

3.6.4 Paracrine Propagation

In many cases the intercellular wave is propagated by the diffusion of an extracellular messenger, usually (but not always) ATP or glutamate. Unlike waves that propagate through gap junctions, these paracrine waves do not pause at cell boundaries, although the average speed remains approximately 10–20 μm s^{-1}. Such similarity in wave speeds is presumably because ATP, glutamate, and IP_3 all diffuse at similar speeds. Furthermore, paracrine waves can cross regions where there are no cells (Hassinger et al, 1996).

To model a paracrine mechanism, one need only add one more equation to the model discussed above, and make minor adjustments to the equations for c and p. If we let A denote the concentration of the extracellular messenger (which, for convenience, we shall here assume to be ATP), then

$$\frac{\partial A}{\partial t} = D_A \nabla^2 A + f_1(c) f_2(p) - \frac{V_A A}{K_A + A}, \tag{3.148}$$

where f_1 and f_2 model the rate of ATP formation as functions of $[Ca^{2+}]$ and $[IP_3]$, respectively. Note also that ATP secretion is not included explicitly, so this equation considers both intracellular and extracellular processes. There are no jump conditions for A, as ATP does not need to go through gap junctions.

Different authors have used different expressions for f_1 and f_2; since there have been few models of paracrine extracellular waves there is no consensus on which are the most suitable expressions to use for any particular cell type. In their model of an astrocyte network, Stamatakis and Mantzaris (2006) assumed that both f_1 and f_2 were biphasic functions, while Bennett et al (2005), again modelling intercellular Ca^{2+} waves in astrocytes, assumed that ATP production was dependent on p, but was not dependent on c.

Similarly, the effect of ATP on the production of IP_3 has been modelled in at least two different ways. Bennett et al (2005) constructed a detailed model of the GPCR cascade leading to production of IP_3, and thus ATP was modelled explicitly as a ligand binding to its receptor. Stamatakis and Mantzaris (2006) took the simpler approach of assuming that the rate of production of IP_3 was an increasing algebraic function of A, to give

$$\frac{dp}{dt} = \nu_4 \left(\frac{A}{\kappa_4 + A}\right)\left(\frac{c}{\kappa_5 + c}\right) - k_p p. \tag{3.149}$$

Hassinger et al (1996) showed experimentally that a cell-free lane of width $80\,\mu$m causes a delay in the wave of 18 s, while the model of Bennett et al (2005) showed a 14 s delay for a cell-free lane of width $125\,\mu$m. The model wave thus goes rather faster than physiological ones, suggesting that effects of gap junctions may be more important than assumed in the model.

By combining extracellular ATP diffusion with intercellular diffusion of IP_3 through gap junctions, one can construct a model that propagates a wave by a combination of both mechanisms. If ATP, IP_3, and Ca^{2+} are all assumed to be defined on the same numerical grid, then the model equations can all be solved on a single grid; in this case p and c will have internal jump conditions but ATP will not. A slightly different numerical approach was taken by Edwards and Gibson (2010), who avoided the use of jump conditions, but their different solution method is unlikely to result in qualitative differences. They showed that, depending on the weighting given to the extracellular and the intercellular mechanisms, the waves could have quite different qualitative properties, either purely diffusive, partially regenerative, or fully regenerative.

3.7 Connecting the Cytosol to the Membrane

In every cell the cytosolic concentration of Ca^{2+} is modulated, to a greater or lesser extent, by the membrane potential. For a start, not only can changes in the membrane potential open and close voltage-gated Ca^{2+} channels, but such changes will also affect the driving force for Ca^{2+} entry. Furthermore, in many cell types the cytosolic Ca^{2+} is, in turn, an important modulator of the membrane potential, via the activation of K^+ and Cl^- channels. This close intertwining of Ca^{2+} and the membrane potential results in a complex coupled system; in many cases, if both cytosolic Ca^{2+} and membrane potential can oscillate independently, the resultant system of coupled oscillators can give behaviour of enormous complexity. For instance, electrical bursting and excitation-secretion coupling in hypothalamic and pituitary neurons and in pancreatic β cells (Sections 8.7 and 8.4) are closely dependent on the interaction of cytosolic Ca^{2+} with the membrane potential, as is excitation-secretion coupling in exocrine epithelia (Section 6.3). Indeed, most of the models in Chapters 7 and 8 are of this type.

From the classic cell circuit diagram of the membrane potential used by Hodgkin and Huxley (1952) (for a more detailed presentation see Keener and Sneyd (2008)), we have that the membrane potential, V, satisfies the equation

$$C_m \frac{dV}{dt} + \sum I_{\text{ionic}} = 0, \tag{3.150}$$

where C_m is the capacitance of the cell membrane, and the sum is over all the ionic currents across the cell membrane. By convention, V is defined to be the internal potential minus the external potential, and I_{ionic} is defined to be positive when positive charge leaves the cell.

For simplicity, let us consider the simple case of a cell that has a single Na^+ current, a single Ca^{2+}-sensitive K^+ current, and a voltage-gated Ca^{2+} channel. Then

$$C_m \frac{dV}{dt} = -I_{\text{Na}}(V) - I_{\text{K,Ca}}(V, c) - I_{\text{Ca}}(V, c), \tag{3.151}$$

where, as usual, c denotes $[Ca^{2+}]$.

In writing down the conservation equation for c we need to ensure that we include the flux through the Ca^{2+} channel. We know that the flux of Ca^{2+} is related to the Ca^{2+} current through

$$I_{\text{Ca}} = -\frac{d}{dt}(2Fwc), \tag{3.152}$$

where F is Faraday's constant (9.648×10^4 C/mol) and w is the volume of the cell. The factor of 2 is included because each Ca^{2+} ion carries two positive charges.

Hence, as long as the volume of the cell is constant, the conservation equation for c will be

$$\frac{dc}{dt} = -\frac{I_{\text{Ca}}}{2Fw} + \sum \text{all other Ca}^{2+} \text{ fluxes}. \tag{3.153}$$

In cases where the volume of the cell is changing one must be careful to include the necessary terms in dw/dt. Furthermore, it is important to note that (3.153) does not take Ca^{2+} buffering into account; it describes the total flow of Ca^{2+} into the cell. Ca^{2+} buffering must be incorporated, either by including the explicit equations of Ca^{2+} buffering, or, if the buffers are fast and linear, by scaling the Ca^{2+} current by the required buffering scale factor, as described in Section 2.9.

Thus, for a cell of fixed volume, we have the coupled equations (3.151) and (3.153), as well as any additional equations used to model the membrane potential, additional Ca^{2+} fluxes, and Ca^{2+} buffers.

The functions $I_{K,Ca}$ and I_{Ca} depend upon the particular channels under consideration. For example, one common choice is to assume that the K^+ current is a linear function of the membrane potential, while the conductance of the channel is an increasing sigmoidal function of c. In this case

$$I_{K,Ca} = g_K(c)(V - V_K), \tag{3.154}$$

where V_K is the Nernst potential of K^+, and

$$g_K(c) = \frac{\alpha_1 c^n}{\alpha_2^n + c^n}, \tag{3.155}$$

for some constants α_1 and α_2, and where n is some small integer (see, for example, Section 8.4.1).

Greater complexity can be introduced into the models of the ion channels, including time-dependent gating or dependence on other agonists, and other electrogenic Ca^{2+} transport mechanisms such as Na^+/Ca^{2+} exchangers can introduce additional current terms into the equation for V. However, the basic structure of the model remains the same.

It is an interesting exercise to contrast pancreatic β cells and parotid acinar cells (Petersen, 2014). Both cell types require a rise in $[Ca^{2+}]$ in order to carry out their function; pancreatic β cells to secrete insulin, parotid acinar cells to transport water. However, because of their different functions, the link between the cytosolic Ca^{2+} and the membrane potential is quite different in the two cell types.

In pancreatic β cells, depolarisation of the cell membrane leads to activation of voltage-gated Ca^{2+} channels, and subsequent Ca^{2+} entry, which then drives the exocytotic machinery to secrete insulin. Thus, the link from V to c is mostly in one direction, with an increase in V driving an increase in c. However, even though water transport by acinar cells also depends on an increase in $[Ca^{2+}]$, simple depolarisation and influx through voltage-gated Ca^{2+} channels would not work. Water flow is driven by net secretion of Cl^- through Ca^{2+}-sensitive channels in the apical membrane. If the membrane potential were allowed to depolarise, the driving force for Cl^- exit would be diminished, Cl^- secretion would stop, as would water flow. To prevent this happening, the rise in Ca^{2+} concentration also opens Ca^{2+}-sensitive K^+ channels, preventing depolarisation of the cell, and ensuring that the driving force for Cl^- exit is maintained.

Hence, in acinar cells it is not a rise in V that stimulates the release of Ca^{2+} from internal stores, it is stimulation by an agonist, with subsequent IP_3-induced release of Ca^{2+} from the ER. This rise in Ca^{2+} then controls V precisely so as to ensure that the cell does not depolarise. Rather than V controlling c, it is now c that controls V.

3.8 The Effects of Buffers

In Section 2.9 we saw how buffers can change the dynamics of a Ca^{2+} model, and how they can change the effective diffusion coefficient. For two reasons, it is of immediate interest to determine the qualitative and quantitative effects of these changes. Firstly, Ca^{2+} concentrations are observed using fluorescent Ca^{2+} dyes, which are themselves Ca^{2+} buffers. In other words, to measure $[Ca^{2+}]$, one must add an additional buffer, which might change the behaviour of the responses. We need to know if our attempts to measure $[Ca^{2+}]$ disrupt the very thing we are trying to measure. Secondly, exogenous buffers such as BAPTA or EGTA are often added as part of experimental procedures (see, for example, Dargan and Parker (2003) or Kidd et al (1999)). It is thus important to be able to predict the effects of such buffer additions.

3.8.1 Qualitative Effects

The qualitative effects of buffers on Ca^{2+} waves have been investigated in detail in a series of papers by Je-Chiang Tsai (Tsai and Sneyd, 2005, 2007, 2011; Tsai, 2013), who used the buffered bistable equation and the buffered FitzHugh-Nagumo model as caricature models of Ca^{2+} waves. The conclusions depend to a large extent on the exact properties of the buffer; whether it is immobile, or diffuses quickly, whether it has a high or low affinity for Ca^{2+}.

If the buffer is immobile, it cannot change the existence, uniqueness, or stability of the Ca^{2+} waves, although it can change the wave amplitude (Kaźmierczak and Volpert, 2007) However, if a buffer is highly mobile, and there is enough of it, waves can be eliminated. Nevertheless, if the waves exist they remain unique and stable. These qualitative results, although derived from caricature models, are consistent with the available experimental evidence. Nuccitelli et al (1993) have shown that waves can eliminated by the addition of enough exogenous buffer, but use of a dextran-bound Ca^{2+} dye, with a very low diffusion coefficient, has no appreciable effect on waves (Lechleiter and Clapham, 1992).

Tsai et al (2012) also showed that, under some conditions, a mobile buffer can increase the wave speed. This is due to facilitated diffusion, whereby a buffer can pick up a Ca^{2+} ion, diffuse quickly with it, then release it, resulting in an increase in the effective speed of Ca^{2+} diffusion (Thul and Falcke, 2004a; Rüdiger et al, 2010a). This theoretical result agrees with the numerical results of Jafri and Keizer (1995), but has yet to receive direct experimental confirmation.

3.8.2 Quantitative Effects

Quantitative effects of buffers have been studied in a large number of detailed models. Early studies were those of Sala and Hernàndez-Cruz (1990), Backx et al (1989), Nowycky and Pinter (1993), and Jafri and Keizer (1995). The results of such studies are crucially dependent on the exact buffering parameters used, and are quantitative rather than qualitative in nature, making it impractical to give summaries of the results, for which the interested reader is referred to the original papers. In addition, studies of Ca^{2+} in skeletal muscle depend to a large extent on detailed models of Ca^{2+} buffers, and the quantitative changes these make to the whole-cell Ca^{2+} responses. This is discussed briefly in Section 7.3.

In Section 3.4.4 we saw how quickly the Ca^{2+} microdomain can form, and its approximate extent. However, the size of the microdomain also depends crucially on the buffer parameters and the extent to which the ER depletes. This has motivated a number of detailed numerical studies of how a stochastic IPR will interact with its local environment, and how this can be modified by buffers and by the presence of neighbouring IPR (Falcke, 2003a; Shuai et al, 2007, 2008; Zeller et al, 2009; Rüdiger et al, 2010a). The computations are intricate and difficult to do, as the fast openings and closings of the IPR must be resolved, as must the quickly-forming steep Ca^{2+} gradients. Thus, small time and space steps are required, and the stochastic IPR must be coupled with a deterministic formulation of the Ca^{2+} diffusion and reaction. Furthermore, nearly all the results to date have been obtained by using the De Young-Keizer IPR model (Section 2.7.1), or slight modifications thereof. It is an open question how much these qualitative results will change (if at all) if more modern IPR models are used instead.

Some of the major results to come out of this work are

- Stationary buffers increase the mean open probability of an isolated IPR. When the channel opens, the rise time of the $[Ca^{2+}]$ is increased as the buffer fills up, and when the channel shuts, the buffers release their bound Ca^{2+}, thus decreasing the rate at which $[Ca^{2+}]$ returns to rest. These effects cause little change to the mean open time, but serve to decrease the mean closed time significantly, as the longer duration of high $[Ca^{2+}]$ around the mouth of the channel promotes channel reopening.

- Mobile buffers have the opposite effect, of decreasing the mean open probability and increasing the rate of decline of $[Ca^{2+}]$ when the IPR shuts. They do this by binding Ca^{2+} and then diffusing away from the channel, an effect called *shuttling*. Buffers such as BAPTA, that bind and release Ca^{2+} quickly, have a greater shuttling effect than do slow buffers such as EGTA.
- When IPR are grouped in clusters, as they are in vivo, immobile buffer has little effect on open probability. This somewhat counterintuitive result is because the $[Ca^{2+}]$ in the cluster is affected much more by the number of open channels rather than by the local buffer dynamics. Ca^{2+} coming from a neighbouring open channel has a greater effect than Ca^{2+} coming off an immobile buffer.
- However, a fast mobile buffer has a significant effect on the open probability of IPR in clusters, mostly via its effect on the $[Ca^{2+}]$ within the cluster. As the concentration of buffer increases, the IPR open probability first increases (from around 0.2 to around 0.4) before falling steeply to around 0.1 when the buffer concentration is 10 mM.
- Slow mobile buffers have little effect on the IPR dynamics within a cluster, but affect the ability of one cluster to excite neighbouring clusters. Thus they can disrupt whole-cell wave propagation.

On a larger spatial scale, Zeller et al (2009) studied the effects of different buffers on the interaction between clusters of IPR, and compared their results to the experimental results of Dargan and Parker (2003) and Dargan et al (2004). At particular issue was the experimental observation that EGTA (a buffer with a slow on rate) and BAPTA (a buffer with a fast on rate) have different effects on the spatiotemporal properties of Ca^{2+} responses in *Xenopus* oocytes. EGTA causes global responses to fracture into separate localised responses, while BAPTA slows the response down and tends to promote global responses. The results of the theoretical study of Zeller et al (2009) are not always easy to reconcile with these data, or with the similar data of Kidd et al (1999), indicating that the precise effects of buffers remain poorly understood, particularly on the scale of multiple IPR clusters.

In whole-cell models, the effects on Ca^{2+} dynamics of fast and slow buffers were studied by Gin et al (2006). They introduced a scaling parameter, λ, such that $\lambda = 1$ corresponded to infinitely fast buffers, $\lambda = 0$ corresponded to infinitely slow buffers. As λ varies from 0 to 1 the important model dynamics changed qualitatively only in a small region for λ between approximately 10^{-2} and 10^{-3}. However, this result was dependent on the particular model chosen for their investigation, and so it is unclear how well it will extrapolate to other models. In particular, since it is difficult to know, a priori, the exact region of buffer speeds where such qualitative changes occur, it is not possible, in general, to conclude much about the effects of buffers on such things as homoclinic orbits or period doubling.

Chapter 4
Hierarchical and Stochastic Modelling

4.1 Introduction

As we have already pointed out (Section 1.3), the basic building block of a Ca^{2+} response is a stochastic event at the level of an individual Ca^{2+} channel, whether an IP_3 receptor, a ryanodine receptor, or a voltage-gated Ca^{2+} channel. The behaviours that we call oscillations, or spiking, or waves, or microdomain transients, are merely emergent properties of the interaction of channel-level stochastic processes.

In other chapters we mostly consider models, based on ordinary or partial differential equations, that address aspects of intracellular Ca^{2+} dynamics which can be described by mean field behaviour and parameter dependencies arising thereof. Stochastic modelling, accounting for the hierarchy of structures (channel, cluster, cluster array) and events (blip, puff, spike), provides further insight into the mechanisms that control Ca^{2+} signalling.

Molecular randomness introduces random behaviour even on the cellular level. In all cell types analysed so far, the standard deviation of interspike intervals is of the same order of magnitude as the average interspike interval. Indeed, it appears that CICR exploits noise rather than being perturbed by it. The question of how such microscopic events can be organised into whole-cell responses thus becomes of immediate relevance.

The random behaviour of molecules (on the microscopic scale) and cells (on the mesoscopic scale) requires the calculation of probability distributions of variables instead of single values. Simulations of time courses or trajectories in phase space, which are the typical result of deterministic models, are individual realisations of a stochastic process. But, due to the randomness in the system, each realisation is different. Thus, complete information on a stochastic process cannot be obtained from single trajectories, but only by considering the distribution of trajectories or distributions of other characteristics of the system like, for example, interspike intervals or amplitudes.

G. Dupont et al., *Models of Calcium Signalling*, Interdisciplinary Applied Mathematics 43, DOI 10.1007/978-3-319-29647-0_4

Hence, in this chapter we study models of a type quite different to those seen elsewhere in this book, models in which the variables are distributions.

The fundamental question that we wish to answer is the following: given the stochastic behaviour of individual Ca^{2+} channels, can we derive an expression for the distribution of the times between successive whole-cell Ca^{2+} spikes? In the deterministic models we have studied so far, this question is answered by studying how the period of the limit cycle solutions depends on the parameters of the model, including the parameters of the individual Ca^{2+} channels. Although analytic results are not usually obtainable, nevertheless the techniques of numerical bifurcation theory and numerical simulation give us ways to answer this question.

However, serious difficulties can arise if one relies too much on purely deterministic approaches, and these are illustrated by one of the earliest theoretical investigations of stochastic Ca^{2+} dynamics. Using a stochastic version of the De Young-Keizer model of the IPR, coupled to the diffusion of Ca^{2+} in two dimensions, Falcke (2003b) showed how, for values of the IP_3 concentration for which there is no limit cycle, the stochastic model exhibits behaviour that looks almost indistinguishable from experimentally observed Ca^{2+} oscillations. In other words, the model appears to oscillate, even when it can be proven that no deterministic oscillations exist.

An intuitive explanation of this behaviour is relatively easily understood. Suppose that an excitable system is driven by noise, so that, every so often, the system is driven over the excitability threshold, leading to a spike. Suppose further that, after the spike, there is a refractory period during which no additional spikes can be stimulated (for example, due to depletion of the ER, although this is by no means the only possible refractory mechanism). After the refractory period, there will be a random waiting time before the next spike is initiated. However, if the noise has a large enough amplitude, this waiting time will be very short, leading to a second spike almost immediately after the refractory period has ended. This will lead to a series of spikes, with period almost the same as the refractory period, and this series of spikes will look almost exactly like an oscillation, even though, in this case, there is no underlying limit cycle.

Thus warned, we realise that the emergent behaviour of an ensemble of stochastic Ca^{2+} channels is not only vital for studying Ca^{2+} dynamics, it is also more complex than might appear at first. It is this emergent behaviour that we study in this chapter, and to do so, we take the approach of hierarchical modelling. Although stochastic modelling is important for understanding how all Ca^{2+} channels work, they can all be studied using only minor variants on the basic theory. For this reason, and partly for convenience, we shall focus our discussion here on stochastic modelling of IP_3 receptors.

4.1.1 Hierarchical Modelling Across Different Structural Levels

We can illustrate the basic motivating idea of hierarchical modelling by considering cars in traffic. A high-level description of the car characterises the basic task of going from point A to point B. There are just four parameters: the position and velocity of the car on the two-dimensional road network. At lower levels, the number of parameters describing the car is larger. The engine can be described by the rotation velocity of the crankshaft, the gear box by the current gear, the brakes by the torque they exert on the wheels, the steering system by the position of the steering wheel, and so on. The number of parameters required to describe a car at yet lower levels (at the level, say, of nuts and bolts) is enormous.

So, although the number of parameters and degrees of freedom increases dramatically while going from the highest to the lowest level, in the end, all the parameters of the lower levels control only the four parameters that describe the car as an elemental unit of the traffic flow. How can we take advantage of these observations in terms of modelling? We can describe flow by using a description of cars by position and velocity. We may even introduce some interaction between individual cars affecting the velocity. However, as long as we know all possible velocities and interactions between cars, in order to describe traffic flow we do not need to know how the lower-level parameters affect the velocity of the car.

Splitting the description of a system into structural levels is useful whenever one of those levels can be described by a small set of parameters. This is the case in the above example. It also applies to many physical systems, where statistical physics tells us that the astronomical number of microscopic degrees of freedom and parameters causes a macroscopic behaviour described often by only a few parameters.

The existence of a description with a small number of parameters cannot be taken for granted, and must be checked for each system. However, we will see that it applies to Ca^{2+} signalling. For example, the puff duration distribution (recall that a puff is the release of Ca^{2+} through a cluster of IPR, while a blip is the release of Ca^{2+} through a single IPR) can be described by two parameters that are essentially equal for all puff sites, and one parameter defining the puff site. Similarly, interpuff interval (IPI) distributions can be well characterised by two parameters only. Hence, the dynamics of the hundreds of microscopic states of the ensemble of channels forming the cluster can be essentially captured by two fixed and three cluster-specific parameters. That is a good starting point for hierarchical modelling. We will also see that, on the cellular level, interspike interval (ISI) distributions can be described by three parameters; two of them are specific to the cell type and pathway, and one characterises the individual cells. The onset of spiking at low stimulation and the transition to over-stimulation appear also to be governed by two parameters only. As yet, spike duration and amplitude distributions have not been investigated as systematically as ISI distributions, and thus we do not know how many parameters characterise them. However, at first glance, spike duration appears to show less variability than the interspike interval.

4.1.2 Distributions, Blips, and Puffs

There is a clear hierarchy of structures in IP_3-induced Ca^{2+} signalling (Fig. 4.1). The smallest structure we are considering is the individual IPR. Typical time scales on that level are milliseconds for individual channel openings, or microseconds for the change of the Ca^{2+} concentration at the channel mouth upon channel opening in a volume of approximately the same size as the channel molecule.

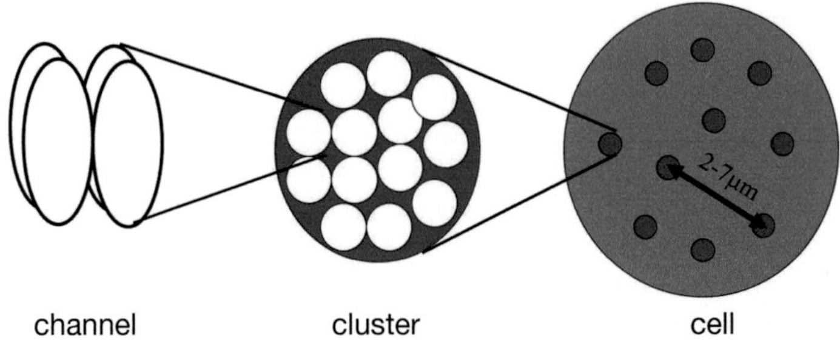

Fig. 4.1 IP_3-induced Ca^{2+} release happens on a hierarchical structure. The individual channels are on the lowest level, and typically operate on a time scale of μs to ms, although some slower processes can have time scales of hundreds of ms. Channels form clusters, and typically operate on a time scale of a few tens to a few hundreds of milliseconds. The array of clusters corresponds to the cell level. The processes on the structural levels are, respectively, single-channel openings, puffs, and cellular spikes.

IPR form clusters. Clusters are regions, with a diameter of approximately 300 nm (Wiltgen et al, 2010), in which there is a high density of IPR. A puff involves the opening of a random number of channels in a cluster (Dickinson and Parker, 2013) and lasts a few tens of milliseconds. Regions that generate puffs exhibit release amplitude distributions with a well-defined maximum, and release sources are localised to a small volume. Both observations indicate the existence of clusters. Recent observations in SH-SY5Y human neuroblastoma cells indicate there are between 1 and about 15 channels per cluster (Dickinson and Parker, 2013). The channels within a cluster are strongly coupled by Ca^{2+} diffusion and Ca^{2+}-induced Ca^{2+} release (CICR).

The coupling between different clusters is also due to CICR but is much weaker. Since the coupling is weak, not every puff initiates puffs at neighbouring sites. Puffs lead to whole-cell release spikes only with a given probability, and thus it can take a few attempts before a cellular spike is initiated. For this reason, ISIs are longer than IPIs. Additionally, we find an absolute refractory phase after each spike, which we do not observe with puffs.

To use hierarchical modelling to study Ca^{2+} signalling is a process requiring multiple steps, not all of which are, as yet, completely understood. We start by calculating the IPI and puff duration distributions, which we do based upon single-channel models and information on the number of channels per cluster. Then, for a given cell type and signalling pathway, we calculate the ISI distributions from the IPI and puff duration distributions as well as from knowledge of global feedbacks. The waiting time distributions, or probabilities per unit time, that are derived in this manner can be used for simulations (Thurley and Falcke, 2011), but their main use is to help explain statistical properties of cells in terms of distributions and relations between their moments.

4.2 Characteristics of Puffs

4.2.1 Interpuff Interval Distributions

The autonomous activity of individual Ca^{2+} puff sites can be measured by suppressing global Ca^{2+} dynamics with EGTA. This Ca^{2+} buffer substantially reduces diffusion of free Ca^{2+} and therefore reduces intercluster coupling, which, in turn, prevents coordinated global Ca^{2+} spikes. Without EGTA, global spikes in SH-SY5Y cells are induced even at weak stimulation, due to diffusive coupling of puff sites. Release at individual Ca^{2+} puff sites can be imaged by TIRF microscopy (Smith et al, 2009). Although this technique only resolves Ca^{2+} puffs that occur near the plasma membrane, we have no reason to assume that these puff sites are not representative. Some data from this cell type are shown in Fig. 4.2.

It would require an impracticably long time series to demonstrate that EGTA completely uncouples puff sites, but the data show that the addition of EGTA is sufficient to eliminate Ca^{2+} spiking and cellular time scales from puff site behaviour. Thus, for these results, it is reasonable to consider the puff sites as statistically independent.

Upon opening of the first channel in a cluster any other channel in the cluster opens with a probability of 2/3 (Dickinson and Parker, 2013). Hence, the probability of observing the opening of only one channel in a cluster (i.e., the probability of observing a blip) goes down with cluster size like $3^{-(N_{\mathrm{ch}}-1)}$, where N_{ch} is the number of channels in the cluster. The number of open channels quickly reaches its maximum with a mean rise time of 34 ms, which, for $N_{\mathrm{ch}} > 2$, is independent of cluster size (Dickinson and Parker, 2013). We call this maximum the puff amplitude.

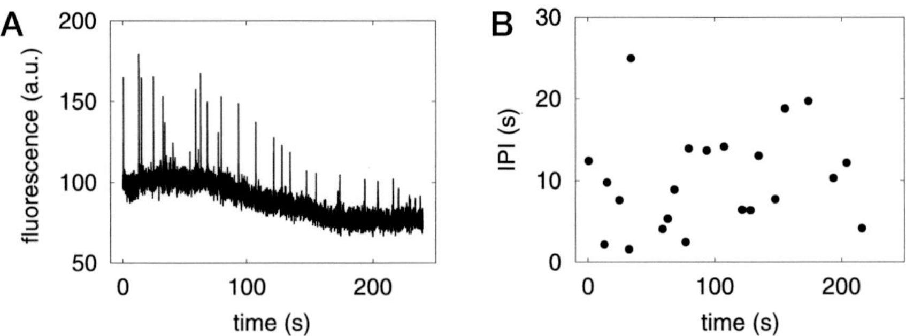

Fig. 4.2 **A:** Fluorescence recorded from a single puff site in a SH-SY5Y cell. EGTA was used to decrease intercluster coupling and thus to prevent coordinated global responses. **B:** The sequence of interpuff intervals (IPIs) derived from the recording in A (Thurley et al, 2011).

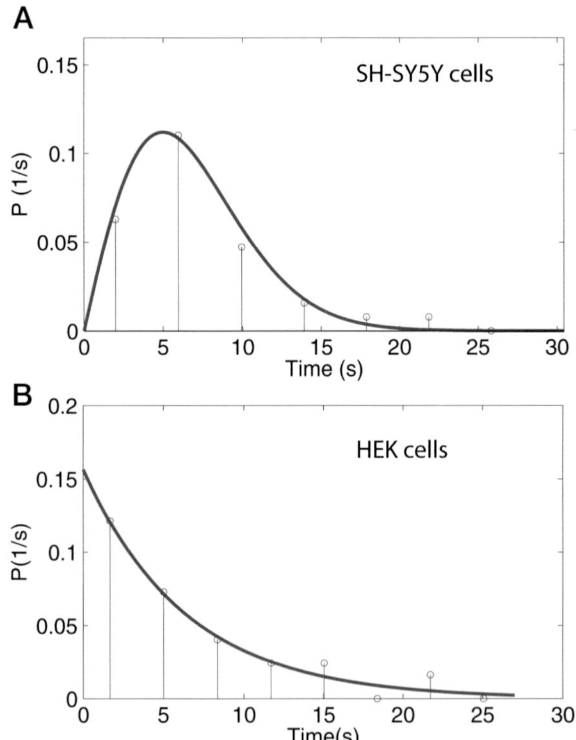

Fig. 4.3 IPI distributions from SH-SY5Y cells (panel A) and HEK 293 cells (panel B). $P(t)$ is the probability per second of observing an ISI of length t. The vertical lines show the measured normalised histograms, the full lines show fits to smooth functions. The distribution in panel A has been fit to (4.2), the one in panel B to an exponential function (Thurley et al, 2011).

The mean amplitude of puffs is mainly determined by the number of channels that have IP_3 bound, and can therefore be activated by Ca^{2+} (Smith et al, 2009; Dickinson et al, 2012). Experiments at maximal stimulation revealed that the amplitude distribution of an individual puff site is uniform between 1 and N_{ch} (Dickinson and Parker, 2013). Smith and Parker (2009) could simulate similar data with a channel open probability proportional to the cube root of the number of open channels.

The closing phase starts when the maximum number of open channels has been reached, and has a mean duration consistent with independent closing of channels with a mean waiting time of 47 ms (Smith and Parker, 2009; Dickinson and Parker, 2013). More recently, coupled coordinated closing of channels has been reported (Wiltgen et al, 2014).

Histograms of IPI measured at individual puff sites (Fig. 4.3 A) show that, in SH-SY5Y cells, very short IPIs are rare, with the most common IPI being around 5 s. One interpretation of this result is that there is a refractory period after a puff, which prevents another puff occurring too soon. The existence of such a refractory period is also suggested by investigations of amplitudes and correlations in puff sequences. After IP_3 photorelease, the amplitude (i.e., the maximal number of open channels) of the first puff is, on average, larger than the mean amplitude of subsequent puffs at the same site (Dickinson and Parker, 2013). This effect increases with cluster size and reaches about 40% for clusters larger than five channels. The IPI following a puff with large amplitude is on average longer than the one following a small puff. There is also a weak positive correlation of the length of an IPI to the amplitude of the following puff. However, amplitudes of subsequent puffs are not correlated in a measurable way (Dickinson and Parker, 2013).

Recovery from inhibition can be described by a process with a probability per unit time for occurrence of a puff given by

$$\Lambda(t) = \lambda(1 - e^{-\xi t}). \tag{4.1}$$

The parameter ξ is the rate of recovery from inhibition; $\xi \ll \lambda$ means very slow recovery, while for large ξ, puff generation is like a Poisson process with constant Λ. It follows that the IPI distribution, ψ_o, which is just the distribution of waiting times before the next puff, is given by

$$\psi_o = \Lambda(t)e^{-\int_0^t \Lambda(s)\,ds}. \tag{4.2}$$

The shape of the IPI distribution shows considerable variability between puff sites of the same cell type. For many sites, particularly in HEK cells (Thurley et al, 2011), the ratio ξ/λ_0 is so large (i.e., recovery from inhibition is so rapid) that the distributions are well fit by exponential distributions (Fig. 4.3 B). In these cases, the probability of a puff occurring is constant and unaffected by the preceding puff. For other puff sites, ξ is smaller (i.e., recovery from inhibition is slow). These distributions have a lesser probability of a puff occurring immediately after a preceding puff, and a pronounced maximum probability at an IPI of the same order of magnitude as the mean IPI.

Although the shape is variable, measured IPI distributions are all well fit by (4.2) (or by (4.27), as discussed in more detail in Section 4.4), with each IPI distribution invariably having only one maximum. Oscillatory behaviour would provide a multimodal distribution with maxima at the period and multiples of it (Verechtchaguina et al, 2007). Such multimodal distributions were not observed and we conclude that, for periods in the range of the mean IPI, there is no periodicity in the IPI sequences.

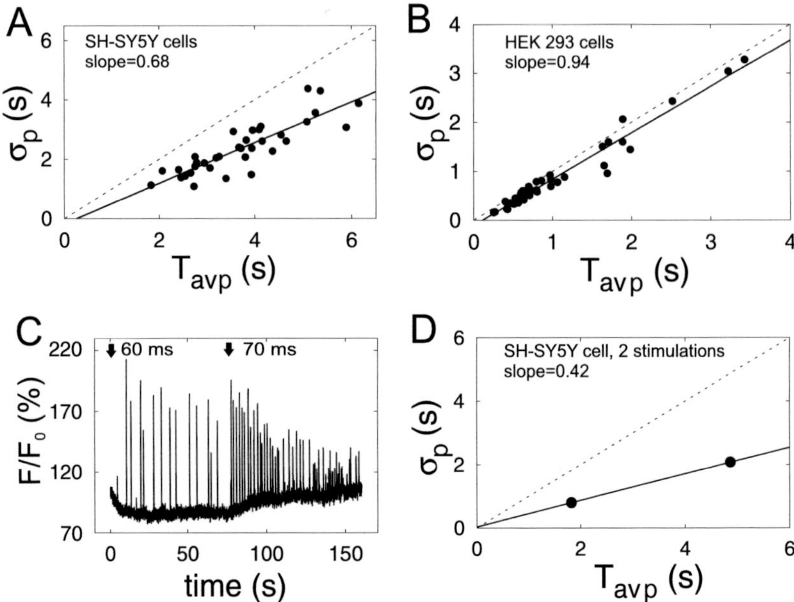

Fig. 4.4 The relationship between standard deviation (σ_p) and mean (T_{avp}) of IPI distributions in SH-SY5Y and HEK 293 cells. **A** and **B:** σ_p and T_{avp} were calculated from IPI sequences similar to those shown in Fig. 4.2. Dashed lines mark the line where $\sigma_p = T_{\text{avp}}$, and solid lines are linear fits to the data. The slope of each fit is shown in the legend. Each solid circle is a data point from a single puff site in either an SH-SY5Y cell (A) or an HEK 293 cell (B). The intercepts with the T_{avp}-axis are 0.26 s (A) and 0.11 s (B). **C:** Fluorescence at a single puff site in an SH-SY5Y cell was recorded during stimulation with two successive UV flashes at the indicated times and with the indicated flash durations. **D:** T_{avp}-σ_p relationship calculated from the data shown in C. Reproduced from Thurley et al (2011), Fig. 3, with permission from Elsevier.

4.2.2 The Coefficient of Variation

The relationship between the mean (T_{avp}) and standard deviation (σ_p) of the IPI is called a cumulant relation, since the mean and standard deviation are the first and second cumulants of a distribution. It can be obtained directly from the data (Fig. 4.4 A and B). For both SH-SY5Y and HEK 293 cells, the

T_{avp}-σ_p relation is linear, with slope less than one and intercept (on the T_{avp} axis) close to zero. Since (4.2) also has an approximately linear relationship between the standard deviation and the mean, this provides additional confirmation that (4.1) is an accurate model of the recovery from inhibition after a puff.

The slope of the T_{avp}-σ_p line may be interpreted as a population-averaged coefficient of variation (CV $= \sigma_p/T_{\mathrm{avp}}$), which is a measure of noise in the underlying stochastic process. CV $= 1$ indicates a Poisson process, which has maximal noise, while CV $= 0$ would be a completely deterministic process. Since the T_{avp}-σ_p line intercepts the horizontal axis so close to zero, this indicates that there is no absolute refractory period following the puff (or, at most, only a very short one). However, the fact that the line has slope less than one indicates that inhibition after the puff reduces the CV compared to a pure Poisson process. Such inhibition is more obvious in SH-SY5Y cells than HEK 293 cells, since the slope of the T_{avp}-σ_p relation is smaller in SH-SY5Y than in HEK cells (Fig. 4.4 A and B). This corresponds to the observation that the ratio ξ/λ_0 is also smaller in SH-SY5Y than in HEK cells.

Hitherto, T_{avp}-σ_p relationships have been derived from data points from many different puff sites, each providing a single σ_p and T_{avp} (Fig. 4.4 A and B). To allow the relationship to be analysed for a single puff site, SH-SY5Y cells were successively stimulated with flashes of increasing intensity and the puff sequences were recorded from the same site after each stimulus (Fig. 4.4 C). Two values of σ_p and T_{avp} from these sequences are sufficient to determine a linear relation, yielding the relationship for an individual puff site (Fig. 4.4 D). From seven independent experiments, the mean slope of the single-site T_{avp}-σ_p relationship was 0.49 ± 0.14, shallower than that of the population relationship (which has a slope of 0.68, as seen in Fig. 4.4 A), and the mean intercept was -0.11 s $\pm$ 0.73 s. In both the population and the single-site analyses, the intercepts were almost 0 s, and the slopes less than 1, indicating inhibition after the puff, but without an absolute refractory period.

4.2.3 Puffs are not Periodic

In oscillatory reaction-diffusion systems, the global period (the average interspike interval, or ISI, in our context) is set by the time scale of the local dynamics (i.e., the time scale of recovery from inhibition after a puff). However, IPI distributions have maxima at less than 5 s for SH-SY5Y cells and less than 1 s for HEK 293 cells, and recovery times ($1/\xi$) are shorter than 3 s in both cell types (Thurley et al, 2011). Thus, at submaximal stimulation, the mean IPI in SH-SY5Y cells is considerably shorter than the mean ISI ($\sim$100 s) (Van Acker et al, 2002; Skupin et al, 2008). Even when the stimulation intensity is increased, the mean IPI remains shorter than the ISI (Smith et al, 2009). In HEK 293 cells too, ISIs evoked by carbachol are much longer

($>$ 20 s) than the IPIs reported here (Skupin et al, 2008). Likewise with *Xenopus* oocytes, IPIs are shorter than ISIs (Marchant and Parker, 2001). We conclude that in each of these three systems, *Xenopus* oocytes, SH-SY5Y cells, and HEK 293 cells, the period of the Ca^{2+} spikes (ISI) is not set by the time scale of Ca^{2+} puffs (IPI).

Another possibility is that there is an oscillatory process on a time scale of several IPIs modulating the puff sequences. If puff sequences were periodic, with period similar to an ISI, the points in a plot of the nth IPI vs. the $(n+1)^{\text{st}}$ IPI (i.e., the return map) would lie on a closed curve, at least approximately. Thus, we can investigate possible periodicity in puff sequences by plotting the return map for a time period at least as long as two or three ISIs, to see whether the sequence of points in the return map repeats.

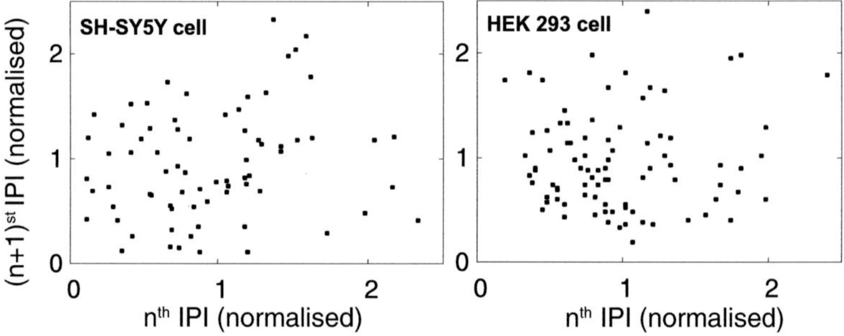

Fig. 4.5 Successive interpuff intervals are uncorrelated, as shown by plots of the $(n+1)^{\text{st}}$ IPI against the nth IPI in SH-SY5Y and HEK 293 cells. Here, we call this plot a "return map". IPIs have been normalised to the mean IPI of the site. The lack of any clear repetition, or pattern, in this return map suggests that there is no periodicity in IPI sequences, at least on the time scale of the ISI (Thurley et al, 2011).

Such an analysis has been done for traces containing more than 40 (SH-SY5Y cells) or 60 (HEK 293 cells) successive puffs (Thurley et al, 2011). Each data set was collected over a time period equivalent to at least two ISIs. It is very difficult to measure longer puff sequences. Figure 4.5 shows typical examples, whence it can be seen that the return map shows no obvious repetition. It remains possible that puff sequences are periodic on time scales much greater than the ISI, but such a periodicity would, presumably, be unrelated to the mechanism generating the ISI, and thus of no interest to us here. Such long periodicity is, in any case, unlikely.

The combination of T_{avp}-σ_p relations, IPI distributions, and the fact that successive IPIs are uncorrelated, provides strong support for the conclusion that Ca^{2+} puffs are not periodic – neither on the time scale of the mean IPI nor on the time scale of the mean ISI – and have recovery times that are much faster than global Ca^{2+} spikes.

This conclusion is supported by theoretical investigations. Local $[Ca^{2+}]$ inside clusters is in the range of 20–200 μM, much larger than the dissociation constants of the regulatory binding sites on the IPR (Falcke, 2004; Thul and Falcke, 2004a; Bentele and Falcke, 2007). Hence, in a deterministic model, these binding sites are either empty at resting $[Ca^{2+}]$ or saturated as soon as a channel opens. The system is monostable or bistable for almost all parameter values (Thul and Falcke, 2004b, 2007). In the tiny parameter range where oscillations occur, their amplitude is too small and their period too short in comparison to Ca^{2+} spikes. The large local concentration changes turn the channel state dynamics essentially into a bistable system, which exhibits state changes only if it is stochastic. Hence, there is no oscillatory regime of the local dynamics of CICR that can explain Ca^{2+} spiking. This result does not depend on the details of channel state dynamics, and consequently applies to all models with dissociation constants for Ca^{2+} binding sites much smaller than the local $[Ca^{2+}]$ at the mouth of an open channel.

The mean IPI of an ensemble of puff sites depends on the number of channels in each cluster, and this dependence can be well fit by a single linear function. Hence, the number of channels is the parameter that most strongly affects the IPI. Other parameters that distinguish individual puff sites in SH-SY5Y cells do not strongly affect the mean IPI (Dickinson et al, 2012).

Dickinson et al (2012) also characterised how puffs depend on $[IP_3]$, which they controlled by releasing caged IP_3 by UV flashes. It is generally assumed that $[IP_3]$ is proportional to the flash duration. Fig. 4.6 shows their results, together with fits to functions which are purely heuristic, not based on underlying biophysical mechanisms. The maximal number of open channels in a puff depends on the flash duration in a cooperative manner, and the relationship can be well fit by a Hill function with Hill coefficient 4. This could possibly be explained by the tetrameric structure of the IPR and is a fit of satisfying quality and simplicity.

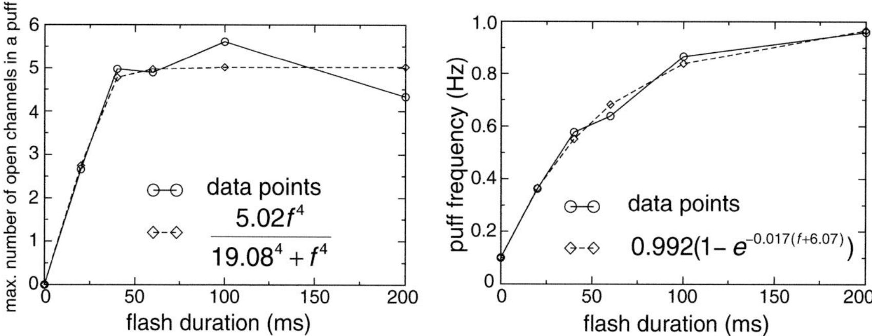

Fig. 4.6 Puff characteristics in SH-SY5Y cells measured by Dickinson et al (2012). $[IP_3]$ is set by the duration of UV flashes, which uncage photoreleasable IP_3. Solid lines and circles show the experimental results, diamonds and dashed lines show the fits to the equations in the legends, where f denotes the flash duration in ms. The offset in flash duration in the right panel improves the fit, and may correspond to a basal $[IP_3]$.

4.3 Properties of Sequences of Cellular Spikes

Many different cell types exhibit sequences of Ca^{2+} spikes (Rooney et al, 1989; Berridge, 1990; Capite et al, 2009; Dupont et al, 2011a; Falcke, 2004). Figure 4.7 shows examples from five different cell types and six different pathways. The interspike interval (ISI) is the time difference between the maxima of two consecutive spikes. The sequences of ISIs are also shown in Fig. 4.7.

For each of the cell types or pathways in Fig. 4.7, we plot the standard deviation of ISI sequences, σ, against the mean, T_{av}. Results are shown in Fig. 4.8. There are a few prominent features common to all the results:

1. σ is of the same order of magnitude as T_{av} for many cells.
2. σ increases with T_{av}. The relation can be well approximated by a linear function (Skupin et al, 2008)

$$\sigma = \alpha\,(T_{\mathrm{av}} - T_{\mathrm{min}}) + \sigma_{\mathrm{min}}. \tag{4.3}$$

   The slope α of the linear relation depends on the cell type and the agonist used to stimulate the spiking.
3. There is a minimal mean ISI, T_{min}, corresponding to an absolute refractory period, dividing the ISI into a deterministic part, T_{min}, and a stochastic part, ISI $- T_{\mathrm{min}}$.

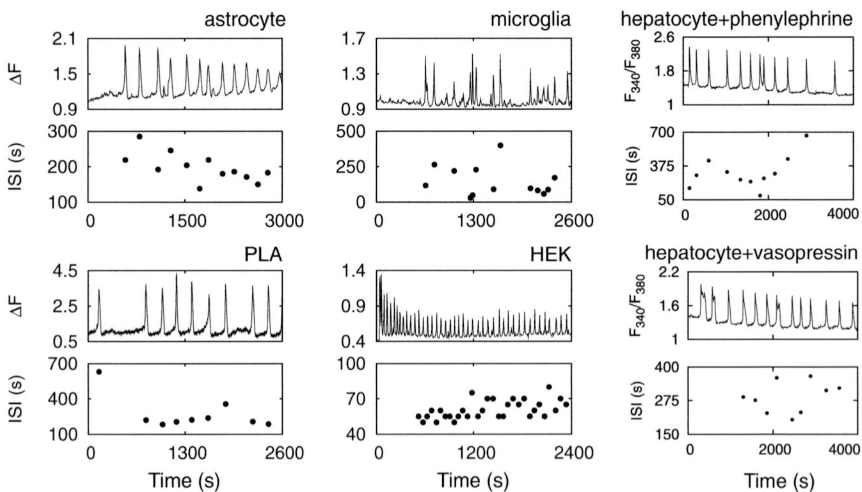

Fig. 4.7 Spike sequences for a variety of cell types and two different pathways in hepatocytes. The upper panel of each pair shows either the fluorescence change or the fluorescence ratio, thus indicating the cytosolic $[Ca^{2+}]$. The lower panel of each pair shows the ISI following the spike that occurs at the same time as the symbol.

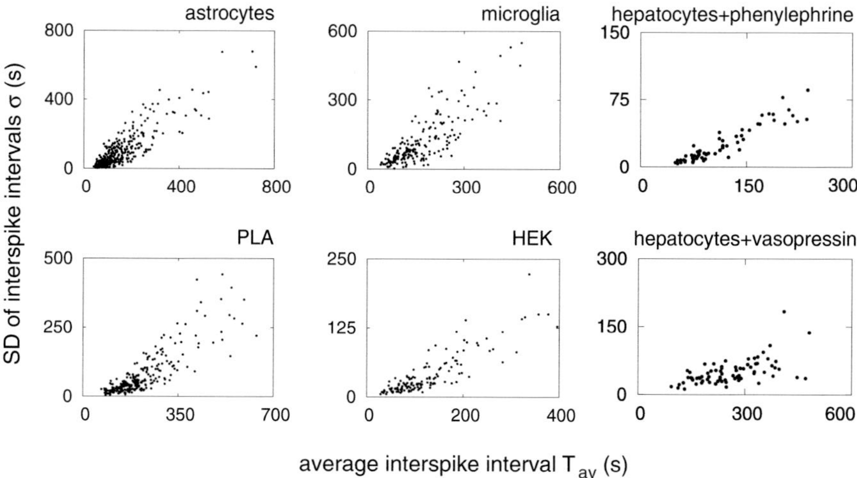

Fig. 4.8 The relation between the mean ISI, T_{av}, and the standard deviation σ. Each data point is the result of the analysis of one spike sequence.

Each data point in Fig. 4.8 corresponds to one cell. The individual cells have diverse values for such parameters as cell size, protein concentrations, resting $[\mathrm{Ca}^{2+}]$, and the number of IPR clusters. Nonetheless, we find a relation between σ and T_{av}, one that is robust with respect to changes of many parameters.

The existence of this robust relationship also suggests that spiking is a stochastic process. The above cases with a slope equal to 1 correspond to a Poisson process, while a slope smaller than 1 could arise from an inhomogeneous Poisson process. If spiking were oscillatory dynamics perturbed by noise, we would expect that at least some of the properties setting the period were independent of the standard deviation, or other properties, of the noise. If such were the case, we would not expect to find such a well-defined relationship between standard deviation and mean.

4.3.1 Wave Nucleation

The mean ISI is several times longer than the mean IPI, and so how does the cellular time scale of the ISI emerge from puff dynamics? The generation of a cellular concentration spike has been observed in detail in *Xenopus* oocytes (Marchant et al, 1999; Marchant and Parker, 2001). Not every puff causes a global spike but several channel clusters have to open within a short time in order to activate all other IPR clusters. Ca^{2+} diffusing from the puff site that opens first may activate neighbouring clusters, in which case Ca^{2+} release can sweep like a wave through the whole cell. However, for this to occur there must be a sufficient number of puffs sufficiently frequently, and this happens only with some probability, not with certainty. The smaller this probability is, the more attempts are required to cause a global spike and the larger is the ISI/IPI ratio.

The initiation of such a wave of Ca^{2+} release is called wave nucleation. The minimal number of open clusters required to cause a global spike with very high probability is called the critical nucleus. Wave nucleation has also been observed as the spike generation mechanism in cells smaller than *Xenopus* oocytes (Rooney et al, 1990; Bootman et al, 1997b). In addition to nucleation, a small rise of the spatially averaged Ca^{2+} concentration may contribute to an increase of the IPR open probability and thus prime the cell for wave nucleation, or may cause global release which is not spatially coordinated (Marchant et al, 1999; Bootman et al, 1997b; Smith et al, 2009). In large cells, a rise of the basal $[Ca^{2+}]$ by increased puff frequency in a wave nucleation region is likely to be part of the nucleation process (Marchant et al, 1999). In small cells, the volume comprising the critical number of puff sites might even be comparable to the cell size. Ca^{2+} buffers decrease the spike generation probability in both cases, since they reduce diffusion of free Ca^{2+} and slow down the rise of the basal $[Ca^{2+}]$.

The spike generation probability is the product of two probabilities: the probability per unit time for the occurrence of a puff (puff probability) and the probability that this puff opens sufficiently many clusters for a global spike to occur (coupling probability). The puff probability is proportional to the number of clusters in the cell. It could therefore be very large in large cells (Falcke, 2003b) and render spike generation an essentially deterministic process. However, this is not what has been observed. The most likely reasons are that in large cells the cluster density is too low for wave nucleation to occur except at certain nucleation sites (Rooney et al, 1990; Marchant et al, 1999), while in small cells there are too few clusters (Smith et al, 2009). One possible exception to this general rule is Ca^{2+} oscillations in airway smooth muscle, which occur with high frequency, and without any apparent Ca^{2+} puffs (Perez and Sanderson, 2005b; Cao et al, 2014). It is hypothesised that such behaviour is the result of a cell with a high density of IPR clusters, all tightly coupled, so that the firing of a single puff site will nearly always initiate a whole-cell response. In such cases, the whole-cell response is governed by the cluster dynamics.

4.3.2 The Effects of Buffers on Wave Nucleation

If spike generation is an inhomogeneous Poisson process determined by the spike generation probability, decreasing that probability should increase to a similar extent both the mean ISI and the standard deviation. If we reduce the spike generation probability by decreasing the coupling probability, we can also investigate the role of wave nucleation versus the basal $[Ca^{2+}]$. If buffers control the spike generation probability via the basal $[Ca^{2+}]$ and the puff probability, time scales should be proportional to the total buffer capacity according to the rapid buffer approximation (Section 2.9). However, the coupling probability depends sensitively on the strength of the spatial coupling by Ca^{2+} diffusion, and thus if buffers act on the spike generation probability by reducing diffusion of free Ca^{2+}, the response could be substantially more sensitive (Thul and Falcke, 2004a).

Results of experiments changing the coupling probability by additional buffer loading are shown in Fig. 4.9. After initial loading with Ca^{2+} indicators, which also act as buffers with fast on and off rates for Ca^{2+}, a first time series was recorded, thus providing reference values for T_{av} and σ. Then, additional buffer was loaded as indicated.

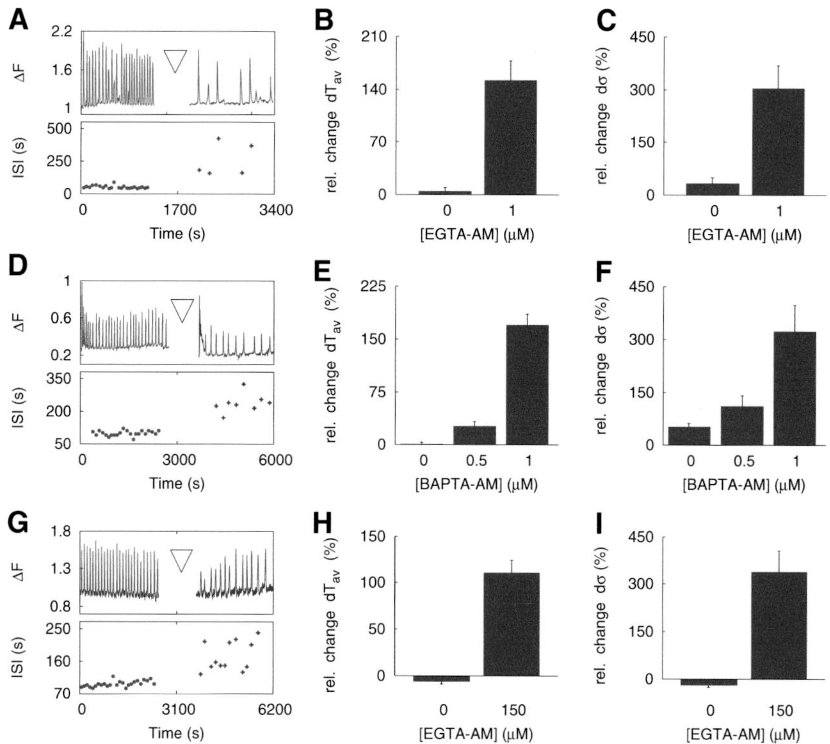

Fig. 4.9 Ca^{2+} buffers decrease the spike generation probability. Results show Ca^{2+} spikes in astrocytes (A-C), HEK 293 cells (D-F), and PLA cells (G-I). Red curves are before loading with additional buffer, blue curves are after. For the 50% of cells that resumed Ca^{2+} spiking, changes in T_{av} and σ are shown relative to the values obtained before incubation with the Ca^{2+} buffer. For all three cell types, the relative increase of σ is larger than for T_{av}, since T_{min} was not excluded from the calculation of the relative change in T_{av}. Reproduced from Skupin et al (2008), Fig. 6, with permission from Elsevier.

A second spike sequence was measured after the additional loading. The additional buffer loading caused a substantial increase of both T_{av} and σ in all three cell types (Fig. 4.9 B, C, E, F, H, I), and this increase was dependent on the amount of buffer loading (Fig. 4.9 E, F). The mean ISI depends more sensitively on buffer concentration than could be explained by its effect on basal $[Ca^{2+}]$ alone (Skupin et al, 2008). Hence, wave nucleation appears to be essential for spike generation also in small cells like HEK cells.

Experiments like those in Fig. 4.9 provide information on the T_{av}-σ relation of individual cells in which the data points are from experiments with the same cell. Table 4.1 compares the slopes of the population relations, in which the data points come from multiple cells under identical conditions, with the population average of the slope of the individual relations. All three values agree well. Hence, the additional buffer moves individual cells along the T_{av}-σ relation to larger T_{av} and σ values. This is consistent with spiking being caused by an inhomogeneous Poisson process the cumulant relation of which does not depend on the buffer concentration.

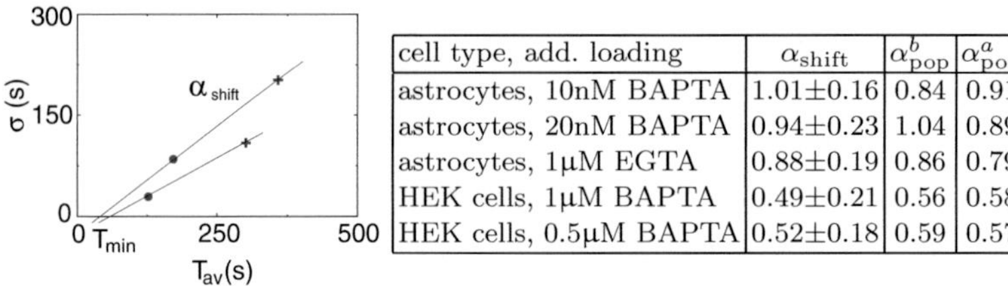

cell type, add. loading	α_{shift}	α^b_{pop}	α^a_{pop}
astrocytes, 10nM BAPTA	1.01±0.16	0.84	0.91
astrocytes, 20nM BAPTA	0.94±0.23	1.04	0.89
astrocytes, 1μM EGTA	0.88±0.19	0.86	0.79
HEK cells, 1μM BAPTA	0.49±0.21	0.56	0.58
HEK cells, 0.5μM BAPTA	0.52±0.18	0.59	0.57

Table 4.1 The T_{av}-σ relation can be calculated for individual cells, if time series have been measured with two different conditions. The red dot marks T_{av} and σ before additional buffer loading, the blue cross after it. The table shows a comparison between the average shifting slope α_{shift}, the population slopes α^b_{pop} before and α^a_{pop} after additional buffer loading.

The existence of the cumulant relation and the results of the buffer experiments suggest that the slope of the cumulant relation should be robust against cell-to-cell variability and consequently against changes of cellular parameters. The experiments reported in Fig. 4.10 investigate its robustness against three different pharmacological perturbations of the Ca^{2+} spiking pathway. All three perturbations affect T_{av} but do not significantly change α, the slope of the linear relation.

4.3.3 Information Content and Signal Encoding

The α values (i.e., the slope of the T_{av}-σ line) of different pathways are significantly different. A statistical measure of the possibility for Ca^{2+} effectors to discriminate between ISI sequences of different pathways is the Kullback

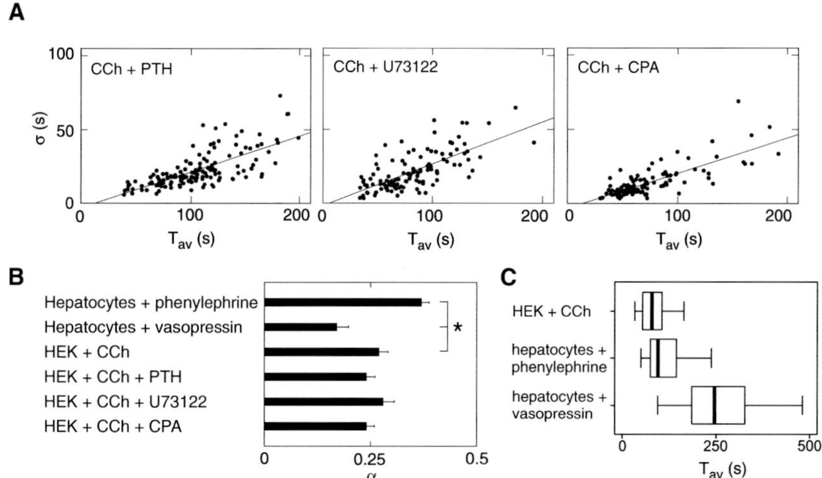

Fig. 4.10 Robust cumulant relations of stochastic Ca^{2+} signals. **A:** T_{av}-σ relations for HEK 293 cells stimulated with CCh in the presence of PTH (which increases the IPR sensitivity to IP_3), U73122 (which blocks PLC) or CPA (which blocks the SERCA pumps). **B:** The slopes of the T_{av}-σ relation differ between cell types and stimuli, but with CCh as the stimulus, PTH, U73122, or CPA do not significantly change the slopes, even though all three perturbations make a significant change to T_{av}. **C:** T_{av} varies considerably from cell to cell. Box plots of T_{av} from HEK 293 cells stimulated with CCh, hepatocytes stimulated with phenylephrine or vasopressin. Bold lines indicate medians, boxes show interquartile ranges, and whiskers show minima and maxima.

entropy. It also measures the information content of a given time series in comparison to a reference series. The value of α of ISI sequences of spontaneously spiking cells like astrocytes and microglia is 1, while stimulated spiking shows values smaller than 1. That suggests the choice of the homogeneous Poisson process of spontaneous spiking as the reference process. Skupin and Falcke (2007) show that stimulated spike sequences are more easily distinguishable from spontaneous ones (and can transmit more information) the smaller the value of α is. This is as one would intuitively expect, given that $\alpha = 0$ denotes a purely deterministic process, while $\alpha = 1$ denotes a random process.

If there is any functional relevance of the mean spike frequency for Ca^{2+} effectors, α^{-1} can be considered as the signal-to-noise ratio or at least determining the signal-to-noise ratio. Hence, stimulated spike sequences have a larger signal-to-noise ratio than spontaneous sequences. Information transmission and the signal-to-noise ratio illustrate why a small value of α is valuable for cells and why it is protected by this extraordinary robustness.

Temporal randomness of spikes is not the only impediment effectors face when reading Ca^{2+} signals. They also have to deal with large cell-to-cell variability of T_{av}, as shown in Fig. 4.10 C. The spread in T_{av} values can be almost one order of magnitude. It is also illustrated by the range of T_{av} values in Fig. 4.8. Hence, there is no consistent relation between T_{av} and stimulus concentration. Consequently, T_{av} and the mean frequency are not good signal carriers.

The parameters σ and T_{av} are two different time scales of an individual cell. The robustness of the ratio $\sigma/(T_{\mathrm{av}} - T_{\mathrm{min}})$ strongly suggests that the time scales of individual cells are distinguished by a single factor and ratios of two time scales of a cell are not cell specific and therefore a good signal carrier. That applies also to values of $T_{\mathrm{av}} - T_{\mathrm{min}}$ measured at different stimulus concentrations. Figure 4.11 A shows the spike sequence from a paired stimulation experiment providing values of T_{av} at two different stimulations (T_{av1}, T_{av2}). We analyse the ratio

$$\beta = \frac{[T_{\mathrm{av1}} - T_{\mathrm{min}}] - [T_{\mathrm{av2}} - T_{\mathrm{min}}]}{T_{\mathrm{av1}} - T_{\mathrm{min}}} = \frac{T_{\mathrm{av1}} - T_{\mathrm{av2}}}{T_{\mathrm{av1}} - T_{\mathrm{min}}}. \tag{4.4}$$

This equation describes the relative change of the mean stochastic part of the mean ISI, and is more conveniently written as

$$T_{\mathrm{av1}} - T_{\mathrm{av2}} = \Delta T_{\mathrm{av}} = \beta \left[T_{\mathrm{av1}} - T_{\mathrm{min}}\right]. \tag{4.5}$$

We call (4.5) an encoding relation. Examples of this relation for hepatocytes and HEK 293 cells are shown in Fig. 4.11 B and C. The quality of the linear fit is assessed by Pearson's correlation coefficient and explained uncertainty (Fig. 4.11 D and E; see Tippmann et al (2012) for details). Both are equal to 1 for a perfect fit. The linear relation (4.5) is confirmed for all stimulation steps except the smallest one. Here, to be determined sufficiently precisely, the small change of T_{av} at these parameters requires longer spike sequences than can be obtained in the paired stimulation experiments. Nevertheless the data point fits very well into the concentration-response curve below.

The information carrier – the variable encoding the stimulus concentration – should ideally be equal for all cells. The property which is the same for all cells in a given paired stimulation experiment is the value of β. Figure 4.11 E shows relative deviations as coefficients of variation of the data points from the population average, T_{pop2}, of T_{av2}, $\mathrm{CV}(\mathrm{T}_{\mathrm{pop2}})$, and the encoding relation $\mathrm{CV}(\beta)$. The latter was calculated as the root-mean-squared orthogonal distance of the data points from the form $T_{\mathrm{av2}} = (1-\beta)T_{\mathrm{av1}} + \beta T_{\mathrm{min}}$ of the encoding relation divided by T_{av2}. The relative deviation from the encoding relation is smaller than from $\mathrm{T}_{\mathrm{pop2}}$, i.e., β represents the behaviour of the cell population better than T_{pop}. The experimental results show that the relative change is the same for all cells in the experiment.

Another prerequisite for being the information carrier is a unique dependence on the stimulation step. In general, β may depend on the starting value of [CCh], as well as on the concentration step, Δ[CCh]. The relation between β and Δ[CCh] is shown in Figure 4.11 F. It is clearly unique. For greater generality, we now denote the stimulus concentration by S, and the

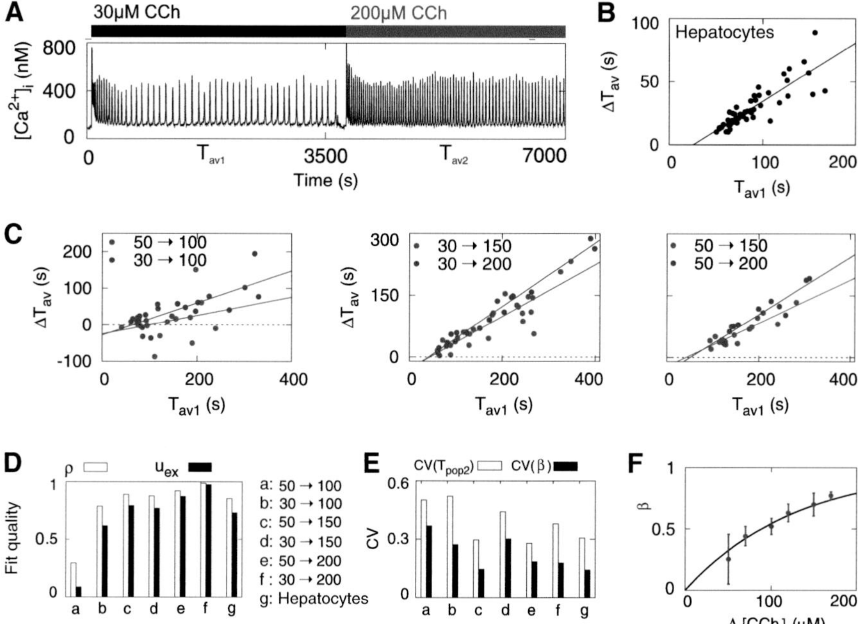

Fig. 4.11 Paired stimulation experiments. **A:** Ca^{2+} spikes in a HEK 293 cell. Similar experiments were used to establish T_{av1} and ΔT_{av} for (**B**) hepatocytes stimulated with 0.6 µM and then 1 µM phenylephrine and (**C**) HEK cells stimulated with the indicated CCh concentrations (µM). **D:** Pearson's correlation coefficients (ρ) and explained uncertainties (u_{ex}) for $T_{\text{av1}} - \Delta T_{\text{av}}$ relations for HEK cells stimulated with the indicated steps in CCh concentration (µM), or hepatocytes stimulated as in B. The code applies to **D** and **E**. **E:** Comparison of the average deviation of individual cell behaviour from the population average T_{pop} of $T_{\text{av}}(\text{CV}(T_{\text{pop}}))$ and the encoding relation ($\text{CV}(\beta)$) for the paired stimulation protocols. **F:** Relative changes calculated from slopes of $T_{\text{av1}} - \Delta T_{\text{av}}$ relations for all steps in CCh concentration, colour-coded to indicate initial CCh concentration (red 30 µM, blue 50 µM). Line shows the exponential relationship between the relative change and Δ[CCh], with γ being the only fit parameter.

concentration step by ΔS, and slightly rewrite (4.4) to get

$$\begin{aligned} &[T_{\text{av}}(S+\Delta S) - T_{\text{min}}] - [T_{\text{av}}(S) - T_{\text{min}}] \\ &= -\beta(S,\Delta S)\,[T_{\text{av1}} - T_{\text{min}}] \\ &\approx -\left[\beta\,(S,0) + \left.\frac{\partial\beta\,(S,\Delta S)}{\partial \Delta S}\right|_{\Delta S=0} \Delta S\right] [T_{\text{av1}} - T_{\text{min}}] \end{aligned} \tag{4.6}$$

The first term in the Taylor expansion of β is 0 according to (4.4). In the limit as $\Delta S \to 0$ we obtain

$$\frac{d\,[T_{\text{av}}(S) - T_{\text{min}}]}{dS} = -\gamma(S)\,[T_{\text{av}}(S) - T_{\text{min}}]\,, \tag{4.7}$$

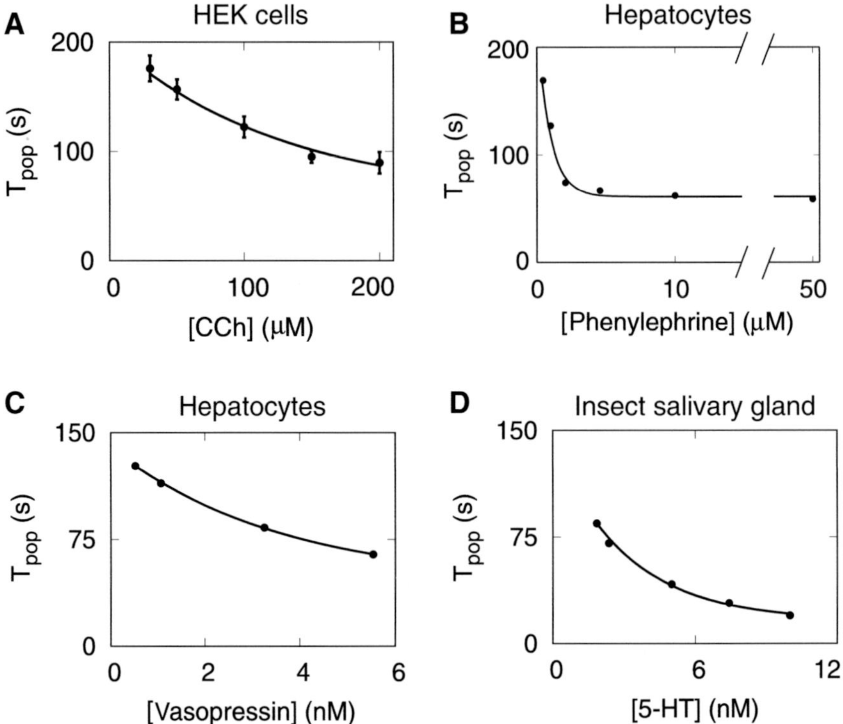

Fig. 4.12 Universal exponential concentration-response relation. **A:** Population average (T_{pop}) of T_{av} for HEK 293 cells at each CCh concentration (data from Fig. 4.11 C). Line drawn using the parameter $\gamma = 7.84\ \mathrm{mM}^{-1}$ (value from the fit in Fig. 4.11 F), with $T_{\mathrm{min}} = 57\,\mathrm{s}$ (average value from Fig. 4.11 C), but with no additional curve fitting. **B-D:** Hepatocytes (Rooney et al, 1989) stimulated with phenylephrine (B) or vasopressin (C), and insect salivary gland (Rapp and Berridge, 1981) stimulated with 5-HT (D) also show an exponential relation between stimulus intensity and T_{pop}. Lines are best fits in parameters T_{min} and γ to (4.10).

$$\gamma(S) = \left.\frac{\partial \beta\,(S, \Delta S)}{\partial \Delta S}\right|_{\Delta S=0}. \tag{4.8}$$

The solution of this equation is

$$\begin{aligned} T_{\mathrm{av}}(S) &= \left[T_{\mathrm{av}}\left(S_{\mathrm{ref}}\right) - T_{\mathrm{min}}\right] e^{-\int_{S_{\mathrm{ref}}}^{S} dS' \gamma(S')} + T_{\mathrm{min}}, \\ \beta(S_{\mathrm{ref}}, \Delta S) &= 1 - e^{-\int_{S_{\mathrm{ref}}}^{S} dS' \gamma(S')}. \end{aligned} \tag{4.9}$$

We introduced the reference concentration S_{ref} here, and $\Delta S = S - S_{\mathrm{ref}}$. Equation (4.9) describes the type of concentration-response relation resulting from a relative change like (4.4). Since β is the same for all cells in the experiment, γ is also. T_{min} was also found to be independent of the stimulation S (Thurley et al, 2014). The large cell-to-cell variability enters via the pre-factor of the exponential function in (4.9). This equation implies also that the population average exhibits the same concentration dependency as each individual cell, merely with a different pre-factor.

Red data points in Fig. 4.11 F represent steps starting at 30 μM [CCh], and blue points started at 50 μM. All data points are on the same curve. Hence, the dependence of β on Δ[CCh] is not affected by the starting value of the stimulation step and β does not depend on [CCh]. Consequently, γ is a constant and the same, not only for all cells in the experiment, but also for all cells of the same type and stimulated with the same agonist. It thus follows that

$$T_{\mathrm{av}}(S) = \left[T_{\mathrm{av}}\left(S_{\mathrm{ref}}\right) - T_{\mathrm{min}}\right] e^{-\gamma(S-S_{\mathrm{ref}})} + T_{\mathrm{min}}, \tag{4.10}$$

$$\beta(\Delta S) = 1 - e^{-\gamma \Delta S}. \tag{4.11}$$

Equation (4.11) is shown in Fig. 4.11 F and fits the data well. Remarkably, it is the average stochastic part $T_{\mathrm{av}} - T_{\mathrm{min}}$ of the mean ISI which is controlled by stimulation.

On the basis of the properties of the concentration-response relation of the population average mentioned above, we can deduce the type of single-cell dependency from population averages. If the population average can be fitted by a single exponential, all individual cells in the sample obey an exponential concentration-response curve.

Fits of single exponentials to experimental data are shown in Fig. 4.12. They fit the data well. We conclude that the cell types and pathways shown in Fig. 4.12 obey an exponential concentration-response curve and consequently also show relative change encoding of stimulus concentration steps both for individual cells and the population average. This is despite the biochemical differences of the pathways upstream from PLC, differences that show up in the value of γ, which is pathway specific.

4.3.4 Summary

In summary, IP_3-induced spike sequences show an absolute refractory period T_{min} followed by a random waiting time till the next spike. Usually, the stochastic part of the ISI is of the same order of magnitude as the refractory period, but at high stimulation it is short and sequences appear to be regular with an ISI approximately equal to T_{min}. The mean ISI, T_{av}, exhibits large cell-to-cell variability within a population of cells under identical conditions. The relation between the standard deviation of ISI and T_{av} is linear, its slope, α, is pathway specific and robust against cell-to-cell variability and perturbations of many parameters.

Ca^{2+} spiking encodes stimulus concentration steps by relative changes of the average stochastic part of the ISI. The relative change does not depend on the initial concentration of the step. That entails an exponential concentration-response relation, both for individual cells and the population average. Cell-to-cell variability is reflected by the pre-factor of the exponential. The factor of the stimulus concentration in the exponent γ is pathway specific, but the same for all individual cells of a given type using the pathway.

Remarkably, within the temporal randomness and cell-to-cell variability we find three conserved properties: $T_{\min}$, α, and γ. Two of them are related to stochasticity; α describes how fluctuations grow with the mean, and γ describes how the spike probability is controlled by stimulation.

The fact that the information in the Ca^{2+} signal is carried by a relative change in the ISI can perhaps be most easily appreciated by the analogy to music. When we listen to a tune, our ears and brain detect, not the frequency of the notes per se, but the relative changes in frequency as the tune is played; if the tune starts at 440 Hz, the melody sounds the same to us as if it started on 450 Hz, as the relative changes in frequency are the same, even though the actual frequencies are different. Indeed, most people would be unable to place the starting frequency on an absolute scale. To a large extent, we are simply not designed to detect absolute frequencies, only relative changes. The same is true of the visual system, which is designed to detect, not absolute light levels but changes in light intensity.

It is intriguing that Ca^{2+} signalling seems also to be designed to transmit, not the "pitch" of the signal, but only the "melody".

4.4 Appendix: An Incomplete Theory of Calcium Spiking

In this appendix we present a partial theoretical explanation for how the ISI distribution can be derived, step by step, from underlying, low-level mechanisms. Although it is necessarily at a much higher level of mathematical difficulty than the rest of this book, it can simply be skipped by readers less interested in the technical details, without affecting their ability to read the remainder of the book. Those readers more interested in the general conclusions of the partial theory, but unwilling to work through the myriad details, should simply read the summary in Section 4.4.5.

A complete theory of Ca^{2+} spiking starts from the molecular properties of IPR, the ideas on cluster size, cluster distributions, and cytosolic and luminal Ca^{2+} concentrations, from which it calculates the ISI distribution. In the spirit of hierarchical modelling, it first calculates puff duration, amplitude, and IPI distributions from IPR properties and then ISI distributions from puff properties. Since the response of the local $[Ca^{2+}]$ is much faster than channel state changes (Thul and Falcke, 2004a; Bentele and Falcke, 2007), concentration dynamics can be eliminated by a quasi-steady approximation when considering the cluster level. The master equation for the cluster states is the starting point for the calculation of distributions of puff characteristics (Thul and Falcke, 2004b, 2007; Higgins et al, 2009; Cao et al, 2013).

Here, we will start on the level of channel clusters. Puff property distribution functions are the input of our theory. A stochastic description is particularly simple if the system exhibits Markovian transition probabilities, which are determined only by the current state of the system. However, the open state of a cluster as well as the closed state consists of many substates. That leads to a violation of the Markov property (Thul and Falcke, 2004b, 2007; Higgins et al, 2009). We thus have to take into account that neither the closing probability per unit time nor the opening probability is constant but depend on the time elapsed since the beginning of a puff (closing) and the end of the last puff (opening).

4.4.1 Semi-Markov Processes

We use the concept of semi-Markov processes to formulate a theory for spike generation. By definition, the stochastic process $X_n(t)$ has the finite sample space $\Omega = \{0, 1, ..., S\}$ consisting of all states the process can visit. By $T_n \in \mathbb{R}^+$ we denote the sequence of transition times between these states with $T_0 < T_1 < ... < T_n$. We can now formulate transition probabilities according to

$$Q_{i,j}(t) = \mathrm{P}\{X_{n+1} = j, T_{n+1} - T_n \leq t | X_n = i\}. \tag{4.12}$$

The process is temporally homogeneous since $Q_{i,j}(t)$ is independent of n, and we also assume $Q_{i,j}(0) = 0$ for $j \neq i$.

A process which is governed by (4.12) is called a semi-Markov process (Cinlar, 1969, 1975; Schlicht and Winkler, 2008; Cox, 1970), with $Q = \{Q_{i,j}(t); i, j \in \Omega, t \in \mathbb{R}^+\}$ forming a so-called semi-Markovian kernel. We will briefly summarise the reasoning behind that terminology. By defining

$$C_{i,j} = \lim_{t\to\infty} Q_{i,j}(t) \tag{4.13}$$

one gets the transition probabilities for the embedded Markov chain with normalisation condition $\sum_j^{N_i} C_{i,j} = 1$. In case of a Markov process with unconditioned constant transition rates $\lambda_{i,j}$ $j = 1, \ldots, N_i$

$$C_{i,j} = \frac{\lambda_{i,j}}{\sum_k^{N_i} \lambda_{i,k}} \tag{4.14}$$

holds. Now one can define the following probability distributions:

$$G_{i,j}(t) = \frac{Q_{i,j}(t)}{C_{i,j}} = \mathrm{P}\{T_{n+1} - T_n \leq t | X_n = i, X_{n+1} = j\}. \tag{4.15}$$

The successive visits of the process X_n form a Markov chain with transition probabilities $C_{i,j}$, whereas the length of the sojourn time intervals $[T_n, T_{n+1})$ are given by the distribution functions $G_{i,j}(t)$. If these distributions can be written as $G_{i,j}(t) = 1 - \mathrm{e}^{-\sum_j q_{i,j} t} \equiv G_i(t)$, than the process is a pure Markov process with rates $q_{i,j}$, the sojourn times are exponentially distributed and independent of the next state. In that case the process is memoryless, which means that

$$\mathrm{P}\{T_{n+1} - T_n > s + t | T_{n+1} > t\} = \mathrm{P}\{T_{n+1} - T_n > s\}, \tag{4.16}$$

for $s, t > 0$ applies.

In conclusion, a semi-Markov process (X_n, T_n) still fulfils the Markov property with respect to the subsequent state transitions, but allows arbitrary (with respect to $\sum_j \lim_{t\to\infty} Q_{i,j}(t) = 1$) sojourn time distributions and hence loses its memorylessness with respect to the transition times $T_{n+1} - T_n$. It is therefore the ideal framework to apply the desired non-exponential waiting times often found experimentally. They are now exactly defined by

$$\frac{d}{dt} Q_{i,j}(t) dt = \Psi_{i,j} dt = \mathrm{P}\{t < T_{n+1} - T_n < t + dt | X_{n+1} = j, X_n = i\}. \tag{4.17}$$

By noting that the transition probabilities of the embedded Markov chain are given by $p_{i,j} = \int_0^\infty \Psi_{i,j}(t) dt$, the semi-Markov process is completely defined by a set of conditioned waiting time densities. Also note that by definition of the transition probabilities in (4.12), the time variable t does not correspond to a systems time, but describes the wait after the last transition T_n.

4.4.2 Interpuff Interval Distributions, Puff Duration Distributions and Their Dependencies on Cellular Parameters

After the maximum number of open channels during a puff is reached at time t_{rise}, the individual channels in a cluster close with rate μ, which does not depend on the Ca^{2+} concentration and is a molecular property of the IPR (Smith and Parker, 2009). Here, we follow Smith and Parker (2009) and Dickinson and Parker (2013) and assume independent closing. The waiting time density ψ_c for closing of an IPR cluster with, on average, N_p channels involved in a puff follows from the independence as

$$\psi_c(t) = \begin{cases} 0, & t \le t_{\mathrm{rise}}, \\ N_p \mu e^{-\mu(t - t_{\mathrm{rise}})} (1 - e^{-\mu(t - t_{\mathrm{rise}})})^{N_p - 1}, & t > t_{\mathrm{rise}}. \end{cases} \tag{4.18}$$

The channel closing rate μ has recently been determined by TIRF microscopy in SH-SY5Y cells as $59\,\mathrm{s}^{-1}$ (Smith and Parker, 2009) and $47\,\mathrm{s}^{-1}$ (Dickinson and Parker, 2013). It may depend on the IPR subtype, but this has not yet been measured.

Due to CICR, the individual opening probabilities per unit time depend on the local Ca^{2+} concentration, which is a function of the number of already open clusters. Moreover, ψ_o depends on the number, N_{ch}, of IPR per cluster, and on [IP_3].

ψ_o was quantified in HEK 293 cells and SH-SY5Y cells by TIRF microscopy (Thurley et al, 2011) for resting cytosolic Ca^{2+}. The mean IPIs were between 2 s and 5 s (SH-SY5Y cells) and between 0.5 s and 2 s (HEK cells). IPI histograms are well approximated by two-parameter probability distributions, indicating recovery times between successive opening events. Starting from the assumption that the probability of puff generation $\Lambda(t)$ relaxes from 0 just after a puff to an asymptotic value λ with rate ξ, it follows that

$$\Lambda(t) = \lambda\left(1 - e^{-\xi t}\right) \tag{4.19}$$

where t is the waiting time since the last puff. That leads to

$$\psi_o(t) = \Lambda(t) e^{-\int_0^t \Lambda(\theta) d\theta}. \tag{4.20}$$

The mean interpuff interval T_{avp} obeys

$$T_{\mathrm{avp}} = \frac{1}{\lambda} e^{\frac{\lambda}{\xi}} \left(\frac{\lambda}{\xi}\right)^{1-\frac{\lambda}{\xi}} \left[\Gamma\left(\frac{\lambda}{\xi}\right) - \Gamma\left(\frac{\lambda}{\xi}, \frac{\lambda}{\xi}\right)\right] \tag{4.21}$$

$$= \frac{1}{\lambda} F\left(\frac{\lambda}{\xi}\right). \tag{4.22}$$

The factor in square brackets is well approximated by the Γ-function for $\lambda/\xi > 4$.

$$\begin{aligned} T_{\mathrm{avp}} &= \frac{1}{2\lambda} e^{\frac{\lambda}{\xi}} \left(\frac{\lambda}{\xi}\right)^{1-\frac{\lambda}{\xi}} \Gamma\left(\frac{\lambda}{\xi}\right) \\ &= \frac{1}{2\lambda} e^{\frac{\lambda}{\xi}} \left(\frac{\lambda}{\xi}\right)^{-\frac{\lambda}{\xi}} \left(\frac{\lambda}{\xi}\right)! \\ &= \frac{1}{2\lambda} e^{\frac{\lambda}{\xi}} \left(\frac{\lambda}{\xi}\right) \sqrt{2\pi \frac{\lambda}{\xi} e^{-\frac{\lambda}{\xi}}} \left(\frac{\lambda}{\xi}\right)^{\frac{\lambda}{\xi}} \\ &= \frac{1}{\lambda} \sqrt{\frac{\pi}{2} \frac{\lambda}{\xi}}. \end{aligned} \tag{4.23}$$

We also find that $F(0) = 1$. Hence, we supplement the above approximation for $\lambda/\xi > 4$ by a constant term to get

$$T_{\mathrm{avp}} = \frac{1}{\lambda}\sqrt{1 + \frac{\pi}{2}\frac{\lambda}{\xi}}. \tag{4.24}$$

The second moment of the distribution is

$$\begin{aligned} T^2 &= \frac{1}{\lambda^2} e^{\frac{\lambda}{\xi}} {}_2F_2 \left[\left(\frac{\lambda}{\xi}, \frac{\lambda}{\xi} \right), \left(1 + \frac{\lambda}{\xi}, 1 + \frac{\lambda}{\xi} \right), -\frac{\lambda}{\xi} \right] \\ &\approx \frac{2}{\lambda^2} \left(1.0 + 1.1 \frac{\lambda}{\xi} \right), \end{aligned} \tag{4.25}$$

with ${}_2F_2$ being the generalised hypergeometric function. With these approximations we get

$$\begin{aligned} \sigma &= \sqrt{T^2 - T^2_{\mathrm{avp}}} \\ &\approx \frac{1}{\lambda}\sqrt{1 + \left(2.2 - \frac{\pi}{2}\right)\frac{\lambda}{\xi}}. \end{aligned} \tag{4.26}$$

The good quality of the approximations is shown in Fig. 4.13.

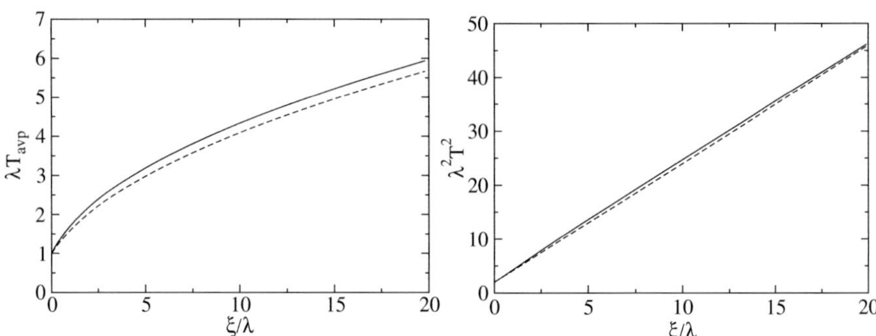

Fig. 4.13 The approximations for the moments in (4.24) and (4.25). The solid lines show the exact values, the dashed lines the approximations.

A drawback of this relaxation ansatz is that it exhibits only coefficients of variations (CVs) of IPI larger than 0.6329, the asymptotic limit for $\xi \to 0$. Hence, we also use generalised exponential (GE) distributions (Gupta and Kundu, 2007), because they are more convenient for operations with Laplace transforms and exhibit CVs of 0.19 and larger:

$$\psi_o(t) = n\lambda(1 - e^{-\lambda t})^{n-1} e^{-\lambda t}, \tag{4.27}$$

where $n \geq 1$ is the shape parameter and $\lambda > 0$ is the scale parameter. The mean and standard deviation of the generalised exponential distribution obey (Gupta and Kundu, 2007)

$$T_{\text{avp}} = \frac{1}{\lambda}\left[\psi(n+1) - \psi(1)\right], \tag{4.28}$$

$$\sigma = \frac{1}{\lambda}\sqrt{\psi'(1) - \psi'(n+1)}, \tag{4.29}$$

where ψ denotes the polygamma function and ψ' its derivative. The generalised exponential distribution has the additional advantage that its coefficient of variation for the interpuff interval depends only on n, which is convenient for determining n from experimental data. The larger n is, the later the generalised exponential function reaches its maximum. Hence, large n is analogous to slow recovery from negative feedback, i.e., it is analogous to small values of ξ.

Both distributions (4.20) and (4.27) were chosen for mathematical convenience, as they exhibit (almost) linear cumulant relations between the standard deviation and the mean IPI. In some cases we measured simple exponential distributions. That corresponds to $n = 1$ in (4.27) and to $\xi \gg \lambda_0$ in (4.20).

Recent data from TIRF microscopy revealed key properties of $\psi_o(t)$ (Thurley et al, 2011; Dickinson et al, 2012), but are not sufficient for a closed mathematical model, because the dependencies on $[Ca^{2+}]$ and $[IP_3]$ have not yet been completely established. Opening probabilities ψ_o of a cluster of IPR can be computed from single-channel models by solution of a master equation for about 2000 cluster states (Higgins et al, 2009) or simulations (Cao et al, 2013). We used this type of simulation to fit parameter dependencies of IPI distributions. In particular, we adopt an implementation of the De Young-Keizer model (De Young and Keizer, 1992; Higgins et al, 2009), a model that is presented in detail in Section 2.7.1.

We use the generalised exponential distribution to describe the dependency of IPI distribution parameters on $[Ca^{2+}]$ and $[IP_3]$. Based on these simulations, we assume simply for convenience that the transition from 0 to 1 open cluster obeys a pure exponential distribution with mean interpuff interval λ_0 as recorded for SH-SY5Y cells for typical values of cellular parameters. That assumption is at least not in contradiction with experimental results. Hence, for all clusters closed, $\alpha = 1$ in (4.27). For subsequent cluster openings ($1 \to 2$, $2 \to 3$,... open clusters), $\alpha > 1$ applies.

We are now in a position to write down the $\Psi_{i,j}$ for an array of clusters. The starting configuration S_l has N_o open clusters and N_{cl} closed clusters. The set of integers $\{h_j\}$, $j = 1, \ldots, N_o$, indexes the open clusters and the set $\{i_j\}$, $j = 1, \ldots, N_{\text{cl}}$, indexes the closed ones. In general, each cluster has its specific opening and closing waiting time distribution $\psi_o^i(t)$ and $\psi_c^i(t)$. Let k denote one of the closed clusters. The transition to the configuration with cluster k open (configuration S_m) and all other channels in the same state as in S_l obeys

$$\Psi_{l,m}(t) = \psi_o^k(t) \prod_{j=1, i_j \neq k}^{N_{cl}} \left(1 - \int_0^t d\tau \psi_o^{i_j}(\tau)\right) \prod_{j=1}^{N_o} \left(1 - \int_0^t d\tau \psi_c^{h_j}(\tau)\right). \tag{4.30}$$

It is the product of the probabilities that cluster k opens at t, none of the other closed clusters opens and none of the open clusters closes. For a linear chain with identical clusters, where the configurations are indexed by the number of open clusters, the expressions between neighbouring transitions are

$$\Psi_{N_o,N_o+1}(t) = N_{cl}\psi_o(t)\left(1-\int_0^t d\tau\psi_o(\tau)\right)^{N_{cl}-1}\left(1-\int_0^t d\tau\psi_c(\tau)\right)^{N_o},$$
$$\Psi_{N_o,N_o-1}(t) = N_o\psi_c(t)\left(1-\int_0^t d\tau\psi_o(\tau)\right)^{N_{cl}}\left(1-\int_0^t d\tau\psi_c(\tau)\right)^{N_o-1}.$$

Here, we took into account that the linear chain does not distinguish between individual clusters and therefore a configuration transition $S_{N_o} \to S_{N_o+1}$ happens by opening of any of the closed clusters.

4.4.3 Detailed Derivation of the First Passage Time Density

In a simplified view, the probability for a cellular spike is the probability for the occurrence of a puff multiplied by the probability that this puff causes a spike. This may happen by a variety of scenarios. If the removal of Ca^{2+} from the cytosol is slow, its concentration increases globally with each puff. The puff frequency may increase due to rising Ca^{2+} till it is so large that we observe global release. In this scenario, different puffs may not happen concurrently but still cause a spike. We could model puffs as point processes (van Kampen, 2001) (vanishing puff duration) and the onset of a spike as a critical value of the puff frequency. Unfortunately, this scenario has not been investigated yet except in simulations (Falcke, 2003b,a), and hence we will not discuss it here. It may be the scenario of spike initiation in some cell types (Marchant et al, 1999).

If the removal of Ca^{2+} from the cytosol is fast, the first puff will cause more puffs only during its existence. Puffs happen concurrently. A puff causes a spike if it causes a critical number N_{cr} of open clusters. Once this critical number has been reached, the probability for the transition to a cellular spike is close to 1. The waiting time distribution $F_{0,N}$ for the transition from no puff to N puffs, $N = N_{\mathrm{cr}}$, is then an approximation for the distribution of the stochastic part of the ISI. The critical number of open clusters might be much smaller than the total number of clusters in the cell. N_{cr} also depends on the $[IP_3]$, strength of spatial coupling by Ca^{2+} diffusion and other parameters.

The two scenarios are not mutually exclusive. A steady rise of puff frequency may be part of the emergence of a spike once N_{cr} has been reached. Or, even if the frequency rises due to puffs, it may still require a supercritical number of simultaneous puffs to cause a spike.

4.4.3.1 Calculations Based on the Laplace Transform of Waiting Time Distributions

An elegant way of obtaining an expression for the first passage time density, $F_{i,j}$, for a state transition $i \to j$ is the renewal equation (van Kampen, 2001). It formulates the general identity that the probability to start in state i at time 0 and to be in state j at time t with $i \neq j$ is equal to the probability to arrive there for the first time at any time between 0 and t and be again there at time t:

$$P_{i,j}(t) = \int_0^t d\tau F_{i,j}(t-\tau)P_{j,j}(\tau). \tag{4.31}$$

We solve this equation for $F_{i,j}(t)$. A standard technique for such integral equations with a convolution kernel uses the Laplace transform $\mathcal{L}\{f(t)\} = \int_0^\infty e^{-st} f(t)dt = \tilde{f}(s)$, with the convolution theorem

$$\mathcal{L}\{f(t) * g(t)\} = \mathcal{L}\{f(t)\}\mathcal{L}\{g(t)\}. \tag{4.32}$$

We obtain for the Laplace transform $\tilde{F}_{i,j}(s)$ of $F_{i,j}(t)$

$$\tilde{F}_{i,j}(s) = \frac{\tilde{P}_{i,j}(s)}{\tilde{P}_{j,j}(s)}. \tag{4.33}$$

The moments of the first passage time distribution are given by van Kampen (2001):

$$\langle t^n \rangle = (-1)^n \frac{\partial^n}{\partial s^n} \tilde{F}_{i,j}(s)|_{s=0}. \tag{4.34}$$

This equation yields for $n = 1$ the mean first passage time. Hence, we need to calculate $\tilde{P}_{i,j}(s)$ and $\tilde{P}_{j,j}(s)$.

The master equation for $P_{i,j}(t)$ is

$$\frac{dP_{i,j}(t)}{dt} = \sum_{l \neq j}^{N} I^i_{l,j}(t) - \sum_{l \neq j}^{N} I^i_{j,l}(t). \tag{4.35}$$

$I^i_{l,j}(t)$ denotes the probability flux due to transitions from state l to j under the condition that the process started at state i at time 0. The Laplace transform of the master equation is

$$s\tilde{P}_{i,j}(s) - \delta_{ij} = \sum_l \tilde{I}^i_{l,j}(s) - \sum_l \tilde{I}^i_{j,l}(s), \tag{4.36}$$

containing the Laplace transformed probability fluxes $\tilde{I}^j_{il}$. That entails the Laplace transform of the FPT density

$$\tilde{F}_{i,j}(s) = \frac{\sum_l \tilde{I}^i_{l,j}(s) - \sum_l \tilde{I}^i_{j,l}(s)}{\sum_l \tilde{I}^j_{l,j}(s) - \sum_l \tilde{I}^j_{j,l}(s) + 1}. \tag{4.37}$$

The fluxes obey (Prager et al, 2007):

$$I_{l,j}^{i}(t) = \int_0^t \Psi_{l,j}(t-\tau) \sum_{k}^{N_{in}} I_{k,l}^{i}(\tau) d\tau + f_{l,j}^{i}(t), \tag{4.38}$$

which is a convolution of the conditioned waiting time to go from state l to j and all incoming fluxes to the state l. The $f_{l,j}^{i}$ are the initial fluxes with $f_{l,j}^{i}(t) \equiv 0$ for $i \neq l$.

Equation (4.38) does not reflect the full complexity of spike dynamics. The equation assumes that transition probabilities depend only on the time since arrival in a state. Since state changes correspond to opening and closing of clusters, (4.38) assumes that the $\Psi_{l,j}(t)$ are determined by the time scale of puff dynamics. However, there is the additional time scale of recovery from negative feedback since the last spike. Including this time scale in (4.38) renders the derivation of algebraic equations for the Laplace transform of the FPT density difficult. Hence, we neglect it for the remaining calculations of this section. In section 4.4.3.2 we consider the complementary approximation, neglecting the recovery from the last puff in the IPI distribution but taking recovery from the last spike into account.

By writing the set of fluxes as a vector $\mathbf{I}^j(t)$ and by writing the conditioned waiting time densities in an appropriate matrix $\mathbf{\Psi}(t)$, the system of integral equations can be written as

$$\mathbf{I}^j(t) = \int_0^t \mathbf{\Psi}(t-\tau)\mathbf{I}^j(\tau)d\tau + \mathbf{f}^j(t), \tag{4.39}$$

with $\mathbf{f}^j(t)$ as vector of the initial functions. Equation (4.39) reads in Laplace space:

$$(\mathbb{1} - \tilde{\mathbf{\Psi}})\tilde{\mathbf{I}}^j = \tilde{\mathbf{f}}^j. \tag{4.40}$$

This is an inhomogeneous linear system of equations for the Laplace transformed fluxes $\tilde{I}_{il}$ which can be solved exactly by standard algebraic methods. The solution yields all Laplace transformed fluxes as functions of the Laplace transformed waiting time distributions. As can be seen in (4.37) for the FPT $\tilde{F}_{i,j}(s)$, there are two sets of fluxes needed, reflecting the initial states i or j, respectively. Inserting these fluxes into (4.37) gives the Laplace transformed FPT density $\tilde{F}_{i,j}(s)$.

In case of a linear chain with $N+1$ states (see Fig. 4.14), we write for the flux vector

$$\mathbf{I}^j = (I_{01}^j, I_{10}^j, I_{12}^j, ..., I_{N-1,N}^j, I_{N,N-1}^j), \tag{4.41}$$

and the matrix elements $\mathbf{\Psi}_{il}$ are determined by the set of equations given in (4.39) (see also the section below for an example). Equation 4.37 for the FPT simplifies to

$$\tilde{F}_{0,N} = \frac{\tilde{I}_{N-1,N}^0 - \tilde{I}_{N,N-1}^0}{\tilde{I}_{N-1,N}^N - \tilde{I}_{N,N-1}^N + 1}. \tag{4.42}$$

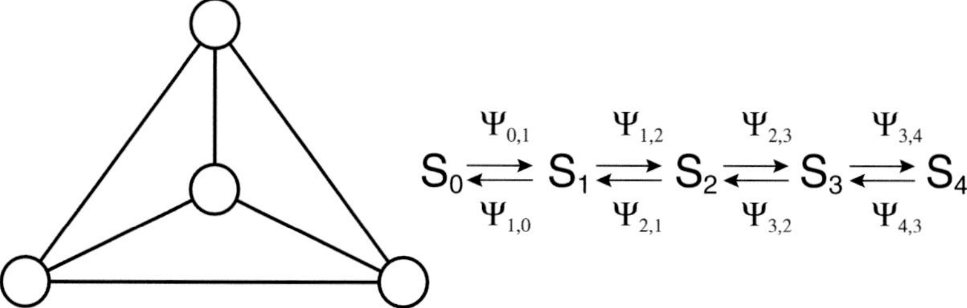

Fig. 4.14 The tetrahedron model of a cluster array. Clusters are arranged on the vertices of a tetrahedron and are identical. All configurations with n open clusters, $n = 0, \ldots, 4$, are equivalent. Hence, the state of the array is completely captured by the number of open clusters and the state graph is a linear chain. An array with more than four clusters having the same property does not exist. When we use a five or six cluster array in the calculation of the first passage time or splitting probability, we approximate the coupling strength in order to regain equivalence of states with equal number of open clusters.

For the stochastic tetrahedron cluster model (Fig 4.14) with $N = 4$, $\tilde{\mathbf{\Psi}}$ is

$$\tilde{\mathbf{\Psi}} = \begin{bmatrix} 0 & \tilde{\Psi}_{0,1} & 0 & 0 & 0 & 0 & 0 & 0 \\ \tilde{\Psi}_{1,0} & 0 & 0 & \tilde{\Psi}_{1,0} & 0 & 0 & 0 & 0 \\ \tilde{\Psi}_{1,2} & 0 & 0 & \tilde{\Psi}_{1,2} & 0 & 0 & 0 & 0 \\ 0 & 0 & \tilde{\Psi}_{2,1} & 0 & 0 & \tilde{\Psi}_{2,1} & 0 & 0 \\ 0 & 0 & \tilde{\Psi}_{2,3} & 0 & 0 & \tilde{\Psi}_{2,3} & 0 & 0 \\ 0 & 0 & 0 & 0 & \tilde{\Psi}_{3,2} & 0 & 0 & \tilde{\Psi}_{3,2} \\ 0 & 0 & 0 & 0 & \tilde{\Psi}_{3,4} & 0 & 0 & \tilde{\Psi}_{3,4} \\ 0 & 0 & 0 & 0 & 0 & 0 & \tilde{\Psi}_{4,3} & 0 \end{bmatrix} . \tag{4.43}$$

The vectors for the right-hand side of (4.40) are $\tilde{\mathbf{f}}^0 = \left\{\delta_{i0}\tilde{\Psi}_{0,1}\right\}$ and $\tilde{\mathbf{f}}^4 = \left\{\delta_{i8}\tilde{\Psi}_{4,3}\right\}$, $i = 1, \ldots, 8$. Using these vectors in (4.40), solving for $\mathbf{I}^0$ and $\mathbf{I}^4$ and inserting them into (4.42) yields the Laplace transform of the FPT:

$$\tilde{F}_{0,4} = \frac{\tilde{\Psi}_{0,1}\tilde{\Psi}_{1,2}\tilde{\Psi}_{2,3}\tilde{\Psi}_{3,4}}{1 - \tilde{\Psi}_{0,1}\tilde{\Psi}_{1,0} - \tilde{\Psi}_{1,2}\tilde{\Psi}_{2,1} - \tilde{\Psi}_{2,3}\tilde{\Psi}_{3,2} + \tilde{\Psi}_{0,1}\tilde{\Psi}_{1,0}\tilde{\Psi}_{2,3}\tilde{\Psi}_{3,2}} . \tag{4.44}$$

The results for $N = 2, 3$, and 5 are

$$\tilde{F}_{0,2} = \frac{\tilde{\Psi}_{0,1}\tilde{\Psi}_{1,2}}{1 - \tilde{\Psi}_{0,1}\tilde{\Psi}_{1,0}}$$

$$\tilde{F}_{0,3} = \frac{\tilde{\Psi}_{0,1}\tilde{\Psi}_{1,2}\tilde{\Psi}_{2,3}}{1 - \tilde{\Psi}_{0,1}\tilde{\Psi}_{1,0} - \tilde{\Psi}_{1,2}\tilde{\Psi}_{2,1}}$$

$$\tilde{F}_{0,5} = \frac{\tilde{\Psi}_{0,1}\tilde{\Psi}_{1,2}\tilde{\Psi}_{2,3}\tilde{\Psi}_{3,4}\tilde{\Psi}_{4,5}}{\begin{array}{l}1 - \tilde{\Psi}_{0,1}\tilde{\Psi}_{1,0} - \tilde{\Psi}_{1,2}\tilde{\Psi}_{2,1} - \tilde{\Psi}_{2,3}\tilde{\Psi}_{3,2} - \tilde{\Psi}_{3,4}\tilde{\Psi}_{4,3} \\ \quad + \tilde{\Psi}_{0,1}\tilde{\Psi}_{1,0}\left(\tilde{\Psi}_{2,3}\tilde{\Psi}_{3,2} + \tilde{\Psi}_{3,4}\tilde{\Psi}_{4,3}\right) + \tilde{\Psi}_{1,2}\tilde{\Psi}_{2,1}\tilde{\Psi}_{3,4}\tilde{\Psi}_{4,3}\end{array}}.$$

All of these expressions have the structure

$$F_{0,N} = \frac{\aleph(s)}{\Upsilon(s)}. \tag{4.45}$$

For example, we have with $N = 4$

$$\begin{aligned}\aleph &= \tilde{\Psi}_{0,1}\tilde{\Psi}_{1,2}\tilde{\Psi}_{2,3}\tilde{\Psi}_{3,4}, \\ \Upsilon &= 1 - \tilde{\Psi}_{0,1}\tilde{\Psi}_{1,0} - \tilde{\Psi}_{1,2}\tilde{\Psi}_{2,1} - \tilde{\Psi}_{2,3}\tilde{\Psi}_{3,2} + \tilde{\Psi}_{0,1}\tilde{\Psi}_{1,0}\tilde{\Psi}_{2,3}\tilde{\Psi}_{3,2}.\end{aligned}$$

In general, both $\aleph$ and Υ are linear in the $\tilde{\Psi}_{i,j}$ and the second derivatives of $\aleph$(s) and Υ(s) with respect to $\tilde{\Psi}_{i,j}$ vanish, $\aleph(0)=\Upsilon(0)$ since $\tilde{F}_{0,N}(0)=1$, and $\partial\aleph/\partial\tilde{\Psi}_{i,j} = \aleph/\tilde{\Psi}_{i,j}$. We obtain for the mean first passage time $T_{0,N}$

$$\begin{aligned}T_{0,N} &= \left[-\frac{1}{\Upsilon}\left(\sum_{i=0}^{N-1}\frac{\partial\aleph}{\partial\tilde{\Psi}_{i,i+1}}\tilde{\Psi}'_{i,i+1} - \frac{\aleph}{\Upsilon}\sum_{i=0}^{N-1}\sum_{j=0}^{N-1}\frac{\partial\Upsilon}{\partial\tilde{\Psi}_{i,j}}\tilde{\Psi}'_{i,j}\right)\right]\Bigg|_{s=0} \\ &= -\left(\sum_{i=0}^{N-1}\frac{\tilde{\Psi}'_{i,i+1}(0)}{\tilde{\Psi}_{i,i+1}(0)} - \frac{1}{\Upsilon(0)}\sum_{i=0}^{N-1}\sum_{j=0}^{N-1}\frac{\partial\Upsilon}{\partial\tilde{\Psi}_{i,j}}\Bigg|_{s=0}\tilde{\Psi}'_{i,j}(0)\right),\end{aligned} \tag{4.46}$$

and with $N = 4$

$$\begin{aligned}T_{0,4} = -\frac{1}{\Upsilon}\Bigg[&\left(\tilde{\Psi}_{1,2}\tilde{\Psi}_{2,3}\tilde{\Psi}_{3,4} - \tilde{\Psi}_{0,1}\tilde{\Psi}_{1,2}\tilde{\Psi}_{2,3}\tilde{\Psi}_{3,4}\frac{\tilde{\Psi}_{1,0}\tilde{\Psi}_{2,3}\tilde{\Psi}_{3,2} - \tilde{\Psi}_{1,0}}{\Upsilon}\right)\tilde{\Psi}'_{0,1} \\ &- \tilde{\Psi}_{0,1}\tilde{\Psi}_{1,2}\tilde{\Psi}_{2,3}\tilde{\Psi}_{3,4}\frac{\tilde{\Psi}_{0,1}\tilde{\Psi}_{2,3}\tilde{\Psi}_{3,2} - \tilde{\Psi}_{0,1}}{\Upsilon}\tilde{\Psi}'_{1,0} \\ &+ \left(\tilde{\Psi}_{0,1}\tilde{\Psi}_{2,3}\tilde{\Psi}_{3,4} + \tilde{\Psi}_{0,1}\tilde{\Psi}_{1,2}\tilde{\Psi}_{2,3}\tilde{\Psi}_{3,4}\frac{\tilde{\Psi}_{2,1}}{\Upsilon}\right)\tilde{\Psi}'_{1,2} \\ &+ \tilde{\Psi}_{0,1}\tilde{\Psi}_{1,2}\tilde{\Psi}_{2,3}\tilde{\Psi}_{3,4}\frac{\tilde{\Psi}_{1,2}}{\Upsilon}\tilde{\Psi}'_{2,1} \\ &+ \left(\tilde{\Psi}_{0,1}\tilde{\Psi}_{1,2}\tilde{\Psi}_{3,4} - \tilde{\Psi}_{0,1}\tilde{\Psi}_{1,2}\tilde{\Psi}_{2,3}\tilde{\Psi}_{3,4}\frac{\tilde{\Psi}_{0,1}\tilde{\Psi}_{1,0}\tilde{\Psi}_{3,2} - \tilde{\Psi}_{3,2}}{\Upsilon}\right)\tilde{\Psi}'_{2,3} \\ &- \tilde{\Psi}_{0,1}\tilde{\Psi}_{1,2}\tilde{\Psi}_{2,3}\tilde{\Psi}_{3,4}\frac{\tilde{\Psi}_{0,1}\tilde{\Psi}_{1,0}\tilde{\Psi}_{2,3} - \tilde{\Psi}_{2,3}}{\Upsilon}\tilde{\Psi}'_{3,2} + \tilde{\Psi}_{0,1}\tilde{\Psi}_{1,2}\tilde{\Psi}_{2,3}\tilde{\Psi}'_{3,4}\Bigg]\Bigg|_{s=0}.\end{aligned} \tag{4.47}$$

We can substantially simplify (4.47) with $C_{i,j} = \tilde{\Psi}_{i,j}(s=0)$, $C_{0,1} = 1$, $C_{i,i-1} + C_{i,i+1} = 1$ and the mean conditioned waiting time $T_{i,j} = \tilde{\Psi}'_{i,j}(s=0)$ to reach

$$\Upsilon(0) = C_{1,2}C_{2,3}C_{3,4} \tag{4.48}$$

and

$$\begin{aligned} T_{0,4} = \frac{1}{\Upsilon(0)} & \left[\left(\Upsilon(0) + C_{1,0}\left(1 - C_{2,3}C_{3,2}\right)\right) T_{0,1} \right. \\ & + \left(1 - C_{2,3}C_{3,2}\right)\left(T_{1,0} + T_{1,2}\right) + C_{1,2}\left(T_{2,1} + T_{2,3}\right) \\ & \left. + \, C_{1,2}C_{2,3}\left(T_{3,2} + T_{3,4}\right)\right]. \end{aligned} \tag{4.49}$$

Note that we did not have to specify the explicit form of the conditioned waiting times for the entire calculation up to here. The mean ISI is a weighted sum of mean conditioned dwell times.

$T_{0,1}$ is two orders of magnitude larger than all the other $T_{i,j}$. Hence, we continue by taking only terms with $T_{0,1}$ into account. We can rewrite the expression for T_{av} now like

$$\begin{aligned} T_{0,N} & \approx - \left[\left(\frac{1}{\Upsilon} \frac{\partial \aleph}{\partial \tilde{\Psi}_{0,1}} - \frac{\aleph}{\Upsilon^2} \frac{\partial \Upsilon}{\partial \tilde{\Psi}_{0,1}} \right) \tilde{\Psi}'_{1,0} \right]_{s=0} \\ & = \left(1 - \frac{1}{\Upsilon(0)} \frac{\partial \Upsilon}{\partial \tilde{\Psi}_{0,1}}(0) \right) T_{0,1} \end{aligned} \tag{4.50}$$

$$T_{0,4} \approx \left(1 + \frac{C_{1,0}\left(1 - C_{2,3}C_{3,2}\right)}{\Upsilon(0)} \right) T_{0,1}. \tag{4.51}$$

The second moment $T^{(2)}_{0,N}$ of the ISI can be derived with the same approximation to give

$$T^{(2)}_{0,N} = \left[\left(\frac{1}{\Upsilon} \frac{\partial \aleph}{\partial \tilde{\Psi}_{0,1}} - \frac{\aleph}{\Upsilon^2} \frac{\partial \Upsilon}{\partial \tilde{\Psi}_{0,1}} \right) \left(\tilde{\Psi}''_{1,0} - \frac{2}{\Upsilon} \frac{\partial \Upsilon}{\partial \tilde{\Psi}_{0,1}} \tilde{\Psi}'^2_{1,0} \right) \right]_{s=0}. \tag{4.52}$$

We define $\mathrm{CV}_{0,1}$ as the coefficient of variation of $\Psi_{0,1}$. We obtain for the CV of ISI

$$\begin{aligned} \mathrm{CV} & = \sqrt{ \left. \frac{ \mathrm{CV}^2_{0,1} - \frac{2}{\Upsilon} \frac{\partial \Upsilon}{\partial \tilde{\Psi}_{0,1}} + \frac{\aleph}{\Upsilon} \frac{\partial \Upsilon}{\partial \tilde{\Psi}_{0,1}} }{ 1 - \frac{1}{\Upsilon} \frac{\partial \Upsilon}{\partial \tilde{\Psi}_{0,1}} } \right|_{s=0} } \\ & = \sqrt{ \frac{ \mathrm{CV}^2_{0,1} - \left. \frac{1}{\Upsilon(0)} \frac{\partial \Upsilon}{\partial \tilde{\Psi}_{0,1}} \right|_{s=0} }{ 1 - \left. \frac{1}{\Upsilon(0)} \frac{\partial \Upsilon}{\partial \tilde{\Psi}_{0,1}} \right|_{s=0} } } \\ & = \sqrt{ \frac{ \mathrm{CV}^2_{0,1} + \frac{T_{0,N}}{T_{0,1}} - 1 }{ \frac{T_{0,N}}{T_{0,1}} } }. \end{aligned} \tag{4.53}$$

The CV does not depend on the details of the conditioned waiting time distributions $\Psi_{i,j}$. It depends on $\mathrm{CV}_{0,1}$ and the ratio of the stochastic part of the ISI $\mathrm{T}_{0,N}$ to the cellular IPI $\mathrm{T}_{0,1}$, if $\mathrm{T}_{0,1}$ is much larger than the other $\mathrm{T}_{i,j}$. The cellular IPI is approximately a single-site ISI divided by the number of clusters in the cell. Hence, $\mathrm{T}_{0,N}/\mathrm{T}_{0,1}$ is typically larger than 10 and the CV is close to 1 (see also Fig. 4.15). That applies to all values of $\mathrm{CV}_{0,1}$ and a range of N values and corresponds to the robustness of the CV observed in experiments. The CV of ISI is also 1 if $\mathrm{CV}_{0,1}$ is equal to 1. Interpuff interval distributions are often close to an exponential distribution (Thurley et al, 2011). Hence, $\mathrm{CV}_{0,1} \approx 1$ is true for many puff sites.

We look again at the example with $N = 4$

$$\mathrm{CV} = \sqrt{\frac{\mathrm{CV}_{0,1}^2 + \frac{C_{1,0}}{C_{1,2}}\left(1 + \frac{1}{C_{3,4}}\left(\frac{1}{C_{2,3}} - 1\right)\right)}{1 + \frac{C_{1,0}}{C_{1,2}}\left(1 + \frac{1}{C_{3,4}}\left(\frac{1}{C_{2,3}} - 1\right)\right)}}. \tag{4.54}$$

Despite the approximation of using the derivatives of $\tilde{\Psi}_{0,1}$ only, some information on the other transitions enters via the $C_{i,j}$. If the clusters are strongly coupled, almost each puff will cause the other sites to open also. Consequently, $C_{1,0}/C_{1,2} \ll 1$ and $T_{0,N} \approx T_{0,1}$ hold, and $\mathrm{CV} \approx \mathrm{CV}_{0,1}$ applies (Fig. 4.15). $C_{1,0}/C_{1,2}$ is much larger than 1 and $\mathrm{T}_{0,N}$ is much larger than $T_{0,1}$ for weak cluster coupling, which entails $\mathrm{CV} \approx 1$.

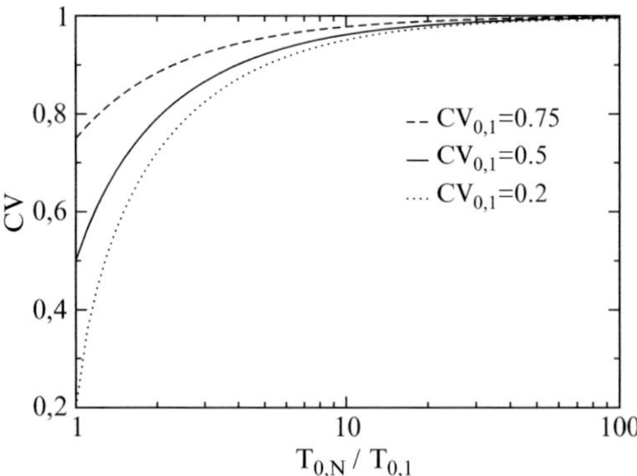

Fig. 4.15 The relation between the $\mathrm{CV}_{0,1}$ of IPIs ($\mathrm{T}_{0,1}$) and the CV of cellular ISIs ($\mathrm{T}_{0,N}$) in the approximation of (4.53).

The quality of the approximation in (4.50) and (4.53) can be checked by comparing it with exact calculations for the cases $N = 2, 3$, and 4. The formulas for the moments of $F_{0,N}$ are given in Section 4.4.4. We need to specify the $\Psi_{i,j}$ to that end. The distributions $\psi_o(t)$ and $\psi_c(t)$ obey (4.27) and (4.18), respectively, with the parameters λ, n, N_p, and μ. The data in Fig. 4.6 A can be used to calculate the number of maximally open channels

N_p during the puff for a given [IP$_3$]. The channel closing rate μ has been measured as $59\,\mathrm{s}^{-1}$ in SH-SY5Y cells (Smith and Parker, 2009). We used $\mu = 50\,\mathrm{s}^{-1}$ as the standard value.

The cluster opening rate λ is described by

$$\lambda = \lambda_0\left([\mathrm{IP}_3]\right) V_c^{\max} \frac{[\mathrm{Ca}^{2+}]^4}{k_c^4 + [\mathrm{Ca}^{2+}]^4} \tag{4.55}$$

$$V_c^{\max} = \left(\frac{(0.5k_c)^4}{k_c^4 + (0.5k_c)^4}\right)^{-1} \tag{4.56}$$

$$[\mathrm{Ca}^{2+}] = C_0 + N_o N_p C_1 \tag{4.57}$$

The function $\lambda_0\left([\mathrm{IP}_3]\right)$ is the fit of the puff frequency to experimental data in Fig. 4.6 B. The choice of $V_c^{\max}$ guarantees that λ coincides with these experimental data for $[\mathrm{Ca}^{2+}] = 0.5k_c$. Equation (4.57) describes the coupling by the Ca^{2+} concentration profiles around open clusters. The amplitude of these profiles is set proportional to the number of open channels in the cluster (N_p), and the profiles from different open clusters add up (N_o). C_0 is the resting concentration.

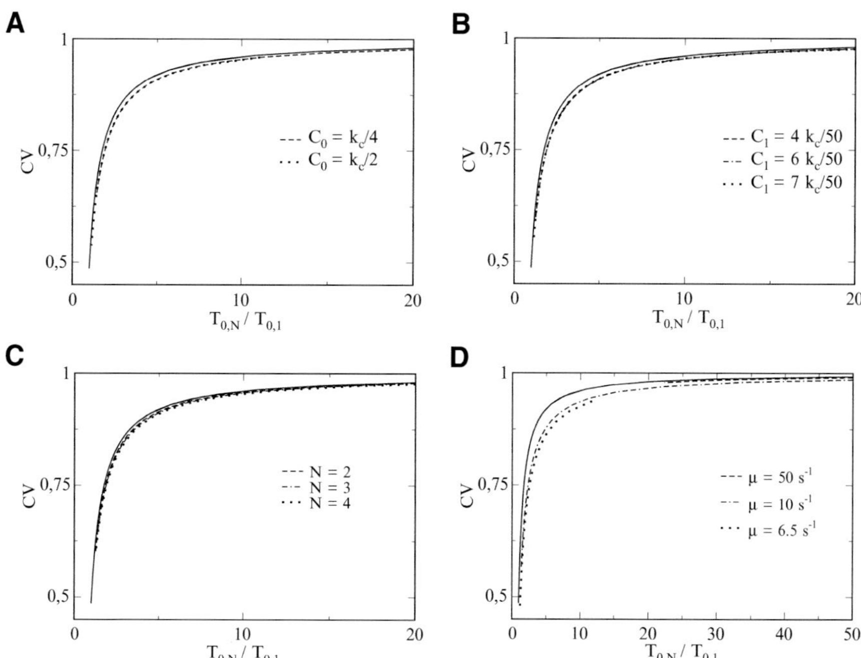

Fig. 4.16 Comparison of the approximation for the CV (4.53) (solid line in all panels) with exact calculations. We varied $\mathrm{T}_{0,N}/\mathrm{T}_{0,1}$ by changing the [IP$_3$] in $\lambda_0\left([\mathrm{IP}_3]\right)$ (4.55). The standard parameter values are $\mathrm{C}_0 = 5k_c/16$, $\mathrm{C}_1 = 6k_c/50$, $\mu = 50\,\mathrm{s}^{-1}$, $N = 4$, $\mathrm{CV}_{0,1} = 0.486$. The legends in the plots specify parameter values deviating from the standard values. We used $\mathrm{C}_0 = k_c/4$, $\mathrm{C}_1 = 3k_c/50$ in panel D.

All relations $CV(T_{0,N}/T_{0,1})$ are very close to (4.53), as shown in Fig. 4.16. Changing parameter values essentially only shifts the relation along the graph of (4.53). The condition $T_{0,1} \gg T_{i,j}$, $i > 0$, for the validity of (4.53) arises from the relation $\mu \gg T_{0,1}^{-1}$. But even if we choose μ about one order of magnitude smaller than the measured value of 47–59 s^{-1}, (4.53) is still a good approximation for the CV.

The coefficient of variation quickly approaches 1 with increasing $T_{0,N}/T_{0,1}$ in all plots in Fig. 4.16. That result is interesting by itself, since it shows that a birth and death process consisting of non-Poissonian individual steps leads in the limit of large $T_{0,N}/T_{0,1}$ to Poissonian first passage time behaviour.

We observed CVs smaller than 1 in most of the experiments. Our results suggest that a decrease of the CV below 1 is caused by processes not included in the above calculations, i.e., by global feedback. Possibly, experimental spike sequences also show the approach of the CV for $T_{0,N}/T_{0,1} \to 1$ to $CV_{0,1}$, but they are not precise and long enough to verify that. In summary, the results of the calculations agree with our findings in experiments and simulations that the CV is set by global feedbacks (not included in the above calculations) and robust against changes of local properties (Thurley et al, 2014; Thurley and Falcke, 2011). But they do not yet explain them exhaustively, since the derivation of the cumulant relation from puff properties with global feedback and the explanation for the concentration-response relation of T_{av} is still lacking.

4.4.3.2 Resampling an IPI Distribution to Obtain an ISI Distribution

The spike probability is the product of the probability for a puff multiplied by the probability that this puff causes a supercritical nucleus. We consider the regime in which puffs cause more puffs only as long as they are open, i.e., the regime with fast Ca^{2+} removal. Since puff closing is fast, reaching of the critical nucleus upon opening of the first cluster happens quickly, if it happens. Since puff duration is much shorter than IPI, we have a timescale separation between the time for the transition from 1 to N_{cr} open clusters and an interpuff interval. In the approximation we apply in this section, we calculate the distribution of the intervals till the first successful puff and neglect the time for the transition from 1 to N_{cr} open clusters. That is in technical terms a resampling of the IPI distribution with the probability for the transition from 1 to N_{cr} open clusters. The resampling can be carried out for a puff rate depending on the time elapsed since the last spike, but not with an additional dependence on the time since the last puff. For example, if we use the relaxation ansatz (4.19), the time in this equation is the time since the last spike.

We go through the ideas and calculations first by picking a specific site. The probability for a puff at this site between t and $t+dt$ is then $\lambda(t)dt$. The time t is the time till the end of the absolute refractory period of the last spike. We denote by $\Pi_{1,N_{\mathrm{cr}}}(t)$ the probability that the transition from 1 to N_{cr} open clusters occurs before all clusters close. More generally, we let $\Pi_{i,N}(t)$ denote the probability of reaching N open clusters before all clusters are closed when starting with i open clusters.

We derive now the probability $p(n,t)$ that the nth puff of a specific site initiates a spike in $(t, t+dt)$. The times of occurrence of the $n-1$ preceding puffs are $t^{(0)}, \ldots, t^{(n-1)}$. Simply for convenience, we set $t^{(0)} = 0$.

$$\begin{aligned}
p(n,t) &= \prod_{i=0}^{n-1} \int_{t^{(i)}}^{t} dt^{(i+1)} e^{-\int_{t^{(i)}}^{t^{(i+1)}} d\theta \lambda(\theta)} \left(1 - \Pi_{1,N_{\mathrm{cr}}}\left(t^{(i+1)}\right)\right) \lambda\left(t^{(i+1)}\right) \times \\
&\qquad e^{-\int_{t^{(n-1)}}^{t} d\theta \lambda(\theta)} \Pi_{1,N_{\mathrm{cr}}}(t)\lambda(t) \\
&= e^{-\int_{t^{(0)}}^{t} d\theta \lambda(\theta)} \Pi_{1,N_{\mathrm{cr}}}(t)\lambda(t) \prod_{i=0}^{n-1} \int_{t^{(i)}}^{t} dt^{(i+1)} \left(1 - \Pi_{1,N_{\mathrm{cr}}}\left(t^{(i+1)}\right)\right) \lambda\left(t^{(i+1)}\right) \\
&= e^{-\int_{t^{(0)}}^{t} d\theta \lambda(\theta)} \Pi_{1,N_{\mathrm{cr}}}(t)\lambda(t) \frac{\left[\int_{t^{(0)}}^{t} d\theta \left(1 - \Pi_{1,N_{\mathrm{cr}}}(\theta)\right) \lambda(\theta)\right]^{n}}{n!}
\end{aligned} \tag{4.58}$$

The product is over all $n-1$ time points. In the integrand, we first have the probability for no puff occurring between $t^{(i)}$ and $t^{(i+1)}$, times the probability that the puff occurring at $t^{(i+1)}$ does not open N_{cr} clusters, times the probability density that a puff occurs in $(t^{(i+1)}, t^{(i+1)} + dt^{(i+1)})$. This is multiplied with the probability that a successful puff occurs at t. The total probability $p(t)$ that an individual puff site causes a spike at point in time t is the sum over all possible numbers of failing puffs between 0 and t

$$\begin{aligned}
p(t) &= e^{-\int_{t^{(0)}}^{t} d\theta \lambda(\theta)} \Pi_{1,N_{\mathrm{cr}}}(t)\lambda(t) \sum_{n=0}^{\infty} \frac{\left[\int_{t^{(0)}}^{t} d\theta \left(1 - \Pi_{1,N_{\mathrm{cr}}}(\theta)\right) \lambda(\theta)\right]^{n}}{n!} \\
&= \Pi_{1,N_{\mathrm{cr}}}(t)\lambda(t)\ e^{-\int_{t^{(0)}}^{t} d\theta \lambda(\theta)} e^{\int_{t^{(0)}}^{t} d\theta (1-\Pi_{1,N_{\mathrm{cr}}}(\theta))\lambda(\theta)} \\
&= \Pi_{1,N_{\mathrm{cr}}}(t)\lambda(t)\ e^{-\int_{t^{(0)}}^{t} d\theta \Pi_{1,N_{\mathrm{cr}}}(\theta)\lambda(\theta)}.
\end{aligned} \tag{4.59}$$

Hence, resampling the stochastic process defined by (4.19) by $\Pi_{1,N_{\mathrm{cr}}}(t)$ provides the probability for spike initiation by an individual site at time t. The exponential in (4.59) is the probability that a spike is not initiated before t. In order to go to an array of N clusters we first introduce the index j marking the jth cluster: $p^j(t)$, $\Pi^j_{1,N_{\mathrm{cr}}}$, λ^j. The probability for spike generation by the array is the probability that none of the clusters has generated a spike before time t and one of them initiates a spike at t. It is the interspike interval distribution $p_{\mathrm{ISI}}(t)$

$$\Psi_{\mathrm{ISI}}^{st}(t) = \sum_{j=1}^{N} \Pi_{1,N_{\mathrm{cr}}}^{j}(t)\lambda^{j}(t) \prod_{i=1}^{N} e^{-\int_{t(0)}^{t} d\theta \Pi_{1,N_{\mathrm{cr}}}^{i}(\theta)\lambda^{i}(\theta)} \tag{4.60}$$

$$= \sum_{j=1}^{N} \Pi_{1,N_{\mathrm{cr}}}^{j}(t)\lambda^{j}(t) e^{-\int_{t(0)}^{t} d\theta \sum_{i=1}^{N} \Pi_{1,N_{\mathrm{cr}}}^{i}(\theta)\lambda^{i}(\theta)} \tag{4.61}$$

$$= e^{-\int_{t(0)}^{t} d\theta \sum_{i=1}^{N} \Pi_{1,N_{\mathrm{cr}}}^{i}(\theta)\lambda^{i}(\theta)} \sum_{j=1}^{N} \Pi_{1,N_{\mathrm{cr}}}^{j}(t)\lambda^{j}(t). \tag{4.62}$$

Hence, the rate of spike generation (probability per unit time irrespective of events at other times) $\psi_{\mathrm{ISI}}(t)$ is

$$\psi_{\mathrm{ISI}}(t) = \sum_{j=1}^{N} \Pi_{1,N_{\mathrm{cr}}}^{j}(t)\lambda^{j}(t), \tag{4.63}$$

i.e., the sum over the rates of all individual clusters. If we take into account that the puff rate $\lambda^{j}(t)$ depends linearly on N_{ch} (Dickinson et al, 2012) we obtain with the single-channel opening rate $p_o(t)$

$$\psi_{\mathrm{ISI}}(t) = \sum_{j=1}^{N} \Pi_{1,N_{\mathrm{cr}}}^{j}(t) N_{\mathrm{ch}}^{j} p_o(t). \tag{4.64}$$

To simplify the calculation of $\Pi_{1,N_{\mathrm{cr}}}^{j}$ we assume again identical clusters and a linear chain of the configurations with 1 to N_{cr} open clusters. The probability $\Pi_{i,N}$ of going from i to N open clusters is calculated by averaging over all possible trajectories from i to N which avoid 0. The trajectories start with probability $C_{i,i+1}$ with a step to $i+1$ and with probability $C_{i,i-1}$ with a step to $i-1$:

$$\Pi_{i,N} = C_{i,i+1}\Pi_{i+1,N} + C_{i,i-1}\Pi_{i-1,N}. \tag{4.65}$$

Additionally, $\Pi_{N,N} = 1$ and $\Pi_{0,N}^{0} = 0$. More specifically, for $i = 1$ and $N = 3$ or $N = 4$ we obtain

$$\begin{pmatrix} 1 & -C_{1,2} \\ -C_{2,1} & 1 \end{pmatrix} \begin{pmatrix} \Pi_{1,3}^{0} \\ \Pi_{2,3}^{0} \end{pmatrix} = \begin{pmatrix} 0 \\ C_{2,3} \end{pmatrix}, \tag{4.66}$$

$$\begin{pmatrix} 1 & -C_{1,2} & 0 \\ -C_{2,1} & 1 & -C_{2,3} \\ 0 & -C_{3,2} & 1 \end{pmatrix} \begin{pmatrix} \Pi_{1,4}^{0} \\ \Pi_{2,4}^{0} \\ \Pi_{3,4}^{0} \end{pmatrix} = \begin{pmatrix} 0 \\ 0 \\ C_{3,4} \end{pmatrix}. \tag{4.67}$$

The solutions for $N = 2, \cdots, 5$ are

$$\Pi^0_{1,2} = C_{1,2}, \tag{4.68}$$

$$\Pi^0_{1,3} = \frac{C_{1,2}C_{2,3}}{1 - C_{1,2}\left(1 - C_{2,3}\right)}, \tag{4.69}$$

$$\Pi^0_{1,4} = \frac{C_{1,2}C_{2,3}C_{3,4}}{1 - C_{1,2}\left(1 - C_{2,3}\right) - C_{2,3}\left(1 - C_{3,4}\right)}, \tag{4.70}$$

$$\Pi^0_{1,5} = \frac{C_{1,2}C_{2,3}C_{3,4}C_{4,5}}{1 - C_{1,2}\left(1 - C_{2,3}\right) - C_{2,3}\left(1 - C_{3,4}\right) - C_{3,4}\left(1 - C_{4,5}\right) + C_{1,2}\left(1 - C_{2,3}\right)C_{3,4}\left(1 - C_{4,5}\right)}. \tag{4.71}$$

Here we used $C_{i,i-1} + C_{i,i+1} = 1$.

From (4.19) we obtain for the resampled rate

$$\Lambda(t) = \begin{cases} 0, & t \leq T_{\text{min}}, \\ \Pi^0_{1,N_{\text{cr}}}\lambda\left(1 - e^{-\xi(t - T_{\text{min}})}\right) & t \geq T_{\text{min}}, \end{cases} \tag{4.72}$$

where t is the waiting time since the last spike and T_{min} the absolute refractory period after a spike. The ISI distribution $P(t)$ is then

$$P(t) = \begin{cases} 0, & t \leq T_{\text{min}}, \\ e^{\int_{T_{\text{min}}}^{t} d\theta \Pi^0_{1,N_{\text{cr}}}\lambda\left(1 - e^{-\xi(\theta - T_{\text{min}})}\right)} \Pi^0_{1,N_{\text{cr}}}\lambda\left(1 - e^{-\xi(t - T_{\text{min}})}\right), & t \geq T_{\text{min}}. \end{cases} \tag{4.73}$$

4.4.4 Some Formulae

This is a collection of formulae used previously in this chapter. The first and second moments of $F_{0,N}(t)$ for $N = 2, 3, 4$ are ($T_{0,4}$ is given by (4.47) and (4.49))

$$T_{0,2} = \left(1 + \frac{C_{1,0}}{\Upsilon(0)}\right) T_{0,1} + \frac{1}{\Upsilon}\left(T_{1,0} + T_{1,2}\right), \tag{4.74}$$

$$\begin{aligned} T^{(2)}_{0,2} = \frac{2}{\Upsilon(0)} &\left[C_{1,0}\left(1 + \frac{C_{1,0}}{\Upsilon(0)}\right) T^2_{0,1} + 2\left(1 + \frac{C_{1,0}}{\Upsilon(0)}\right) T_{0,1}T_{1,0} \right. \\ &+ \left.\left(1 + \frac{C_{1,0}}{\Upsilon(0)}\right) T_{0,1}T_{1,2} + \frac{1}{\Upsilon(0)} T^2_{1,0} + \frac{1}{\Upsilon(0)} T_{1,0}T_{1,2}\right] \\ &+ \left(1 + \frac{C_{1,0}}{\Upsilon(0)}\right) T^{(2)}_{0,1} + \frac{1}{\Upsilon}\left(T^{(2)}_{1,0} + T^{(2)}_{1,2}\right), \end{aligned} \tag{4.75}$$

$$
\begin{aligned}
T_{0,3} = & -\frac{1}{\Upsilon}\left[\left(\tilde{\Psi}_{1,2}\tilde{\Psi}_{2,3} + \frac{\tilde{\Psi}_{0,1}\Psi_{1,2}\tilde{\Psi}_{2,3}}{\Upsilon}\tilde{\Psi}_{1,0}\right)\tilde{\Psi}'_{0,1}\right. \\
& + \frac{\tilde{\Psi}^2_{0,1}\Psi_{1,2}\tilde{\Psi}_{2,3}}{\Upsilon}\tilde{\Psi}'_{1,0} \\
& + \left(\tilde{\Psi}_{0,1}\tilde{\Psi}_{2,3} + \frac{\tilde{\Psi}_{0,1}\Psi_{1,2}\tilde{\Psi}_{2,3}}{\Upsilon}\tilde{\Psi}_{2,1}\right)\tilde{\Psi}'_{1,2} \\
& \left.\left. + \frac{\tilde{\Psi}_{0,1}\Psi^2_{1,2}\tilde{\Psi}_{2,3}}{\Upsilon}\tilde{\Psi}'_{2,1} + \tilde{\Psi}_{0,1}\Psi_{1,2}\tilde{\Psi}'_{2,3}\right]\right|_{s=0}
\end{aligned} \tag{4.76}
$$

$$
= \frac{1}{\Upsilon(0)}\left[\left(C_{1,2}C_{2,3} + C_{1,0}\right)T_{0,1} + T_{1,0} + T_{1,2} + C_{1,2}\left(T_{2,1} + T_{2,3}\right)\right] \tag{4.77}
$$

$$
= \left(1 + \frac{C_{1,0}}{\Upsilon(0)}\right)T_{0,1} + \frac{1}{\Upsilon}\left(T_{1,0} + T_{1,2}\right) + \frac{C_{1,2}}{\Upsilon}\left(T_{2,1} + T_{2,3}\right), \tag{4.78}
$$

$$
\begin{aligned}
T^{(2)}_{0,3} = & \frac{1}{\Upsilon(0)}\left[2C_{1,0}\left(1 + \frac{C_{1,0}}{\Upsilon(0)}\right)T^2_{0,1} + 4\left(1 + \frac{C_{1,0}}{\Upsilon(0)}\right)T_{0,1}T_{1,0}\right. \\
& + 2\left(1 + \left(1 + C_{2,1}\right)\frac{C_{1,0}}{\Upsilon(0)}\right)T_{0,1}T_{1,2} \\
& + 2C_{1,2}\left(1 + 2\frac{C_{1,0}}{\Upsilon(0)}\right)T_{0,1}T_{2,1} + 2C_{1,2}\left(1 + \frac{C_{1,0}}{\Upsilon(0)}\right)T_{0,1}T_{2,3} \\
& + \frac{2}{\Upsilon(0)}T^2_{1,0} + \frac{2}{\Upsilon(0)}\left(1 + C_{2,1}\right)T_{1,0}T_{1,2} + 4\frac{C_{1,2}}{\Upsilon(0)}T_{1,0}T_{2,1} \\
& + 2\frac{C_{1,2}}{\Upsilon(0)}T_{1,0}T_{2,3} + 2\frac{C_{2,1}}{\Upsilon(0)}T^2_{1,2} + 4\left(1 + \frac{C_{1,2}C_{2,1}}{\Upsilon(0)}\right)T_{1,2}T_{2,1} \\
& + 2\left(1 + \frac{C_{1,2}C_{2,1}}{\Upsilon(0)}\right)T_{1,2}T_{2,3} + 2\frac{C^2_{1,2}}{\Upsilon(0)}T^2_{2,1} + 2\frac{C^2_{1,2}}{\Upsilon(0)}T_{2,1}T_{2,3} \\
& \left. + \left(C_{1,2}C_{2,3} + C_{1,0}\right)T^{(2)}_{0,1} + T^{(2)}_{1,0} + T^{(2)}_{1,2} + C_{1,2}\left(T^{(2)}_{2,1} + T^{(2)}_{2,3}\right)\right],
\end{aligned} \tag{4.79}
$$

$$
\begin{aligned}
T^{(2)}_{0,4} = & \frac{2}{\Upsilon(0)}\left[C_{1,0}\left(1 - C_{2,3}C_{3,2}\right)\left(1 + \frac{C_{1,0}\left(1 - C_{2,3}C_{3,2}\right)}{\Upsilon(0)}\right)T^2_{0,1}\right. \\
& + 2\left(1 - C_{2,3}C_{3,2}\right)\left(1 + \frac{C_{1,0}\left(1 - C_{2,3}C_{3,2}\right)}{\Upsilon(0)}\right)T_{0,1}T_{1,0} \\
& + \left(1 - C_{2,3}C_{3,2}\right)\left(1 + \frac{C_{1,0}\left(1 + C_{2,1}\right)\left(1 - C_{2,3}C_{3,2}\right)}{\Upsilon(0)}\right)T_{0,1}T_{1,2} \\
& + C_{1,2}\left(1 + 2\frac{C_{1,0}\left(1 - C_{2,3}C_{3,2}\right)}{\Upsilon(0)}\right)T_{0,1}T_{2,1} \\
& + \left(C_{1,2} - C_{1,0}C_{3,2} + \frac{C_{1,2}C_{1,0}\left(1 + C_{3,2}\right)\left(1 - C_{2,3}C_{3,2}\right)}{\Upsilon(0)}\right)T_{0,1}T_{2,3}
\end{aligned}
$$

$$
\begin{aligned}
&+C_{2,3}\left(C_{1,2}-C_{1,0}+2\frac{C_{1,2}C_{1,0}\left(1-C_{2,3}C_{3,2}\right)}{\Upsilon(0)}\right)T_{0,1}T_{3,2}\\
&+C_{1,2}C_{2,3}\left(1+\frac{1-C_{2,3}C_{3,2}}{\Upsilon(0)}\right)T_{0,1}T_{3,4}+\frac{\left(1-C_{2,3}C_{3,2}\right)^2}{\Upsilon(0)}T_{1,0}^2 \qquad (4.80)\\
&+\left(1-C_{2,3}C_{3,2}\right)\left(C_{2,3}C_{3,4}+2\frac{C_{2,1}}{\Upsilon(0)}\right)T_{1,0}T_{1,2}\\
&+2\frac{C_{1,2}\left(1-C_{2,3}C_{3,2}\right)}{\Upsilon(0)}T_{1,0}T_{2,1}\\
&+\left(\frac{C_{1,2}\left(1+C_{3,2}\right)\left(1-C_{2,3}C_{3,2}\right)}{\Upsilon(0)}-C_{3,2}\right)T_{1,0}T_{2,3}\\
&+C_{2,3}\left(2\frac{C_{1,2}\left(1-C_{2,3}C_{3,2}\right)}{\Upsilon(0)}-1\right)T_{1,0}T_{3,2}\\
&+\frac{C_{1,2}C_{3,2}\left(1-C_{2,3}C_{3,2}\right)}{\Upsilon(0)}T_{1,0}T_{3,4}\\
&+\frac{C_{2,1}\left(1-C_{2,3}C_{3,2}\right)}{\Upsilon(0)}T_{1,2}^2+2\left(1+\frac{C_{1,2}C_{2,1}}{\Upsilon(0)}\right)T_{1,2}T_{2,1}\\
&+\left(1+\frac{C_{1,2}C_{2,1}\left(1+C_{3,2}\right)}{\Upsilon(0)}\right)T_{1,2}T_{2,3}\\
&+C_{2,3}\left(1+2\frac{C_{1,2}C_{2,1}}{\Upsilon(0)}\right)T_{1,2}T_{3,2}+C_{2,3}\left(1+\frac{C_{1,2}C_{2,1}}{\Upsilon(0)}\right)T_{1,2}T_{3,4}\\
&+\frac{C_{1,2}^2}{\Upsilon(0)}T_{2,1}^2+\frac{C_{1,2}^2\left(1+C_{3,2}\right)}{\Upsilon(0)}T_{2,1}T_{2,3}+2\frac{C_{1,2}^2C_{2,3}}{\Upsilon(0)}T_{2,1}T_{3,2}\\
&+\frac{C_{1,2}^2C_{2,3}}{\Upsilon(0)}T_{2,1}T_{3,4}+\frac{C_{1,2}^2C_{3,2}}{\Upsilon(0)}T_{2,3}^2+2C_{1,2}\left(1+\frac{C_{1,2}C_{2,3}C_{3,2}}{\Upsilon(0)}\right)T_{2,3}T_{3,2}\\
&\left.+C_{1,2}\left(1+\frac{C_{1,2}C_{2,3}C_{3,2}}{\Upsilon(0)}\right)T_{2,3}T_{3,4}+\frac{C_{1,2}^2C_{2,3}^2}{\Upsilon(0)}T_{3,2}^2+\frac{C_{1,2}^2C_{2,3}^2}{\Upsilon(0)}T_{3,2}T_{3,4}\right]\\
&+\left(1+\frac{C_{1,0}\left(1-C_{2,3}C_{3,2}\right)}{\Upsilon(0)}\right)T_{0,1}^{(2)}+\frac{1-C_{2,3}C_{3,2}}{\Upsilon(0)}\left(T_{1,0}^{(2)}+T_{1,2}^{(2)}\right)+\\
&\frac{C_{1,2}}{\Upsilon(0)}\left(T_{2,1}^{(2)}+T_{2,3}^{(2)}\right)+\frac{C_{1,2}C_{2,3}}{\Upsilon(0)}\left(T_{3,2}^{(2)}+T_{3,4}^{(2)}\right).
\end{aligned}
$$

Additional splitting probabilities are

$$\Pi_{2,5}^0=\frac{C_{2,3}C_{3,4}C_{4,5}}{Z}, \qquad (4.81)$$

$$\Pi_{3,5}^0=\frac{C_{3,4}C_{4,5}\left(1-C_{1,2}\left(1-C_{2,3}\right)\right)}{Z}, \qquad (4.82)$$

$$\Pi_{4,5}^0=\frac{C_{4,5}\left(1-C_{1,2}\left(1-C_{2,3}\right)-C_{2,3}\left(1-C_{3,4}\right)\right)}{Z}, \qquad (4.83)$$

where

$$Z = 1 - C_{1,2}\,(1 - C_{2,3}) - C_{2,3}\,(1 - C_{3,4}) - \\ C_{3,4}\,(1 - C_{4,5}) + C_{1,2}\,(1 - C_{2,3})\,C_{3,4}\,(1 - C_{4,5})\,. \qquad (4.84)$$

On the basis of the sequence $\Pi_{i,N}$ for $i = 1, \ldots, N-1$ we can define the size $i^{(\varepsilon)}_{\text{crit}}$ of the critical nucleus by $\Pi_{i^{(\varepsilon)}_{\text{crit}},N} \geq 1-\varepsilon$ and $\Pi_{i^{(\varepsilon)}_{\text{crit}}-1,N} < 1-\varepsilon$. Here, N has to be chosen large enough for representing with certainty a supercritical nucleus.

4.4.5 Summary

We defined the task of modelling as the calculation of the ISI distribution from the IPI distribution, puff duration distributions, and properties of global feedbacks. How far have we come? We used simple models to understand the basic observations theoretically, since the distributions we would like to calculate are simple and are characterised by few parameters only, and since simplicity often implies broad applicability. Excluding global feedbacks, we could calculate the moments for the ISI distribution, if the spike generation process can be mapped to a birth and death process (fast removal of Ca^{2+} from the cytosol). With this approach we find $\alpha = 1$ for not too small T_{av}, even with non-Poissonian waiting time distributions for puff occurrence and termination. Hence, they do not suffice to generate non-Poissonian cumulant relations, confirming our conclusion from experiments that global negative feedback is required to obtain $\alpha < 1$.

Calculation of the ISI distribution, including global feedback, faces the problem of a system time in the convolution of probability fluxes with the puff generation distribution. That has not yet been solved. We have taken into account global feedbacks in the resampling approach of Section 4.4.3.2 and distribution ansatzes only. Neither have we yet calculated the ISI distribution for the case of accumulation of cytosolic Ca^{2+} on the time scale of several IPI. That causes correlations across several puff openings and therefore cannot be mapped to a birth and death process with waiting time distributions depending on the current state only (semi-Markov property).

More theoretical understanding of the role of global feedbacks would augment our understanding of Ca^{2+} spiking. Experiments have shown that different cell types and different pathways exhibit different values of the slope of the cumulant relation. We also know that the time scale of ISI is not in the local puff dynamics (Thurley et al, 2011) and that α does not respond to static changes of local properties, i.e., its value is fixed by global feedback processes. One way to test this hypothesis would be to change experimentally the time scale of recovery from negative feedback, to check whether that changes α. However, this experiment has not been done yet. Also, calculation of the puff duration and IPI distribution is still technically demanding, although interesting approaches have been suggested (Thul and Falcke, 2006, 2007). A theory more easily applicable to channel state models, resulting from patch clamp experiments or similar measurements, is desirable.

The concentration-response relation should also be the result of a stochastic theory of IP_3-induced Ca^{2+} spiking. That has not been accomplished yet. Hence, there is still much interesting and challenging theory to be developed.

Chapter 5
Nonlinear Dynamics of Calcium

The focus of the earlier chapters of this book has been on the construction of mathematical models based on information about the underlying physiology, and on comparison of model predictions with experimental data. A recurrent theme has been that a relatively small number of broad principles are applicable in the construction of a wide variety of different Ca^{2+} models in different physiological contexts. This chapter has a different focus, and instead looks at some mathematical methods that have proved useful for the analysis of Ca^{2+} models. However, there is an analogous theme, which is that a relatively small number of mathematical methods underlie much of current practice in model analysis.

The main purpose for a mathematical model is to help explain the behaviour of the system being modelled. Given a detailed model like those discussed in the earlier chapters, along with values for all model parameters and for the initial configuration of the system, it is reasonably straightforward to use standard numerical methods to perform numerical integration or simulation of the model and, hence, to extract predictions about model behaviour. While direct numerical solution is a relatively quick way to determine, for instance, steady-state values of certain variables or the period of an oscillation, there are two main reasons why we frequently need more than this for answering modelling questions.

First, many of the parameter values in a model will be known only approximately, if at all, and the initial configuration for the model will be, at best, a crude approximation of the initial configuration in the system being modelled. Thus, model predictions based on a specific choice of parameter values and initial conditions may bear little direct relationship to experimental data. Even when parameter values have been carefully fitted to experimental data such as a measured time series for a specific physiological quantity, different values might be obtained for a second time series obtained under "identical" experimental conditions. For a model to be useful, therefore, it

G. Dupont et al., *Models of Calcium Signalling*, Interdisciplinary Applied Mathematics 43, DOI 10.1007/978-3-319-29647-0_5

is important to understand how robust model predictions are to changes in parameter values and initial conditions, and this cannot easily be deduced from numerical solution of the model at specified values.

Second, a common aim of modelling is to understand which parts of a model are important for producing a particular type of behaviour. For instance, we might be interested in how the details of an IPR model affect the overall Ca^{2+} dynamics. A related, but broader, question is whether certain component models are needed at all. For instance, if some reactions are known to occur on much faster or much slower time scales than others, or to involve only tiny variations in overall concentration, can these reactions be ignored in the model without destroying the behaviour being studied? Determining which mathematical mechanisms underlie the model behaviour of interest is crucial to answering these types of question, and is something that cannot usually be achieved by numerical solution of a model at isolated parameter values and for one or two choices of initial conditions.

Thus, it is important to have a broader set of techniques for the analysis of models than just numerical solution. In this chapter we focus on techniques that have proved useful for the analysis of the various types of deterministic Ca^{2+} model discussed in the earlier chapters, and in particular on dynamical systems methods for investigating the robustness of models to changes in model structure or parameter values and for investigating the sensitivity of model behaviour to changes in initial conditions. Most of the techniques we discuss are applicable to a wide variety of biophysical models besides those involving Ca^{2+}, and also to models from other disciplines, such as physics and chemistry.

5.1 An Illustrative Model: The Hybrid Model

For convenience, we restate the equations for a model of Ca^{2+} oscillations that we shall be using extensively in this chapter to illustrate the theoretical discussion. The model we use is based on the gating IPR model of Atri et al (1993), converted into a hybrid model by Domijan et al (2006). It was briefly discussed in Section 3.2.7.1, and forms the basis for the work of Harvey et al (2010), where more details can be found.

The model equations are

$$\frac{dc}{dt} = J_{\mathrm{IPR}} - J_{\mathrm{serca}} + \delta(J_{\mathrm{in}} - J_{\mathrm{pm}}), \tag{5.1}$$

$$\frac{dc_t}{dt} = \delta(J_{\mathrm{in}} - J_{\mathrm{pm}}), \tag{5.2}$$

$$\frac{dp}{dt} = \nu\left(1 - \frac{\alpha k_4}{c + k_4}\right) - \beta p, \tag{5.3}$$

$$\frac{dy}{dt} = \frac{1}{\tau}\left(\frac{k_2^2}{k_2^2 + c^2} - y\right), \tag{5.4}$$

where

$$\begin{aligned}
J_{\mathrm{IPR}} &= \left[k_{\mathrm{flux}}\gamma\left(\mu_0 + \frac{\mu_1 p}{k_\mu + p}\right) y \left(b + \frac{V_1 c}{k_1 + c}\right)\right](c_t - (1 + 1/\gamma)c),\\
J_{\mathrm{serca}} &= \frac{V_e c}{K_e + c},\\
J_{\mathrm{pm}} &= \frac{V_p c^2}{k_p^2 + c^2},\\
J_{\mathrm{in}} &= \alpha_1 + \alpha_2 \nu.
\end{aligned}$$

The parameter values are given in Table 5.1.

When $\alpha = 0$ and $\tau \neq 0$, the model is a Class I model, but increasing α from 0 introduces Class II terms. Alternatively, fixing $\alpha \neq 0$ and taking the limit as $\tau \to 0$ effectively removes the IPR dynamics from the model, converting the model to Class II.

Finally, note that ν denotes agonist concentration. We thus anticipate that ν will be an important bifurcation parameter in the study of this model.

parameter	value	parameter	value	parameter	value
b	0.111	k_2	0.7 μM	k_{flux}	6 s^{-1}
δ	0.01	k_4	1.1 μM	V_p	24.0 μM s^{-1}
γ	5.405	k_p	0.4 μM	V_e	20.0 μM s^{-1}
μ_0	0.567	k_e	0.06 μM	α_1	1.0 μM s^{-1}
μ_1	0.433	k_1	1.1 μM	α_2	2.5
V_1	0.889	k_μ	4.0 μM	β	0.08 s^{-1}

Table 5.1 Parameter values for the hybrid model of (5.1)–(5.4) (Domijan et al, 2006).

5.2 Bifurcation Analysis for ODE Models

The most common way to investigate parameter dependence of model behaviour is the construction of a numerical bifurcation diagram, using bifurcation and continuation software such as XPPAUT (Ermentrout, 2002) or AUTO (Doedel, 1981; Doedel et al, 1998). Bifurcation diagrams typically show how the location and stability of steady-state and oscillatory solutions vary with a single parameter of the system; if the chosen bifurcation parameter corresponds to a system parameter that can be measured experimentally, then comparison between the qualitative features of the bifurcation diagram and experimental data can be made.

A bifurcation diagram for the Li-Rinzel simplification of the De Young-Keizer model of the IPR was shown in Fig. 3.2, where equilibrium and periodic solutions of the model were plotted as a function of p, the IP_3 concentration, assumed to be a parameter of the model. The bifurcation diagram has a "Hopf bubble", a feature common to many models of intracellular Ca^{2+} dynamics

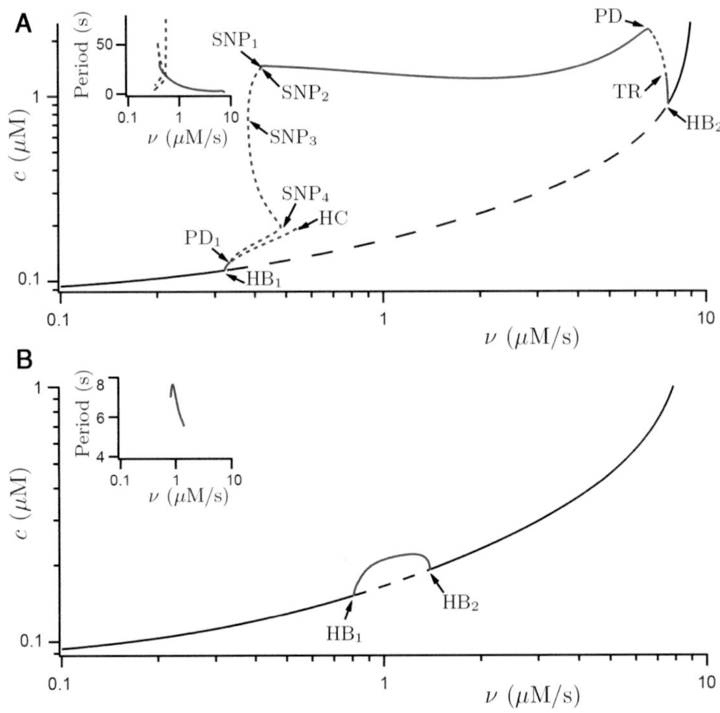

Fig. 5.1 Partial bifurcation diagrams for the hybrid model, (5.1)–(5.4), plotting the cytosolic $[Ca^{2+}]$, c, versus the maximal rate of IP_3 formation, ν. **A**: the bifurcation diagram for the hybrid version of the model ($\tau = 2$, $\alpha = 1$). **B**: the analogous diagram for the Class II version ($\tau = 0$, $\alpha = 1$). The black curves are steady-state solutions (dashed for unstable, solid for stable). The red and blue curves indicate the maximum amplitudes of stable and unstable periodic orbits, respectively. Bifurcations are labelled HB (Hopf), PD (period-doubling), SNP (saddle-node of periodic orbits), HC (homoclinic), and TR (torus bifurcation). The insets in each panel show the periods of the periodic orbits.

where $[Ca^{2+}]$ is plotted as a function of p or agonist concentration. Specifically, there are no oscillations for low or high values of p, but oscillations occur for intermediate values of p, in the region between a pair of Hopf bifurcations. This occurs as a result of a very general modelling assumption based on experimental observation: the steady-state open probability curve for the IPR is bell-shaped as a function of $[Ca^{2+}]$ so the IPR is closed at both high and low $[Ca^{2+}]$ (Fig. 2.14). Since resting $[Ca^{2+}]$ is an increasing function of p in this model and since oscillations arise through the sequential opening and closing of the IPR, this assumption leads directly to the occurrence of the Hopf bubble; the IPR is unlikely to open at high or low values of p and so there will be no oscillations in those parameter regimes.

In other models the details might be different, but the overall qualitative pattern of oscillations occurring at intermediate values of the parameter is robust across a wide range of models. This is illustrated in Fig. 5.1, where we show bifurcation diagrams for two versions of the hybrid model described in the previous section. In the hybrid version of the model (Fig. 5.1 A), oscillations occur for intermediate values of ν, the maximal rate of IP_3 production, but now the branches of periodic orbits have a much more complicated structure. Although periodic orbits still appear and disappear at Hopf bifurcations (labelled HB_1 and HB_2), and thus we technically still have a Hopf bubble, between the Hopf bifurcations we now have a much more complex structure, with period-doubling bifurcations, multiple saddle-node of periodics bifurcations, and disconnected branches of stable orbits. However, the bifurcation diagram of the Class II version of the model (Fig. 5.1 B) has the simple Hopf bubble seen in the Li-Rinzel model (Fig. 3.2). Thus we see that, even in a single model, changes in parameters can have dramatic effects on the bifurcation structure, which can be of great complexity.

Sometimes, one of the variables of a model (rather than a system parameter) is chosen as the bifurcation 'parameter', and the position and stability of steady-state and oscillatory solutions are computed with respect to changes in this variable. This technique can be useful when the variable in question is known to evolve much more slowly than other variables in the system, and is discussed at more length in Section 5.4.

A limitation of standard bifurcation diagrams is that all but one of the parameters must be fixed before the diagram is computed. By comparing bifurcation diagrams obtained using different choices of the fixed parameters, or by choosing an alternative bifurcation parameter, it may be possible to infer how qualitative changes in the bifurcation diagrams arise, but a more efficient approach might be feasible. Specifically, it is often possible to compute numerically a two-parameter bifurcation set, which shows the position of bifurcations such as folds of steady states or Hopf bifurcations as two parameters of the system are varied and, importantly, identifies points or regions in the two-parameter space where qualitative changes in the bifurcation structure of the model occur.

The ability to compute two-parameter bifurcation sets opens up the possibility of using bifurcation theory to investigate how certain terms in a model influence the overall dynamics. The idea is to introduce a so-called *homotopy* parameter to the system as a multiplicative factor in front of the terms of interest. When this parameter is equal to 1, the terms are present as normal in the model, but they decrease in amplitude and then disappear completely as the parameter is decreased to zero. By plotting a two-parameter bifurcation set using a regular bifurcation parameter and the homotopy parameter as the two distinguished parameters, it is possible to deduce how the bifurcation diagram for the system depends on the presence of the terms of interest.

An illustration of this approach is provided by the hybrid model, where either α or τ can be used as a homotopy parameter. Plotting the two-parameter bifurcation sets using either α and ν or τ and ν as the bifurcation parameters gives valuable clues about the mathematical mechanisms underlying the differences between the dynamics of the different forms of the hybrid model.

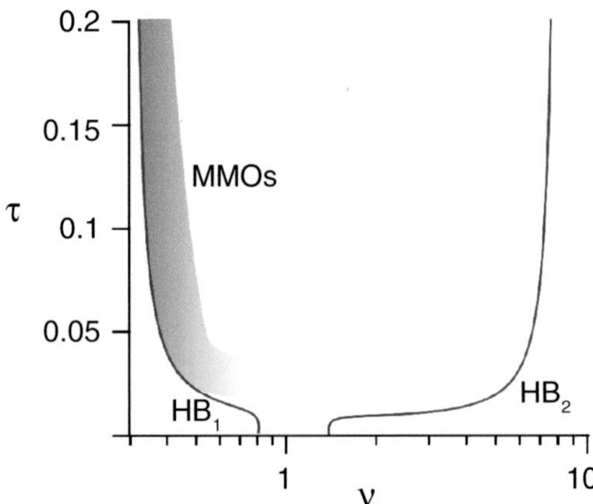

Fig. 5.2 Partial two-parameter bifurcation set for the hybrid model, (5.1)–(5.4), using τ and ν as the bifurcation parameters. The curves HB_1 and HB_2 correspond to the left and right Hopf bifurcations on the Hopf bubble, and the shaded region indicates the approximate region in which complex oscillations known as mixed-mode oscillations (MMO) occur (Section 5.4.2). The bifurcation diagrams shown in Fig. 5.1 A and B correspond to taking slices through this bifurcation set, at $\tau = 2.0$ and $\tau = 0$, respectively. The loci of the bifurcations marked SNP, TR, and PD in Fig. 5.1 A are not indicated.

For example, Fig. 5.2 shows a partial bifurcation set obtained by varying τ and ν in the hybrid model. Increasing ν corresponds to increasing agonist. For a fixed τ, increasing agonist first leads to oscillations, as the HB_1 curve is crossed, and then leads to disappearance of the oscillations, as the HB_2 curve is crossed. As τ decreases, the IPR dynamics become faster, and the hybrid model begins to behave more like a Class II model, in which the oscillations are governed entirely by the Ca^{2+} control of IP_3 production. The region between the curves of Hopf bifurcations narrows markedly at $\tau \approx 0.02$, and as the model becomes more like a Class II model, the oscillatory region shrinks (Harvey et al, 2011). This can also be seen clearly in Fig. 5.1; oscillations in the Class I hybrid model (panel A) occur over a much wider region (i.e., for a much larger interval of ν) than in the Class II hybrid model (panel B).

The complicated sequence of bifurcations seen close to HB_1 in Fig. 5.1 A gives rise to analogous complications in the two-parameter bifurcation diagram of Fig. 5.2. Instead of tracking these all individually as functions of τ, which would result in an incomprehensible diagram, we simply shade the approximate region where these bifurcations occur (the shaded region labelled MMOs). In this region we see a variety of complex oscillations, known as mixed-mode oscillations, behaviour that is discussed in more detail in Section 5.4.2.

5.3 Model Reduction

Careful selection of the processes to include in a model is crucial if a useful model is to be developed. Computational power is readily available and relatively cheap, meaning that large numbers of variables and parameters can be accommodated by numerical methods for investigating model solutions. There is, therefore, a temptation to include everything in a model that might possibly be relevant. However, if the aim is to understand the mathematical mechanisms underlying the dynamics of a model, smaller models are generally more tractable than larger ones and so it is often helpful to try to simplify a model as much as possible before launching into too much analysis. The aim is to remove variables or terms that model processes that are not relevant for the particular issue being investigated, even if they are otherwise vital to the overall functioning of the system.

At the most basic level, terms can be removed by trial and error, by computing time series or bifurcation diagrams for two different versions of a model, one with the term(s) of interest and one without, to see if there is a qualitative difference. Alternatively, a two-parameter bifurcation set computed using a homotopy parameter as described above, might be helpful. However, more systematic approaches to model reduction can be useful, particularly those based on arguments about time scales for various processes. For instance, it is common during modelling to conclude that certain variables evolve so fast relative to other variables that they can be assumed to be in quasi-steady state, hence justifying replacing a differential equation for the evolution of that variable by an algebraic equation; such an approach was used repeatedly, for instance, in the derivations of IPR models in Section 2.7, or in the reduction of Markov models for Ca^{2+} pumps and exchangers (Sections 2.2 and 2.3). Alternatively, it is sometimes the case that certain variables evolve so slowly that they can be assumed to be constant over the typical time scale of the process being considered. Reduction based on these types of argument is intuitively appealing, but can lead to erroneous conclusions and warrants a closer look.

5.3.1 Identifying Time Scales

The first step in performing model reduction based on timescale separation is to identify correctly the time scales in the model. In many situations, physiological intuition based on experimental observation provides useful information. For instance, Fig. 8.16 shows experimentally obtained time series for the membrane current and cytosolic $[Ca^{2+}]$ for a bursting GnRH neuron. It is usual to interpret time series of this type as showing fast oscillations of current (or voltage) on a raised baseline accompanied by a slower increase in $[Ca^{2+}]$, interspersed with periods of slow decay and then quiescence in voltage and slow decrease in [Ca^{2+}]. This commonly leads to the assertion that voltage changes over a shorter time scale than [Ca^{2+}] and, hence, that voltage is fast and Ca^{2+} is slow in this system; this idea is built into many models used to describe the qualitative behaviour of bursting pancreatic β cells or neuroendocrine cells (Section 8.7). In other systems, common indications that multiple time scales might be important include time series showing mixed-mode oscillations, which have small, subthreshold oscillations interspersed with large amplitude spikes, and bifurcation diagrams in which branches of periodic orbits have an amplitude that changes very rapidly as the bifurcation parameter is varied over a small interval (as seen in the left end of the Hopf bubble in Fig 5.1 A). However, observations of this type, based on qualitative properties of time series or bifurcation diagrams, need to be backed up with quantitative identification of distinct time scales in the corresponding mathematical models if they are to be mathematically convincing and useful.

Quite apart from any difference in overall speed of evolution between voltage and [Ca^{2+}] inferred from the time series in Fig. 8.16, it appears that there are at least two different time scales at play in the voltage time series: relatively fast evolution during the bursting, and slower decay, then quiescence between bursts. This suggests there must be different physiological processes that operate on distinct time scales contributing to the dynamics of the voltage variable. In general, a focus on the time scales of the processes contributing to the dynamics of a system rather than on the overall time scales of the observed variables can be useful.

One way to check for distinct time scales in a model is to employ a type of non-dimensionalisation, in which there are two main steps:

1. algebraically manipulating the model equations to aggregate parameters, and
2. multiplying variables by characteristic scales, which take the units of a physical quantity, and rescaling the corresponding variables so that the maximum magnitude of each rescaled variable is about 1.

We can apply this procedure to the hybrid model as follows. We begin by introducing new dimensionless variables, (C, C_t, P, t_1), with

$$c = Q_c C\,, \quad c_t = Q_c C_t\,, \quad p = Q_p P\,, \quad t = T t_1\,, \tag{5.5}$$

where $Q_c = 1\,\mu$M and $Q_p = 10\,\mu$M are typical concentration scales for Ca^{2+} and IP_3, respectively, and $T = Q_c/(\delta V_P) = 100/24$ s is a typical time scale for the c_t dynamics (Harvey et al, 2011). The variable y was already dimensionless in the original model, taking on values between 0 and 1, and so does not require rescaling. This change of variables then yields the rescaled equations

$$\delta \frac{dC}{dt_1} = \bar{J}_{\mathrm{IPR}} - \bar{J}_{\mathrm{serca}} + \delta(\bar{J}_{\mathrm{in}} - \bar{J}_{\mathrm{pm}}), \tag{5.6}$$

$$\frac{dC_t}{dt_1} = \bar{J}_{\mathrm{in}} - \bar{J}_{\mathrm{pm}}, \tag{5.7}$$

$$\frac{dy}{dt_1} = \frac{1}{\hat{\tau}}\left(\frac{k_2^2}{k_2^2 + Q_c^2 C^2} - y\right), \tag{5.8}$$

$$\frac{dP}{dt_1} = \hat{\nu}\left(1 - \frac{k_4 \alpha}{k_4 + Q_c C}\right) - \hat{\beta} P, \tag{5.9}$$

with dimensionless parameters

$$\hat{\tau} = \frac{\delta V_p}{Q_c}\tau, \quad \hat{\nu} = \frac{Q_c}{Q_p}\frac{\nu}{\delta V_p}, \quad \hat{\beta} = \frac{Q_c}{\delta V_p}\beta, \tag{5.10}$$

and corresponding dimensionless versions of the fluxes: $\bar{J}_{\text{IPR}}$, $\bar{J}_{\text{serca}}$, $\bar{J}_{\text{pm}}$, and $\bar{J}_{\text{in}}$.

Using the parameter values given in Table 5.1, and with ν values corresponding to oscillatory solutions, it can then be seen that the right-hand sides of (5.6), (5.7), and (5.9) are $O(1)$, while the right-hand side of (5.8) is $O(1/\hat{\tau})$. Noting the position of δ in the C equation and its value, we can then deduce the speeds of evolution for the variables: $O(10^2)$ for C, $O(1)$ for C_t and P, and order $O(1/\hat{\tau})$ for y. Thus, if $\hat{\tau}$ is $O(1)$, this system has one fast variable and three slow variables, while if $\hat{\tau}$ is $O(\delta)$, there are two fast variables and two slow variables (since $\delta = 0.01$).

In straightforward cases like this one, non-dimensionalisation will show that certain model variables, corresponding to physiologically identifiable (and, preferably, experimentally measurable) quantities, have a rate of evolution that is clearly separated from the rate of evolution of other variables. A timescale separation of around two orders of magnitude is desirable, but a separation of at least one order of magnitude might be acceptable; any less would not normally be regarded as clear separation and variables separated by less would commonly be treated as occurring on the same time scale.

Difficulties in identifying timescale separation may arise in cases where at least one slow process in a model is not aligned with a measurable quantity. A particularly relevant example of this is the total Ca^{2+} load in a cell, which we called c_t in the models in Section 3.2.5 and in (5.2). This quantity is difficult to measure experimentally to any degree of accuracy, and it is not obvious from a physiological perspective why it would be so important in controlling Ca^{2+} oscillations. Nevertheless, the concept of total Ca^{2+} as a slow variable has proven to be extremely useful in the development of our understanding of how Ca^{2+} influx affects oscillations and waves, and has been used previously to make a number of predictions that have been confirmed experimentally (Sneyd et al, 2004).

There are undoubtedly other quantities with similar status, i.e., that are not directly physically measurable but align directly with distinguished time scales in a model, and identifying them is crucial to formulating a model in a form where the timescale structure is explicit and can be exploited to aid the analysis of the model. Unfortunately, such identification can appear to be more a matter of luck than a systematic process, and has to proceed on a case by case basis.

5.3.2 Reduction Based on Timescale Separation

When a clear separation between time scales in a model can be identified, it may be possible to exploit the timescale separation to simplify the model. Perhaps the most common instance of this is quasi-steady-state reduction (QSSR). This technique is commonly used to simplify Markov models of receptors, pumps, and exchangers, is widely used in enzyme kinetics, and is now ubiquitous in biophysical modelling. Indeed, the method has been used many times in the derivation of models in the earlier chapters of this book. The idea behind this method is to assume that one or more variables in a model evolve so fast relative to other variables that they can be assumed to equilibrate instantaneously and, thus, that ordinary differential equations for the evolution of such variables can be replaced by algebraic equations, thereby reducing the dimension of the model.

QSSR can be effective when transient dynamics is not important and when the system has a globally attracting equilibrium solution. In this case, it can be shown that application of the method amounts to using geometric singular perturbation theory, a mathematically rigorous approach that will be discussed more in Section 5.4. However, significant difficulties can arise when QSSR is applied to systems with oscillations (Pedersen et al, 2008; Erneux and Goldbeter, 2006; Flach and Schnell, 2006; Zhang et al, 2011), which is frequently the case in models of Ca^{2+} dynamics; a detailed discussion of this case is contained in Boie et al (2015). The main issues can be illustrated by seeing how the technique works for two well-known biophysical models: the Hodgkin-Huxley equations and the Hindmarsh-Rose model.

The bifurcation diagram for the Hodgkin-Huxley system, using the applied current, I, as the bifurcation parameter, is shown in Fig. 5.3 (black and blue curves). It is well established that the voltage, v, and activation of the sodium channel, m, have fast kinetics relative to inactivation of the sodium channel (h) and activation of the potassium channel (n), and it is common to simplify the Hodgkin-Huxley equations using QSSR by assuming that m equilibrates instantly. Numerical integration of the resulting system of equations produces the bifurcation diagram shown in Fig. 5.3 (black and red curves). Comparison of the two versions of the bifurcation diagram shows that the bifurcation diagram is qualitatively the same in the two cases, although the positions of the Hopf bifurcations and the amplitude of the branch of periodic orbits are modified by QSSR. In a similar way, it is possible to use QSSR to remove the differential equation for v from the system while retaining the dynamics of m; doing so produces another qualitatively similar bifurcation diagram (black and green curves). The Hodgkin-Huxley equations thus provide an example where QSSR appears to be a reasonable way to simplify the model (Boie et al, 2015).

Applying QSSR does not always give such a satisfactory result, however. The Hindmarsh-Rose model is

$$\dot{v} = I - v^3 + 3v^2 + m - n, \tag{5.11}$$

$$\dot{m} = 1 - 5v^2 - m, \tag{5.12}$$

$$\dot{n} = 0.001(4(v + 1.6) - n), \tag{5.13}$$

where v is the membrane potential, m and n model ionic currents, and I is the applied current. The bifurcation diagram for this system, using I as the bifurcation parameter, is shown in Fig. 5.4 (black and blue curves). In this model, n evolves much more slowly than v and m. If we assume that m equilibrates rapidly, then m can be replaced by $m_\infty(v) \equiv 1 - 5v^2$ in (5.11), yielding the reduced system

$$\dot{v} = I - v^3 - 2v^2 + 1 - n, \tag{5.14}$$

$$\dot{n} = \epsilon(4(v + 1.6) - n). \tag{5.15}$$

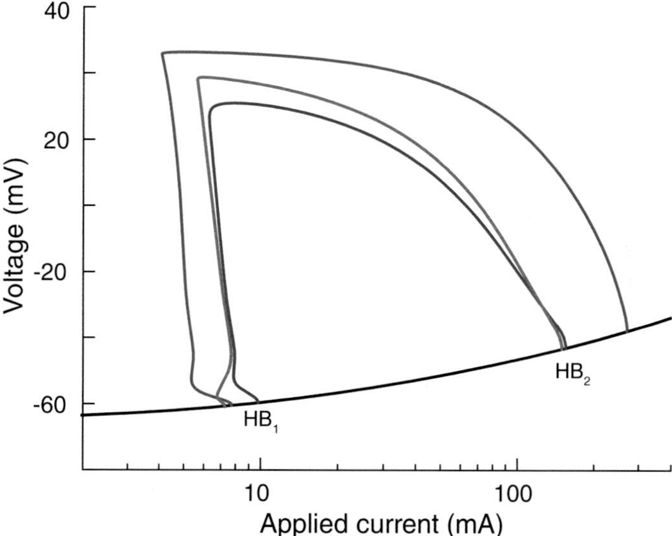

Fig. 5.3 Bifurcation diagrams for the full Hodgkin-Huxley model and the models obtained after removal of v or m via QSSR. The black curve shows the equilibria, and is common to all versions of the model. The coloured curves show the maximum voltages for the family of periodic orbits in each model: orbits of the full model are shown in blue, while orbits of the reduced model with m (respectively v) removed are shown in red (respectively green). Stability of solutions is not indicated. The branches of periodic orbits arise through Hopf bifurcations, labelled HB_1 and HB_2 in the case of the full model.

This version of the model has significantly different dynamics to the full model, having just two Hopf bifurcations rather than four as in the full model. As a result, the full model has a much larger range of values of I for which oscillations are possible. This system thus provides an example where QSSR destroys a significant component of the dynamics.

These two examples show that the application of QSSR will sometimes minimally disrupt the dynamics of a model but in other cases will cause significant changes. It would be useful to know before attempting QSSR which scenario is likely, but this is not currently possible. However, some partial results have been established. Boie et al (2015) classifies Hopf bifurcations according to whether they involve fast or slow variables (or both), and shows that Hopf bifurcations involving just fast variables may be removed by QSSR (as was the case in the HR model discussed above) but other types of Hopf bifurcation are preserved by the reduction method (as was the case in the Hodgkin-Huxley model). In particular, Hopf bifurcations involving both fast and slow variables, specifically those known as singular Hopf bifurcations, are preserved by QSSR. These occur commonly in biophysical models, which explains why QSSR is often useful for models with oscillations. Unfortunately, it can be difficult to tell which sort of Hopf bifurcation occurs in a particular model without significant numerical or theoretical investigation of the model.

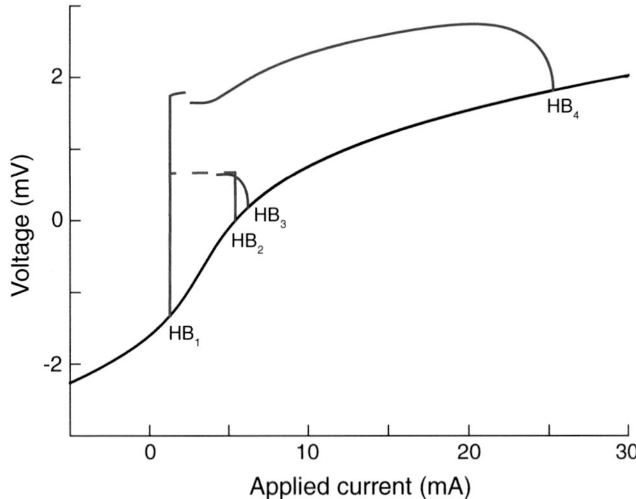

Fig. 5.4 Bifurcation diagrams for the full Hindmarsh-Rose equations and the model obtained after removal of m via QSSR. The black curve shows the equilibria, and is common to both versions of the model. The coloured curves show the maximum voltages for the family of periodic orbits in each model: orbits of the full model are shown in blue while orbits of the reduced model are shown in red. Stability of solutions is not indicated. In the full model, the branches of periodic orbits arise through Hopf bifurcations, labelled HB_1 - HB_4. In the reduced model, Hopf bifurcations occur very close to HB_1 and HB_2, but HB_3 and HB_4 no longer exist.

In addition to removing Hopf bifurcations from the dynamics of a model, QSSR can introduce other significant changes, such as changing the stability of branches of periodic solutions or introducing or removing homoclinic bifurcations (Boie et al, 2015). Essentially nothing systematic is known about the circumstances under which these changes might occur.

As a rule of thumb, it is probably safe to use QSSR on non-oscillatory models where there is a clear timescale separation and in cases where the transient dynamics is not important; all other models should be approached with caution. If QSSR is to be attempted on a model with oscillations, it is important to check that reduction does not destroy important model dynamics before relying on predictions from the reduced model; a systematic procedure for checking is discussed in Boie et al (2015).

Another way in which timescale separation might be exploited to reduce the complexity of a model is by assuming that certain variables are so slow relative to other variables that their evolution can be ignored and they can be treated as constant. However, care needs to be taken in adopting this approach, since there are many examples where very slow processes are crucial to the overall dynamics. For instance, in a model of GnRH neurons (Lee et al, 2010; Duan et al, 2011), it was found that inclusion of a slow Ca^{2+} afterhyperpolarisation current was vital if the bursting observed in experiments was to be reproduced by the model (Section 8.7.2). In this case, and many others, ignoring the very slow process would result in a model that does not capture the dynamics of interest. From a mathematical point of view, the issue seems to be that a variable identified as 'slow' might evolve much slower than other variables over part of the phase space but at a comparable speed in other regions of the phase space (because the 'fast' variables slow down significantly in those regions). In such a case, global removal of a slow variable from the model can destroy the interesting dynamics. Once again, it is important to check that reduction does not destroy important features of the model dynamics before using a reduced model to make predictions about full model dynamics.

5.4 Analysis Based on Timescale Separation

A variety of techniques, both theoretical and numerical, have been developed in recent years for the analysis of models with multiple time scales. In many cases, the theoretical methods used have been ad hoc methods, based on observations of the particular model of interest, but two theoretical methods stand out for being more generally applicable: freezing slow variables and geometric singular perturbation theory.

5.4.1 Freezing Slow Variables

In systems where one variable evolves much more slowly than the others, the origin and nature of complex oscillations can sometimes be understood by treating the slow variable as a constant (i.e., freezing the slow variable), constructing a bifurcation diagram using the slow variable as the bifurcation parameter, and then comparing the phase portrait of the complex oscillation with this bifurcation diagram. Rinzel (1985) introduced this method in his study of bursting electrical oscillations in pancreatic beta cells, and the method has been widely used since then for the analysis of biophysical models. Analysis of the Chay-Keizer model of bursting via this method is discussed in some detail in Section 8.4.1.

While this approach has been used successfully to explain the dynamics of many systems, particularly systems with bursting solutions of various types (Izhikevich, 2000), there are limits to its applicability. First, in its most common form the method relies on there being a single slowest variable with a speed of evolution well separated from the speeds of the other variables. In practice, even if a model has a single identifiable slow variable, the timescale separation between it and the other variables might not be sufficient for the method to be useful. For instance, Fig. 5.5 shows the bifurcation diagram for the Class II version of the hybrid model, i.e., (5.1) and (5.3) with $y(c) = \frac{k_2^2}{k_2^2+c^2}$ and with c_t treated as the bifurcation parameter. Superimposed on the bifurcation diagram are three orbits of the full (unfrozen) Class II model for three different choices of δ corresponding to three different choices of the timescale separation between the model variables: $\delta = 0.0001$ (red), $\delta = 0.001$ (blue), and $\delta = 0.01$ (green). Without performing a careful non-dimensionalisation of the model along the lines discussed in Section 5.3.1, it is not clear what timescale separation results from each choice of δ, but it is apparent from Fig. 5.5 that $\delta = 0.01$ is not small enough for the method to be useful (since the orbit with this value of δ does not follow branches of the frozen

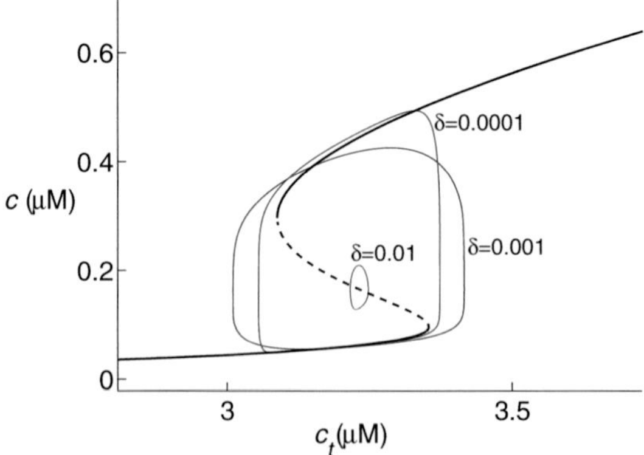

Fig. 5.5 Bifurcation diagram of the frozen Class II hybrid model, (5.1) and (5.3) with $y(c) = \frac{k_2^2}{k_2^2+c^2}$ and c_t treated as the bifurcation parameter. Parameter values are $\nu = 1.0$, $\alpha = 1$, and other constants as in Table 5.1. The black curve indicates steady states of the frozen system, solid for stable, and dotted for unstable. The red, blue, and green curves show solutions of the full Class II system for $\delta = 0.0001$, $\delta = 0.001$, and $\delta = 0.01$, respectively.

bifurcation diagram) and that $\delta = 0.001$ is of marginal use. More generally, a full system orbit is expected to be better approximated by (parts of) the frozen system bifurcation diagram as the timescale separation increases but there is no a priori way to know when the timescale separation is large enough for the frozen system bifurcation diagram to be a good guide to the dynamics of the full system.

A second issue is that many systems have more than one variable evolving on the slowest time scale. It is possible to adapt the freezing method to the case of multiple slow variables, but the procedure rapidly becomes unwieldy since multiple parameters will need to be varied to compute the bifurcation diagram for the frozen system. Attempts are sometimes made to freeze just one of the various slow variables, which amounts to approximating the higher-dimensional bifurcation diagram by a one-dimensional slice, but whether this will provide insight can depend on which variable is frozen, or, more precisely, the orientation of the slice relative to important features of the bifurcation diagram.

The limitations discussed in the paragraphs above arise if the method is used in situations for which it was not designed. A different kind of difficulty results from assumptions implicit in the formulation of the method. A full system orbit is nicely explained by comparison with a frozen system bifurcation diagram in cases, such as for $\delta = 0.0001$ in Fig 5.5, in which the orbit follows an attracting branch of a bifurcation diagram until the branch becomes unstable or disappears, then makes a fast jump to another, coexisting branch of the bifurcation diagram. However, transitions from slow to fast segments of an orbit can be more subtle than this, and the method of freezing may not give accurate information. The classic example is provided by canard solutions, which in this context can be regarded as full system solutions that follow the frozen bifurcation diagram from an attracting branch to a repelling branch, without immediately jumping at the transition point. Freezing the slow variable does not provide any insight into this phenomenon. The difficulties arise because the distinction between fast and slow variables is lost at points where branches of the frozen bifurcation diagram change stability; at such points, one or more 'fast' variables are actually evolving on a time scale comparable with that of the 'slow' variable, possibly resulting in subtle and complex interactions between the variables. A rigorous explanation of this type of phenomenon is provided by geometric singular perturbation theory, which we discuss next.

5.4.2 Geometric Singular Perturbation Theory

Geometric singular perturbation theory (GSPT) is a mathematically rigorous approach that uses limiting cases of a model to make predictions about full model dynamics. The main idea is that a model with two or more time scales will contain one or more small quantities associated with the timescale separation. Letting these quantities tend to zero in appropriate ways produces singular limit systems which are effectively lower dimensional. If the original system is a small enough perturbation of the limiting cases, results from analysis of the limiting systems can then be combined to give information about the mechanisms underlying complex dynamics in the full system. Of particular interest is the explanation of dynamics that involves both fast and slow time scales; dynamics that is purely fast might be explained by a model in which all slow variables are set to some appropriate constant value, and dynamics that is purely slow might be captured by a system in which all fast variables have equilibrated, but an explanation of the mechanisms behind solutions with both fast and slow phases requires information from both fast and slow parts of a model.

There are clear similarities between GSPT and the method of freezing a slow variable; in fact, as we shall see below, the frozen bifurcation diagrams used in the last section are closely related to one of the limiting systems we use in GSPT in the case where a system has two time scales with just one slow variable. However, the freezing method is essentially empirical, with intuition being invoked to explain the transition from slow to fast segments of an orbit (the orbit 'falls off' the frozen bifurcation diagram), meaning only the simplest transitions can be captured. By contrast, in GSPT the details of the transitions are explained using mathematical rigour, in the process of which a great deal of complex and important dynamics can be uncovered. General information about GSPT is contained in Fenichel (1979), Jones (1995), Szmolyan and Wechselberger (2001), and Wechselberger (2005) with further details about the application of GSPT to a selection of Ca^{2+} models being contained in Harvey et al (2010, 2011).

Application of GSPT is not just a matter of being careful in the process of obtaining results that are more easily derived with less rigorous methods; use of GSPT can reveal interesting features of model dynamics that are not otherwise obvious. To illustrate this point, we look at the hybrid model, starting with the non-dimensionalised version given by (5.6)–(5.10). Introducing a small parameter ϵ, we can rewrite the model as

$$\epsilon\frac{dC}{dt} = \bar{J}_{\text{IPR}} - \bar{J}_{\text{serca}} + \epsilon(\bar{J}_{\text{in}} - \bar{J}_{\text{pm}}), \tag{5.16}$$

$$\frac{dC_t}{dt} = \bar{J}_{\text{in}} - \bar{J}_{\text{pm}}, \tag{5.17}$$

$$\frac{dy}{dt} = \frac{1}{\hat{\tau}}\left(\frac{k_2^2}{k_2^2 + Q_c^2 C^2} - y\right), \tag{5.18}$$

$$\frac{dP}{dt} = \hat{\nu}\left(1 - \frac{k_4\alpha}{k_4 + Q_c C}\right) - \hat{\beta}P. \tag{5.19}$$

Decreasing ϵ has the effect of speeding up C relative to the other variables, i.e., increasing the timescale separation in the system. In the discussion below we assume that $\hat{\tau}$ is $O(1)$ and other parameters are as in Table 5.1, in which case C is fast and the remaining three variables are slow. In writing (5.16)–(5.19), it appears that we have just renamed the small quantity δ as ϵ, but note that δ appears in each of the dimensionless parameters in (5.10); these are all $O(1)$ and do not depend on the timescale separation, and will not change their values as $\epsilon \to 0$. Thus, the distinction between ϵ and δ is important.

Alternatively, one can rescale time by making the substitution $T = t/\epsilon$ in (5.16)–(5.19), resulting in the system

$$\frac{dC}{dT} = \bar{J}_{\text{IPR}} - \bar{J}_{\text{serca}} + \epsilon(\bar{J}_{\text{in}} - \bar{J}_{\text{pm}}), \tag{5.20}$$

$$\frac{dC_t}{dT} = \epsilon(\bar{J}_{\text{in}} - \bar{J}_{\text{pm}}), \tag{5.21}$$

$$\frac{dy}{dT} = \frac{\epsilon}{\hat{\tau}}\left(\frac{k_2^2}{k_2^2 + Q_c^2 C^2} - y\right), \tag{5.22}$$

$$\frac{dP}{dT} = \epsilon\left[\hat{\nu}\left(1 - \frac{k_4\alpha}{k_4 + Q_c C}\right) - \hat{\beta}P\right]. \tag{5.23}$$

The systems (5.16)–(5.19) and (5.20)–(5.23) are equivalent for $\epsilon \neq 0$, but taking the limit as $\epsilon \to 0$ results in different, non-equivalent singular systems. Specifically, letting $\epsilon \to 0$ in (5.20)–(5.23) gives the singular limit called the fast subsystem or layer problem,

$$\frac{dC}{dT} = \bar{J}_{\text{IPR}} - \bar{J}_{\text{serca}}, \tag{5.24}$$

$$\frac{dC_t}{dT} = \frac{dy}{dT} = \frac{dP}{dT} = 0, \tag{5.25}$$

which amounts to an evolution equation for the fast variable, C, with each of the slow variables, C_t, y, and P, held constant. On the other hand, taking the limit $\epsilon \to 0$ in (5.16)–(5.19) yields the reduced problem

$$0 = \bar{J}_{\text{IPR}} - \bar{J}_{\text{serca}}, \tag{5.26}$$

$$\frac{dC_t}{dt} = \bar{J}_{\text{in}} - \bar{J}_{\text{pm}}, \tag{5.27}$$

$$\frac{dy}{dt} = \frac{1}{\hat{\tau}}\left(\frac{k_2^2}{k_2^2 + Q_c^2 C^2} - y\right), \tag{5.28}$$

$$\frac{dP}{dt} = \hat{\nu}\left(1 - \frac{k_4\alpha}{k_4 + Q_c C}\right) - \hat{\beta}P, \tag{5.29}$$

which consists of evolution equations for the three slow variables along with an algebraic constraint which determines the surface on which the reduced problem evolves. Note that in the context of GSPT, the phrase 'reduced problem' has the very specific meaning given just above, and it is this we mean when we use this phrase in the discussion below. By contrast, when discussing

QSSR earlier, we sometimes referred to the reduced system, meaning the lower-dimensional system obtained by application of QSSR. It is important not to confuse these two similar terms.

A geometric structure important to both the layer problem and the reduced problem is the critical manifold, which is the manifold of equilibria of the layer problem, i.e., the three-dimensional set of points in the (C, C_t, y, P) space such that

$$\bar{J}_{\text{IPR}} - \bar{J}_{\text{serca}} = 0. \tag{5.30}$$

In the context of the reduced problem, the critical manifold is the surface, corresponding to the algebraic constraint, on which the dynamics of the reduced problem evolves. Fig 5.6 shows the critical manifold of (5.16)–(5.19) in the (C, C_t, y) space for the case $\hat{\nu} = 0.317$ and with fixed $P = 0.95$. As can be seen, the critical manifold is folded with respect to the fast variable, C. The inner sheet of the manifold is repelling within the layer problem, and the outer sheets are attracting.

The reduced and layer problems can be used to construct singular orbits consisting of alternating fast and slow segments; fast segments are solutions to the layer problem and slow segments are solutions to the reduced problem. Fig 5.7 shows schematically a singular periodic orbit consisting of fast segments, which are simple jumps between attracting sheets of the critical manifold, and slow segments, which lie on the critical manifold and exactly join the fast segments to make a closed loop. In the simplest cases, a singular periodic orbit will perturb in a straightforward way when we move away from the singular limit (for example, by increasing ϵ from 0 in the hybrid model), producing a relaxation oscillation that is qualitatively similar to the singular periodic orbit; an example of such an orbit is shown in Fig 5.8 C. In other

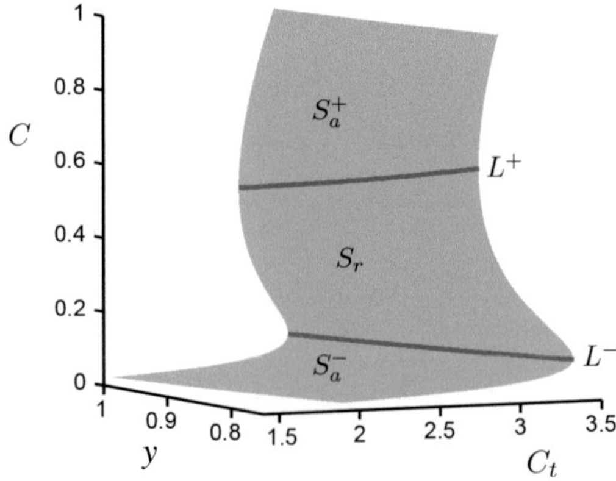

Fig. 5.6 The critical manifold of the hybrid model, (5.16)–(5.19), with $\hat{\nu} = 0.317$ and with fixed $P = 0.95$. The blue curves labelled L^+ and L^- mark the folds of the critical manifold with respect to C, and separate the two attracting sheets of the manifold (S_a^- and S_a^+) from the repelling sheet (S_r).

cases, a singular periodic orbit perturbs to something more complicated such as a mixed-mode oscillation (MMO) like the one shown in Fig. 5.8 A and B. MMOs like this have a number of smaller (subthreshold) oscillations between the larger amplitude relaxation-type spikes; the MMO shown in this example has very small subthreshold oscillations, but their occurrence has a strong influence, substantially lengthening the period of the overall oscillation.

A great deal is known about the appearance of MMOs in Ca^{2+} models, and other biophysical models. Frequently, they are associated with the existence of two or more slow variables in the model, in which case careful examination of the associated singular limit systems using GSPT enables predictions to be made about the types of MMO that will be observed (for example, the number of subthreshold oscillations between each large amplitude spike and the order in which different MMOs appear as the bifurcation parameter is varied) (Szmolyan and Wechselberger, 2001; Wechselberger, 2005). Note that this kind of information is not available from standard frozen system analysis discussed in the previous section.

When it works well, GSPT can be useful for explaining the origin of complicated dynamics in a multiple time scale model; an example of this is contained in the next section. However, as for freezing, there can be problems if the separation between different time scales is not sufficiently large. For instance, in (5.6)–(5.9), if $\hat{\tau}$ is $O(10^{-1})$ then the y variable is neither as fast as C nor as slow as C_t and P, and there is not enough of a separation between the speeds of evolution of y and the other variables to define a new intermediate time scale. In such a case, the correspondence between the full system dynamics and predictions of GSPT based on a singular limit obtained

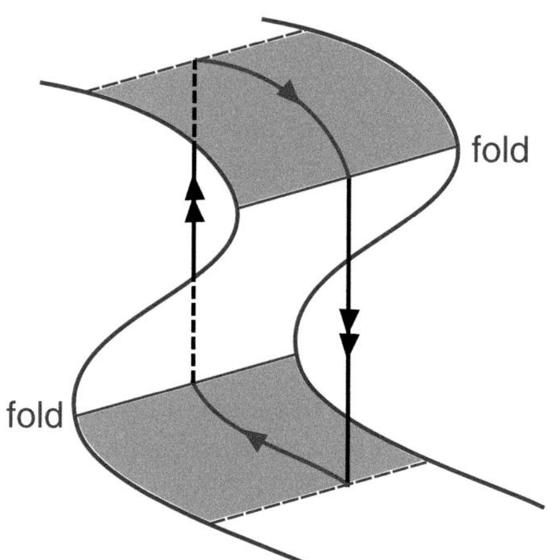

Fig. 5.7 Schematic diagram illustrating the construction of a singular periodic orbit (Nan et al, 2015). The cubic-shaped surface indicated by the blue curves and shading is the critical manifold. The singular orbit consists of alternating slow (red) and fast (black) segments.

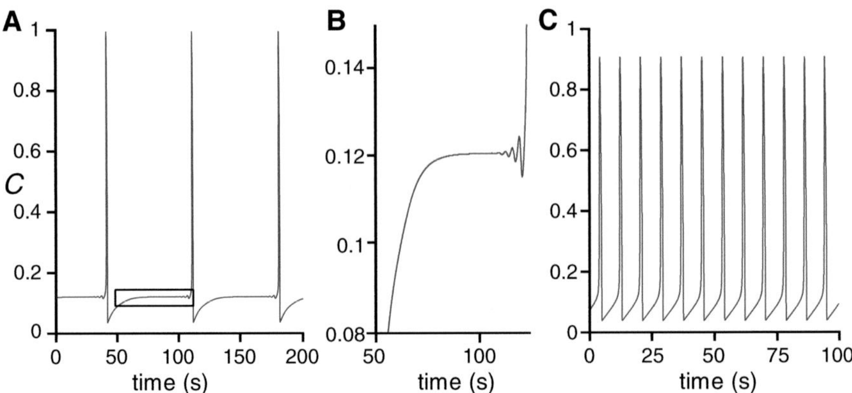

Fig. 5.8 Attracting oscillations in the hybrid model, (5.16)–(5.19), with $\alpha = 0$, $\hat{\tau} = 0.833$, $\epsilon = 0.01$ and other parameters as in Table 5.1. **A:** Mixed-mode oscillations for $\hat{\nu} = 0.167$. **B:** Enlargement of boxed section of panel A. **C:** Relaxation oscillations for $\hat{\nu} = 0.333$. Reproduced from Harvey et al (2011), Fig 2.

by assuming there exist two time scales might be weak or unhelpful. On the other hand, a comprehensive theory enabling GSPT methods to be applied to a system with three or more well-separated time scales does not yet exist, although some partial results are known (Krupa et al, 2008, 2012; Vo et al, 2013; Nan et al, 2015). Models with three or more time scales are particularly relevant for coupled voltage and Ca^{2+} models, where the voltage and Ca^{2+} submodels each have two time scales, and coupling between the submodels typically results in a full model with at least three time scales.

Even though a complete theory for the study of these types of model is not available, some progress can be made with GSPT methods. For example, to understand the dynamics of (5.6)–(5.9) in the case that y is intermediate in speed, one might look at two different approximate models: one with two fast and two slow variables (with y treated as a fast variable) and the other with one fast and three slow variables (with y treated as a slow variable). Standard GSPT analysis of one or other of these models might provide insight into the dynamics of the original model, even if neither is close enough to the original model for predictions to be rigorously justified. Some discussion of this type of approach is contained in Vo et al (2013).

Although GSPT analysis can result in detailed and useful predictions about model dynamics, the procedure for manipulating a model into a convenient form for the application of GSPT can be complicated and lengthy. Models rarely start out in one of the standard fast-slow forms such as

$$\frac{dx}{dt} = f(x, y), \tag{5.31}$$

$$\frac{dy}{dt} = \epsilon g(x, y), \tag{5.32}$$

or

$$\epsilon \frac{dx}{dt} = f(x, y), \tag{5.33}$$

$$\frac{dy}{dt} = g(x, y), \tag{5.34}$$

with $x \in R^n$, $y \in R^m$, f and g $O(1)$ functions, and $0 \leq \epsilon \ll 1$, and identification of appropriate coordinate changes that will put the model into one of these forms can involve as much luck as skill. Sometimes, physiological intuition suggests shortcuts to the process, but care must be taken if misleading results are to be avoided. For instance, in models like the hybrid model, a closed-cell version of the model is obtained by setting c_t, the total Ca^{2+} load of the cell, to a constant or, alternatively, by letting the variable c_t get slower and slower in the open-cell model. This process is somewhat similar to taking a singular limit and it might be tempting to think of the closed-cell model as a singular limit (i.e., the fast subsystem) for the open-cell model and hence to assume that the dynamics of the open-cell model will be a smooth perturbation of the dynamics of the closed-cell model. Although some features of the dynamics might perturb in this simple manner, there are a number of things that can go wrong. For instance, c_t might not be the only slow variable in the full model, in which case the limit in which c_t alone is fixed would not be the fast subsystem, and might not be helpful. Even if the closed-cell model is the fast subsystem, information obtained from analysis of this alone might be misleading. For instance, a Hopf bifurcation can be supercritical in the fast subsystem but subcritical in the full system (or vice versa) no matter how close the full system is to the limiting case (Zhang et al, 2011), so the full system dynamics cannot be deduced from knowledge of one limiting case by itself.

It is worth noting that GSPT methods are applicable to multiple timescale systems that are not in one of the standard fast-slow forms (Fenichel, 1979) but there are, as yet, few examples of physiological models in which this more general approach has been successfully applied.

5.5 Understanding Transient Dynamics

When analysing a model, the focus is often on determining the long-term dynamics, with the existence of stable equilibria or oscillatory solutions being of particular interest. The rationale for this focus is an assumption that the system being modelled either is in, or will rapidly settle to, an attracting state. Within this context, it is still important to locate unstable solutions, since these can be vital for understanding the structure of phase space. For instance, the location and orientation of unstable solutions of saddle type is crucial to understanding the long-term behaviour of typical solutions in systems where there is bistability. Furthermore, keeping track of both stable and unstable solutions is necessary if bifurcation diagrams and other indications of the robustness of a model are to be computed. Nevertheless, the main aim in much model analysis is to identify any attractors of the system.

An important exception to this arises when trying to understand the outcome of experiments in which a system is perturbed, by, for instance, releasing an exogenous pulse of IP_3 or Ca^{2+}, as discussed in Section 3.2.7.2. In such a situation, the focus is on interpreting the transient response of the system to the perturbation. The idea behind such an approach is that the characteristics of the transient response might provide information about dynamic mechanisms in the system that are not obvious from the steady-state behaviour alone, perhaps thereby allowing one to distinguish between different competing hypotheses about the underlying physiological mechanisms. From a mathematical point of view, the perturbation sets a new initial configuration for the model, away from the dominant attractor, and observing the transient response allows one to explore the region of phase space between that initial configuration and the attractor.

The phase space features most important for determining the transient response of a model are likely to be invariant manifolds of various types, such as the stable and unstable manifolds of saddle equilibria and periodic orbits, or, in systems in which there is a timescale separation, critical manifolds and their perturbations. An example that shows the importance of these is provided by an analysis of the transient response of the hybrid model to an exogenous pulse of IP_3 (Domijan et al, 2006; Harvey et al, 2010). An IP_3 pulse can be modelled by the inclusion of a term

$$S(t) = MH(t - t_0)\, H(t_0 + \Delta - t) \tag{5.35}$$

in the right-hand side of the equation for P in (5.16)–(5.19). Here M denotes the pulse magnitude and H is the Heaviside function

$$H(x) = \begin{cases} 0, \text{ if } x < 0, \\ 1, \text{ if } x \geq 0. \end{cases}$$

The pulse is applied at $t = t_0$ and has a duration Δ. Fig. 5.9 shows the model response under various conditions. Panel A shows the Class I model response when $\hat{\nu} = 0.40$: a temporary increase in oscillation frequency is seen immediately after the pulse, but then the time series rapidly returns to its pre-pulse form. Panel C shows the Class II model response when $\hat{\nu} = 0.417$: the pulse is followed by a relatively long quiescent period before the oscillations in Ca^{2+} concentration resume. The different model responses in these two cases would seem to be explained by the heuristic argument given in Section 3.2.7.2, but the actual situation is more complicated, as can be seen by examination of Panel B, which shows a Class I model (with $\hat{\nu} = 0.233$) responding in a manner closer to that expected from Class II; the pulse is followed by a number of faster oscillations followed by a long quiescent period before oscillations resume.

The difference in model response can be explained by recourse to GSPT. As established in Section 5.4.2, the Class I version of the hybrid model can be regarded as having one fast and three slow variables while the Class II version reduces to a three-dimensional model with one fast and two slow variables. In the Class II case the critical manifold is a two-dimensional surface in the three-dimensional (C, C_t, P) phase space, and is folded with respect to the fast variable, C; there are two fold curves for $P < 0.45$ with the folds merging in a cusp at $P \approx 0.45$, and no folds for $P > 0.45$, as shown in Fig. 5.10. For the values of ϵ and $\hat{\nu}$ used in that figure, the attracting orbit is a relaxation

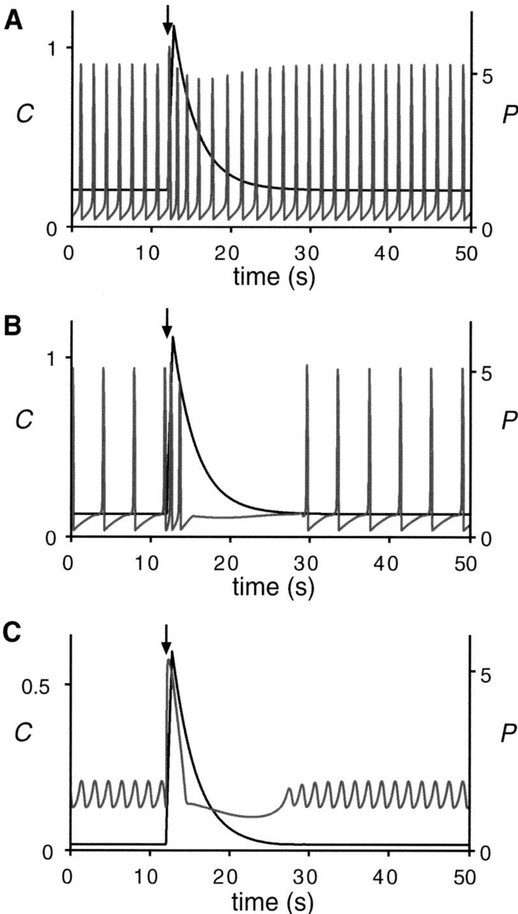

Fig. 5.9 Some responses of (5.16)–(5.19) with $\epsilon = 0.01$ to a pulse of IP_3. The form of the pulse is given by (5.35) with $M = 8.33\dot{3}$, $t_0 = 12$, and $\Delta = 0.72$. Each panel shows time series for the concentrations of Ca^{2+} (red curve) and IP_3 (black curve). The arrow indicates the time of application of the pulse. **A:** Class I: $\alpha = 0$, $\hat{\tau} = 0.48$ for $\hat{\nu} = 0.40$ and other parameter values as in Table 5.1. **B:** Class I: as in panel A except with $\hat{\nu} = 0.233$. **C:** Class II: $\alpha = 1$, $\hat{\tau} = 0$ for $\hat{\nu} = 0.417$.

oscillation that resides in the region where the critical manifold is folded. A sufficiently large pulse of IP_3 resets the P variable to a high value, past the cusp. Since ϵ is small, the orbit first moves quickly towards the critical manifold then evolves on the critical manifold, following the slow flow until a fold curve is reached, after which a fast jump in C could occur. Since there are no folds in the critical manifold for large P, immediately after a large pulse there will be no jumps and hence no oscillations. Eventually P will become small enough that the orbit enters the region in which the critical manifold is folded, and then the oscillations resume. Thus, the quiescent period in the Class II response to pulsing seen in Fig. 5.9 C occurs because there can be no oscillations in the transient orbit until P has decayed to a small enough value.

The characteristic features of the Class I response to pulsing result from quite different mechanisms than for Class II. The Class I model is four-dimensional and has a three-dimensional critical manifold. As for Class II, the critical manifold is folded with respect to C, but in this case the folds occur on two-dimensional surfaces that extend to all values of P, as shown in Fig. 5.11; there is no analogue to the cusp of fold points that was vital to the pulse response for the Class II model. Most points on the fold surfaces are regular jump points. If an orbit reaches the fold surface at one of these points it will make a fast jump away from the fold surface towards another attracting sheet of the critical manifold. However, one of the fold surfaces contains a one-dimensional curve of distinguished fold points, called *folded singularities*, which are parametrised by P (see Fig. 5.11); the nature of the singularity depends on $\hat{\nu}$ and P. These folded singularities are crucial to understanding the transient response of the Class I model to pulsing in IP_3.

For the case shown in Fig. 5.9 A the attracting orbit in the absence of pulsing is a relaxation oscillation which meets each fold surface once per

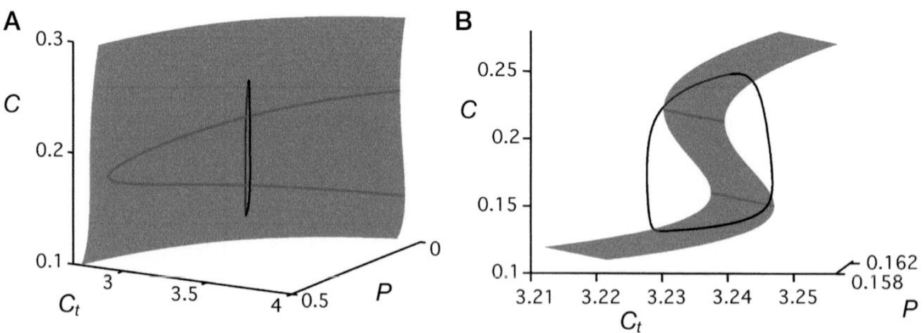

Fig. 5.10 The critical manifold for the Class II hybrid model and the attracting relaxation oscillation that occurs when $\epsilon = 0.001$ and $\hat{\nu} = 0.417$. **A:** The critical manifold, shown in pink, has two fold curves for $P < 0.45$ (blue curves), which merge in a cusp point at $P \approx 0.45$. The relaxation oscillation is shown in black. **B:** An enlargement of part of panel A from a different view point. Reproduced from Harvey et al (2010), Fig. 11, with permission from AIP Publishing.

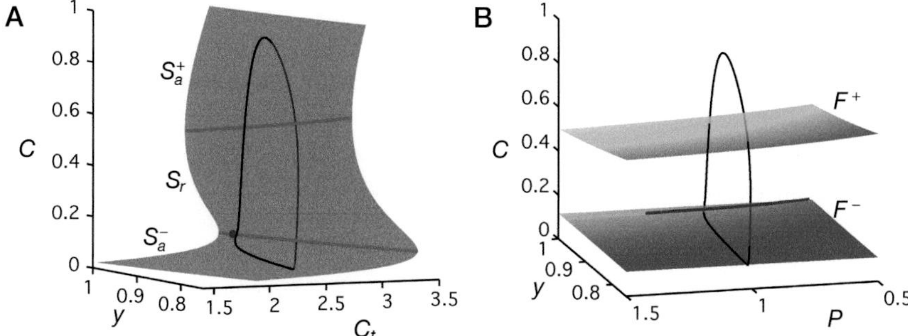

Fig. 5.11 Geometric structures important for the analysis of transient dynamics in the Class I hybrid model. The surfaces shown were computed with $\epsilon = 0.001$ and $\hat{\nu} = 0.317$, but qualitatively similar results are seen for other choices of parameters. **A:** The pink surface is a slice through the critical manifold, with the slice defined by fixing P at its equilibrium value, $P = \hat{\nu}/\hat{\beta} = 0.95$. The blue curves are the intersections of the fold surfaces with the slice. The black curve shows the relaxation oscillation that is the attracting solution at these parameter values. The blue dot marks the folded singularity that occurs for this value of P. **B:** the fold surfaces projected on to (C, y, P) coordinates. The C_t-coordinate is represented as a colour gradient (dark to light as C_t increases). The green and dark blue curve is the locus of folded singularities; blue for folded focus and green for folded node. Reproduced from Harvey et al (2010), Fig. 2, with permission from AIP Publishing.

period, at a regular jump point in each case. A pulse of IP_3 resets the P variable to a higher value, from which point the orbit moves quickly towards the critical manifold then moves slowly along the critical manifold until a fold surface is reached, all the while with P decaying. For the choice of $\hat{\nu}$ used for Panel A, each crossing of the fold surface made by the transient orbit occurs at a regular jump point, resulting in a sequence of relaxation-type spikes, with decaying P, until the original relaxation orbit is reached.

By contrast, in the case shown in Fig. 5.9 B the unpulsed orbit is an MMO (with subthreshold oscillations that are almost invisible on the scale of Fig. 5.9). This orbit crosses each fold surface once per period, but does so at a folded singularity on one surface and a regular jump point on the other; passage near the folded singularity (which is of *folded node* type) produces the subthreshold oscillations. When a pulse of IP_3 is applied, the orbit makes several regular jumps, with P decaying, before it encounters the first folded singularity, which is of folded saddle type. Because of the particular orientation of the stable and unstable manifolds of that folded saddle, the transient orbit is then forced to follow a one-dimensional curve of folded saddles as P decays further, resulting in a lengthy delay before reestablishment of oscillations. When P has decreased sufficiently, the curve of folded saddles turns into a curve of folded nodes and MMO-type oscillations resume, with the orbit eventually settling back to the attracting MMO.

Thus, while there is superficial similarity between the delays seen in the pulse responses of the Class I and II models shown in Fig. 5.9 B and C, the mathematical mechanisms underlying these delays are quite different, being a consequence of the shape of the critical manifold in the Class II case and a result of passage near a particular type of folded singularity in the Class I case. Further detail, including discussion of other types of transient response occurring at different values of $\hat{\nu}$, is contained in Harvey et al (2010). Discussion about the type of response that might be expected from a hybrid model in which both Class I and II mechanisms are present is also contained in Harvey et al (2010).

A heuristic argument was given in Section 3.2.7.2 about likely responses of Class I and Class II models to IP_3 pulsing. It was postulated that a Class I model would respond with a temporary increase in oscillation frequency followed by resumption of regular oscillations, while a Class II model would respond with a delay before resumption of oscillations. In light of the detailed GSPT analysis described above, we now see that the situation is more complicated, and that there is in fact no clear-cut distinction between the responses of these two model types to a pulse in IP_3. This suggests that using IP_3 pulsing to determine whether Class I or Class II mechanisms underlie Ca^{2+} oscillations in a particular cell type might not be as straightforward as first thought.

In any case, understanding of the mechanisms behind the observed responses in the hybrid model depends crucially on a careful identification of the timescale structure of the model and then on exploitation of this structure using GSPT, thus providing a clear example of the utility of GPST methods.

5.6 Coupled Voltage and Calcium Models

In many cell types, the dynamics of cytosolic Ca^{2+} and the cell membrane potential are inextricably intertwined, as discussed in Section 3.7, and models that couple Ca^{2+} and the membrane potential are common. Coupled voltage and Ca^{2+} models are particularly important for the study of neuroendocrine cells (Bertram et al, 2006; Bertram and Sherman, 2000; Bertram et al, 2010; Fletcher and Li, 2009; LeBeau et al, 1997; Lee et al, 2010; Li et al, 1994; Roper et al, 2004; Tsaneva-Atanasova et al, 2010a,b; Zhang et al, 2003), and appear in many places throughout this book, particularly in Chapter 8.

Models of this type generally have multiple time scales, since oscillations in the membrane potential typically occur with a time scale of milliseconds, which is orders of magnitude faster than typical Ca^{2+} oscillations. In fact, since coupled voltage and Ca^{2+} models are frequently built by coupling a two (or more) timescale membrane potential model with a two (or more) timescale Ca^{2+} model, the resulting system will typically have three or more time scales, depending on the relative speeds of slower variables in the membrane potential model and faster variables in the Ca^{2+} model.

An oscillation pattern frequently associated with such coupled models is electrical bursting, discussed in the context of pancreatic β cells in Section 8.4. Some features of bursting oscillations can be explained by assuming a model has two time scales and then freezing the slow variables of the model, as in Section 5.4.1; other examples of this type of analysis are contained in Hindmarsh and Rose (1984), Izhikevich (2000) and Keener and Sneyd (2008).

However, not all instances of bursting can be explained in this way. For example, in their model of bursting in GnRH neurons, Lee et al (2010) found that an appropriate interburst interval could only be obtained by inclusion of currents that evolved on a third, very slow, time scale (Section 8.7.2). A more complete explanation of bursting, and of other solution features such as control of the number of spikes in a burst, requires the use of GSPT and the consideration of evolution of variables on more than two time scales. Vo et al (2013) contains a detailed discussion of the role of Ca^{2+} dynamics in bursting in a pituitary lactotroph model, and in particular shows how GSPT methods developed for two-timescale systems can be used to give information about the dynamics of a model with three time scales, while Krupa et al (2012) and Nan et al (2015) consider the use of GSPT methods for three-timescale models based on coupled GnRH neuron models. More commonly, however, coupled models with three or more time scales are investigated using numerical methods, with more detailed analysis awaiting the development of new theoretical techniques.

5.7 Calcium Waves

Phenomena associated with propagation of waves in Ca^{2+} models are described elsewhere in this book, most notably in Section 3.5 and in Chapter 6. Direct numerical simulation is useful for establishing the existence and stability of waves in a particular wave model, but some more general principles can be uncovered by the use of bifurcation analysis. As discussed in Section 3.5.2, in many models it is sufficient to consider wave propagation in one spatial dimension, and here we restrict ourselves to this case.

If we are interested in solitary or periodic travelling waves with constant wave speed, it is helpful to use a moving frame by introducing a travelling wave coordinate and then work with the travelling wave equations. The travelling wave equations are convenient because they are ordinary differential equations, meaning that bifurcation analysis and the other techniques of the previous sections of this chapter are relatively straightforward to apply. Solitary and periodic travelling waves in the original partial differential equation correspond to homoclinic (or heteroclinic) and periodic solutions, respectively, in the moving frame, and so we will be interested in locating homoclinic or heteroclinic bifurcations and Hopf bifurcations of the travelling wave equations, as both a bifurcation parameter of the original model and the wave speed are varied. In the partial differential equation formulation of the model, the wave speed is selected by the dynamics and is not a parameter of the equations, but in the travelling wave equations we treat the wave speed as a bifurcation parameter.

Here we briefly illustrate the method using a particularly simple model of Ca^{2+} waves (Tsai et al, 2012). In this model we omit any dynamic behaviour of the IPR, and assume instead that the IPR open probability merely follows a steady-state bell-shaped curve (as a function of c; see Fig. 2.14). Thus

$$P_o = \frac{c^2}{\phi_1^2 + c^2} \cdot \frac{\phi_2}{c + \phi_2}, \tag{5.36}$$

for some constants ϕ_1 and ϕ_2. We omit any dependence on the cytosolic IP_3 concentration, p, for reasons that will become clear shortly. The release flux from the ER is then given by

$$J_{\mathrm{ER}} = (k_f P_o + \alpha)(c_e - c), \tag{5.37}$$

for some constant α which denotes a background leak from the ER, while the ATPase pump fluxes are assumed to take the simplest possible, linear, forms. Thus

$$J_{\mathrm{serca}} = k_s c, \tag{5.38}$$

$$J_{\mathrm{pm}} = k_m c. \tag{5.39}$$

Hence the model is

$$\frac{\partial c}{\partial t} = D_c \frac{\partial^2 c}{\partial x^2} + J_{\mathrm{ER}} - J_{\mathrm{serca}} + \delta(J_{\mathrm{in}} - J_{\mathrm{pm}}), \tag{5.40}$$

$$\frac{\partial c_e}{\partial t} = \gamma(J_{\mathrm{serca}} - J_{\mathrm{ER}}). \tag{5.41}$$

For convenience we have assumed that Ca^{2+} does not diffuse within the ER, an assumption which is certainly not true. However, it is useful for illustrating the basics of travelling wave approaches.

In contrast to models discussed previously, here we use J_{in} as the bifurcation parameter, for a fixed value of p. This is why p does not appear explicitly in the model equations. Since J_{in} can be easily manipulated experimentally this is a useful point of view, amenable to making verifiable predictions and simple to compare to data. This different approach illustrates the fact that there are many different ways of analysing Ca^{2+} models, each of which is useful in different circumstances.

In the absence of diffusion (setting $D_c = 0$ in (5.40)), the model has a unique equilibrium point for each choice of J_{in}, with a branch of periodic orbits existing for intermediate values of J_{in}, as shown in Fig. 5.12.

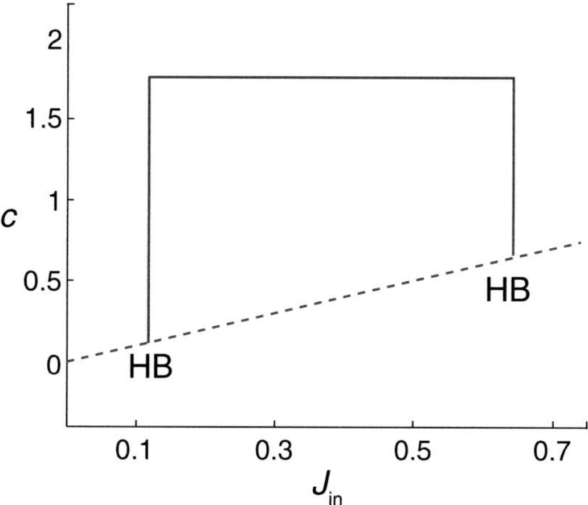

Fig. 5.12 Bifurcation diagram of the model of Ca^{2+} wave propagation in the absence of diffusion, (5.40)–(5.41) with $D_c = 0$ and other parameter values as in Table 5.2. The dashed purple curve indicates equilibrium solutions and the solid blue curve shows the maximum value of c on the branch of periodic solutions. Stability is not indicated. HB denotes a Hopf bifurcation.

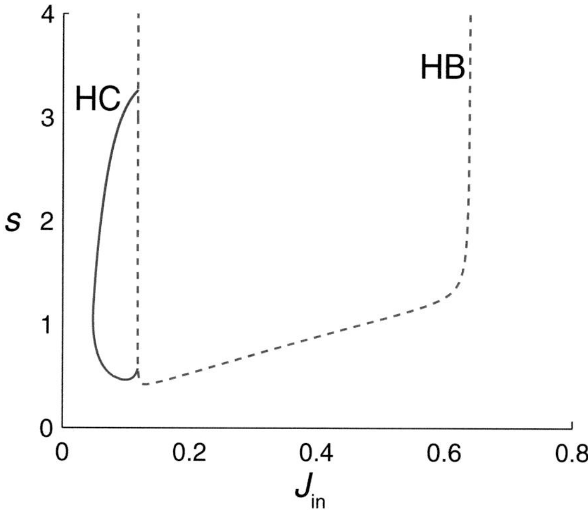

Fig. 5.13 Partial bifurcation set of the travelling wave equations of the model of Ca^{2+} wave propagation (5.40)–(5.41). HC denotes a curve of homoclinic bifurcations (the solid blue curve); HB denotes a curve of Hopf bifurcations (the dashed red curve). Computed using the parameter values in Table 5.2.

$D_c = 25\,\mu\text{m}^2\text{s}^{-1}$	$\alpha = 0.05\,\text{s}^{-1}$
$k_s = 20\,\text{s}^{-1}$	$k_f = 20\,\text{s}^{-1}$
$k_m = 0.1\,\text{s}^{-1}$	$\phi_1 = 2\,\mu\text{M}$
$\phi_2 = 1\,\mu\text{M}$	$\delta = 0.001$
$\gamma = 5.0$	

Table 5.2 Parameters of the simple model of Ca^{2+} waves (Tsai et al, 2012).

We obtain the travelling wave equations for this model using the procedure outlined in Section 3.5.2. Specifically, we change variables by introducing the travelling wave coordinate $\xi = x + st$, where s is the wave speed, and rewrite (5.40) and (5.41) as

$$c' = d, \tag{5.42}$$

$$d' = \frac{1}{D_c}(sd - J_{\text{ER}} + J_{\text{serca}} - \delta(J_{\text{in}} - J_{\text{pm}})), \tag{5.43}$$

$$c_e' = \frac{\gamma}{s}(J_{\text{serca}} - J_{\text{ER}}), \tag{5.44}$$

where a prime denotes differentiation with respect to ξ, and $d = c'$. The result of a two-parameter numerical bifurcation analysis of these equations, with s and J_{in} as the bifurcation parameters, is shown in Fig. 5.13. The behaviour as $s \to \infty$ is that of the model in the absence of diffusion, as expected from the general theory (Maginu, 1985) and shown in Fig. 5.12. Thus, for large values of s there are two Hopf bifurcations. If we track these Hopf bifurcations in the (s, J_{in}) plane they form a U-shaped curve; this marks the locus of onset of periodic travelling waves in the original partial differential equations, with periodic waves occurring in the region inside the U. A C-shaped branch of homoclinic bifurcations lies to the left of the U-shaped Hopf bifurcation curve. For values of s and J_{in} on this branch, the original equations have travelling wave solutions. At some point there is a transition from single travelling pulses to periodic waves, but exactly how this type of transition occurs is not completely understood (Champneys et al, 2007).

The curves in Fig. 5.13 are by no means the only bifurcations in the model, although they serve to outline the major wave features. In particular, periodic solutions of the travelling wave equations can be created by mechanisms other than Hopf or homoclinic bifurcations (for example, by saddle-node bifurcations of periodic orbits or by heteroclinic bifurcations involving equilibria or periodic solutions (Zhang et al, 2012)). Furthermore, an additional curve of homoclinic bifurcations occurs to the right of the Hopf U, as will be discussed further in Section 5.8. While the details of these additional bifurcations may be interesting mathematically, they are in most cases irrelevant from the point of view of physiology.

Working in the moving frame allows the existence of travelling waves in the original model to be established, but gives no information about stability of these solutions. Stability can easily be checked numerically by direct simulation of the partial differential equations or, in some cases, can be computed using Evans functions or other theoretical methods (for an example of this in the context of Ca^{2+} waves, see Romeo and Jones (2003)). For the model discussed above, it can be shown numerically that the upper section of the C-shaped curve of homoclinic orbits corresponds to stable travelling wave solutions. This branch of homoclinic orbits denotes travelling waves with speeds of around 1–3 μm s^{-1}, which is considerably slower than the physiological values of around 10–15 μm s^{-1}. This suggests that the model is too simple to be a reliable quantitative guide to Ca^{2+} waves. Nevertheless, it has the correct qualitative features. For example, as J_{in} increases, so does the wave speed, consistent with what has been observed experimentally.

It is natural to ask which of the periodic solutions inside the U-shaped Hopf curve correspond to stable periodic travelling waves in the original model, i.e., to ask which wave speed will be selected by the dynamics of the partial differential equation. For a particular model, this could again be checked by direct simulation, but no general answer is known; it is believed to depend on the precise boundary and initial conditions for the partial differential equation. The issue was investigated numerically in Simpson et al (2005) for a model of the basal region of pancreatic acinar cells, and it was shown that extremely complicated, seemingly chaotic travelling solutions can occur inside the corresponding U-shaped Hopf curve; presumably, similarly complicated solutions can also occur in models such as the one considered above.

The basic structure of the bifurcation diagram shown in Fig. 5.13, with a C-shaped branch of homoclinic orbits and a U-shaped branch of Hopf bifurcations, seems to be common to many models of excitable systems. The wave versions of the FitzHugh-Nagumo equations and the Hodgkin-Huxley equations have the same basic C-U structure, as do all models of Ca^{2+} wave propagation for which this question has been studied. This seems to be a consequence of the general shape and relative position of nullclines in these models, which follows from the underlying physiology. Specifically, as argued in Champneys et al (2007), many excitable models have two important nullclines, one N-shaped and one roughly linear, and these move relative to one another as a bifurcation parameter is varied. For instance, in Ca^{2+} models such as (5.42)–(5.44), the nullcline for $d(= c')$ is N-shaped while the nullcline for c_e is roughly linear. These nullclines move across each other as a bifurcation parameter (for example, J_{in} in (5.42)–(5.44)) is varied so that they intersect first on the left branch of the N-shaped nullcline, then, in turn, on the middle and right branches. This relative motion of the nullclines typically results in the occurrence of a homoclinic bifurcation followed by two or more Hopf bifurcations, as is consistent with a C-U structure (Champneys et al, 2007; Tsai et al, 2012). In all the cases studied, stable travelling pulses are associated with the upper section (higher s values) of the C-shaped curve, although the branch can change stability towards its rightmost end. More complicated solitary waves, with two or more pulses within the wave packet, may also occur in excitable systems of this type (Champneys et al, 2007; Carter and Sandstede, 2014).

Consideration of different time scales for cellular processes is as much of an issue for models of Ca^{2+} waves as it is for spatially independent models, and many of the ideas discussed in Section 5.4 can also be used for the analysis of wave models. Ideas from GSPT have been successfully applied to analysis of the wave version of the FitzHugh-Nagumo equations (Bell and Deng, 2002; Carter and Sandstede, 2014; Jones, 1984; Krupa et al, 1997) but, until recently, have been used less for the analysis of Ca^{2+} wave models. One recent approach has been to look for a singular analogue of the C-U bifurcation structure, and to demonstrate that bifurcations of the full, non-singular, system are perturbations of this. Tsai et al (2012) investigated a singular (closed-cell) version of (5.42)–(5.44), finding a singular C-U structure and showing numerically that the C-U structure for the full (open-cell) system lies near to the singular C-U structure (see Fig. 5.14). Work still remains to be done to show rigorously how the non-singular C-U structure arises as a perturbation of the singular case.

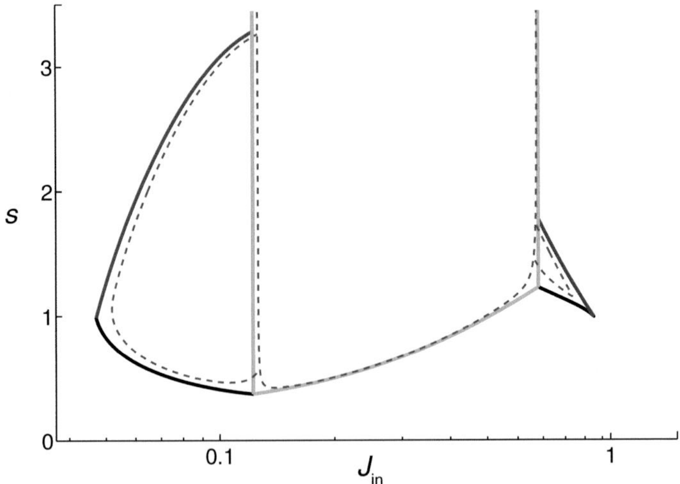

Fig. 5.14 Partial bifurcation set for singular ($\delta = 0$) and non-singular ($\delta = 0.001$) versions of (5.42)–(5.44), with parameter values as in Table 5.2 (Tsai et al, 2012). Bifurcations for the non-singular system are plotted with dashed curves; the left and right dashed curves are the loci of homoclinic bifurcations and the middle (U-shaped) curve is the locus of Hopf bifurcations. Bifurcations for the singular system are plotted with solid curves; the red and blue curves show the locus of travelling fronts, the black curves indicate the locus of travelling pulses, and the green curve shows the locus of Hopf bifurcations producing periodic travelling waves. Information about the stability of solutions is not included in the diagram.

5.8 Calcium Excitability and the FitzHugh-Nagumo Equations

Calcium excitability, which occurs when the release of a small quantity of Ca^{2+} induces the release of a larger quantity of Ca^{2+}, is an important feature of many models of Ca^{2+} dynamics. As discussed in Section 3.2.4, much of the current theoretical understanding of Ca^{2+} excitability is based on analogy with the theory of electrical excitability, and in particular on results about the dynamics of the canonical excitable model, the FitzHugh-Nagumo equations. However, there are some important structural differences between Ca^{2+} models and the FitzHugh-Nagumo equations which lead to differences in dynamics.

The FitzHugh-Nagumo equations can be written as

$$\frac{\partial u}{\partial t} = D\frac{\partial^2 u}{\partial x^2} + u(u-\alpha)(1-u) - w + I, \tag{5.45}$$

$$\frac{\partial w}{\partial t} = \epsilon(u - \gamma w), \tag{5.46}$$

where I denotes the applied current, and will be the main bifurcation parameter in the following discussion. The parameter ϵ is normally assumed to satisfy $0 \leq \epsilon \ll 1$, in which case the model is a two-timescale model. The parameter D is the diffusion constant, γ is a positive constant and $\alpha \in (0, \frac{1}{2})$. The corresponding travelling wave equations are

$$\frac{du}{dz} = v,$$

$$\frac{dv}{dz} = \frac{1}{D}\left(sv - u(u-\alpha)(1-u) + w - I\right), \tag{5.47}$$

$$\frac{dw}{dz} = \frac{\epsilon}{s}(u - \gamma w). \tag{5.48}$$

If there is no diffusion, there are strong similarities between the dynamics of the FitzHugh-Nagumo equations and the dynamics of typical Ca^{2+} models. For instance, Fig. 5.15 shows the bifurcation diagram for (5.45)–(5.46) with $D = 0$; the similarities to the bifurcation diagram for the diffusion-free version of (5.40)–(5.41) shown in Fig. 5.12 are clear. This is a consequence of structural similarities between the models; both model types have two time scales, with a fast variable (v for the FitzHugh-Nagumo model and c for a typical Ca^{2+} model) with a roughly cubic-shaped nullcline, and a slow variable (w for FitzHugh-Nagumo and c_t for a typical Ca^{2+} model) with a roughly linear nullcline. Differences in the details of the dynamics can occur, particularly since Ca^{2+} models frequently have more dependent variables than the FitzHugh-Nagumo equations, but the overall dynamics is broadly similar.

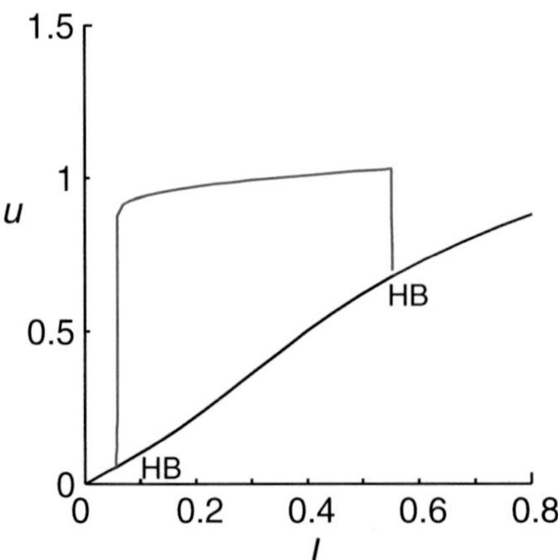

Fig. 5.15 Bifurcation diagram for the FitzHugh-Nagumo model, (5.45)–(5.46), without diffusion ($D = 0$), and with $a = 0.1$, $\gamma = 1.0$, $\epsilon = 0.1$. The black curve shows the position of the steady-state solution and the red curve indicates the maximum amplitudes of periodic orbits. Hopf bifurcations are labelled HB. Stability is not indicated.

Even when diffusion is included, there are still significant similarities between the dynamics of the two types of model. In particular, as noted in the previous section, the travelling wave equations of the FitzHugh-Nagumo equations have a C-U bifurcation structure very similar to that seen in Ca^{2+} models; again, this is a consequence of the structural similarities between the models. However, closer investigation shows that there are important differences between the mechanisms underlying the dynamics in the different models.

Perhaps the most significant difference comes from the way diffusion acts relative to the time scales in the model. In the FitzHugh-Nagumo model, diffusion has an effect on the evolution of the fast variable only and does not enter the definition or the evolution equation for the slow variable. By contrast, in Ca^{2+} models, diffusion in the cytoplasm affects both fast and slow variables. To see this, we note that an appropriate slow variable for many Ca^{2+} wave models is $c_t = s(c_e/\gamma + c) - D_c d$; this definition of c_t is the wave model analogue of the earlier definition of c_t (introduced in Section 3.2.5 and discussed in Section 5.3.1). That this combination of original variables is slow follows from rewriting (5.42)-(5.44) by replacing c_e with c_t, to get

$$c' = d, \tag{5.49}$$

$$D_c d' = sd - (k_f P_0 + \alpha)\left(\frac{\gamma}{s}(c_t + D_c d - sc) - c\right) + J_{\text{serca}} - \delta(J_{\text{in}} - J_{\text{pm}}), \tag{5.50}$$

$$c_t' = \delta(J_{\text{in}} - J_{\text{pm}}), \tag{5.51}$$

where a prime indicates differentiation with respect to the travelling wave coordinate, $\xi = x + st$, and other quantities are as for (5.40) and (5.41). It can now be seen that when δ is sufficiently small, c and d are fast variables and c_t is slow, and the explicit appearance of D_c in the definition of c_t reflects the fact that diffusion affects the slow variable.

Diffusion introduces nonlinear coupling between the fast variables of (5.49)–(5.51), via the term in $P_0 d$ in the d' equation (since P_0 is a function of c). This type of coupling arises from the structure of the equations and will thus be present in typical Ca^{2+} models. The difference between this and the simpler coupling found in the FitzHugh-Nagumo equations is seen in the nature of the Hopf bifurcations in these models. In the FitzHugh-Nagumo equations, the Hopf bifurcations corresponding to the near vertical arms of the Hopf U are degenerate in the singular limit, meaning that the first Lyapunov coefficient of the bifurcation is zero and the bifurcation is neither supercritical or subcritical (Tsai et al, 2012). This degeneracy is removed by the more generic coupling in the Ca^{2+} models. It is not known exactly how this difference in the singular limit influences the dynamics in the non-singular models and more work on this issue is needed.

Note, however, that in both the FitzHugh-Nagumo equations and typical Ca^{2+} models the Hopf bifurcations are singular Hopf bifurcations, so that the frequency of the oscillations at onset tends to zero as $\delta \to 0$ and the amplitude of the branch of oscillations grows very rapidly with respect to the bifurcation parameter near the Hopf bifurcation. The difference between supercritical and subcritical (or even degenerate) Hopf bifurcations may therefore be hard to observe. In all cases, the bifurcation diagram is likely to have a steep branch of periodic solutions near the Hopf bifurcation, followed by a plateau where the oscillations are of relaxation type (as, for instance, in Figs. 5.15 and 5.12); whether the steep branch corresponds to stable or unstable periodic orbits may be of interest mathematically but has less relevance from a physiological point of view since the branch is confined to a relatively small range of parameter values.

Another important structural difference between typical Ca^{2+} models and the FitzHugh-Nagumo equations is that the FitzHugh-Nagumo equations are symmetric, with the form of the equations being unchanged by the following transformation (Champneys et al, 2007):

$$u \to \frac{2}{3}(1+\alpha) - u, \quad v \to -v, \quad w \to \frac{2}{3\gamma}(1+\alpha) - w, \tag{5.52}$$

$$I \to \frac{2}{3}(1+\alpha)\left[\frac{1}{\gamma} - \frac{(2-\alpha)(1-2\alpha)}{9}\right] - I. \tag{5.53}$$

One consequence is that the C-U bifurcation structure is symmetric about the mid-line of the Hopf U curve for the FitzHugh-Nagumo equations, and there are two C-shaped curves of homoclinic bifurcations, one on each side of the Hopf U. A second consequence is that travelling pulses can arise either as a perturbation of two symmetry-related singular travelling fronts (corresponding to symmetry-related heteroclinic orbits in the moving frame) or as a perturbation of a single singular travelling pulse (corresponding to a homoclinic orbit in the moving frame) (Tsai et al, 2012). On the other hand, Ca^{2+} models typically do not have this type of symmetry, and so the C-U bifurcation set is unlikely to be symmetric; Fig. 5.14 illustrates this point for (5.42)–(5.44). However, Tsai et al (2012) have shown that, even in the absence of symmetry, travelling pulse solutions to (5.42)–(5.44) can arise as perturbations of either two singular travelling fronts or a single singular travelling pulse, in an analogous way to the case for the FitzHugh-Nagumo equations. This issue does not appear to have been considered in other Ca^{2+} models, and it remains to establish what the typical case is.

Overall, we can see that, in many respects, the excitability associated with the FitzHugh-Nagumo equations provides a reasonable qualitative guide to the dynamics of Ca^{2+} excitability, but there are some structural differences between the models that arise because the FitzHugh-Nagumo equations are too simple to be generic; these lead to some important differences in model dynamics, and we conclude that Ca^{2+} excitability is of a somewhat different nature to excitability in the FitzHugh-Nagumo equations.

Part II
Specific Models

Chapter 6
Nonexcitable Cells

In the remainder of this book we present a small number of examples, from a range of specific cell types, in more depth. Of course, it is not possible to give a complete overview of all the Ca^{2+} models that have been constructed, and so our choice here is based on a mixture of convenience and personal preference. However, the selected examples cover a broad range of modelling styles, in many qualitatively different types of cells, and should give a reasonably comprehensive picture of the types of models that have proved useful.

In this chapter each of our examples comes from nonexcitable cells, which tend to have quite different Ca^{2+} dynamics from excitable cells. In excitable cells the Ca^{2+} dynamics of the cytoplasm is usually strongly linked to Ca^{2+} entry through voltage-gated channels, and thus to the membrane potential. Hence excitable cells often have a cytosolic Ca^{2+} oscillator linked to a membrane potential oscillator, potentially giving rise to highly complex behaviours; a number of such models are discussed in Chapter 8. However, in nonexcitable cells such as epithelial cells, exocrine epithelial cells, oocytes, and hepatocytes, cytosolic Ca^{2+} dynamics tends to occur largely independently of the cell membrane potential, being governed instead by the internal dynamics of Ca^{2+} release through IPR and RyR.

6.1 *Xenopus* Oocytes

Xenopus laevis is an aquatic frog, native to sub-Saharan Africa, which, thanks to the large size of its egg cells, or oocytes, has been widely used in experimental investigations. The oocyte diameter is of the order of 1-1.2 mm, i.e., 50–100 times larger than an average cell. Furthermore, an albino lineage of this frog is easily available, which greatly facilitates imaging. In the field of Ca^{2+} signalling, it is one of the most widely used experimental models for the study of Ca^{2+} blips, puffs , oscillations, and intracellular waves. Very low $[IP_3]$

G. Dupont et al., *Models of Calcium Signalling*, Interdisciplinary Applied Mathematics 43, DOI 10.1007/978-3-319-29647-0_6

induces Ca^{2+} blips and puffs (see Chapter 4), while Ca^{2+} oscillations are observed in immature oocytes in response to stimulation by acetylcholine (ACh) or by flash photolysis of sufficient amounts of caged IP_3. These oscillations correspond to the repetitive propagation of intracellular waves, which can take the spectacular form of spirals as illustrated in Chapter 1 (see Fig. 1.7). These waves propagate at about $20\,\mu m\,s^{-1}$, with a frequency of about 1 per minute.

Upon maturation, oocytes become eggs. Upon fertilisation of the egg, sperm-egg fusion produces a different type of Ca^{2+} response, in that a single, slow Ca^{2+} wave invades the egg. Given that the speed of this fertilisation wave is about $5\,\mu m\,s^{-1}$, it takes about 4 minutes to traverse the egg. After passage of the wave, a sustained high $[Ca^{2+}]$ is maintained for 5–6 minutes. This high $[Ca^{2+}]$ is necessary for the prevention of polyspermy and for resumption of the embryonic cell cycle.

Besides the abundance of experimental data, the *Xenopus* oocyte is well suited to a modelling approach because of the limited number of elements of the Ca^{2+} toolbox necessary to provide a good description of the observed Ca^{2+} dynamics. Stable oscillations are maintained in the absence of external Ca^{2+}, which suggests that Ca^{2+} exchange with the extracellular medium does not play a significant role in the oscillatory mechanism (Yao and Parker, 1994). Although mitochondria clearly affect the Ca^{2+} wave speed and frequency (Section 6.1.2), they are not essential, as changes in the rate of mitochondrial Ca^{2+} transport can neither induce nor abolish the oscillations (Jouaville et al, 1995). Furthermore, the only Ca^{2+}-releasing channels in the ER membrane are the IPR, as RyR are not expressed.

6.1.1 A Heuristic Model for Calcium Oscillations and Waves

The earliest model of Ca^{2+} dynamics in *Xenopus* oocytes is that of Atri et al (1993). The model includes only Ca^{2+} release from the ER (J_{IPR}), Ca^{2+} pumping back into the ER (J_{serca}), and a small Ca^{2+} entry from the outside (β), which is not necessary to get oscillations. Thus, letting c denote $[Ca^{2+}]$, we have

$$\frac{dc}{dt} = J_{\text{IPR}} - J_{\text{serca}} + \beta. \tag{6.1}$$

Since the PMCA pumps are assumed to behave in the same way as the SERCA pumps, and since the ER $[Ca^{2+}]$ is not modelled explicitly, the flux through the PMCA can be amalgamated into the SERCA flux, which explains why it does not appear in the equation explicitly.

This heuristic model is based on the gating model of the IPR described in Section 2.7.3. The steady-state Ca^{2+} flux through the IPR is a bell-shaped function of c, determined by fitting to the data of Parys et al (1992). This bell-shaped curve is the macroscopic counterpart of the biphasic curves that represent the IPR open probability shown in Fig. 2.14. However, the dynamics of IPR inhibition by Ca^{2+}, which is assumed to occur on a slower time scale than IPR activation by Ca^{2+} and IP_3, is not determined directly from experimental data.

Pumping back into the ER is modelled by a simplified version of (2.28) where, for simplicity, cooperativity in Ca^{2+} binding is not included. Thus,

$$J_{\text{serca}} = \frac{V_{\text{max}} c}{K + c}. \tag{6.2}$$

In (6.1), Ca^{2+} buffers are assumed to be fast and in excess (see Section 2.9), and so all fluxes should be interpreted as effective fluxes.

The model defined by (2.122), (2.123), (2.137), (6.1), and (6.2) can reproduce Ca^{2+} oscillations, and, despite its simplicity, can also reproduce experimental results about Ca^{2+} changes in response to successive IP_3 pulses.

The model can be extended to simulate Ca^{2+} waves, by adding the diffusion term $D_c \nabla^2 c$ to the right-hand side of (6.1). Because Ca^{2+} buffers are not included explicitly, D_c here is the effective diffusion coefficient, as described in Section 2.9. In experiments, Ca^{2+} waves are mostly initiated by the photorelease of a bolus of IP_3 in the cell cytoplasm. Thus, $[IP_3]$ (denoted by p) cannot be treated as a constant. Instead, p evolves according to

$$\frac{\partial p}{\partial t} = D_p \nabla^2 p - k_p p, \tag{6.3}$$

which is the simplest possible form as it ignores endogenous IP_3 synthesis (supposed to be small because GPCR have not been activated) as well as the enzymatic nature of the reactions that metabolise IP_3. If k_p is small, IP_3 will persist for a long time, allowing the emergence of complex spatiotemporal behaviour that lasts until the IP_3 has all been degraded.

To simulate Ca^{2+} waves, the model equations are solved numerically on a square two-dimensional domain with no-flux boundary conditions (Fig. 6.1). Flash photolysis of IP_3 is simulated through the initial conditions for $[IP_3]$: $p(x, 0)$ is set at a high value in a restricted region of the domain corresponding to the illuminated part of the egg. This leads to a rapid release of Ca^{2+} in the region of the flash. Both IP_3 and released Ca^{2+} diffuse into the rest of the cell and sequentially activate Ca^{2+} release from neighbouring regions. As IP_3 diffuses quickly ($D_p = 300\,\mu\text{m}^2\,\text{s}^{-1}$) compared to Ca^{2+} (Ca^{2+} buffering results in an effective diffusion coefficient for Ca^{2+} of around $20\,\mu\text{m}^2\,\text{s}^{-1}$, as discussed in Section 2.9), the speed of the wave is mainly dependent on the square root of D_c, the Ca^{2+} diffusion coefficient. As D_c is varied from $20\,\mu\text{m}^2\,\text{s}^{-1}$ to $50\,\mu\text{m}^2\,\text{s}^{-1}$, the speed of the model wave varies from about $7\,\mu\text{ms}^{-1}$ to about $11\,\mu\text{ms}^{-1}$, which agrees well with the experimentally measured wave speed of between 5 and $25\,\mu\text{ms}^{-1}$ (Girard et al, 1992; Girard and Clapham, 1993).

In addition, Atri et al (1993) showed that the appearance of the wave depends on the shape of the domain in which IP_3 is elevated initially, i.e.,

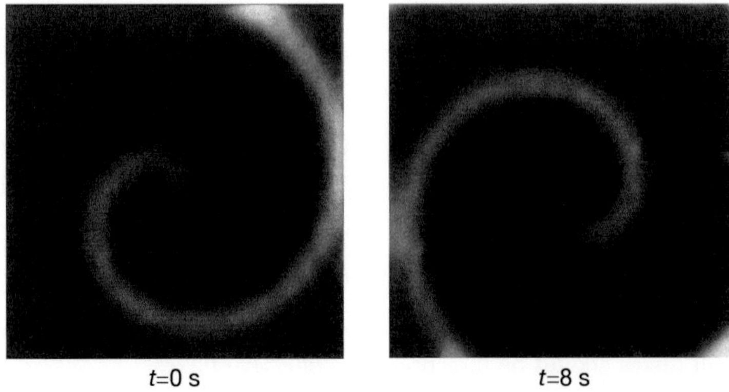

Fig. 6.1 Spiral Ca^{2+} waves in the model of Atri et al (1993). The domain is 600 μm square. In the experiments (Fig. 1.7), the wave is initiated by the photorelease of caged IP_3. In the model, the wave is initiated by raising IP_3 above its steady-state concentration in a portion of the domain, and then letting IP_3 diffuse. Lighter shades denote higher $[Ca^{2+}]$. Reproduced from Atri et al (1993), Fig. 11, with permission from Elsevier.

the spatial extent of the IP_3 bolus. A sharp, linear front originates from a line-like IP_3 increase, while spirals require a global IP_3 stimulation together with some heterogeneities in basal Ca^{2+}, allowing for the breakdown of circular fronts into spirals. In simulations, spirals can also be obtained when taking into account that both the rate of IP_3 synthesis and the IPR density are not spatially uniform (Dupont, 1998).

Finally, we note that Ca^{2+} waves occur only in the cortex of the cell, which is a thin region well approximated by a two-dimensional sheet, and thus scroll Ca^{2+} waves, or other three-dimensional phenomena, have not been observed in *Xenopus*.

6.1.2 Mitochondria and Spiral Wave Stability

Mitochondrial transport of Ca^{2+} has a significant effect on wave propagation in *Xenopus* oocytes (Jouaville et al, 1995). As shown in Fig. 6.2, addition of pyruvate/malate, which increases the mitochondrial membrane potential difference and thus increases the rate of Ca^{2+} uptake, converts multiple small Ca^{2+} waves into a smaller number of larger waves. When the Ca^{2+} response is viewed as a function of time (Fig. 6.2 B) the changes in frequency and amplitude are clear.

It is not immediately clear why such changes in wave properties should result from increased mitochondrial Ca^{2+} uptake. One possible explanation was provided by Falcke et al (1999) who showed that changes in mitochondrial uptake could affect the stability of wave fronts, which could in turn result in waves of lower frequency and higher amplitude, as seen experimentally.

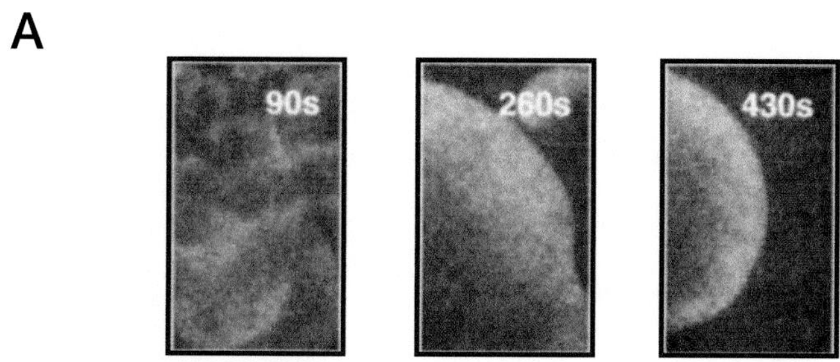

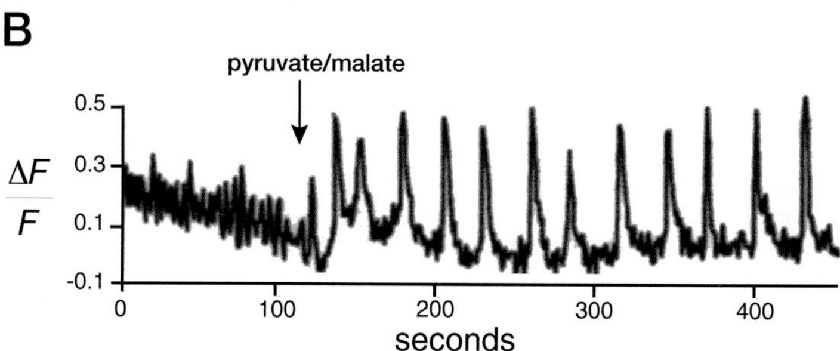

Fig. 6.2 Ca^{2+} waves in *Xenopus* oocytes before and after mitochondrial energisation at approximately 110 seconds. **A:** Ca^{2+} fluorescence as a function of space, measured at the indicated times. The leftmost panel (90 s) is before mitochondrial energisation, the other two after. **B:** Ca^{2+} fluorescence measured at a fixed position, and plotted as a function of time. Reproduced from Jouaville et al (1995), Fig. 1a, c, with permission from the Nature Publishing Group.

To show this, Falcke et al (1999) used a closed-cell model of Ca^{2+} dynamics (Section 3.2.5) based on the IPR model of Tang et al (1996) – a model that is identical in structure to the gating models discussed in Section 2.7.3 – and with the SERCA pumps modelled using the Hill function of (2.28). However, the major innovation of the model of Falcke et al (1999) was to include transport of Ca^{2+}, in simplified form, to and from the mitochondria. Uptake of Ca^{2+} by the uniporter was modelled in the simplest fashion by the Hill function of (2.76),

$$J_{\text{uni}} = \frac{V_{\text{uni}} c^2}{K_d^2 + c^2}, \tag{6.4}$$

ignoring any explicit dependence on mitochondrial membrane potential, while the NCLX was modelled by (2.82), assuming that n_c is constant. Energisation of the mitochondria was modelled simply by an increase in V_{uni}, which increases the rate at which Ca^{2+} is transported into the mitochondria. The resultant model has three variables; cytosolic $[Ca^{2+}]$ (c), mitochondrial $[Ca^{2+}]$ (c_m), and the inactivation variable of the IPR.

Fig. 6.3 A shows a typical wave in the model, with and without energised mitochondria. Mitochondrial transport sharpens the peak of the wave, as the uniporter removes Ca^{2+} more quickly, but the wave peak is then followed by a pronounced shoulder, as Ca^{2+} is released from the mitochondria through the NCLX. The waves take the form of circular plane waves (Fig. 6.3 B), and the change in period can be seen clearly in the time series (Fig. 6.3 C).

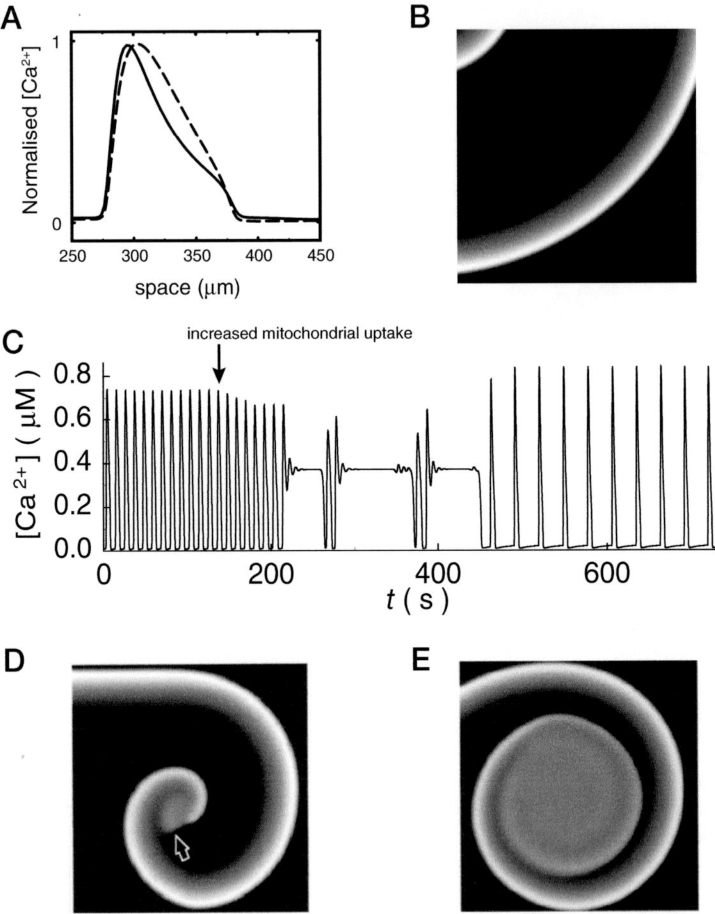

Fig. 6.3 Simulations of the model of Falcke et al (1999), showing how mitochondrial transport changes wave shape and stability. **A:** Waves (travelling from right to left) in the model with (solid line) and without (dashed line) mitochondrial Ca^{2+} uptake. The mitochondria cause a sharper wave peak, as Ca^{2+} is taken up by the uniporter, followed by a pronounced shoulder, as Ca^{2+} is released again through the NCLX. **B:** In the presence of mitochondrial transport, the waves take on the form of circular plane waves with decreased period (compared to the waves in the absence of mitochondrial transport). **C:** Time series taken from a fixed position in a two-dimensional simulation, showing how energisation of the mitochondria (by increasing V_{uni}) changes the amplitude and period of the waves. **D:** In the presence of mitochondrial uptake the core of the spiral becomes unstable, as the trailing shoulder of the wave (see A) starts to propagate a secondary wave, thus destabilising the spiral core and turning the spiral wave into a circular wave, as shown in E.

Fig. 6.3 D and E show how the trailing wave shoulder destabilises the spiral core, transforming the spiral waves into circular plane waves, thus changing the large-scale spatial structure of the waves.

But why should the destabilisation of spirals lead to decreased wave frequency? The answer to this lies in the fact that, in *Xenopus*, spiral waves have a higher frequency than circular wave fronts and target patterns (Lechleiter et al, 1998). Furthermore, spatial structures of higher frequency always take over the domain, as they progressively push out waves of lower frequency. Thus, in *Xenopus* under normal conditions, high-frequency spirals predominate, and the Ca^{2+} responses take the form of multiple high-frequency spiral foci. However, once the spirals lose stability they disappear, thus allowing the lower frequency circular waves to appear over large parts of the domain.

A more detailed theoretical analysis of wave stability and mitochondrial transport was done by Falcke et al (2000).

6.1.3 Bistability and the Fertilisation Calcium Wave

In contrast with the circular and spiral Ca^{2+} waves observed in immature oocytes, fertilisation of mature *Xenopus* eggs triggers a single, high-amplitude, and long-lasting wave of increased $[Ca^{2+}]$ that starts at the point of sperm-egg fusion and crosses the egg in a few minutes. After passage of this wave, often referred to as the "Ca^{2+} tide", cytoplasmic $[Ca^{2+}]$ remains high for at least 5 minutes. The wave starts as a thin crescent near the site of sperm-egg fusion and then propagates with a concave shape (in fact, nearly flat) until it reaches the opposite side of the egg. This means that the speed of propagation through the cortex is larger than in the centre. One of the aims of modelling is to provide a plausible physiological explanation for this unexpected behaviour.

The molecular mechanisms responsible for the fertilisation Ca^{2+} wave in *Xenopus* oocytes and its peculiarities have been investigated in a series of joint experimental and modelling papers (Fontanilla and Nuccitelli, 1998; Wagner et al, 1998; Bugrim et al, 2003; Wagner et al, 2004; Fall et al, 2004). The observation that the wave involves an abrupt transition of $[Ca^{2+}]$ from a low value (about $0.2\,\mu$M) to an elevated value (about $1.5\,\mu$M) that is maintained for an extended period of time strongly suggests that the underlying dynamics is bistable. In other words, during maturation, the egg somehow acquires the capacity to support two alternative steady states; a low concentration typical of the pre-fertilisation state and a high concentration behind the wave. In agreement with this hypothesis, cytoplasmic Ca^{2+} additions by flash photolysis of caged Ca^{2+} allow the passage from low to high concentration. There is even a threshold $[Ca^{2+}]$ below which additions of Ca^{2+} fail to initiate a wave. This threshold likely corresponds to an unstable steady state acting as a separatrix between the stable steady states.

The question next arises as to the nature of the changes, occurring during maturation of the egg, that transform the cytoplasm from an oscillatory medium with a single stable limit cycle to a bistable medium with no limit cycle and two stable fixed points. Using the Li-Rinzel model of the IPR ((2.117)–(2.118); also see Section 3.2.1), Wagner et al (1998) proposed an elegant mechanism that can account for this transition.

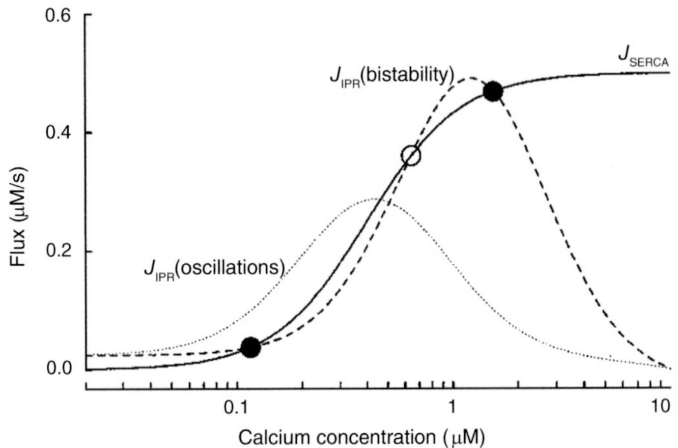

Fig. 6.4 Passage from oscillations to bistability induced by a change in the balance between Ca^{2+} fluxes in and out of the ER. Specifically, y_∞ and K_5 are twice as large for bistability than for oscillations. For the parameters that give oscillations, only one steady state exists, but for the parameters that give bistability there are three steady states. In the latter case the solid circles denote stable steady states, while the open circle denotes an unstable steady state. Reproduced from Wagner et al (1998), Fig. 1, with permission from Elsevier.

The passage from oscillations to bistability can be viewed as a shift in the balance between the fluxes governing the Ca^{2+} exchanges between the cytoplasm and the ER, i.e., IP_3-mediated Ca^{2+} release (J_{IPR}) and Ca^{2+} pumping by the SERCA pumps (J_{serca}). Since there is assumed to be no transport across the plasma membrane, the equation for c is

$$\frac{dc}{dt} = J_{\text{IPR}} - J_{\text{serca}}, \tag{6.5}$$

where J_{IPR} and J_{serca} are given by (3.10) and (3.12). There is an associated equation for y (the fraction of IPR that have been inactivated by Ca^{2+}), given by (2.115). Hence the steady-state value for c is given by $J_{\text{IPR}} = J_{\text{serca}}$, when y is set at its equilibrium value, and p is considered as a constant parameter. This equation is most easily solved graphically, as shown in Fig. 6.4. For parameters corresponding to a single steady state, the two curves (dotted and solid) intersect only once. Numerical integration shows that this steady state is unique, unstable, and surrounded by a stable limit cycle. Upon changes in the parameter values that characterise $J_{\rm IPR}$ (dashed curve), the relative position of this curve with respect to the $J_{\rm serca}$ curve (solid curve) can change drastically, which can result in the existence of three intersection points, two of which are stable and one of which is unstable. At the end of this section, we will discuss in greater detail the possible characteristics of IP_3-mediated Ca^{2+} release that are modified during maturation.

In this framework, the fertilisation wave would correspond to the passage from the low steady state to the high one in response to an appropriate perturbation in the form of an increase in either $[IP_3]$ or $[Ca^{2+}]$, induced by sperm-egg fusion. The aim of the subsequent studies was to infer both the exact nature of the perturbation and the spatial characteristics of the IP_3-mediated Ca^{2+} release in the oocyte from a detailed comparison of the characteristics of the wave seen experimentally and in the model.

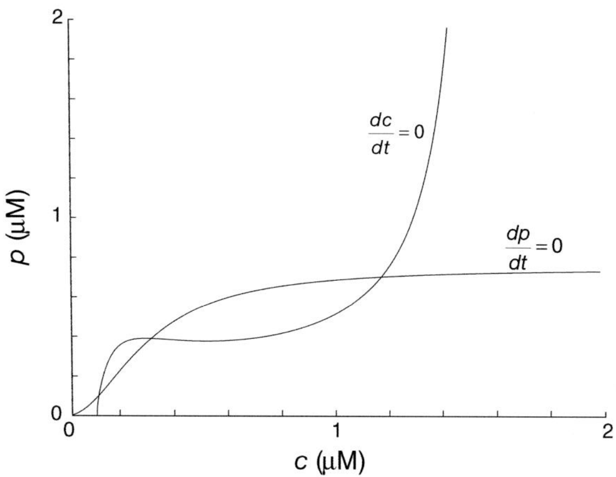

Fig. 6.5 Bistability arising from the interplay between Ca^{2+} and IP_3 dynamics. Curves represent the projections on the c, p plane of the nullclines for Ca^{2+} (6.5) and IP_3 (6.6), obtained by setting $y = y_\infty$.

Answering this question was not a simple task. Three-dimensional simulations suggested that the shape of the wave could only be reproduced when imposing strict conditions on the ER density, on the steady-state distribution of IP_3 before fertilisation, and on the shape of the IP_3 bolus resulting from sperm binding (Wagner et al, 1998; Bugrim et al, 2003). Although none of these hypotheses was unrealistic from a physiological point of view, the precision of the required arrangement suggested some lack of robustness that is not expected for a process as important as fertilisation.

On the other hand, experimental evidence in favour of an active role of IP_3 dynamics in the fertilisation wave was accumulating. Subsequent measurements strongly suggested that IP_3 is produced in a wave-like manner throughout the egg (Wagner et al, 2004). In the above mechanism (Fig. 6.4),

IP_3 was considered as a parameter that was not supposed to change during the fertilisation wave as both steady states correspond to the same value of $[IP_3]$. Subsequent models relaxed this assumption, and included an additional equation for IP_3 dynamics (Fall et al, 2004),

$$\frac{\partial p}{\partial t} = V_{\mathrm{PLC}} \frac{c^2}{K_{\mathrm{PLC}}^2 + c^2} - k_d p. \tag{6.6}$$

The stimulation of PLC activity by Ca^{2+} was found to be necessary to account for experimental results, in agreement with observations performed in some cell types. Although purely hypothetical when the model was developed, this assumption is now corroborated by the observation that the sperm factor is a Ca^{2+}-activated PLC (PLCζ; Swann et al (2004)). The exact nature of the sperm factor in frogs still remains to be fully established. More sophisticated descriptions of IP_3 degradation, as described in Section 2.10, have also been adopted in the framework of the *Xenopus* fertilisation Ca^{2+} wave (Bugrim et al, 2003; Wagner et al, 2004).

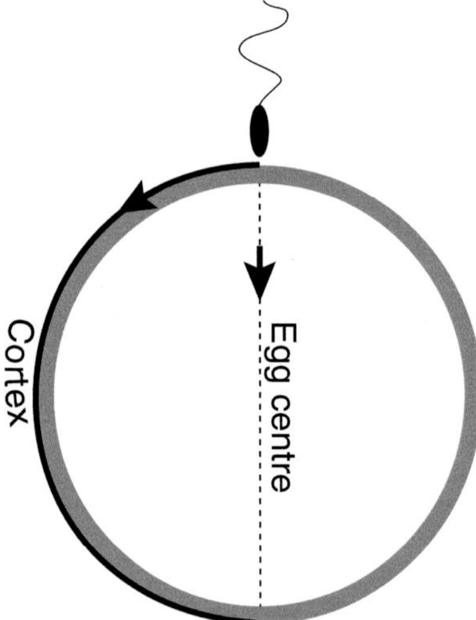

Fig. 6.6 Schematic representation of one-dimensional paths used to investigate the different characteristics of the Ca^{2+} waves propagating through the cortex (solid line) or through the centre of the egg (dotted line). The grey shaded region indicates the region where IP_3 is synthesised from its membrane-bound precursor.

The simplest model consists of three ODEs; for c (6.5), y (2.115), and p (6.6). By putting y at its steady state, y_∞, and then projecting the c and p nullclines on the (c, p) phase plane, one can see that bistability can also result from the interaction between IP_3 and Ca^{2+} regulation (Fig. 6.5). In this interpretation, the fertilisation Ca^{2+} wave still corresponds to the passage from one steady state to the other but it is now accompanied by an IP_3 wave. Interestingly, when Ca^{2+} and IP_3 diffusions are included, this mechanism naturally leads to a front that is always concave, as seen in vivo. This can be understood in the following way. A two-dimensional cross-section of the oocyte is sketched in Fig. 6.6. The solid and dotted lines denote the paths followed by the wave to travel along the cortex or through the bulk of the egg, respectively. Given that IP_3 is synthesised from its membrane-bound precursor PIP_2, the grey region indicates the region where IP_3 is produced. Thus V_{PLC} is space-dependent, taking its usual value in the cortex of the cell, but being zero in the interior.

Propagation in the cortical region can be investigated by simulations in one spatial dimension, with the line of propagation corresponding to a half perimeter of the circle shown in Fig. 6.6. Along this path, V_{PLC} is set at a constant, non-zero value. The upper panel of Fig. 6.7 shows the $[Ca^{2+}]$ (solid line) and $[IP_3]$ (dashed line) profiles obtained by numerical simulations: $[IP_3]$ increases just ahead of the Ca^{2+} wave that travels at about $5\,\mu m\,s^{-1}$. These profiles differ strikingly from those simulating one-dimensional propagation through the centre of the egg, where V_{PLC} is non-zero only at the extremities of the simulation line (Fig. 6.7, lower panel). As expected intuitively, propagation is much slower through the centre of the egg, as IP_3 diffuses only passively. Furthermore, $[Ca^{2+}]$ and $[IP_3]$ gradients extend into the cell, and these gradients are present both before and after fertilisation, as observed experimentally.

Thus, the wave propagates faster in the cortex but has further to go. It is not implausible that these effects compensate in such a way that the increased $[Ca^{2+}]$ initiated by the production of IP_3 at the point of sperm-egg fusion will progress vertically at about the same rate. To address this question properly, three-dimensional simulations in a spherical geometry were performed by Wagner et al (2004), who confirmed that the longer path taken by the wave along the cortex compensates for the faster propagation. Simulations also predict that active propagation is limited to the cortex of the egg, a prediction that remains untested, as changes in $[Ca^{2+}]$ can only be observed up to $300\,\mu m$ from the plasma membrane.

6.1.4 Increased IP_3 Sensitivity During Egg Maturation

In contrast to the oocyte, the mature egg shows bistability as seen in the preceding section. To try to understand what causes this "Ca^{2+} signalling differentiation" during oocyte maturation, Machaca (2004) and Ullah et al (2007) investigated aspects of Ca^{2+} signalling that have different dynamics in eggs than in oocytes. In particular, they investigated the duration of the $[Ca^{2+}]$ increases induced by the localised photorelease of small amounts of IP_3 in the presence of BAPTA. In the presence of this rapid Ca^{2+} buffer, the $[Ca^{2+}]$ increase remains localised and does not trigger a global fertilisation wave. It was found that the $[Ca^{2+}]$ increase in response to a given amount of IP_3 is much longer in eggs than in immature oocytes. They also compared the small-scale, spontaneous $[Ca^{2+}]$ increases (called puffs, see Chapter 4) in both states of the egg. Intriguingly, there they found the opposite result, that the average puff duration is smaller in eggs than in oocytes. These results are summarised in Fig. 6.8.

To understand this paradoxical result, modelling was used to find out which parameters can account for the change in duration of the responses to IP_3 uncaging. The modelling identified two possibilities. The change in behaviour could be due to an increase in the affinity of the IPR for Ca^{2+} or for IP_3, as both changes increase the amount of Ca^{2+} released through the IPR. To discriminate between the two possibilities, stochastic simulations of a small number of IPR were performed. A shorter average puff duration can only be obtained with an increase in IP_3 affinity. A closer look at the distribution of puff durations reveals that this average effect is due to an increased number of short events ($< 20\,\text{ms}$), while the duration of longer puffs is barely

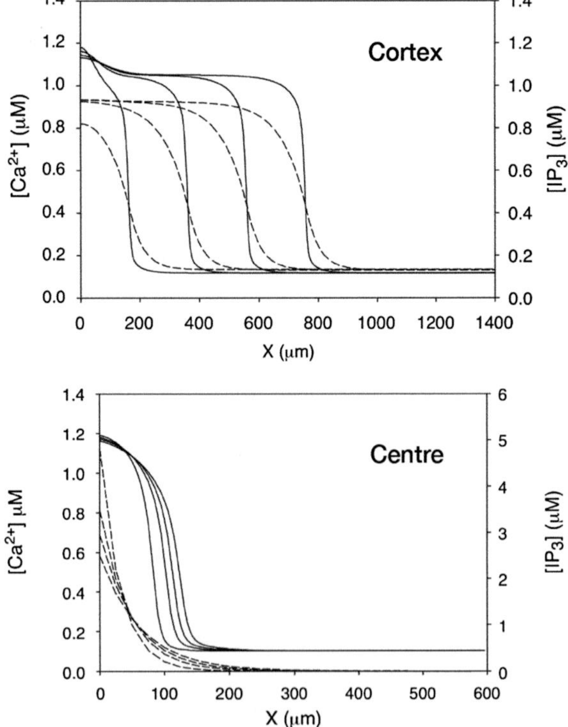

Fig. 6.7 Cortical (upper) and radial (lower) one-dimensional propagation of the fertilisation Ca^{2+} wave. Profiles are plotted at four successive times (50, 100, 150, and 200 s). Reproduced from Fall et al (2004), Figs. 2 and 4, with permission from Elsevier.

affected by the change in affinity. The conclusion is that by increasing the IPR affinity for IP_3, the Ca^{2+} signalling machinery is switched to a new state during *Xenopus* oocyte maturation. This is signified both microscopically by the modulated puff dynamics and macroscopically by the Ca^{2+} response to IP_3. A highly plausible molecular explanation for this prediction would be a change in the respective densities of the three IPR subtypes.

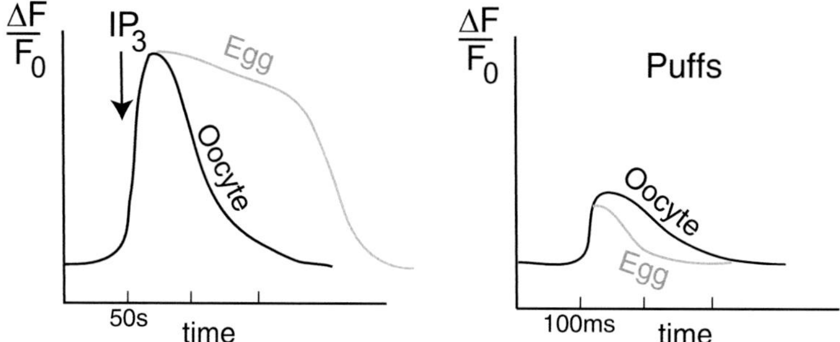

Fig. 6.8 Schematic representation of the "Ca^{2+} signalling differentiation" during oocyte maturation into a mature egg. Maturation is associated with both an increase in the duration of the $[Ca^{2+}]$ rises induced by the photorelease of IP_3 (left panel) and a decrease in the average duration of spontaneous Ca^{2+} puffs (right panel).

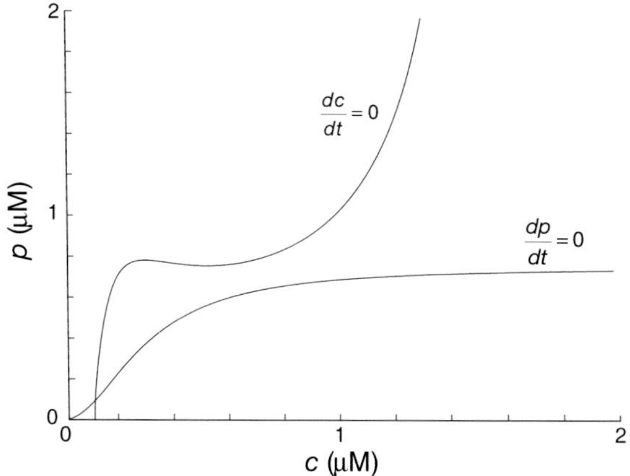

Fig. 6.9 Disappearance of bistability upon decreasing the affinity of the IPR for IP_3. Curves have been obtained as for Fig. 6.5, except that the parameter K_1 in the IPR model has been increased by a factor of 2.

It is tempting to draw a link between this prediction about an increased IPR receptor affinity for IP_3 and the prediction made by Wagner et al (2004) about the onset of bistability during maturation. In the Li Rinzel IPR model, as used above in the *Xenopus* oocyte model, the affinity of the IP_3 receptor for IP_3 is measured by the parameter K_1 (Section 2.7.2). An increase in affinity corresponds to a decrease in K_1. Fig. 6.9 shows the relationship between the c and p nullclines with a value of K_1 twice as large as that in Fig. 6.5. This situation would correspond to an immature oocyte. As compared to Fig. 6.5, the increase in K_1 shifts the c nullcline upwards, while the p nullcline remains unchanged, as K_1 does not appear in (6.6). In consequence, there is only one remaining steady state. Such agreement between various computational results suggests that the available models provide a reliable description of Ca^{2+} dynamics in *Xenopus* oocytes and eggs.

6.2 Hepatocytes

The liver is a vital organ that performs a wide spectrum of physiological functions, and plays a major role in metabolism. For example, it stores glycogen, the multimeric form of glucose that serves as a store to be released between meals, and secretes biochemical effectors such as the bile that is necessary to digest fats. All these processes are under the control of Ca^{2+} signalling. For example, glycogenolysis (the process through which the stored glycogen is broken down into glucose) is governed by the liver phosphorylase kinase, a Ca^{2+}-activated enzyme.

Hepatocytes are the most abundant cells in the liver. Ca^{2+} responses in these cells have been studied for many years, and it was in hepatocytes that hormone-induced Ca^{2+} oscillations were observed for the first time (Woods et al, 1986). The agonists used most frequently to stimulate these cells experimentally are vasopressin (Vp), noradrenaline (Nor), and ATP, each of which induces a distinct pattern of oscillations (Fig. 6.10).

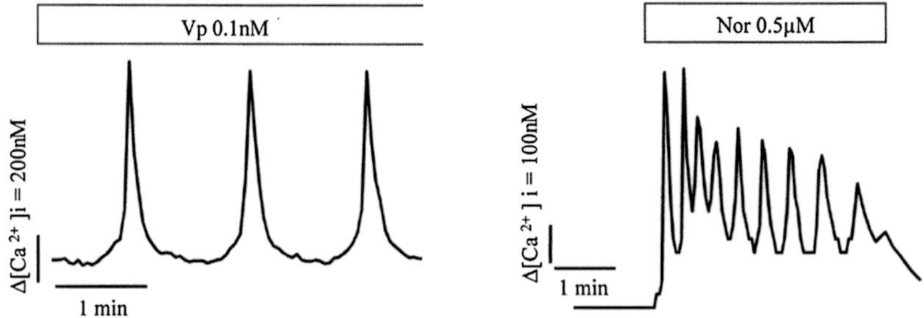

Fig. 6.10 Ca^{2+} oscillations in isolated rat hepatocytes after stimulation by different agonists. The pattern of oscillations depends on the nature of the stimulus. Not shown in this figure is the fact that the frequency of oscillations increases with the concentration of the extracellular agonist, as in other cell types. Reproduced from Dupont et al (2000a), Fig. 2, with permission from Elsevier.

To generate these repetitive spikes, hepatocytes use the classical elements of the Ca^{2+} toolbox (Chapter 2); PLC-mediated IP_3 synthesis, IPR, and SERCA pumps. For most purposes, Ca^{2+} exchange with the extracellular medium can be neglected, as oscillations persist for a long time in the absence of extracellular Ca^{2+}. As in many other cell types, each spike corresponds to the propagation of a Ca^{2+} wave at around $20\,\mu m\,s^{-1}$. When observed in the intact liver, these intracellular waves appear to form larger-scale intercellular waves that propagate throughout an entire lobule, which represents the functional unit of the liver (Gaspers and Thomas, 2005). In this section, we will focus on some characteristics of liver Ca^{2+} dynamics that have been successfully studied by a combination of experiment and modelling.

6.2.1 Effect of IP_3 Metabolism on Calcium Oscillations

Early models of Ca^{2+} signalling in hepatocytes suggested that, in response to stimulation, both $[IP_3]$ and $[Ca^{2+}]$ oscillate in synchrony (Cuthbertson and Chay, 1991; Larsen et al, 2004). As we have seen in Section 2.10, IP_3 oscillations could arise through two plausible mechanisms. The first mechanism is a Ca^{2+}-dependent regulation of IP_3 synthesis that can itself have two different causes. Protein kinase C (PKC), a Ca^{2+}- and DAG-dependent kinase, can exert a negative feedback on the receptor/G protein complex that activates PLC (Woods et al, 1987; Kummer et al, 2000). Alternatively, Ca^{2+} can stimulate PLC activity directly, which would provide positive feedback from Ca^{2+} to IP_3 synthesis (Meyer and Stryer, 1988). However, neither of these effects, that depend on the precise isoforms of PKC and PLC expressed in a given cell type, seems to be at play in hepatocytes (Renard et al, 1987; Bird et al, 1997).

The second plausible mechanism involves Ca^{2+}-regulated IP_3 catabolism (Section 1.2.4). IP_3 can be transformed either by 5-phosphatase-mediated dephosphorylation to give IP_2 or by an IP_3 3-kinase to yield inositol 1,4,5-tetrakisphosphate (IP_4). The activity of the 3-kinase is Ca^{2+}-dependent, and the binding of Ca^{2+}/calmodulin (Ca^{2+}/CaM) to the 3-kinase enhances its activity to a variable extent. The A isoform of the enzyme is stimulated 2- to 3-fold by Ca^{2+}/CaM, whereas the B isoform is stimulated up to 10-fold (Takazawa et al, 1990).

As expected intuitively, when models for intracellular Ca^{2+} dynamics take the stimulation of IP_3 catabolism by Ca^{2+} into account, IP_3 oscillations accompany the Ca^{2+} oscillations (Dupont and Erneux, 1997), but what really matters here is to know if these IP_3 oscillations have some consequences for Ca^{2+} signalling. One could conceive of two effects of these catabolism-induced IP_3 oscillations. Firstly, they could provide an explanation for the observation that the Ca^{2+} signal following uncaging of poorly metabolised IP_3 analogues decays more slowly than the signal following IP_3 uncaging (Chatton et al, 1998). Secondly, and more importantly, one could argue that in the course of agonist-induced oscillations the enhanced degradation of IP_3 following a

Ca^{2+} spike (due to Ca^{2+}-enhanced activity of the IP_3 3-kinase) plays a role in determining the relatively large period of Ca^{2+} oscillations in hepatocytes (of the order of 1 minute). Such large periods cannot be explained by the kinetic properties of the IPR, which are characterised by time scales on the order of seconds (see Section 2.7). In this view, each Ca^{2+} spike would cause a decrease in $[IP_3]$ to values too low to allow Ca^{2+} release through the IPR. Consequently, the period of oscillations would correspond to the time necessary to increase $[IP_3]$ again, up to a value that can reactivate the IPR. As phenomena such as secretion or glycogen breakdown are sensitive to the frequency of Ca^{2+} oscillations, the possible identification of a process that regulates the frequency is an important topic. In the following, we will see how modelling, by proposing a way to test this hypothesis experimentally, has helped to demonstrate that it is incorrect (Dupont et al, 2003).

The model consists of four differential equations; one each for the cytosolic Ca^{2+}, IP_3, and IP_4 concentrations (c, p, and Z), coupled to the gating IPR model described by (2.120) and (2.121). As the cell is assumed to be closed (see Section 3.2.5) the equation for c takes into account only fluxes in and out of the ER. Hence

$$\frac{dc}{dt} = J_{\mathrm{ER}} - J_{\mathrm{serca}}, \tag{6.7}$$

where

$$J_{\mathrm{ER}} = (k_f P_o + J_{\mathrm{leak}})(c_e - c), \tag{6.8}$$

$$J_{\mathrm{serca}} = \frac{V_s c^2}{K_s^2 + c^2}. \tag{6.9}$$

Since the cell is closed, $\gamma c + c_e = c_t$, where c_t is constant, and γ is the ratio of the cytoplasmic volume to the ER volume. Here, k_f controls the maximal Ca^{2+} flux through the IPR, while J_{leak} allows for a basal (small) efflux of Ca^{2+} from the ER into the cytosol in the absence of IP_3. The SERCA pump is modelled as a Hill function, with coefficient 2 (Section 2.2). The model is completed by the inclusion of equations for p and Z, which are given by (2.176) and (2.177).

6.2.1.1 Simulation Results

Fig. 6.11 A shows the simulation of a control situation corresponding to in-phase Ca^{2+} and IP_3 oscillations in response to stimulation by an agonist. The idea is to decrease the amplitude of the IP_3 oscillations to see if this would have an effect on the observed pattern of Ca^{2+} oscillations. As IP_3 oscillations are due to Ca^{2+} stimulation of IP_3 catabolism (V_{3K} in (2.175)), it can be expected that their amplitude would be much reduced if the relative importance of the other degradation pathway (V_{5P} in (2.175)) were increased. Thus, increasing the ratio of 5-phosphatase to 3-kinase should lead to the disappearance of IP_3 oscillations. This can be done experimentally by injecting 5-phosphatase into hepatocytes. As shown in Fig. 6.11 B, where the concentration of 5-phosphatase is increased 25-fold with respect to the control, the model suggests that this would abolish Ca^{2+} oscillations. There is a low steady-state [Ca^{2+}], consistent with the observed reduction of [IP_3]. Interestingly, although not surprisingly, the model suggests that if the external stimulation is then increased, Ca^{2+} oscillations are recovered, but now in the presence of a nearly constant [IP_3] (Fig. 6.11 C). In this case, the activity of the phosphatase exceeds that of the kinase 30 times, while both activities are, on average, roughly the same as in the control situation corresponding to Fig. 6.11 A. The characteristics of the repetitive spikes (shape, amplitude, and order of magnitude of the period) remain similar to those of the control. Thus, the model suggests that IP_3 oscillations driven by Ca^{2+}-activated IP_3 degradation do not significantly modulate Ca^{2+} oscillations in hepatocytes. The shape of the spikes is not affected because receptor inactivation is much faster than IP_3 metabolism. Also, the period is not significantly altered because the minimal [IP_3] during the course of oscillations remains above the threshold required for oscillatory behaviour.

6.2.1.2 Testing the Model Predictions

The model predicts that IP_3 oscillations have little effect on Ca^{2+} oscillations. To test this prediction experimentally, one can take advantage of the fact that it is possible to find two cells that exhibit practically identical Ca^{2+} oscillations. In an intact liver, hepatocytes are tightly coupled by gap junctions. Small groups of cells can be isolated in such a way that they remain connected, as in the intact liver. These groups, most of the time containing two or three cells, are called *doublets* or *triplets*, respectively. A striking feature of the responses in these connected cells is the quasi-identical pattern of Ca^{2+} increases in the coupled cells, with only a slight phase shift between the Ca^{2+} spikes in adjacent cells (Tordjmann et al, 1997, 1998). Although such similarity in response might simply be due to tight coupling of two cells with quite different intrinsic properties, this appears not to be the case. Rather, the similar oscillations appear to be a result of close similarity between the cells, an issue which we shall discuss in more detail in Section 6.2.3.

A doublet of connected hepatocytes thus provides an ideal tool to assess the role of IP_3 metabolism in the regulation of Ca^{2+} dynamics. By injecting 5-phosphatase – which is too large to diffuse through gap junctions – into one cell of the doublet, and stimulating both cells with the same dose of agonist, one can directly compare the conditions simulated in Figs. 6.11 A and B.

The results of the experiments are shown in Fig. 6.12. As shown in the upper panel (Fig. 6.12 A), control injection in one cell of inactivated 5-phosphatase did not result in any difference between the two cells in response to noradrenaline. In contrast, Ca^{2+} signals in the two cells were different when active 5-phosphatase was injected in one cell of the doublet, whatever

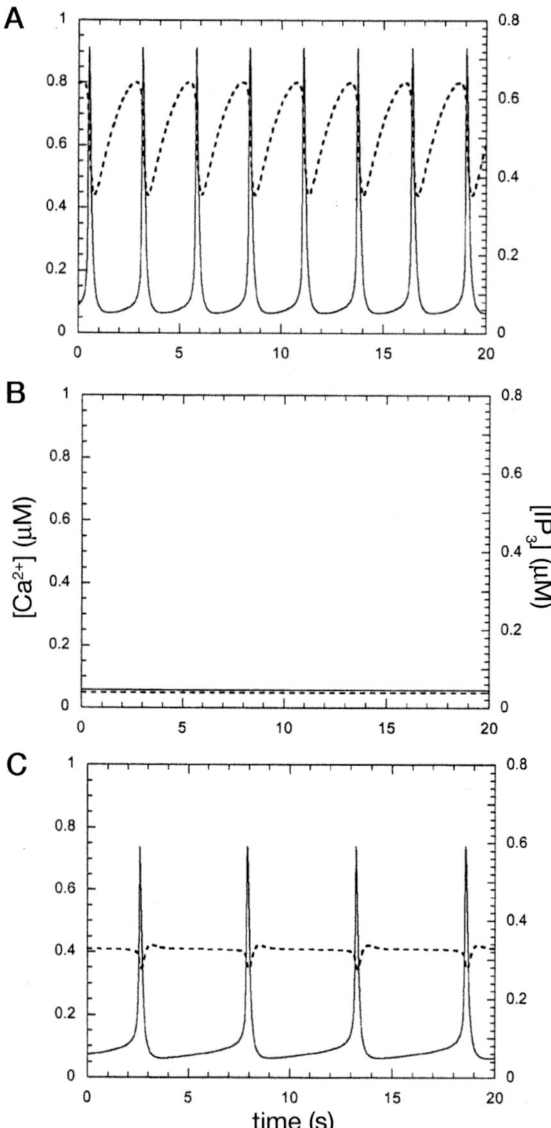

Fig. 6.11 Modelling predictions as to the possible role of Ca^{2+}-simulated IP_3 catabolism. **A:** Oscillations in Ca^{2+} (solid line) and IP_3 (dashed line) in a cell stimulated with a submaximal dose of agonist. **B:** The amount of 5-phosphatase in the simulated cell has been multiplied by 25 as compared to its value in A, and this makes the oscillations disappear. **C:** Ca^{2+} oscillations can reappear if the cell is stimulated with a high dose of agonist. Reproduced from Dupont et al (2003), Fig. 2, with permission from Elsevier.

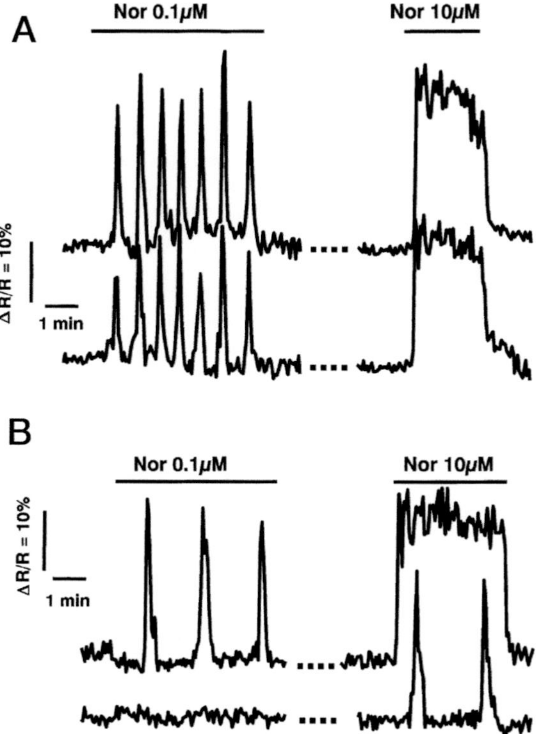

Fig. 6.12 Effect of 5-phosphatase on IP_3-dependent agonist-induced Ca^{2+} oscillations. One cell of the doublet (lower trace) was microinjected with fura-2 and either with inactive (A) or active 5-phosphatase (B). Fura-2 can flow through gap junctions, but not 5-phosphatase because of its high molecular weight. Then, hepatocyte doublets were challenged with noradrenaline (Nor) for the time shown by the horizontal bar. Reproduced from Dupont et al (2003), Fig. 3, with permission from Elsevier.

the concentration of the agonist (Fig. 6.12 B), consistent with the reduction of $[IP_3]$ in the injected cell anticipated by the model (Fig. 6.11 B). At high concentrations of noradrenaline ($10\,\mu$M), the non-injected cell shows a high sustained $[Ca^{2+}]$, reflecting a high $[IP_3]$, but the injected cell displays low-frequency Ca^{2+} oscillations, typical of an intermediate $[IP_3]$ (Fig. 6.12 B, right panel).

Thus, as predicted by the model (Fig. 6.11 C), a hepatocyte that has been made silent by injection of 5-phosphatase can become responsive again by increasing the concentration of the agonist. The critical observation is that oscillatory Ca^{2+} signals resembling those observed in control conditions can be observed at high enough agonist concentrations, despite the fact that IP_3 catabolism is mainly carried out by a Ca^{2+}-independent pathway. We conclude that the kinetics of IP_3 degradation does not control the period of Ca^{2+} oscillations in hepatocytes.

6.2.2 Deterministic Versus Stochastic Aspects of Calcium Oscillations

We have seen in Chapter 4 that, in many cell types, subthreshold stimulation by IP_3-generating agonists induces localised and short-lived Ca^{2+} increases called Ca^{2+} blips and Ca^{2+} puffs. As they rely on a low number of channels, these events are stochastic. Interestingly, Ca^{2+} blips and puffs have until now not been observed in hepatocytes, even at low agonist (or IP_3) dose. This might simply be due to technical limitations (the fact that something is not observed does not mean that it does not exist!). Another plausible explanation is related to the density of IPR. The latter is 100 times higher in hepatocytes than in *Xenopus* oocytes (Parys and Bezprozvanny, 1995). The average distance between clusters of IPR is thus expected to be smaller in hepatocytes than in other cell types such as *Xenopus* oocytes or HeLa cells where blips and puffs have been reported. Thus the Ca^{2+} released by the activity of one puff site would nearly always activate the release of Ca^{2+} from all the other IPR of the hepatocyte.

One might intuitively expect that the distinction between a stochastic and a deterministic description is not clear-cut, and that some intermediate behaviour should be observed. When only a small number of IPR are involved, Ca^{2+} signalling can only be described in stochastic terms. If there are many IPR (and the coupling between them very effective), fluctuations arising from the molecular noise, proportional to $1/\sqrt{N}$, cancel out on average, and a deterministic description is adequate. Between these two extremes there should be a regime where a deterministic description is acceptable, although the effect of internal fluctuations would be visible. Thus, even if blips and puffs are not visible in hepatocytes, Ca^{2+} signals in this cell type may exhibit such a borderline behaviour.

To test this possibility, one must assess the regularity of Ca^{2+} oscillations generated at a low stimulation level (because then fewer receptors are bound to IP_3 and thus able to participate). The level of irregularity can be quantified by the coefficient of variation (CV, defined as the standard deviation divided by the period; see Section 4.2.2). Analysis of Ca^{2+} time series of noradrenaline-stimulated hepatocytes reveals that the CV depends on the concentration of extracellular hormone and varies between 10 and 15% (Dupont et al, 2008). Higher CVs are associated with lower frequencies. One has to keep in mind, however, that the robustness of Ca^{2+} oscillations also depends on the agonist used to stimulate the cell (Perc et al, 2008), because different agonists stimulate different oscillatory mechanisms. For noradrenaline-induced responses, Ca^{2+} oscillations are rather regular, but still intrinsically noisy. Starting from a deterministic description of oscillations, how can we understand this irregular character?

Computational methods are available to assess the impact of internal fluctuations on chemical oscillations. One of the most popular is due to Gillespie (1976), who developed an algorithm that allows the simulation of a random walk that exactly represents the distribution of the master equation. This algorithm, which is heavily used in computational systems biology, associates a probability to each kinetic transition considered in the reaction scheme. This probability is a function of the specific stochastic reaction rate and of a combinatorial term that depends on the stoichiometry of the reaction. Originally, non-elementary reactions involving more than two molecules were modelled as a sequence of binary reactions but this step can be bypassed (see, for example, Gonze et al (2002)). Gillespie's method can be used to assess if the irregular character of Ca^{2+} oscillations in hepatocytes can be explained by the internal fluctuations due to the limited number of IPR present in this cell (even though there are enough IPR to prevent blips or puffs).

In applying Gillespie's method to a model, the evolution of the model variables is described by a set of transitions each of which corresponds to a given term in the deterministic evolution equations. Furthermore, the variables are no longer concentrations, but numbers of particles of the various species of the system. To provide a stochastic description of Ca^{2+} dynamics in hepatocytes, one ascribes a so-called *propensity* to the various transitions shown in Fig. 6.13. In this scheme, the italic R_{ij} represents the number of IPR in state R_{ij}, with i and j denoting, respectively, the number of Ca^{2+} ions bound to the activating and inhibiting site of the IPR. Each site can accommodate two Ca^{2+} ions, and cooperativity in Ca^{2+} binding at both sites is accounted for by assuming that $k_{a2+} \gg k_{a1+}$, $k_{a2-} \ll k_{a1-}$, $k_{i2+} \gg k_{i1+}$, $k_{i2-} \ll k_{i1-}$. IP_3 binding and unbinding are not simulated explicitly as these

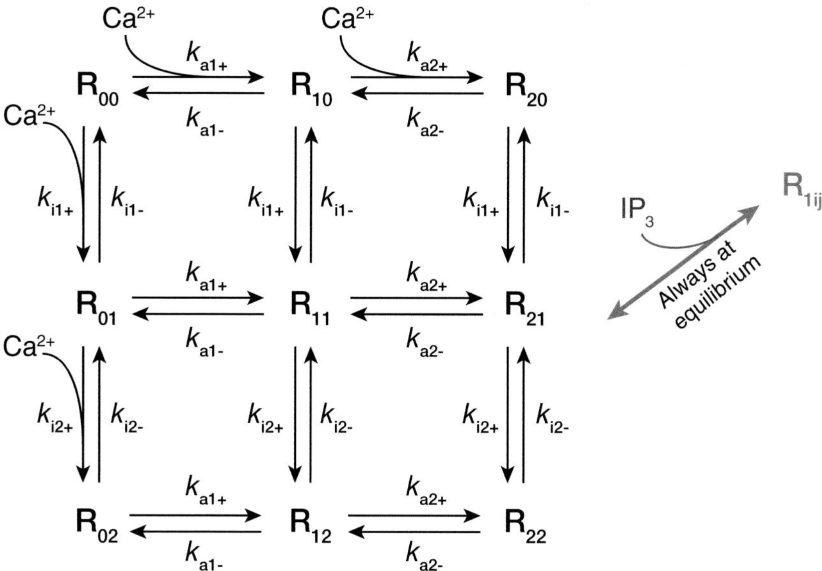

Fig. 6.13 Schematic representation of the IPR model used to simulate stochastic dynamics of the IPR (Dupont et al, 2008).

processes are assumed to be faster than Ca^{2+} binding and unbinding and thus at quasi-equilibrium.

In Gillespie's method, binding of one Ca^{2+} ion to the activating site of the receptor in the state R_{00} occurs with probability

$$P(\mathrm{R}_{00} \rightarrow \mathrm{R}_{10}) = \frac{k_{a1+} C R_{00}}{\Omega}, \tag{6.10}$$

where C represents the number of Ca^{2+} ions (i.e., the concentration × the volume of the hepatocyte) and Ω is a factor that permits the modulation of the total number of particles in the system. Given the entire set of reaction propensities, the algorithm determines the time of the next reaction and which reaction it will be. Then the number of particles is changed according to the transition that just happened. For example, if the transition (6.10) occurs, the number of particles is updated in the following way:

$$R_{00} \rightarrow R_{00} - 1; \qquad R_{10} \rightarrow R_{10} + 1; \qquad C \rightarrow C - 1. \tag{6.11}$$

C increases if the selected reaction is Ca^{2+} release through an open IPR, the probability of which is given by

$$P(C_{\mathrm{ER}} \rightarrow C) = k_1 \frac{R_{20}}{R_T} \frac{p}{K_{\mathrm{IP}} + p} \Omega, \tag{6.12}$$

where k_1 is the maximal flux through all the IPR and R_T is the total number of IPR considered in the stochastic simulation. Conversely, C decreases if two Ca^{2+} ion are removed by the SERCA pumps, which are described by a stochastic version of (6.9).

R_T is a crucial parameter in this algorithm, as we expect the stochastic character of Ca^{2+} oscillations to be related to the limited number of IPR. The density of IPR in hepatocytes has been estimated to be about 190 fmol/mg protein, which amounts to about 6000 receptors in a single cell (Spät et al, 1986). For comparison, a basal $[Ca^{2+}]$ of 100 nM in the cytoplasm corresponds to about 30,000 free Ca^{2+} ions in the entire cell. Another factor that has to be considered is the clustering of IPR. As described in Chapters 1 and 4, IPR are not distributed homogeneously in the cytoplasm but rather in groups of closely packed channels, called clusters. The accurate description of this spatial arrangement is a demanding computational problem. A simple way of dealing with this problem – that is, however, not expected to provide exact quantitative theoretical predictions – is to assume that a cluster of n channels can be modelled as one channel with an n-fold larger conductance than an isolated receptor. Indeed, if the receptors inside the cluster are assumed to be in close contact, the $[Ca^{2+}]$ in their vicinity can be considered as being roughly the same, and thus, to a first approximation, they will all open and close simultaneously, at least as compared to the characteristic time of oscillations. This implies that, in the simulations, R_T represents the total number of IPR divided by the number of channels per cluster.

Fig. 6.14 shows the results of simulations when assuming 25 channels per cluster. The Ca^{2+} spikes, shown over approximately 100 s, look noisy. The histogram of the interspike interval, however, indicates that the distribution of the periods remains narrow, with a CV of 14%. It can thus be concluded that the noisy character of Ca^{2+} oscillations seen in hepatocytes can be ascribed to the limited number of IPR. However, it also emphasises that the deterministic description is adequate to describe global Ca^{2+} signalling in hepatocytes. In Fig. 6.14, the irregularity of Ca^{2+} oscillations is further assessed by computation of the autocorrelation function (Dupont and Combettes, 2009). As is known to be the case for noisy periodic systems, the envelope of the autocorrelation function decreases exponentially with a decay half-time that is of the order of the period itself. This rapid phase diffusion can be related to the fact that the phase of the Ca^{2+} oscillation does not contain any physiologically relevant information, as it does for other rhythmic phenomena in cellular physiology such as, for example, circadian rhythms (Gonze et al, 2002).

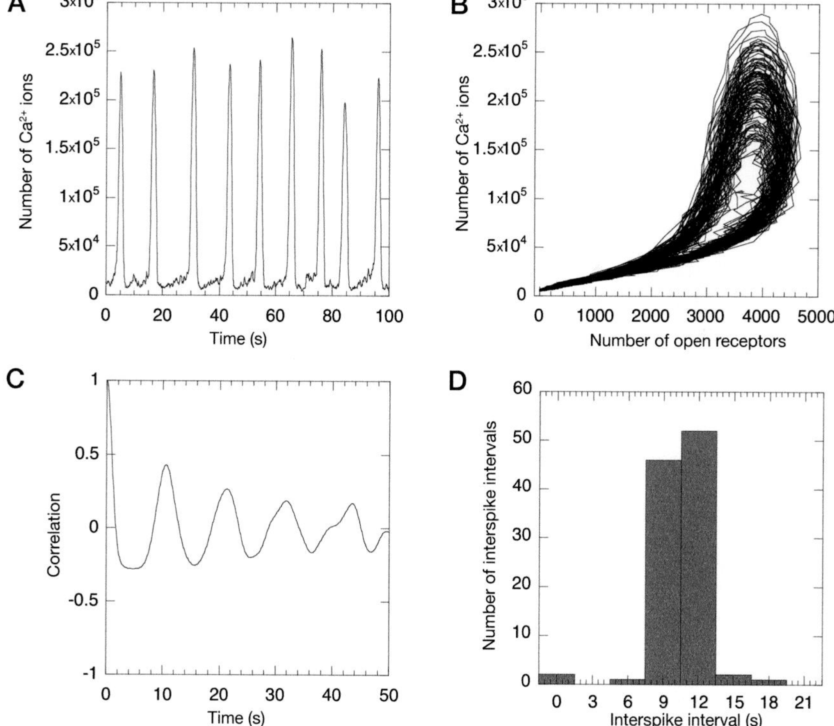

Fig. 6.14 Simulations of stochastic Ca^{2+} oscillations in hepatocytes. Reproduced from Dupont and Combettes (2009), Fig. 1, with permission from AIP Publishing.

Besides putting forward a plausible explanation for the irregularity of the oscillations, simulations also reveal an interesting phenomenon. Sequences of repetitive spikes can be obtained when $[IP_3]$ is below the Hopf bifurcation of the corresponding deterministic system (as we discussed in Section 4.1). As shown in Fig. 6.15 A, these spikes are widely spaced, leading to irregular oscillations (CV $\approx$ 35%). In fact, they are transients in an excitable system that are generated by random fluctuations that, every so often, cross the excitability threshold. The time interval between two successive spikes corresponds to the time interval between two such superthreshold fluctuations, which explains why it is highly variable. In contrast, just at the beginning of the oscillatory domain, i.e., when [IP$_3$] is just above the Hopf bifurcation, oscillations are much more regular (CV $\approx$ 17% for the situation shown in Fig. 6.15 A).

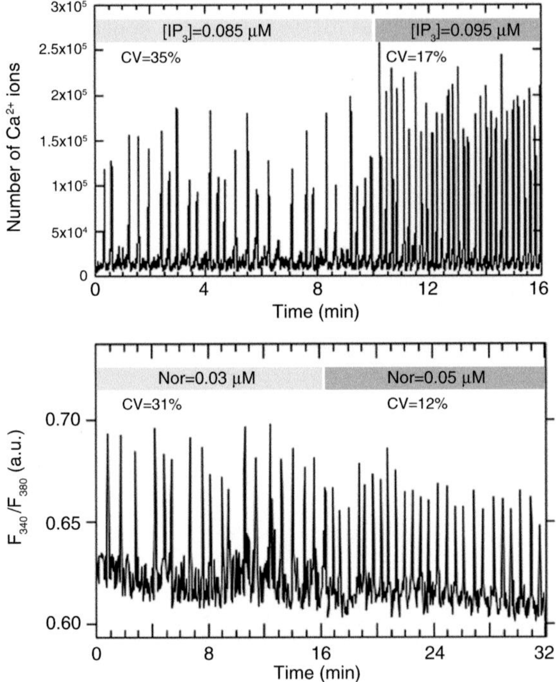

Fig. 6.15 Repetitive Ca^{2+} spikes at subthreshold (left) and superthreshold (right) levels of stimulation in hepatocytes. Model (top panel) and experiments (lower panel). Reproduced from Dupont et al (2008), Fig. 3, with permission from Elsevier.

To see if such behaviour can be observed in reality, a large number of hepatocytes were stimulated with very low doses of noradrenaline (0.03 μM), which, in the vast majority of cases, does not induce any Ca^{2+} increase in the stimulated cell. However, if the experimentalist is patient and careful enough they will notice irregular oscillations occurring in some of the cells. An example of such atypical behaviour is shown in Fig. 6.15 B. Upon increasing slightly the concentration of noradrenaline in the extracellular medium (up to 0.05 μM), regular oscillations resembling those usually observed for normal stimulation levels appear. This shows that the noisy behaviour seen at

0.03 μM noradrenaline does not correspond to anomalous cells in which the Ca^{2+} signalling machinery has been damaged. In contrast, irregular Ca^{2+} oscillations at subthreshold levels of stimulation reveal that in most cases, Ca^{2+} oscillations in hepatocytes result from a deterministic mechanism that is perturbed by noise due to the relatively low number of IPR.

6.2.3 Phase Waves Coordinate Calcium Spiking Between Connected Hepatocytes

In the intact liver, Ca^{2+} oscillations occur nearly synchronously, and imaging the $[Ca^{2+}]$ in a whole lobule reveals the existence of periodic intercellular waves (Gaspers and Thomas, 2005). This coordination of Ca^{2+} spiking is observed at a smaller scale in doublets or triplets, i.e., small groups of two or three hepatocytes that have been isolated in such a way that gap junctions remain intact. Gap junctions are made of the apposition of two hemichannels (also called connexins) located in the plasma membranes of two adjacent cells and allow various molecules and ions to pass between cells. Their permeability is different for each compound and much depends on the molecular weight and the charge of the species. Upon stimulation of a triplet of hepatocytes connected in such a way, a wave propagates in the first cell, and then, after a short pause, invades the second, and then the third one (Fig. 6.16). When plotting the spatially averaged $[Ca^{2+}]$ in each cell, this looks like coordinated Ca^{2+} spiking, with a slight phase shift.

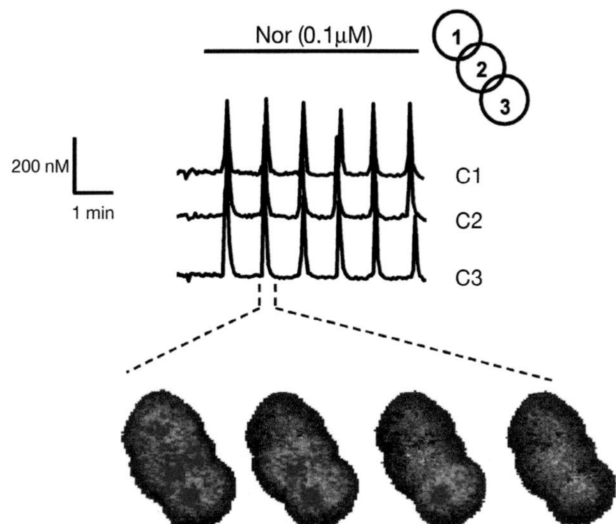

Fig. 6.16 Intercellular Ca^{2+} waves in hepatocytes. In the lowest panel, warmer colours denote higher $[Ca^{2+}]$. Reproduced from Dupont et al (2000a), Fig. 8, with permission from Elsevier.

The most obvious explanation for this phenomenon would be that the Ca^{2+} just propagates from one hepatocyte to the adjacent one through gap junctions. This is in agreement with the observation that inhibitors of gap junctions prevent coordination. In such conditions, asynchronous oscillations arise. However, even with intact gap junctions, coordinated spiking requires that all three cells are continuously stimulated by the agonist; if the agonist is washed away after $[Ca^{2+}]$ has increased in the first two cells, $[Ca^{2+}]$ does not increase in the third cell (Tordjmann et al, 1997, 1998). Also, if only one cell of a doublet or a triplet is stimulated (through careful, local addition of noradrenaline with a micropipette), only the stimulated cell will develop a Ca^{2+} wave. This suggests that the propagation of the wave from one hepatocyte to the other does not correspond to an intercellular flow of Ca^{2+}. In some cell types, such as airway epithelial cells (see Chapters 1 and 3), intercellular waves rely on the flow of IP_3 through gap junctions. This might play a role in the liver as well, but the propagation mechanism clearly differs from that in airway epithelial cells as hepatocytes need the continuous presence of the agonist in order to generate the response, while, in epithelial cells, stimulation of a single cell is able to generate the wave. In addition, the waves are periodic in the liver but not in epithelial cells. At this point, it is important to remember that the liver displays a high level of morphological and functional heterogeneity. Hepatocytes differ according to their position in the liver lobule, or, more precisely, according to their position on the portocentrilobular axis of the liver acinus that joins the periportal (PP, in the centre of the liver) and the perivenous (PV, in the periphery of the liver) zones. Among other things, between the periportal and perivenous zones there is an increasing density of plasma membrane hormone receptors, which accounts for the observed gradient of sensitivity to hormone. The variation of receptor density from one extremity to the other is of the order of 40% (Tordjmann et al, 1998).

To simulate Ca^{2+} dynamics in a triplet of hepatocytes, in each cell one needs to solve (6.7)–(6.9), together with the equations for IP_3 metabolism, (2.176) and (2.177), as well as the equations for the IPR model, (2.120) and (2.121). However, for most purposes one can assume that the concentration of IP_4 remains constant and thus neglect (2.177). The different sensitivities to the extracellular hormone enter the model via different values for the parameter V_{PLC} that represents the rate of IP_3 synthesis by PLC. As each PLC enzyme is activated through a G protein cascade initiated by a hormone-receptor complex, the total PLC activity at the cellular level directly depends on the number of such hormone-receptor complexes. In the model, the relation between both is supposed to be linear. Thus the gradient in V_{PLC} values is estimated directly from the numbers of hormone receptors. As the number of receptors increases by 40% along a liver cell plate that contains about 20 cells, V_{PLC} differs by about 5% between two neighbouring cells. As V_{PLC} dictates the intracellular $[IP_3]$, which in turn controls the frequency of Ca^{2+}

oscillations, this leads to a cell-to-cell variation of about 20% in the period of Ca^{2+} oscillations (Dupont et al, 2000b).

If the same amount of IP_3 is photoreleased in connected cells of a doublet or a triplet, the evolution of intracellular Ca^{2+} is indistinguishable in different hepatocytes. This means that, when the hormone receptor is bypassed, the Ca^{2+} responses are nearly identical in cells that are originally located close to one another in the liver. This suggests that the hepatocytes of the triplet can be modelled by considering all the parameter values, except for V_{PLC}, to be the same for each cell.

Intracellular diffusion of Ca^{2+} and IP_3 are modelled in one spatial dimension, and each hepatocyte is assumed to be $20\,\mu m$ long. Each cell is coupled to its neighbours by appropriate jump conditions that model the diffusion of IP_3 through gap junctions (Section 3.6). As experiments suggest that very little Ca^{2+} moves from one cell to another (Clair et al, 2001), the gap junctions are assumed to be permeable only to IP_3. This jump condition assumes that the flux through the gap junction is proportional to the concentration difference across the gap junction. Thus

$$D_p \frac{\partial p^-}{\partial x} = D_p \frac{\partial p^+}{\partial x} = \mathcal{F}_p (p^+ - p^-), \tag{6.13}$$

where the superscripts + and − denote $[IP_3]$ at the right and left side, respectively, of the border (i.e., the plasma membrane). D_p is the intracellular diffusion coefficient of IP_3, which is around $210\,\mu m^2\,s^{-1}$. $\mathcal{F}_p$ represents the gap junction permeability, and is an unknown parameter that needs to be estimated from experimental observations. Thus, when simulating a doublet of hepatocytes, the value of $\mathcal{F}_p$ must allow for both the coordination of Ca^{2+} spiking when the whole doublet is stimulated and for the absence of Ca^{2+} variations in an unstimulated cell connected to an oscillating one. These two constraints require that $\mathcal{F}_p$ is approximately $0.88\,\mu m/s$. With this value, the steady-state $[IP_3]$ is different in connected cells that are characterised by different rates of IP_3 synthesis (i.e., different values of V_{PLC}), but the gradient in $[IP_3]$ is much less than it would be if the gap junctions were impermeable to IP_3.

Simulation results are shown in Fig. 6.17 A. Ca^{2+} oscillations appear coordinated among the three cells (labelled 1, 2, and 3 in Fig. 6.16). The red cell has the highest rate of IP_3 synthesis (V_{PLC}); less sensitive cells are shown in blue and green. Peaks appear sequentially in cells 1 (red), 2 (blue), and 3 (green), giving the appearance of an intercellular wave. Because no Ca^{2+} is transported from one cell to the other, this wave is a phase wave (Section 3.5.4). The appearance of wave propagation comes from the slight phase shift between the individual oscillators. The phase shift is due to the fact that the latency between stimulation and the onset of the first spike also depends on V_{PLC}. The larger the rate of IP_3 synthesis, the shorter the delay for the first Ca^{2+} spike. Variations in V_{PLC} also cause the oscillation period to be different in each cell. Thus, coordination is progressively lost, as shown in Fig. 6.17 A. This is also seen in experiments (Fig. 6.17 B). The loss of coordination is a genuine phenomenon and is not due to a time-dependent alteration of the experimental preparation. Indeed, resynchronisation can be achieved by washing and restimulation of the same triplet by an identical concentration of vasopressin.

The proposed mechanism accounting for coordination of Ca^{2+} oscillations in hepatocytes in the form of phase waves has been corroborated by various observations such as, for example, the lack of coordination at very low stimulation level or the impermeability of gap junctions to Ca^{2+} (Clair et al, 2001). In contrast to classical wave propagation, the direction of propagation is imposed by the direction of the gradient and the main factor affecting the rate of propagation is the gradient of sensitivity to the external agonist; the lower the gradient, the faster the wave. The gap junction permeability also affects the rate of propagation.

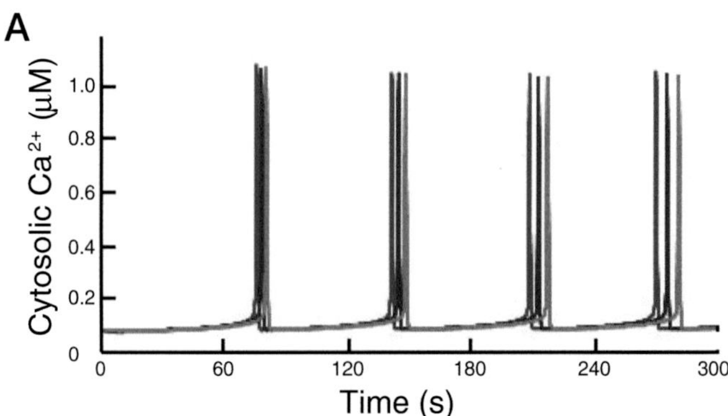

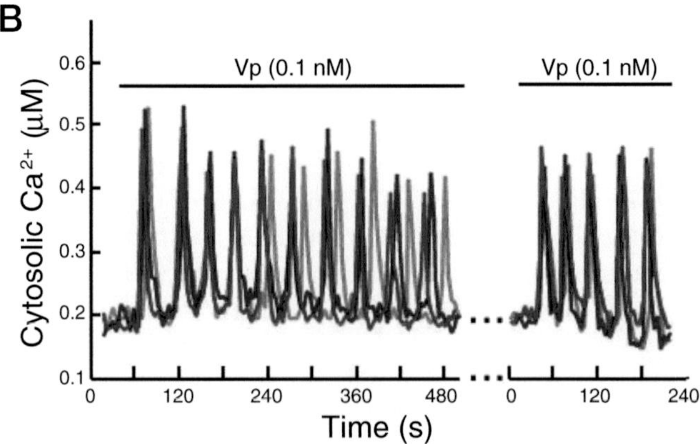

Fig. 6.17 Ca^{2+} oscillations and intercellular phase waves induced by vasopressin in a triplet of rat hepatocytes (Dupont et al, 2000b). **A:** Simulations of the model discussed in Section 6.2.3. **B:** Experimental results. Hepatocytes loaded with fura-2 are stimulated with vasopressin (Vp) for the time shown by the horizontal bars. Between the two periods of stimulation, vasopressin is removed and cells were washed.

6.2.4 Amplitude-Coded Calcium Oscillations in Fish Hepatocytes

In order to assess the universal character of these results and to look for alternative models to mammalian systems, some groups investigated Ca^{2+} signalling in fish cells, and more specifically in fish hepatocytes (Zhang et al, 1992; Nathanson and Mariwalla, 1996; Kroeber et al, 1997; Schweizer et al, 2011). A particularly useful tool in this respect is the fish cell line RTL-W1 that has been derived from primary cultures of rainbow trout liver. In these cells, oscillations of intracellular $[Ca^{2+}]$ are consistently observed in response to all agonists known to activate Ca^{2+} signalling (Schweizer et al, 2011). Most of them are regular, although oscillations of bursting type are observed in response to stimulation by hydrogen peroxide, which operates via plasma membrane depolarisation followed by Ca^{2+} entry. Analysis of the available genetic information reveals that the classical players of Ca^{2+} signalling (PLC, SERCA, IPR, and RyR) in mammalian cells are also present in fish liver cells.

In contrast to what is seen in mammalian cells, where frequency encoding is a basic common property of repetitive Ca^{2+} spiking, in fish hepatocytes Ca^{2+} oscillations appear to be coded by amplitude; upon increasing the concentration of the external stimulus, the amplitude of the oscillations increases. As this is observed for all agonists, the origin of this phenomenon lies in processes downstream of the external receptor. Bifurcation analysis performed in the context of information processing in astrocytes has shown that modification of appropriate parameters can change models from exhibiting frequency encoding to exhibiting amplitude encoding (De Pittà et al, 2009).

In most cases, this change in behaviour is accompanied by a modification in the nature of the bifurcation point leading to oscillations. Subcritical Hopf bifurcations accompany frequency encoding while supercritical Hopf bifurcation points are associated with amplitude encoding. This observation agrees with general properties of supercritical and subcritical Hopf bifurcations. At a subcritical Hopf bifurcation the amplitude of the stable branch remains relatively constant as the bifurcation parameter changes, but, in fast-slow systems at least, the frequency can change much more dramatically. On the other hand, near a supercritical Hopf the amplitude changes significantly as the bifurcation parameter changes, but the frequency remains approximately constant. This concept can be tested with a minimal model for Ca^{2+} oscillations based on a phenomenological description of Ca^{2+}-induced Ca^{2+} release (Goldbeter et al, 1990). This model is discussed in Section 3.2.2, and displays robust frequency encoding.

Schweizer et al (2011) argue that amplitude encoding can be obtained with the same model for different values of the parameters, particularly if these values lead to similar rates of Ca^{2+} transport in and out of the ER. Although it is true that, for some parameters, changing the level of stimulation affects the frequency more than the amplitude of the oscillations, while the opposite is true for a different choice of parameters, it is not obvious how to interpret such results, as, in general, the period and the amplitude both vary when changing the stimulation level.

6.3 Pancreatic and Parotid Acinar Cells

6.3.1 Introduction

Epithelial cells of the exocrine glands have played a major role in studies of Ca^{2+} dynamics. Experiments done in pancreatic acinar cells helped to establish the basic principles of agonist-induced Ca^{2+} release from the ER (Matthews and Petersen, 1973; Streb et al, 1983; Muallem et al, 1985), and Ca^{2+} oscillations in blowfly salivary gland cells were observed as early as 1981 (Rapp and Berridge, 1981). Recent reviews of Ca^{2+} signalling in exocrine epithelial cells are Petersen (2014) and Ambudkar (2014) while an older review is Ashby and Tepikin (2002). Although Ca^{2+} oscillations occur in most exocrine epithelial cell types, our discussion here shall focus on pancreatic and parotid acinar cells, as those are the cell types for which most modelling has been done. Because of the extensive similarities between these two cell types we shall often simply call them acinar cells, distinguishing one cell type from the other only when it is important to do so.

Acinar cells are polarised epithelial cells. The basolateral membrane faces to the extracellular space, while the apical membrane faces into the luminil compartment, which is where the primary saliva (in the case of parotid cells) or digestive enzymes (in the case of pancreatic cells) are secreted. Multiple acinar cells secrete into the same lumen, which has a highly branched structure. The basolateral membrane has a variety of ion exchangers and channels, including the Na^+/K^+ ATPase, a $Na^+/K^+/Cl^-$ cotransporter, and a Ca^{2+}-dependent K^+ channel. On the apical membrane of the parotid cell the most important ion channel is a Ca^{2+}-dependent Cl^- channel, although there are also Ca^{2+}-dependent K^+ channels there.

Upon agonist stimulation $[Ca^{2+}]$ oscillates with a period of around 5–30 seconds (Fig. 1.8). The shape and period of the oscillations depend on the agonist and the cell type. At low agonist concentrations the Ca^{2+} oscillations are restricted to the apical region of the cell, but at higher agonist concentrations they take the form of periodic waves that travel from the apical zone to the basolateral zone (Nathanson et al, 1992; Thorn et al, 1993; Kasai et al, 1993; Fogarty et al, 2000).

Ca^{2+} signalling in acinar cells is governed by the same toolbox components described in Chapter 2. Binding of an agonist to G protein-coupled receptors (P2Y purinergic receptors, for example, or alpha-adrenergic receptors) results in the activation of phospholipase C (PLC) and thus production of IP_3, resulting in the release of Ca^{2+} from the ER. Typical agonists are carbachol (CCh) or cholecystokinin (CCK). Ca^{2+} can also be released from the ER through ryanodine receptors (RyR), and is taken up into the ER by SERCA pumps. Finally, Ca^{2+} is removed from the cell by ATPase pumps on the plasma membrane and enters the cell through a variety of channels (although the modelling of Ca^{2+} entry is often simplified by presuming only a single channel type). Available evidence suggests that Ca^{2+} oscillations in pancreatic acinar cells (and possibly in parotid acinar cells also, although this is not known for sure) are driven by a Class II mechanism (Section 3.2.7), in which oscillations in $[Ca^{2+}]$ and $[IP_3]$ must occur together (Sneyd et al, 2006), and thus Ca^{2+} regulation of IP_3 metabolism will be an important model component. Hence, for example, in the model of Palk et al (2010), the rate of degradation of IP_3 is controlled by Ca^{2+} via its effect on the 3-kinase that converts IP_3 to IP_4.

In both pancreatic and parotid acinar cells, the purpose of the Ca^{2+} oscillations is to stimulate the secretion of proteins, enzymes, and water. Different cell types secrete each in different proportion. In parotid acinar cells there is a relatively high rate of water secretion, as that is needed to make saliva, while in pancreatic acinar cells there is much less water secretion. Instead, the exocrine pancreas is specialised for the secretion of enzymes. However, in both cases it is the Ca^{2+} oscillations that are the controlling mechanism.

The basic mechanism of saliva secretion is illustrated in Fig. 6.18. Stimulation of the cell by agonist results in the release of Ca^{2+} from the ER, and the resultant increased $[Ca^{2+}]$ activates Ca^{2+}-dependent Cl^- channels on the apical membrane. Cl^- thus flows out of the cell, down its electrochemical potential gradient into the luminil space, depolarising the cell. If there were no other ion currents, this Cl^- current would quickly cease as the membrane depolarises. However, simultaneous activation of the Ca^{2+}-dependent K^+ channels on the basolateral membrane maintains the membrane potential at a sufficiently negative voltage to allow for the continued flow of Cl^- out of the cell. Na^+ follows Cl^- into the luminil space, most likely via an extracellular pathway, and water follows via osmosis.

In both pancreatic and parotid acinar cells, it is known that increased $[Ca^{2+}]$ causes the exocytosis of enzyme (just as it causes the exocytosis of neurotransmitter in synaptic terminals) but the details of the mechanism remain far from clear (Low et al, 2010; Messenger et al, 2014).

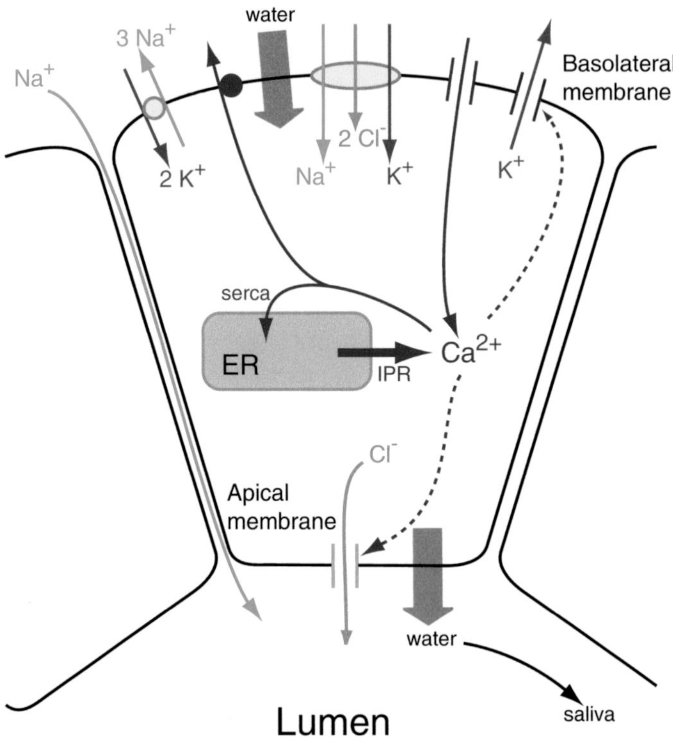

Fig. 6.18 The major ion channels involved in the secretion of saliva, and their control by Ca^{2+}. Although Ca^{2+}-sensitive K^+ channels are also situated on the apical membrane, they are omitted here for clarity.

There have been a number of modelling studies of acinar cells (LeBeau et al, 1999; Sneyd et al, 2000, 2003; Tsaneva-Atanasova et al, 2005; Ventura and Sneyd, 2006; Gin et al, 2007; Palk et al, 2010, 2012). They differ in their choices of how to model each Ca^{2+} flux, and in their treatment of spatial heterogeneity, but all have the same essential structure. The precise details of each model can be found in the original papers, but are of less importance for the present discussion. Instead, we summarise here what has been learned from this succession of models.

6.3.2 Calcium Oscillations and Waves in Acinar Cells

An early model of Ca^{2+} oscillations in pancreatic acinar cells was that of LeBeau et al (1999). Their principal goal was to investigate the possibility that the different types of Ca^{2+} oscillation seen in response to different agonists could be (at least partially) explained by different agonists causing different levels of phosphorylation of the IPR. To test this hypothesis they constructed a new model of the IPR that included an additional phosphorylated state, which was assumed to be inactive (although this is not actually necessary; in principle, the theory would work equally well if phosphorylation activates the IPR). Hence, the IPR could be activated by Ca^{2+} and IP_3, would move spontaneously but more slowly to an inactivated state, but could also move to a second inactivated, phosphorylated, state if the cell had been stimulated by CCK. An experimental test of this hypothesis did indeed find that CCK caused much greater phosphorylation of type 3 IPR than did CCh, confirming this as a possible mechanism for long-period oscillations. Subsequent work from the group of David Yule (Giovannucci et al, 2000; Yule et al, 2003; Straub et al, 2004; Wagner et al, 2008) has shown that type 3 IPR are inhibited by phosphorylation, while type 1 are activated. This variability allows for a complex and precise control of the Ca^{2+} oscillation period.

This is a useful illustration of the oft-quoted fact that models do not have to be correct to be useful. The original IPR model of LeBeau et al (1999) has not withstood the test of time, and has now been entirely superseded by the next generation of IPR models. However, it played a significant role in providing the theoretical framework for a series of experimental papers that have improved our understanding of how the Ca^{2+} oscillation period might be regulated.

The next models of acinar cells were focused on attempting to understand spatial properties of the waves, which are significantly different in parotid and pancreatic versions. In both pancreatic and parotid cells, the apical regions have a greater density of IPR, and so this is where waves are initiated (Kasai et al, 1993; Nathanson et al, 1994). At low $[IP_3]$, Ca^{2+} waves do not propagate out of the apical zone, but at higher $[IP_3]$ apically-induced waves will propagate globally across the cell (Fig. 6.19).

However, global waves in parotid are significantly faster than in pancreas. As can be seen in Fig. 6.19, in pancreas the peak of the Ca^{2+} rise in the basolateral region occurs approximately one second after the peak in the apical region, while in parotid the peaks are separated only by approximately 200 ms. Furthermore, as $[IP_3]$ increases, the wave speed in pancreas at first increases only gradually (Fig. 6.20), while the wave speed in parotid is a much steeper function of $[IP_3]$.

Before we consider the mechanisms underlying such wave properties, it is interesting to note that they may well have functional consequences. Palk et al (2012) have shown that water flow through a parotid cell is maximised when the K^+ and Cl^- channels are activated together (see the discussion below), and thus when $[Ca^{2+}]$ increases simultaneously in the apical and basolateral regions. Hence, the faster wave speed in parotid cells gives a greater secretion of water. Pancreatic cells, although they also secrete a small

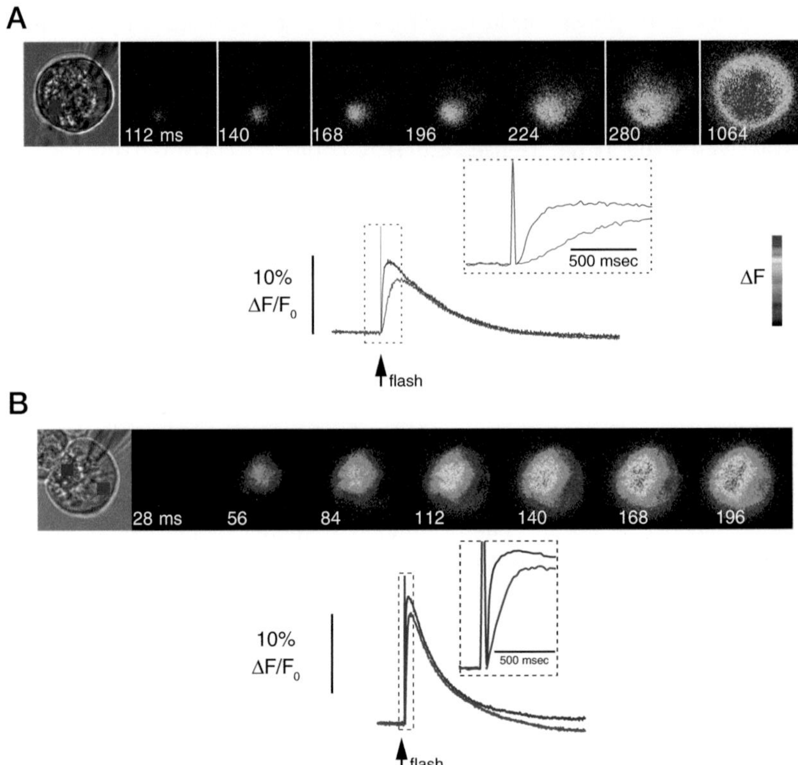

Fig. 6.19 Global Ca^{2+} waves in parotid and pancreatic acinar cells (Giovannucci et al, 2002). Original data provided by David Yule (University of Rochester). **A:** IP_3 was photoreleased across the entire cell by a flash of light at the time indicated by the arrow. $[Ca^{2+}]$ was then measured in the basal and apical regions of each cell, marked, respectively, by the red and blue squares in the image of the cell in the upper panel. The remainder of the upper panel shows the progress of the Ca^{2+} wave across the cell, in a series of still frames taken at the indicated times; warmer colours denote higher $[Ca^{2+}]$. In the lower panel, the blue curve is the time course of $[Ca^{2+}]$ at the blue dot, and similarly for the red curve. The inset shows a magnification of the time courses for the period directly after the light flash. The blue curve increases more quickly than the red curve, showing that $[Ca^{2+}]$ increases more quickly in the apical region (blue square) than in the basal region (red square). **B:** Same as for A, except the results are from a parotid, rather than a pancreatic, acinar cell. Although $[Ca^{2+}]$ in the apical region (blue curve) still increases first, there is less of a delay before $[Ca^{2+}]$ in the basal region (red curve) also rises.

amount of water, are designed rather for the secretion of enzymes, which is (as far as we know) affected only by the basolateral $[Ca^{2+}]$, not by the wave speed. Thus, pancreatic cells work well with a trigger zone to initiate the waves, which are then propagated more slowly across the rest of the cell.

In vivo, the structure of the lumen and the apical region may well play a major role in increasing the wave speed in parotid. Larina and Thorn (2005) have shown that, in submandibular acinar cells, the apical region extends over a greater portion of the cell than it does in pancreatic acinar cells. This, presumably, gives a higher concentration of IPR over a greater region of the cell, resulting in a Ca^{2+} wave that is practically simultaneous in the apical and basolateral regions.

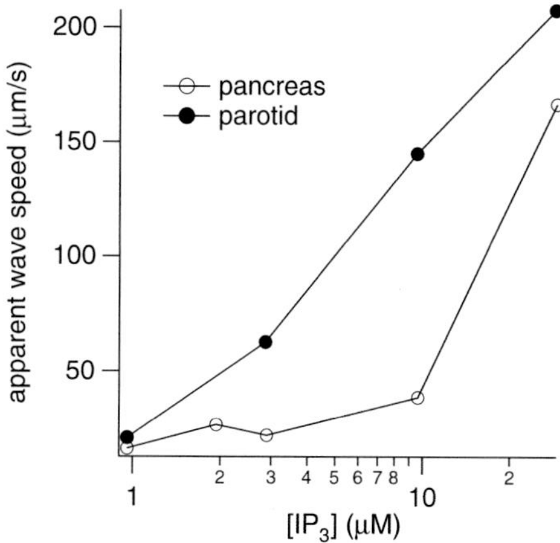

Fig. 6.20 Speed of the global Ca^{2+} waves in parotid and pancreatic acinar cells (Giovannucci et al, 2002).

To return to the mechanism underlying these patterns of wave propagation, there is considerable evidence that, at least in pancreatic acinar cells, the global wave is initiated by IPR in the apical region and then propagated globally by RyR (Straub et al, 2000; Giovannucci et al, 2002). This offers a partial explanation for why, at low $[IP_3]$, the global wave speed is relatively independent of $[IP_3]$ – the wave speed is set by the excitability of the RyR which is unaffected by increased $[IP_3]$. However, the presence of some IPR in the basolateral region means that, once $[IP_3]$ gets large enough, Ca^{2+} release is initiated through the IPR in both regions. The region with a lower IPR density is a little slower to release Ca^{2+}, resulting in a Ca^{2+} phase wave that is driven, not by Ca^{2+} diffusion between excitable release sites, but merely by the difference in the timing of Ca^{2+} release in the apical and basolateral regions (Section 3.5.4).

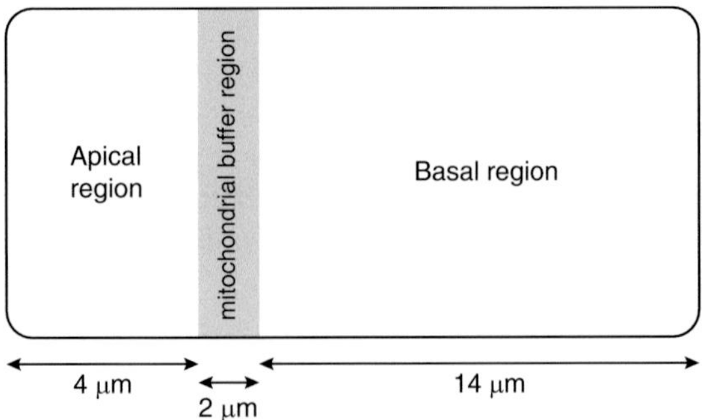

Fig. 6.21 Spatial structure of the model of Ca^{2+} wave propagation in acinar cells (Sneyd et al, 2003).

This mechanism was investigated theoretically by Sneyd et al (2003), who showed that it is possible for a model to display two distinct modes of wave propagation, one in which the wave is driven by the diffusion of Ca^{2+} between excitable release sites and the other kinematic, driven only by differences in the timing of Ca^{2+} release. Their model is based on the following assumptions and choices:

- The cell is modelled as being divided into three parts. The apical end ($4\,\mu$m long) has a high IPR density and a low RyR density. The basolateral end ($14\,\mu$m long) has a higher RyR density and a lower IPR density. The IPR and RyR densities are determined (very approximately) by fitting to experimental data on wave speed as a function of $[IP_3]$. The third region, between the apical and basal regions and $2\,\mu$m wide, contains a high density of mitochondria. This so-called *mitochondrial buffer region* is based on the experimental results of Tinel et al (1999) who showed that uptake of Ca^{2+} by the mitochondria forms a barrier to the formation of global Ca^{2+} waves (Fig. 6.21).
- The IPR model of Sneyd and Dufour (2002) is used, while the RyR are described by the model of Keizer and Levine (1996). These particular choices are of little importance, and use of a wide range of IPR and RyR models would give the same qualitative result.
- The cell is assumed to be thin, so that membrane fluxes appear throughout the regions, not just on the borders.
- Mitochondrial uptake is described using the model of Colegrove et al (2000b). Again, the exact expression for mitochondrial uptake is not important. All that matters is that the mitochondria contribute an additional Ca^{2+} uptake term in the buffer region.
- Both pancreatic and parotid cells have the same spatial structure, and differ only in the densities of IPR and RyR in the three regions.

- Buffering is assumed to be fast and linear, and thus all fluxes are interpreted as effective fluxes (Section 2.9).
- Plasma membrane and ER ATPase pumps are modelled using Hill functions, or minor adaptations thereof.

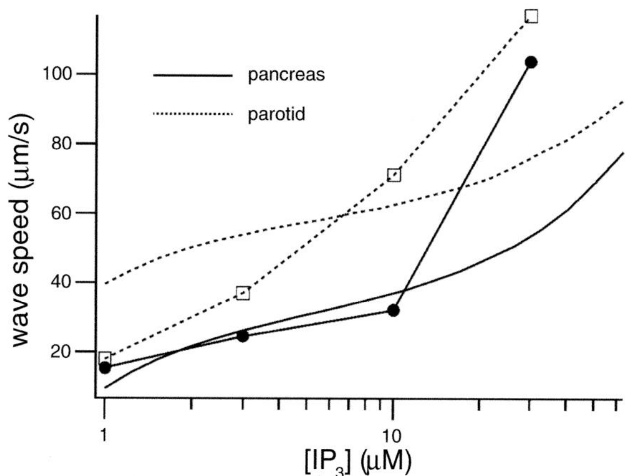

Fig. 6.22 Wave speed as a function of $[IP_3]$ in the acinar cell model (Sneyd et al, 2003), which assumes that both pancreatic and parotid cells have the same spatial structure, differing only in the assumed densities of IPR and RyR in each region. Symbols are experimental data.

Typical results are shown in Fig. 6.22. Although the quantitative agreement is rather poor (despite the parameters being determined by fitting to data!) there are some qualitative features of the results which agree with observation. The parotid waves are faster than the pancreatic waves, and in both cell types the wave speed at low $[IP_3]$ is relatively insensitive to $[IP_3]$. Closer inspection of the model results shows that, at low $[IP_3]$, the wave is propagated across the cell by the RyR, while at high $[IP_3]$ release through RyR becomes less important, as the kinematic mechanism takes over. Thus the model confirms that there are two quite different ways of obtaining a global Ca^{2+} response, and that the actual situation is a complex combination of these two mechanisms.

However, questions remain. For example, the poor quantitative agreement with data has not yet been improved by subsequent modelling, and it remains unclear whether or not global waves in acinar cells can be accurately described by a model involving spatial distributions of IPR and RyR. In addition, detailed investigations of how differences in the three-dimensional structure of the apical region can change wave properties (as suggested by Larina and Thorn (2005)) remain to be done.

6.3.3 Calcium Waves and Water Secretion

Most recently, acinar cell models have focused on the relationship between Ca^{2+} oscillations and water transport in acinar cells of the salivary glands (Palk et al, 2010, 2012). In addition to the equations describing Ca^{2+} and IP_3 dynamics, such models include additional equations modelling the apical and basal membrane potentials, water flux through the cell, and the volume of the cell. It is beyond the scope of this book to discuss these more general models in detail. However, in brief:

- The membrane potentials are modelled by the usual electrical circuit model, as described in Section 3.7, but with the added feature that the basal potential is not the same as the apical potential. For additional simplicity the voltage equations are assumed to be instantaneously at steady state, giving algebraic equations to solve for V_{apical} and V_{basal} in terms of the other variables. If linear current-voltage equations are used, these algebraic equations can be solved explicitly, leading to significant simplification of the model.
- Water flow is via osmosis, and is assumed to be linearly proportional to the difference in total solute concentration across the membrane. Thus, the model tracks the concentrations, not only of Ca^{2+}, but also of K^+, Na^+, and Cl^-. Other ions such as HCO_3^- or H^+ are not yet included in published versions of these models.
- Ions move across the apical and basal membranes through a variety of transport pathways, some of which are shown in Fig. 6.18.
- If the apical water flux does not equal the basal water flux, then the cell volume will change, with consequent changes in all internal ionic concentrations.
- Ion concentrations in the lumen are model variables as well, but the interstitial concentrations are assumed to be held fixed.

This approach gives a set of model equations identical in structure to the classic pump-leak model of cell volume control (Tosteson and Hoffman, 1960; Jakobsson, 1980; Keener and Sneyd, 2008). As described briefly above, changes in $[Ca^{2+}]$ modulate the conductance of the Cl^- channels in the apical membrane and the K^+ channel in the basal membrane, thus leading to overall Cl^- transport into the lumen, whereby Na^+ follows to keep electroneutrality, and water follows by osmosis.

To date, the major results of this model can be summarised as follows:

- The model reproduces well the observed Ca^{2+} dynamics (where, for now, we ignore the problem of wave propagation across the cell). Ca^{2+} oscillations arise for intermediate agonist concentrations, and persist in the absence of extracellular Ca^{2+}. These oscillations are accompanied by oscillations in $[IP_3]$ and in the concentrations of the other ions, as well as small oscillations in cell volume. Physiological concentrations of agonist result in an average cell volume decrease of around 20%, consistent with the available experimental evidence (Foskett and Melvin, 1989).

- Water flow is maximised when approximately 20% of the Ca^{2+}-activated K^+ current is carried by K^+ channels in the apical membrane. Subsequent investigation of this prediction (Almassy et al, 2012) found that there are indeed functional Ca^{2+}-activated K^+ channels in the apical membrane, although it is not known what fraction of the total K^+ current they carry.
- Water flow is maximised when the $[Ca^{2+}]$ at the apical membrane rises simultaneously with that at the basal membrane (or possibly with a slight delay, depending on the exact kinetics of the Ca^{2+}-activated ion channels). This result was obtained, not by solving the model equations in a spatially distributed domain, but by manually imposing a Ca^{2+} rise at the apical and basal ends of the cell. Solution of the spatially distributed model equations results in wave profiles that are quite different in the apical and basal regions, which would confuse the interpretation of the results.
- Water flow is almost entirely insensitive to the frequency of the Ca^{2+} oscillation, being governed instead by the average $[Ca^{2+}]$. This is somewhat in contradiction with much of current dogma, which tends to assume that the frequency of the Ca^{2+} oscillation is an important information carrier.

6.3.4 Detailed Spatial Structure of an Acinus

In all the models discussed so far, the spatial structure of each cell is highly simplified. Even when multiple cells are connected, as in models of intercellular waves in airway epithelial cells (Section 3.6.2) or hepatocytes (Section 6.2.3), not only is each cell a caricature, the multicellular structure is also.

Unsurprisingly, in reality the situation is far more complex, and in any particular case it is not a simple matter to determine how intricate spatial structures may, or may not, influence Ca^{2+} signalling. The importance of spatial structure at the level of microdomains is clear, but the importance of such details at the multicellular level is less so.

An example of an intricate multicellular structure, that might reasonably be expected to influence Ca^{2+} signalling, is shown in Fig. 6.23. This computational reconstruction is based on unpublished data from a mouse parotid gland collected in the laboratory of David Yule. Separate labelling of the Na^+/K^+ ATPase in the basal membrane, and the Ca^{2+}-dependent Cl^- channel in the apical membrane, allows for a three-dimensional reconstruction of multiple cells, together with their associated lumen. A selection of seven cells only is shown in Fig. 6.23, where the basal membranes are shown in red and the apical membranes in green.

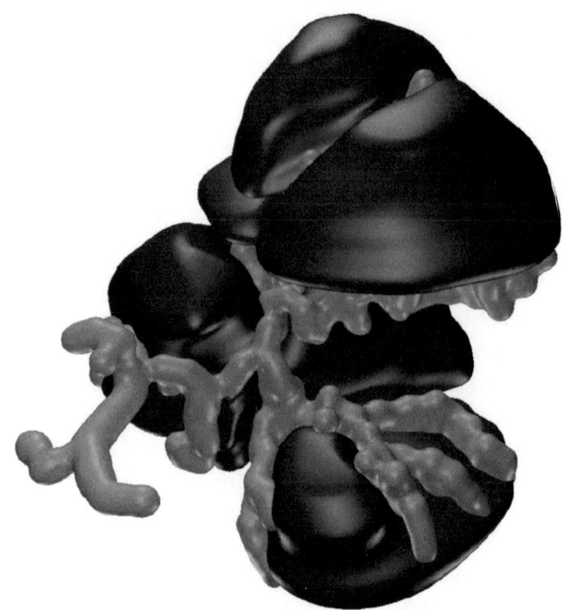

Fig. 6.23 Three-dimensional reconstruction of a group of seven cells in a parotid acinus. One cell has been removed from the diagram so as to gain a better view of the interior of the group. Basal membranes of the cells are shown in red and apical membranes in green. Since the apical membranes surround the lumen, the green surface is effectively a reconstruction of the lumen. Based on unpublished data from the laboratory of David Yule (University of Rochester), with a computational reconstruction by John Rugis (University of Auckland).

The most important point made by this figure is the complex geometry of the acinar lumen. Instead of being a spherical gap between a group of symmetrically positioned cells, the lumen is a highly branched structure that extends "claws" over the surface of each acinar cell. Thus, the apical membrane of each acinar cell has a complex branched structure.

The potential significance of this becomes clear when one recalls that there is a close correlation between spatial structure and the Ca^{2+} toolbox components. IPR, for instance, have a higher density around the apical membrane, while PLC is situated on the basal membrane, which is where IP_3 is necessarily made. It is highly likely that these spatial distributions will have important effects on Ca^{2+} signalling, although, as yet, we have only a very limited understanding of what such effects might be.

6.4 Astrocytes

6.4.1 Introduction

The central nervous system contains two major classes of cell types; neurons and glia. Neurons convey information through electrical signals generated by a variety of rapid and highly regulated ion channels, and are one of the most intensively studied cell types. Glial cells – named for the Greek work for glue – come in a range of different cell types, and play an important supportive role. For example, they supply oxygen and nutrients to neurons and insulate them from one another. However, although the number of glial cells far exceeds that of neurons, they have been studied far less and are not so well understood.

Astrocytes, which are a type of glial cell, are characteristic star-shaped cells in the brain and spinal cord. They are the most abundant cells in our body and, like all types of glial cells, are not electrically excitable. Astrocytes instead display a rich repertoire of temporally and spatially organised Ca^{2+} signals; Ca^{2+} oscillations and intercellular waves are commonly observed in this cell type, and are implicated in important functions of the brain.

Work published over the past 15 years has led to a profound re-evaluation of the roles of glial cells – more specifically of astrocytes – in the mammalian central nervous system. In particular, astrocytes communicate bidirectionally with neurons. Changes in neuronal activity can increase astrocytic Ca^{2+}, which then causes the release of "gliotransmitters" such as glutamate, ATP, or D-serine. These transmitters feed back to pre- and post-synaptic terminals and evoke new responses in adjacent neurons (Hamilton and Attwell, 2010; Navarrete et al, 2013). The mechanistic details of this two-way communication remain poorly understood, but certainly rely on a sophisticated organisation of Ca^{2+} dynamics in astrocytes.

Ca^{2+} changes in astrocytes are either spontaneous (Parri and Crunelli, 2003) or induced by glutamate (Bradley and Challiss, 2012). Astrocytes express both ionotropic and metabotropic glutamate receptors (mGlu receptors). Group I mGlu receptors, which contain both $mGlu_1$ and $mGlu_5$ subtypes, are involved in Ca^{2+} signalling. These receptors couple to $G\alpha_{q/11}$ proteins to stimulate phospholipase C activity and IP_3 formation. Despite this common signal transduction mechanism, the patterns of Ca^{2+} increase induced by $mGlu_1$ and $mGlu_5$ receptors differ (Kawabata et al, 1996); stimulation of the $mGlu_1$ receptor typically causes one large Ca^{2+} peak, followed by damped oscillations on a plateau, whereas stimulation of the $mGlu_5$ receptors most often induces robust, baseline Ca^{2+} spiking. Importantly, Ca^{2+} spiking stimulated by the $mGlu_5$ receptor occurs in parallel with synchronous changes in $[IP_3]$, while stimulation of $mGlu_1$ leads to a non-oscillatory increase in $[IP_3]$ (Nash et al, 2001).

Calcium oscillations triggered by $mGlu_5$ receptor stimulation also display unusual characteristics; they occur over a broad range of stimulation strengths and often fail to transform into a non-oscillatory plateau response at high glutamate concentrations (Kawabata et al, 1996; Nash et al, 2002). On the other hand, Ca^{2+} oscillation frequency and amplitude are little affected by the level of stimulation, in contrast with the usual property of frequency encoding. In this case the density of $mGlu_5$ receptors at the cell surface is the key factor in determining Ca^{2+} oscillation frequency (Nash et al, 2002). The period usually ranges between 10 s and 30 s.

These unusual properties of Ca^{2+} oscillations induced by stimulation of $mGlu_5$ receptors suggest that their mechanism is not simply based on the properties of the IPR (see Section 2.7). To elucidate more precisely how they work, $mGlu_5$ receptors have been expressed in cell lines in which Ca^{2+} signalling can be studied more easily than in astrocytes (which do not easily survive when isolated). The information gained in these experiments can also be useful to understand Ca^{2+} dynamics in neurons that express these receptors in abundance. Such an understanding is particularly important given that modulation of $mGlu_5$ receptor function has been postulated as an approach for the treatment of a number of pathological conditions, including anxiety (Swanson et al, 2005), schizophrenia (Marino and Conn, 2006), Huntington's disease (Ribeiro et al, 2014), and Alzheimer's disease (Lazzari et al, 2014).

6.4.2 Calcium Oscillations Induced by Stimulation of $mGlu_5$ Receptors

All mGlu receptors are trafficked to the plasma membrane and fulfil their signalling function as covalently linked dimers (Pin et al, 2004). The peculiarity of $mGlu_5$ receptors is that they undergo rapid, cyclical phosphorylation along with Ca^{2+} oscillations. Phosphorylation is mediated by PKC and occurs at Ser-839 within the C-terminal tail of the receptor. Such phosphorylation does not occur in $mGlu_1$ receptors (which, as discussed above, generate a peak-plateau response). The difference between these receptor types lies in the amino acid adjacent to the phosphorylation site. In $mGlu_5$ receptors, it is an adjacent Thr-840 that allows PKC-mediated phosphorylation. However, Ser-839 in $mGlu_1$ is not phosphorylated, as the adjacent residue (Asp-854) does not allow that (Kim et al, 2005).

When phosphorylated, the receptor is uncoupled from $G\alpha_{q/11}$/PLC activation. This effect is reversed on receptor dephosphorylation by a Ca^{2+}-independent, protein phosphatase. Such a cyclical change in the transduction mechanism has been called *dynamic uncoupling* (Nash et al, 2002). Receptor phosphorylation is cyclical because PKC activity varies with time. This could be due to diverse factors. The various isoenzymes of PKC are differentially regulated and often fulfil specific roles within the cell. Conventional PKCs are activated by Ca^{2+} and DAG, the membrane-bound ester that is produced together with IP_3 (see Fig. 1.2 in Chapter 1); novel PKCs (nPKC) by DAG, but not by Ca^{2+}; and atypical PKCs are insensitive to both Ca^{2+} and DAG (Gallegos and Newton, 2008). Phosphorylation of the $mGlu_5$ receptor at Ser-839 occurs through PKCϵ which belongs to the nPKC family.

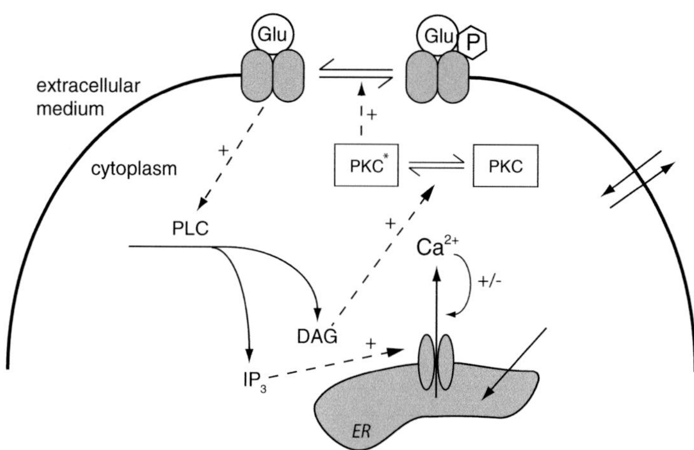

Fig. 6.24 Schematic diagram of the model for $mGlu_5$ receptor-mediated Ca^{2+} oscillations (Dupont et al, 2011b). Glu denotes glutamate, P denotes a phosphate group, and PKC* is the activated form of PKC.

The model of Dupont et al (2011b) is shown schematically in Fig. 6.24. It is assumed that $mGlu_5$ receptors bind glutamate and become active when dimerised. Stimulation of PLC activity, through $G\alpha_{q/11}$ protein activation, generates DAG and IP_3, leading to release of Ca^{2+} from the ER. The specific PLC involved in this pathway is PLCβ (Atkinson et al, 2006), an enzyme that is largely insensitive to $[Ca^{2+}]$ changes within the range occurring during cytoplasmic Ca^{2+} oscillations (Young et al, 2003). DAG stimulates PKC activity, which in turn phosphorylates the $mGlu_5$ receptor at Ser-839. This phosphorylation leads to the uncoupling of the $mGlu_5$ receptor from the G protein signalling cascade. Receptor dephosphorylation occurs through a Ca^{2+}-insensitive protein phosphatase activity.

As well as this phosphorylation/dephosphorylation cycle, the model includes other typical cytoplasmic Ca^{2+} regulatory processes, such as the sequential activation and inhibition of the IPR by cytoplasmic Ca^{2+}, transport into the ER by SERCA pumps, and Ca^{2+} exchange with the extracellular medium. The model variables are

- p ($[IP_3]$), c (cytoplasmic $[Ca^{2+}]$), c_e (ER $[Ca^{2+}]$), and y (fraction of IPR not inactivated by Ca^{2+}),
- R_2 : concentration of $mGlu_5$ dimers,

- M : concentration of ligand-bound $mGlu_5$ dimers,
- D : DAG concentration, and
- P_K : fraction of active protein kinase C.

The equations for these variables correspond to the scheme shown in Fig. 6.24. Binding and unbinding of glutamate (with concentration L, which is assumed to be held fixed) to receptor dimers is described by

$$\frac{dR_2}{dt} = -k_{R+}R_2L^2 + k_{R-}M, \tag{6.14}$$

where k_{R+} and k_{R-} are the kinetic constants of ligand binding and unbinding, respectively. The k_{R-}/k_{R+} ratio, which has the units of concentration squared, represents the affinity (denoted by K_{aff}) of the ligand for the receptor. For simplicity, it is assumed that glutamate binding occurs simultaneously on both glutamate binding sites of the dimer.

Receptor dimers can be phosphorylated by PKC (Kim et al, 2005). The evolution equation for the fraction of active, non-phosphorylated dimer is

$$\frac{dM}{dt} = k_{R+}R_2L^2 - k_{R-}M + V_M\frac{M_p}{K_M + M_p} - P_KV_{\text{PKC}}\frac{M}{K_{\text{PKC}} + M}, \tag{6.15}$$

where M_p is the concentration of phosphorylated receptor dimers. Its value can be obtained by an algebraic equation if it is assumed that the total amount of receptor is fixed (see below). Dephosphorylation is characterised by a maximal rate V_M and a Michaelis constant K_M. The last term in (6.15) represents the PKC-mediated phosphorylation of receptors, with V_{PKC} the maximal rate of this kinase, P_K the fraction of this kinase in its active form, and K_{PKC} the Michaelis constant of $mGlu_5$ receptor phosphorylation by PKC.

The rates of IP_3 and DAG synthesis are assumed to be proportional to the concentration of active receptor dimer. Thus,

$$\frac{dp}{dt} = k_{\text{PLC}}M - k_1p, \tag{6.16}$$

$$\frac{dD}{dt} = k_{\text{PLC}}M - V_D\frac{D}{K_D + D}. \tag{6.17}$$

In these equations k_{PLC} represents the catalytic activity of PLC. For simplicity, metabolism of IP_3 (into IP_4 and IP_2) is combined into a single linear degradation term, with rate constant k_1. For DAG also, a single Michaelis-Menten degradation rate (representing DAG kinase and DAG lipase activities) is used, characterised by a maximal rate V_D and a Michaelis constant K_D.

Phosphorylation of the $mGlu_5$ receptor is most likely mediated by a member (possibly more than one) of the novel PKC family, which is sensitive to

DAG but not Ca^{2+} (Bradley and Challiss, 2011). This process is reversible, and thus

$$\frac{dP_K}{dt} = k_{\text{act}} \frac{D}{K_{AD} + D}(1 - P_K) - k_{\text{des}} P_K. \tag{6.18}$$

The maximal rate of PKC activation (divided by the total concentration of PKC) is k_{act} and K_{AD} corresponds to the concentration of DAG leading to half-maximal activation. The first order rate constant for PKC deactivation is k_{des}. In reality, activation of PKC is accompanied by its translocation to the plasma membrane. However, spatial aspects are not considered in the present phenomenological approach.

The final two differential equations provide a classical description of the activity of the IPR and of cytoplasmic Ca^{2+}, and are similar to those used above to describe Ca^{2+} dynamics in hepatocytes, except for two points. Firstly, Ca^{2+} exchange with the external medium is included. Thus, (6.7) is replaced by

$$\frac{dc}{dt} = \nu_0 + J_{\text{IPR}} - J_{\text{serca}} - k_1 c, \tag{6.19}$$

together with the analogous equation for c_e. Secondly, cooperative binding of IP_3 to the IPR is taken into account. Thus (2.120) is replaced by

$$P_o = \frac{p^2}{K_p^2 + p^2} \frac{c^3}{K_c^3 + c^3} y. \tag{6.20}$$

In order to simplify the model, we assume that ligand binding to and dissociation from the $mGlu_5$ dimer are faster than the other processes described in the model. Within this framework, (6.14) can be replaced by the algebraic equation

$$R_2 = K_{\text{aff}} \frac{M}{L^2}. \tag{6.21}$$

Furthermore, if we let R_1 denote the concentration of $mGlu_5$ monomers, the concentration of phosphorylated dimers can be obtained by the following conservation relation

$$2M_p = R_{\text{tot}} - R_1 - 2R_2 - 2M, \tag{6.22}$$

where R_{tot} is the total intracellular concentration of $mGlu_5$ receptors (in the monomeric form).

The $mGlu_5$ dimerisation reaction is

$$\mathrm{R}_1 + \mathrm{R}_1 \underset{k_{-1}}{\overset{k_1}{\rightleftharpoons}} \mathrm{R}_2, \tag{6.23}$$

and so if we assume this reaction is at equilibrium, we get

$$R_1 = \sqrt{K_{\text{d1}} R_2}, \tag{6.24}$$

where $K_{\text{di}} = k_{-1}/k_1$ is the dimerisation equilibrium constant. In this case, the concentration of phosphorylated dimers is given by

$$2M_p = R_{\text{tot}} - \sqrt{K_{\text{di}}R_2} - 2R_2 - 2M. \tag{6.25}$$

Typical oscillations obtained by simulation of the model are shown in Fig. 6.25. As shown by the evolution of the fraction of active mGlu_5 receptor (panel A, red curve), Ca^{2+} oscillations (panel A, blue curve) result from the repetitive phosphorylation/dephosphorylation of the receptor. Phosphorylation is driven by the periodic activation of PKC by DAG (panel B). The amplitude of the oscillation in active receptor dimers reaches approximately 40 nM, a value which agrees with the experimental estimation of about 50,000 receptors per cell. Agonist-occupied, non-phosphorylated receptor dimers stimulate PLC activity, increasing $[IP_3]$ and inducing Ca^{2+} release from internal stores. This release is modulated through Ca^{2+}-induced Ca^{2+} release and rapidly inhibited by Ca^{2+}-mediated inhibition of the IPR. Thus, $[Ca^{2+}]$ increases are short-lived.

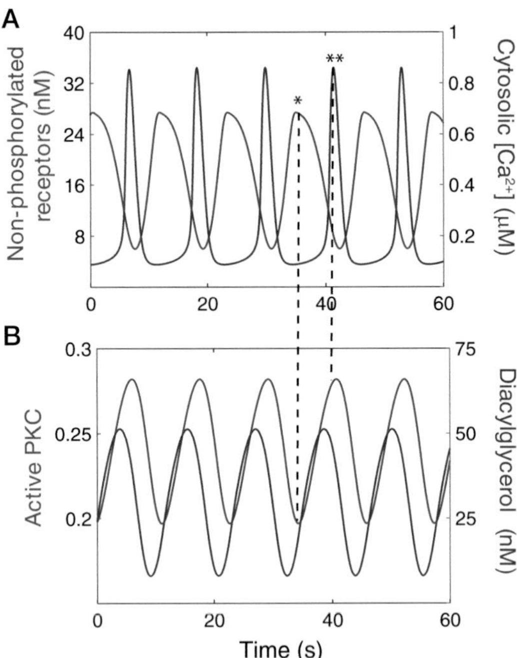

Fig. 6.25 Simulated Ca^{2+} oscillations based on the mechanism of dynamic uncoupling of the mGlu_5 receptor (Dupont et al, 2011b). In this model, the periodic phosphorylation-dephosphorylation of the mGlu_5 receptor dimers drives oscillations of all the variables. The significance of the vertical dashed lines is explained in the text.

The concentrations of DAG and IP_3 increase together, and both second messengers display sinusoidal oscillations. DAG activates PKC, which phosphorylates the $mGlu_5$ receptor, thereby uncoupling it from $G\alpha_{q/11}$. Maximum PKC activity corresponds to a minimum number of active $mGlu_5$ receptors (see the vertical line indicated with a *, spanning panels A and B of Fig. 6.25).

On agonist addition, active, non-phosphorylated $mGlu_5$ receptors peak first, followed by [IP_3] and [DAG] (that peak together), then followed rapidly by [Ca^{2+}], as observed experimentally in Matsu-ura et al (2006), and a bit later by active PKC (see the vertical line indicated with a **). At the peak of $mGlu_5$ receptor-$G\alpha_{q/11}$ uncoupling through PKC-mediated receptor phosphorylation, the [IP_3] is too low to generate a Ca^{2+} peak. In consequence, the dynamic changes in $mGlu_5$ receptor activity determine the timing of Ca^{2+} oscillations. The period of oscillations is imposed by the rates of receptor phosphorylation and dephosphorylation, which have been determined by constraining the model to give oscillation periods matching the experimental observations (12 s in the case of Fig. 6.25).

In the model, the oscillation period is highly sensitive to the rates of the protein kinase and phosphatase. For example, if the maximal rate of the protein phosphatase (V_M) is divided by 5, the period rises to 89 s. This is in agreement with the experimentally observed reduction in frequency in the presence of calyculin A or cantharidin, which inhibit protein phosphatase 1 and 2A, respectively (Nash et al, 2002; Bradley and Challiss, 2011). Similarly, as in the experiments performed in the presence of the PKC activator, phorbol 12,13-dibutyrate (Bradley and Challiss, 2011), an increase in the maximal rate of phosphorylation by PKC (V_{PKC}) leads to a decrease in oscillation frequency. If this rate doubles, the period increases to 23 s. In contrast, as observed experimentally, oscillations are insensitive to increases in the level of glutamate (see below) as modification at the level of the receptor is the rate-limiting step in the overall oscillatory process.

Dynamic uncoupling is only generated in the model at a sufficiently high agonist stimulation level. At the lowest glutamate concentrations, Ca^{2+} oscillations occur, but are based on the IPR dynamics, as the fraction of phosphorylated receptor remains constant. In fact, a stable steady state sometimes coexists with the limit cycle where only the [Ca^{2+}] and the fraction of inactive IPR oscillate. However, the limit cycle, i.e., the oscillatory state, is always reached when simulating an initially unstimulated cell, which corresponds to the experimental protocol. Such Ca^{2+} oscillations occurring independently of dynamic uncoupling are reminiscent of the slower oscillations observed at low stimulation levels in the absence of PKC feedback inhibition (see Fig. 8B in Nash et al (2002)). In this case, the oscillation period is determined by the constant [IP_3] and thus indirectly depends on all the parameters modulating this concentration. For example, decreasing the rate of the protein phosphatase (V_M in (6.15)) leads to an increase in the period, as in the case of dynamic uncoupling, although the level of active dimers does not oscillate.

This is due to an increase in the amount of phosphorylated $mGlu_5$ receptor, leading to a reduced synthesis of IP_3 and thus to slower IPR-based Ca^{2+} oscillations.

Because of the coexistence of two oscillatory mechanisms (dynamic uncoupling and dual regulation of IPR activity by Ca^{2+}), one could theoretically expect complex Ca^{2+} oscillations to occur (Goldbeter et al, 1988). This is indeed the case in the model if IPR regulation by Ca^{2+} occurs on a much faster time scale than dynamic uncoupling. In this case, several cycles of IPR opening and closing can occur while the $mGlu_5$ receptor is in an active state, giving rise to a complex pattern of Ca^{2+} release. However, this situation is not physiologically relevant, as the periods of oscillations induced by stimulation of $mGlu_1$ receptors – and thus IPR based – are very similar to the periods of oscillations based on dynamic uncoupling in response to stimulation of $mGlu_5$, which suggests that the time scales of both processes are of the same order.

It has been shown experimentally that a key factor affecting the period of Ca^{2+} oscillations induced by stimulation of $mGlu_5$ receptors is the receptor density at the cell surface (Nash et al, 2002), and the model reproduces these experimental observations (Fig. 6.26). In the regime of dynamic uncoupling, this behaviour is accompanied by slower cycles of receptor phosphorylation/dephosphorylation, as well as by decreased $[IP_3]$. Moreover, when the amount of receptor is too low, dynamic uncoupling is ineffective, as robust changes in receptor activity are no longer possible; Ca^{2+} oscillations then rely on IPR dynamics. In Fig. 6.26 the frequency of Ca^{2+} oscillations is

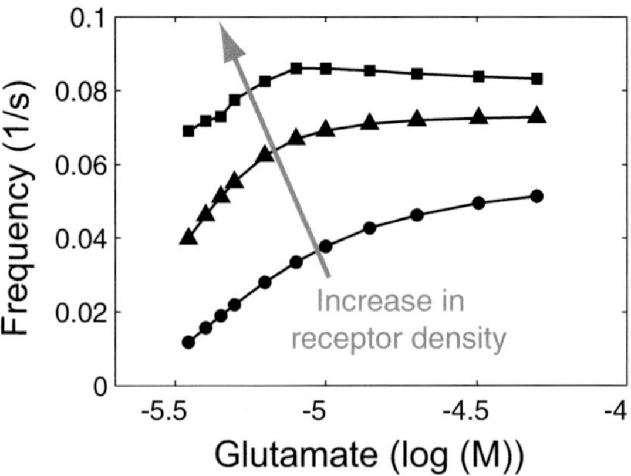

Fig. 6.26 Model prediction of the relationship between the frequency of Ca^{2+} oscillations and the concentration of agonist at different levels of $mGlu_5$ receptor expression (Dupont et al, 2011b). At high receptor densities, oscillation frequency is mainly controlled by the density of receptors and does not vary significantly with the level of stimulation. In contrast, frequency is more sensitive to stimulus strength at low receptor density, where the IPR drives the Ca^{2+} oscillations. The concentration of $mGlu_5$ receptors has been set equal to 0.03 μM (circles), 0.05 μM (triangles), or 0.075 μM (squares).

shown for glutamate concentrations up to 10^{-4} M, but oscillations still occur, with unchanged frequencies, for higher stimulation levels. This maintenance of Ca^{2+} oscillations at very high agonist concentrations is a key characteristic of dynamic uncoupling.

The dynamic uncoupling mechanism is very sensitive to the values of K_M and K_{PKC}. These constants, appearing in (6.15), are the Michaelis constants of the protein phosphatase and kinase for their substrates, i.e., for the $mGlu_5$ receptor. They both need to be small (compared to the concentration of receptors) to allow dynamic uncoupling. This condition implies that both enzymes are always saturated, a phenomenon known as *zero-order ultrasensitivity* (Goldbeter and Koshland, 1981) and characterised by the steep sigmoidal relationship between the amount of unphosphorylated substrate and the rate of the protein kinase at steady state (see the blue dashed curve in Fig. 6.27). If neither enzyme is saturated, this relation becomes smooth and the receptor always remains in the same state (the red dashed curve and dot in Fig. 6.27).

When the dependence is sigmoidal, significant changes in $mGlu_5$ receptor coupling to PLC occur when PKC varies, allowing for robust dynamic uncoupling. The blue cycle in Fig. 6.27 shows the evolution of active receptors

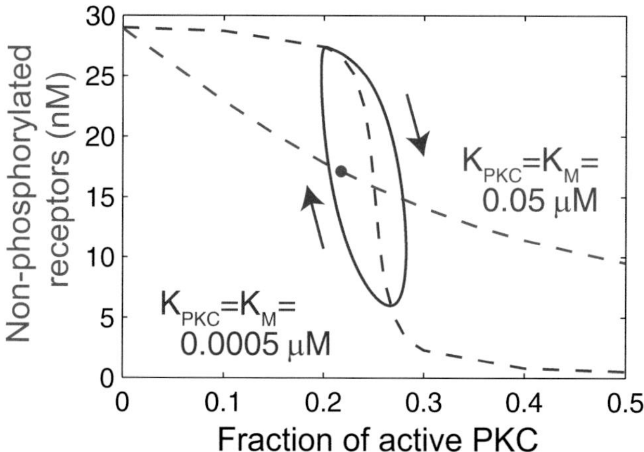

Fig. 6.27 Dependence of the concentration of active $mGlu_5$ receptors on the fraction of active PKC (Dupont et al, 2011b). Oscillations based on dynamic uncoupling require a sharp dependency. The blue curves correspond to a situation where oscillations can occur ($K_{PKC} = K_M = 0.0005\,\mu$M). Under these conditions, active receptors have a sigmoidal dependence on PKC activity at steady state, as indicated by the dashed blue line, which shows the steady-state solution of (6.15). As a consequence, during oscillations (represented by the solid blue line), abrupt transitions in the concentration of active receptor can be obtained upon moderate changes in PKC activity. In contrast, when the concentration of active receptor shows only a weak dependency on PKC activity (red dashed curve, $K_{PKC} = K_M = 0.05\,\mu$M), no oscillations occur and the variables remain at their steady-state values, indicated by the red point.

and active PKC for the oscillations shown in Fig. 6.25. When the level of active PKC lies to the left of the steady-state curve (the blue dashed curve), $mGlu_5$ receptors will rapidly dephosphorylate, because the concentration of non-phosphorylated receptors tends to reach the high steady-state value (approximately 28 nM). Similarly, when PKC increases and lies to the right of the threshold of the steady-state curve, phosphorylation will occur. For higher values of K_M and K_{PKC}, the steepness of the steady-state curve is lost, as shown by the red dashed curve of Fig. 6.27. Under these conditions, oscillations disappear and the system remains in the state indicated by the red point. The sigmoidal steady-state curve shown in Fig. 6.27 also allows us to understand why dynamic uncoupling cannot occur when the number of receptors and/or ligand concentration is too low. In this case the possible range over which the concentration of phosphorylated receptors can change becomes so small that the active receptor concentration cannot be repetitively driven below and above the threshold required to activate Ca^{2+} release.

Zero-order ultrasensitivity has been demonstrated in various signalling cascades (Ferrell and Ha, 2014) but is entirely speculative in the present case. However, in order to get oscillations based on dynamic uncoupling, one must have a sharp dependency of the amount of phosphorylated $mGlu_5$ receptor on PKC activity. This steep function could originate from many other regulatory pathways and feedbacks, which could be as simple as high levels of cooperativity. The existence of such mechanisms is highly plausible in view of the complexity of the G protein signalling cascade involved in coupling $mGlu_5$ receptors to PLC activity (Sorensen and Conn, 2003).

6.4.3 Towards Modelling Calcium Oscillations in Astrocytes

The model just presented describes Ca^{2+} dynamics in cell lines expressing $mGlu_5$ receptors, where Ca^{2+} dynamics rely on these receptors and IPR. As suggested above, astrocytes are in fact much more complex, and an adequate description of their Ca^{2+} dynamics requires the combination of a larger number of modules from the Ca^{2+} toolbox. There have been some attempts to deal with this complexity (Kang and Othmer, 2007; De Pittà et al, 2009). However, computational results are rather difficult to interpret given that these models include many regulatory pathways – whose existence in astrocytes sometimes remains to be demonstrated – giving rise to a much larger variety of behaviours than what can be observed experimentally. Much work remains to be done in this important field. Furthermore, spatial aspects are crucial. Not only are astrocytes polarised (Arizono et al, 2012), but intercellular communication via Ca^{2+} waves also plays an important role in their interplay with neurons (Wallach et al, 2014).

Chapter 7
Muscle

7.1 Introduction

In all three types of muscle cells – skeletal, cardiac, and smooth – Ca^{2+} plays a major role in excitation-contraction (EC) coupling, i.e., the sequence of events that links electrical stimulation to contraction (or, in nonexcitable smooth muscle, the events that link agonist stimulation to contraction). In skeletal and cardiac muscle, an action potential arriving from a neuron is propagated along the membrane of the muscle cell, and penetrates deep into the interior of the cell via invaginations of the cell membrane, called T-tubules. This action potential causes the release of Ca^{2+} from the sarcoplasmic reticulum, which then allows the crossbridge cycle to develop force (Huxley, 1957; Bers, 2001).

Skeletal and cardiac muscle share many similarities. For our purposes, the major difference between these cell types is the way in which EC coupling works. The basic steps of EC coupling in ventricular myocytes are shown in Fig. 7.1. Depolarisation of the T-tubule by the action potential causes the opening of L-type Ca^{2+} channels (also called dihydropyridine receptors, or DHPRs; here we shall call them LCC) and resultant inward flow of Ca^{2+} current. The Ca^{2+} that enters the cell stimulates the release of additional Ca^{2+} from the SR via ryanodine receptors by the process of Ca^{2+}-induced Ca^{2+} release. This Ca^{2+} diffuses through the myoplasm and binds to the myofilaments, causing contraction, before being eventually removed from the myoplasm by ATPases (which pump the Ca^{2+} into the SR or out of the cell) or by the Na^+/Ca^{2+} exchanger (which transports Ca^{2+} out of the cell).

In skeletal muscle the process is slightly different, in that the LCC are directly linked to the RyR; a change in conformation of the channel as a result of depolarisation causes an immediate change in conformation of the RyR and consequent Ca^{2+} release.

G. Dupont et al., *Models of Calcium Signalling*, Interdisciplinary Applied Mathematics 43, DOI 10.1007/978-3-319-29647-0_7

There is a greater diversity in smooth muscle, where the mechanisms of EC coupling can be quite different from those in skeletal and cardiac muscle. This is discussed in more detail in Section 7.4.

There is a vast literature, both experimental and theoretical, on the Ca^{2+} dynamics of muscle cells. It is simply not possible to give here a complete overview of so large a field. Instead, we give only a relatively brief and selective discussion of some of the major questions and models.

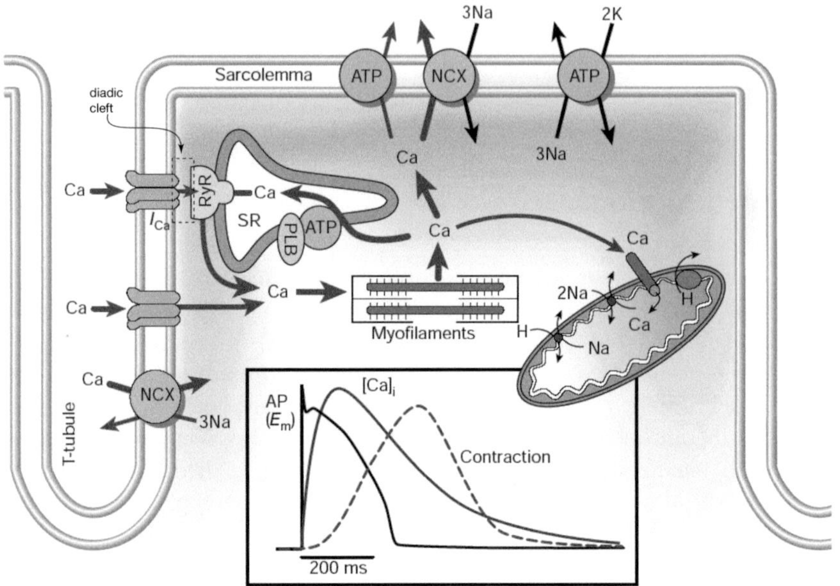

Fig. 7.1 The major Ca^{2+} fluxes underlying excitation-contraction coupling in cardiac ventricular myocytes. NCX is the Na^+/Ca^{2+} exchanger; PLB is phospholamban, a protein that, when unphosphorylated, inhibits the SERCA pump; SR is the sarcoplasmic reticulum. The inset shows the time courses of the action potential (AP), the Ca^{2+} transient, and the contraction. Note that the AP happens first, followed by the Ca^{2+} transient and then by contraction. Bers (2002), Fig. 1: reprinted by permission from Macmillan Publishers Ltd.

7.2 Cardiac Myocytes

7.2.1 Cardiac Excitation-Contraction Coupling

Much of the literature on EC coupling in cardiac myocytes focuses on electrical aspects, enumerating and characterising the large number of different ion channels that interact to form the cardiac action potential (Noble, 1962; Beeler and Reuter, 1977; Luo and Rudy, 1991; Puglisi et al, 2004; Noble,

2008; Williams et al, 2010). However, there has also been a great deal of attention paid to the control of Ca^{2+}, and it is these models that we discuss here.

The most comprehensive review of the experimental literature on cardiac Ca^{2+} is that of Bers (2001), while the modelling literature is most easily accessed through the reviews of Soeller and Cannell (2004), Hinch et al (2006), Williams et al (2010), and Greenstein and Winslow (2011). The shorter review by Bers (2002) is a useful introduction to the field.

The study of Ca^{2+} dynamics in cardiac cells is made more difficult by the spatial aspects of the problem. Both the LCC and the RyR release Ca^{2+} into a small volume, the region between the SR and the sarcolemma. This region, called the diadic cleft (see Fig. 7.1), is only about 15 nm wide with a radius of about 200 nm (although it is not a circular region, it can be reasonably approximated as such), and so has a volume of only about 2×10^{-18} L. Calcium fluxes into such a small volume cause large spatial and temporal gradients that are impossible to measure experimentally and difficult to simulate numerically. Modelling is made even more difficult by the fact that a resting Ca^{2+} concentration of 200 nM (as is typical for the myoplasm of a ventricular myocyte) corresponds to only about 0.2 Ca^{2+} ions in the diadic cleft. In this situation, traditional deterministic and continuous models may not even be applicable (but see Section 7.2.4). In cardiac cells the diadic clefts are separated longitudinally by approximately 2 μm, the length of a sarcomere.

During a single heartbeat, a total of about 70 μmoles of Ca^{2+} per litre cytoplasm enters the cell, with about 1% ending up as unbuffered free Ca^{2+} in the myoplasm, giving a myoplasmic concentration of around 600 nM. Since all this influx comes through the diadic cleft, there are clearly large and rapid changes in concentration there. To understand how the Ca^{2+} transient is controlled it is vital to understand what happens inside the diadic cleft, because it is there that Ca^{2+} modulates the LCC and RyR to control the time course of Ca^{2+} influx and release. Thus, we have the situation in which a full understanding of the macroscopic properties of the cardiac cell (i.e., the myoplasmic Ca^{2+} concentration and its effect on the crossbridges) requires the study of Ca^{2+} dynamics on a much smaller spatial scale.

7.2.2 Common-Pool and Local-Control Models

Two of the most important defining characteristics of Ca^{2+} release in cardiac myocytes are high gain and graded release. High gain means that, in response to a small Ca^{2+} influx through the LCC (J_{LCC}), a much larger amount is released through the RyR (J_{RyR}). In fact, Ca^{2+} influx is about an order of magnitude smaller than Ca^{2+} release from the SR (Cannell et al, 1987; Wier et al, 1994). Graded release means that, if Ca^{2+} influx is smaller, less Ca^{2+} is released from the SR; release is a smooth and continuous function of influx.

This is illustrated in Fig. 7.2. Part A of that figure shows the total LCC flux and the consequent RyR flux as functions of the membrane potential. The first thing to notice is that J_{LCC} is a bell-shaped curve of the potential; at

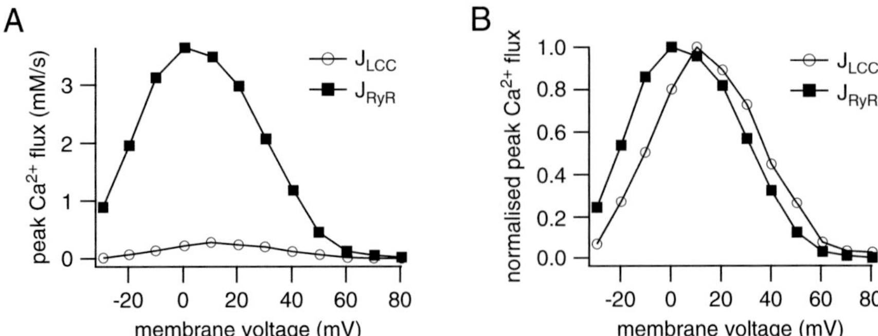

Fig. 7.2 Graded release from rat ventricular myocytes. In response to a voltage step from -40 mV to the indicated level, the peak Ca^{2+} fluxes were measured through the LCC and the RyR. **A:** the lower curve is the flux through the L-type channels (J_{LCC}), the upper curve is the consequent flux through the RyR (J_{RyR}). **B:** the same two curves plotted on a normalised scale. Redrawn from the data in Wier et al (1994), Fig. 3.

low V, the current through each open channel is large, but the probability of opening is small. As V increases, thus increasing the open probability of the channel, the flux increases also. However, as V gets closer to the Ca^{2+} reversal potential of the channel, the current through each open channel begins to fall, leading to a decrease in the total channel flux even though many of the channels are open.

As the LCC flux increases smoothly, so does the RyR flux, as can be seen from the upper curve in Fig. 7.2 A. Interestingly, comparison of the normalised curves (Fig. 7.2 B) shows that the RyR flux decreases at lower voltages than does the LCC flux. An explanation for this is that even while the total LCC current is an increasing function of voltage (due to the increased number of open channels), the flux through each individual channel decreases, leading to less effective coupling with the RyR and a resultant decrease in RyR flux.

The two requirements of high gain and graded release appear at first to be contradictory, and so require further explanation. An excitable system, with a response that is approximately all-or-none, can exhibit high gain without difficulty, but in doing so cannot exhibit a graded release. Conversely, a model that exhibits graded release does not usually exhibit high gain in a stable manner.

Many early models of Ca^{2+} dynamics in muscle are *common-pool* models, in which it is assumed that Ca^{2+} influx and release both occur into the same well-mixed compartment. This well-mixed compartment could be the myoplasm, as in Fig. 7.3 A, or a subcompartment such as the diadic cleft, as in Fig. 7.3 B. Stern (1992b) has shown that such common-pool models, at least in the linear regime, cannot exhibit both high gain and graded release; he introduced, instead, the concept of a *local-control* model, in which the close juxtaposition of the LCC and the RyR leads to the formation of a Ca^{2+} synapse in which the opening of an RyR is controlled by the Ca^{2+} coming

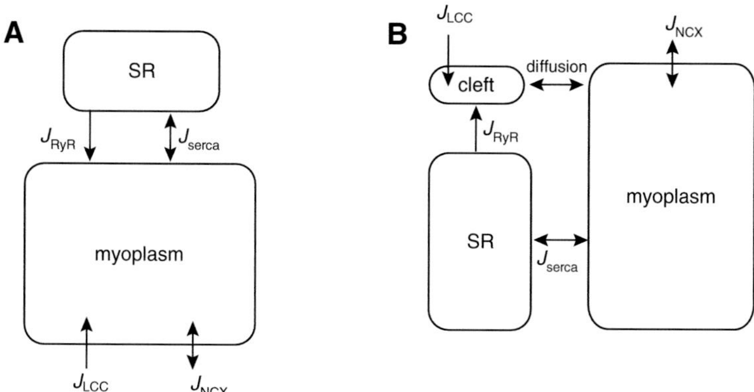

Fig. 7.3 Schematic diagrams of two common-pool models. Each model consists of a number of compartments, each of which is assumed to be well mixed. **A:** a two-compartment model with only the SR and the myoplasm. The LCC current is assumed to flow directly into the myoplasm, whence it stimulates the release of further Ca^{2+} from the SR. **B:** a three-compartment model in which the diadic cleft (with small volume) is included. J_{NCX} is the Na^+/Ca^{2+} exchanger flux. Such common-pool models do not exhibit both a high gain and a graded response.

through a nearby LCC. Although each diadic cleft responds in an all-or-none manner (yielding high gain), the stochastic properties of the population of clefts can give a graded response. Since each diadic cleft is separated from its neighbours by approximately $2\,\mu$m in a longitudinal direction (the length of the sarcomere) and by about $0.8\,\mu$m in a transverse direction, it can respond in a semi-independent fashion. Statistical recruitment from this large pool of locally controlled semi-independent release sites can result in graded whole-cell release, even though each individual release site is all-or-none (Cannell and Soeller, 1999; Soeller and Cannell, 2004).

As with most models of Ca^{2+} dynamics, both common-pool and local-control models of EC coupling are constructed using the Ca^{2+} toolbox, by combining individual models of each of the various Ca^{2+} fluxes, the most important of which are the LCC, the RyR, the SERCA pump, and the Na^+/Ca^{2+} exchanger. Models of all these components are discussed in detail in Chapter 2.

One of the earliest common-pool models that attempted to include the complexity of cardiac Ca^{2+} dynamics was that of Jafri et al (1998).

A schematic diagram of this model is given in Fig. 7.4, whence it can be seen that the model uses four different Ca^{2+} compartments, two in the SR and two in the cytoplasm. The diadic cleft is thus modelled simply as a separate compartment of small volume (a more detailed discussion of microdomains is given in Section 3.4), while the SR is assumed to have an analogous microdomain, the junctional SR. Movement between the compartments is, as usual, linearly proportional to the difference in concentration. The model consists of one ordinary differential equation for the Ca^{2+} concentration in each of the four compartments, as well 26 additional ODEs for the voltage, the other ionic concentrations, and the various fluxes.

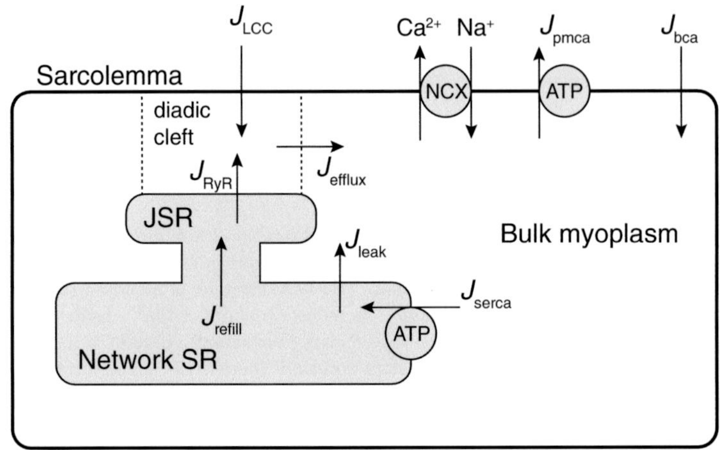

Fig. 7.4 Schematic diagram of the model of Jafri et al (1998). The myoplasm is divided into two regions, the diadic cleft and the bulk myoplasm, while the SR is divided into two analogous regions, the junctional SR (JSR) and the network SR. The J's denote Ca^{2+} fluxes. For clarity, binding of Ca^{2+} to buffers and to the myofilaments has been omitted from the diagram.

A similar common-pool approach has been used in a number of subsequent papers, of mostly increasing complexity. Rice et al (1999) modified the Jafri-Rice-Winslow model to try and reproduce graded responses, and others adapted the model for use with data from a variety of other species (Winslow et al, 1999; Shannon et al, 2004).

However, the common-pool nature of this model meant that, although it has been useful and influential, it has been largely superseded by a new generation of models that incorporate the crucial idea of local control.

7.2.3 Calcium Sparks

The most important discovery underlying the idea of local control was that of Ca^{2+} sparks (Cheng et al, 1993), localised release of Ca^{2+} from RyR that causes the formation of small microdomains of high $[Ca^{2+}]$. Calcium sparks are analogous to Ca^{2+} puffs, which are caused by release through IPR, as discussed in detail in Chapter 4. The discovery of such localised high Ca^{2+} concentrations at the diadic clefts showed conclusively that Ca^{2+} release inside a cardiac cell (or indeed in any other cell type) could not be assumed to be spatially homogeneous, but rather that the whole-cell responses resulted from the summation of a large number of stochastic localised events. However, although there is general consensus about the importance of Ca^{2+} sparks, Ca^{2+} microdomains, and the stochastic behaviour of RyR, there remains much uncertainty about how exactly Ca^{2+} sparks work, and how they are integrated to give whole-cell responses.

One of the major questions around Ca^{2+} sparks is why they turn off. It is relatively easy to understand how CICR can result in an explosive release of Ca^{2+} through RyR into the diadic cleft, but much more difficult to understand the mechanisms that turn this release off, to allow for recovery to the rest state. A number of hypotheses have been proposed, including inactivation by high Ca^{2+} (Rice et al, 1999), stochastic attrition, where stochastic closing of RyR in the cleft leads to termination of release (Stern, 1992b; Hinch, 2004), or depletion of the SR (Sobie et al, 2002), but none has gained unanimous acceptance.

The earliest models to incorporate detailed geometrical information of the diadic cleft were those of Peskoff and Langer (Peskoff et al, 1992; Langer and Peskoff, 1996; Peskoff and Langer, 1998). They modelled the half sarcomere as a thin cylindrical diadic cleft (with a radius of 200 nm and a height of 12 nm) embedded in a cylindrical half sarcomere (radius 500 nm and height 1000 nm). Ca^{2+} release into the diadic cleft resulted in fast large increases in $[Ca^{2+}]$ in this restricted space, with $[Ca^{2+}]$ reaching 10 mM within 10 ms. Outside the diadic cleft, $[Ca^{2+}]$ in the half sarcomere reached only approximately 10 μM within 10 ms, and by 200 ms $[Ca^{2+}]$ in each compartment had returned almost to its resting level. A variation of this basic model, which incorporated uptake by Ca^{2+} pumps by using a homogenised model of the half sarcomere (Higgins et al, 2007), gave similar results, and emphasised the non-homogeneous nature of the $[Ca^{2+}]$ within the diadic cleft.

Similar detailed numerical computations (Soeller and Cannell, 1997; Cannell and Soeller, 1997), which incorporated the effects of the residual charge on the sarcolemma, as well as the kinetics of Ca^{2+} binding to buffers, made similar conclusions. Their results emphasised the effectiveness of the sarcolemmal membrane as a Ca^{2+} sink, as well as the effect of diad size and geometry on the heterogeneity of $[Ca^{2+}]$ within the cleft. One particularly interesting result came from their study of how efficient the LCC are at opening RyR. They defined efficiency approximately by calculating the activation probability of the diadic cleft per unit of Ca^{2+} influx. For a diadic cleft with four RyR, it turns out that the diad works most efficiently when the LCC have a mean open time of around 0.3 ms, consistent with experimentally observed values. For smaller open times the LCC did not allow in enough Ca^{2+} to activate all the RyR, particularly those towards the edges of the diadic cleft, while for larger open times there was more Ca^{2+} than was needed to activate all the RyR.

More recent models have tended towards ever greater computational complexity. The model of Walker et al (2014) has a relatively low spatial resolution of 10 nm, but includes a reasonably detailed geometrical model of the T-tubule and the associated junctional SR (Fig. 7.5 A), while the model of Cannell et al (2013) has a similar model of the T-tubule and junctional SR, but a more detailed model of the surrounding SR (Fig. 7.5 B). Ca^{2+} transport is modelled similarly in both models. Walker et al (2014) use

$$\beta\frac{\partial c}{\partial t} = D_c\nabla^2 c + \sum J, \tag{7.1}$$

where β is a constant that takes into account very fast buffering of Ca^{2+} by the sarcolemma, and the sum of the fluxes, ΣJ, includes fluxes from RyR, SERCA, buffers, and LCC channels. Cannell et al (2013) make no assumption about fast sarcolemmal binding, and incorporate the transport of Ca^{2+} due to the sarcolemmal electric field, but arrive at a similar equation

$$\frac{\partial c}{\partial t} = D_c\nabla(\nabla c + 2c\nabla\Phi) + \sum J, \tag{7.2}$$

where Φ is the electric field within the diadic cleft. As usual, each of the fluxes is modelled by picking the desired model from the available selection in the Ca^{2+} toolbox (Chapter 2). The models thus differ slightly in the specific details of the fluxes. The resultant quantitative differences in model results are important of course, as the modelling of Ca^{2+} sparks is a problem that relies inherently on making distinctions between relatively small quantitative differences, but such fine distinctions are of less importance to us here.

Both models concluded that the most important mechanism governing spark termination was depletion of the junctional SR, with an associated increase in the rate of stochastic attrition. Significantly, direct luminal regulation of RyR open probability played a less important role in both sets of simulations. Both models also investigated the effects of different patterns of RyR positioning within the diad. Cannell et al (2013) concluded that the details of RyR positioning had little effect, while Walker et al (2014) concluded that spark fidelity (i.e., the probability that the opening of a single RyR will trigger a Ca^{2+} spark) was strongly influenced by the positioning and clustering of RyR.

Another detailed computational model is that of Hake et al (2012) who used electron tomograms to construct a diadic cleft based upon

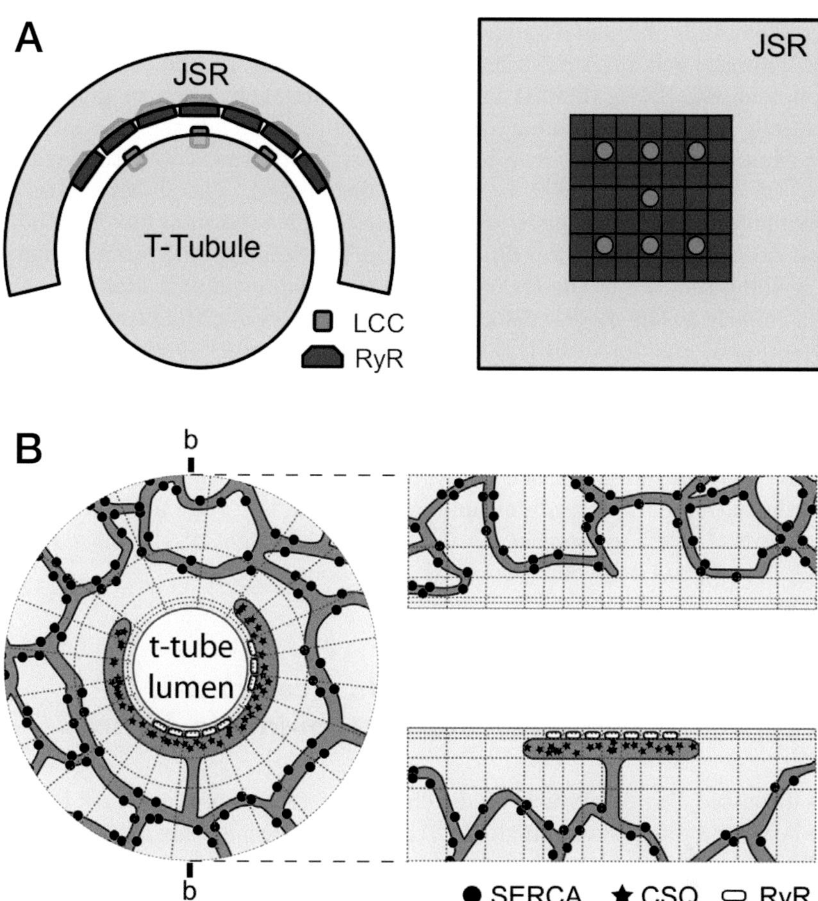

Fig. 7.5 Geometries of two spark models. **A:** In the model of Walker et al (2014) the T-tubule is assumed to be a cylinder of diameter 200 nm, and is partially surrounded by the junctional SR, such that the diadic cleft has a width of 15 nm. The JSR has a diameter of 465 nm. The detailed arrangement of RyR and LCC can be seen in the unfolded geometry on the right. There are seven LCC and 49 RyR, all positioned on a lattice with spacing 31 nm. Reproduced from Walker et al (2014), Fig. 1, with permission from Elsevier. **B:** In the model of Cannell et al (2013) the T-tubule is again a cylinder (of diameter 250 nm) with the JSR wrapping around it such that the diadic cleft has a width of 15 nm. The JSR is shaped like a pancake, 360 nm along the horizontal axis (i.e., the axis projecting out of the plane in the left panel, and running horizontally in the right panels), 26 nm radial thickness, and a circumference of 450 nm. The JSR contains calsequestrin (CSQ) and is connected to the network SR by a tubule along which Ca^{2+} can diffuse. SERCA pumps are uniformly distributed over the network SR. Two side views of the geometry are shown on the right. Here, the T-tubule runs horizontally. The network SR is isotropically distributed to mimic a labyrinth of tubules occupying 3.5% of the cytoplasmic volume. Reproduced from Cannell et al (2013), Fig. 1, with permission from Elsevier.

experimentally measured geometries. Their model is a hybrid model, where the cytosolic domain is modelled by a spatially heterogeneous diffusion equation, but the SR is divided into nine compartments, each modelled by a spatially homogenous ordinary differential equation. Thus the junctional SR, for example, is modelled as a homogeneous compartment, but the cytosolic portion of the diadic cleft is spatially distributed. The fluxes across the boundary of each SR domain are integrated to give the total flux in or out of that domain, and this total flux is used in the ordinary differential equation describing the rate of change of the compartmental concentration.

Similarly to the models discussed above, Hake et al (2012) concluded that depletion of junctional SR is the most important cause of spark termination, and thus having SERCA in the diadic cleft can extend the spark, as Ca^{2+} pumping decreases the rate of SR depletion. This is a somewhat counterintuitive result (one might imagine that increasing the rate of Ca^{2+} removal might decrease the extent of a spark) but becomes plausible when one considers the extreme importance of junctional SR $[Ca^{2+}]$. They also showed that having Na^{+}/Ca^{2+} exchangers in the diadic cleft, right at the place of very high $[Ca^{2+}]$, has little effect on spark behaviour.

7.2.4 The Diadic Cleft Can Be Described by a Continuous and Deterministic Model

As discussed above, a typical diadic cleft has a volume of around 2×10^{-18} L, and contains fewer than a single Ca^{2+} ion at rest. Despite this, Hake and Lines (2008) have shown that, since the (unbuffered) diffusion coefficient of Ca^{2+} is greater than $200\,\mu\text{m}\,\text{s}^{-1}$, a mean-field model of the Ca^{2+} concentration inside the cleft provides an accurate description. Thus, there might be very few Ca^{2+} ions, but they move so fast that the sites where Ca^{2+} binds (buffers, RyR, etc) experience an effectively smoothed mean-field concentration.

This is illustrated in Fig. 7.6, which compares solutions of a continuous model to solutions of a random walk model. For a single run (green line) the average $[Ca^{2+}]$ varies widely during the simulation, but the continuous model (black line) remains an accurate description of the mean-field behaviour, even at low $[Ca^{2+}]$. In particular, the continuous model accurately captures the time course of changes in mean concentration.

7.2.5 Integrative Models

So far we have seen common-pool models that attempt to describe whole-cell behaviour, and Ca^{2+} spark models that concentrate on localised microdomains around the diadic cleft. However, in order to understand the

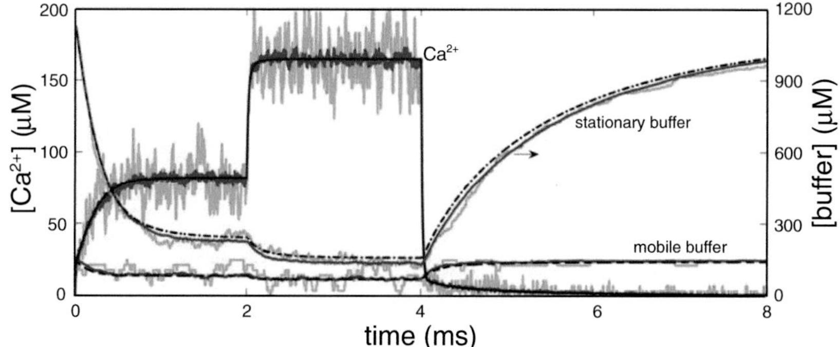

Fig. 7.6 Comparisons of average cleft [Ca^{2+}] from two different models of a diadic cleft (Hake and Lines, 2008). One model was a continuous model of the [Ca^{2+}] (solid black line), while the other model was a simulated random walk of individual Ca^{2+} ions and buffer molecules (coloured lines). One LCC is open initially, and thus [Ca^{2+}] increases. At 2 ms a second LCC opens, leading to an additional increase of [Ca^{2+}], and at 4 ms both LCC were shut, and so [Ca^{2+}] returns to rest. The solid line is [Ca^{2+}], the dashed line is the concentration of the mobile buffer, and the dot-dash line is the concentration of stationary buffer (plotted against the right vertical axis). The green line is from a single run of the random walk model, while the red line is the average of 40 runs of the random walk model. Reproduced from Hake and Lines (2008), Fig. 7, with permission from Elsevier.

macroscopic cellular behaviour in terms of the summation of a large number of Ca^{2+} sparks, it is necessary to build models that can integrate the behaviour of around 20,000 diadic clefts, the approximate number in a single cardiac cell. Clearly, if each diadic cleft was modelled at the level of detail of the complex spark models described above, the computational costs would be prohibitive, and thus there have been numerous attempts to construct simplified models that can explain how multiple stochastic Ca^{2+} sparks could be the building blocks of whole-cell waves.

The basis of these models is the idea of the Calcium Release Unit, or CRU, a combined representation of the LCC, the RyR, and the diadic cleft. Once a CRU model is formulated there are two major ways to combine multiple CRUs into a three-dimensional whole-cell representation.

- Each CRU can be solved independently, the flux out of the CRU calculated, and this flux can be incorporated into a common-pool model of the global [Ca^{2+}]. In this case, the global [Ca^{2+}], $c(t)$, is a function of time only, and satisfies the ODE

$$\frac{dc}{dt} = \sum_i J^i_{\mathrm{CRU}} - \mathrm{efflux}, \tag{7.3}$$

where the sum is over all the fluxes coming out of the individual CRUs. There can, of course, be other Ca^{2+} influx pathways (through voltage-gated channels, for example) but they are omitted here for clarity.

- Each CRU can be incorporated into a spatially distributed model of the cell, in which $c(x,t)$ is now a function of space and time. In this case, c will typically satisfy a reaction-diffusion equation with point sources of Ca^{2+} corresponding to the CRUs. Thus, for example,

$$\frac{\partial c}{\partial t} = D_c\nabla^2 c + \sum_i \delta(x - x_i)J^i_{\mathrm{CRU}} - J_{\mathrm{efflux}}(x), \tag{7.4}$$

where x_i is the position of the ith CRU, and δ is the Dirac delta function. The efflux terms can also be functions of x although this is not necessary.

Early models were those of Keizer et al (1998) and Izu et al (2001), both of who did computations in two dimensions, with each diadic cleft being modelled as a point source of Ca^{2+}. However, as they were more interested in studying the formation of global Ca^{2+} waves from local punctate release of Ca^{2+}, their models of the CRU were rudimentary and lacked the details necessary for incorporating local control.

Greenstein and Winslow (2002) constructed a more detailed model of the CRU and combined 12,500 CRUs into an integrated model of a whole cardiac cell. In their model, $[Ca^{2+}]$ was a global variable, as described above, but each CRU was modelled as a group of five spatially homogeneous compartments: four for the diadic cleft and one for the junctional SR. Each diadic cleft compartment contained one stochastic LCC and five stochastic RyR. Thus, for computational ease, the full spatial heterogeneity of the diad was simplified to a compartmental model. Limited spatial heterogeneity was retained by the use of four compartments for the diadic cleft.

7.2.6 Simplified Approaches

Despite these simplifications, the model of Greenstein and Winslow (2002) remains computationally difficult, as it requires Monte-Carlo simulations of many tens of thousands of individual LCC and RyR. This provides strong motivation for the search for even simpler models of the CRU that could still retain stochastic aspects. The most notable of these efforts are the simplification of Hinch (2004) and the work from the group of Greg Smith (Mazzag et al, 2005; Huertas and Smith, 2007), culminating in the integrative model of Williams et al (2007).

Hinch (2004) first derived a 9-state Markov model for a CRU, consisting of a single RyR combined with a single LCC. The RyR and the LCC can each be in one of three states (closed, open, or inactivated, denoted by C, O, and I, respectively) and thus the combined RyR-LCC model can be in any of nine states. For example, S_{OC} would denote the state where the CRU has an open RyR but a closed LCC. Then, for each CRU state, they derived an analytic expression for the $[Ca^{2+}]$ inside the diadic cleft by assuming that Ca^{2+} in the diadic cleft is at instantaneous equilibrium.

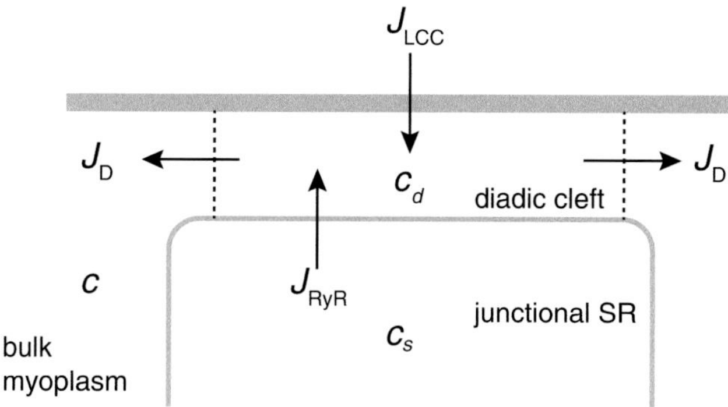

Fig. 7.7 Schematic diagram of the diad model of Hinch (2004).

To illustrate how this simplification works, consider the diad shown in Fig. 7.7. There are only three fluxes into (or out of) the diadic cleft: $J_{\rm RyR}$, $J_{\rm LCC}$, and J_D, the diffusion into the myoplasm. Let c, c_d, and c_s denote $[Ca^{2+}]$ in the myoplasm, diadic cleft, and SR, respectively. For c_d to be at equilibrium we require

$$J_{\rm RyR} + J_{\rm LCC} = 2J_D. \tag{7.5}$$

Then, if we assume that

$$J_D = \frac{g_D}{2}(c_d - c), \tag{7.6}$$

for some constant g_D, we get

$$c_d = c + \frac{J_{\rm RyR} + J_{\rm LCC}}{g_D}. \tag{7.7}$$

Now, for any given state of the CRU ($S_{\rm OC}$, say, where the RyR is open and the LCC is closed) it is possible to calculate the instantaneous value of c_d, and thus avoid having to solve any differential equation for the diadic cleft concentration. Of course the resultant expression for c_d will depend on the models chosen for the RyR and LCC.

For example, if we make the simple assumption that

$$J_{\rm RyR} = J_R(c_s - c_d), \tag{7.8}$$

where J_R is a constant, then, when the CRU is in state S_{OC}, we have $J_{LCC} = 0$, in which case $c_d = c + \frac{J_{RyR}}{g_D}$. When combined with the above equation for J_{RyR}, this gives

$$c_d = \frac{c + \frac{J_R c_s}{g_D}}{1 + \frac{J_R}{g_D}}. \tag{7.9}$$

To solve the equation numerically, now merely needs to do a Monte-Carlo simulation of the 9-state CRU (or, in a global model, of each of 20,000 9-state CRUs). For each change in CRU state, the corresponding value of c_d, and thus the flux into the cytoplasm, can be determined algebraically. Of course, if J_{RyR} and J_{LCC} are complicated nonlinear functions of c_d and c_s, the solution of this algebraic expression for c_d might be nontrivial; the simplicity of the example above required the use of linear flux expressions.

7.2.6.1 The Probability Density Approach

A rather different simplification method was developed by Mazzag et al (2005). This method dispenses with the need for any Monte-Carlo simulations at all by deriving partial differential equations for the probability density functions of c_d jointly distributed with the state of the CRU. This was a modification of a similar approach used by Bertram et al (1999), in the context of synaptic vesicles, and Nykamp and Tranchina (2000), in the context of neuronal populations.

It is easiest to understand the basic theory in a simpler context, so we shall consider the case where a single Ca^{2+} channel, which can be either open or closed, opens into the diadic cleft. This channel is described by the reaction

$$\mathrm{C} \underset{k_{-1}}{\overset{k_1}{\rightleftharpoons}} \mathrm{O}. \tag{7.10}$$

We then define the joint probability distributions

$$\rho_C(c,t) = \mathrm{Prob}[c < c_d < c + dc \quad \text{and} \quad \mathrm{S}(t) = \mathrm{C}], \tag{7.11}$$

$$\rho_O(c,t) = \mathrm{Prob}[c < c_d < c + dc \quad \text{and} \quad \mathrm{S}(t) = \mathrm{O}], \tag{7.12}$$

where S denotes the state of the channel.

We can now write down a conservation law for these joint probability density functions. Note that ρ_C can change in only two ways.

- Firstly, the channel can change state, from C to O, with net rate $k_1\rho_O - k_{-1}\rho_C$, an expression that follows directly from the reaction diagram of the channel.
- Secondly, c_d can move out of the region $(c, c + dc)$. To calculate the rate at which this happens we need to appeal to the assumptions about how c_d can change. Suppose that, when $c_d = c$ and the channel is closed, the

net rate at which Ca^{2+} leaves the diadic cleft is $J_C(c)$. The expression for J can be a sum of multiple fluxes, and arbitrarily complicated, but must contain all the Ca^{2+} fluxes that either decrease or increase c_d. Similarly, $J_O(c)$ is the net flux when the channel is open. Then, when the channel is closed the net rate at which c_d leaves the interval $(c, c+dc)$ is $J_C(c+dc)\rho_C(c+dc,t) - J_C(c)\rho_C(c)$, with a similar expression when the channel is open.

Hence, letting dc tend to zero (and applying the exact same argument to ρ_O) gives the conservation equations

$$\frac{\partial \rho_C}{dt} = -\frac{\partial}{\partial c}(J_C(c)\rho_C) + k_1\rho_O - k_{-1}\rho_C, \tag{7.13}$$

$$\frac{\partial \rho_O}{dt} = -\frac{\partial}{\partial c}(J_O(c)\rho_O) - k_1\rho_O + k_{-1}\rho_C. \tag{7.14}$$

This derivation is essentially the same as the usual derivation of a reaction-diffusion equation for a concentration, as discussed in Keener and Sneyd (2008).

Given a large population of channels, we can now calculate exactly how many are in each state, and how many have a given local $[Ca^{2+}]$. We no longer know the exact state of any individual channel, but we can calculate how the entire population changes over time. From this information we can then calculate the global $[Ca^{2+}]$. Thus, we have replaced multiple Monte-Carlo simulations by the solution of two advection-reaction equations. One might think this was exchanging one difficulty for another, but in practice, once the number of channels gets large enough, the solution of the advection-reaction equations is faster and easier than the Monte-Carlo simulation of the individual channels.

This method is easily extended to more complex systems (Williams et al, 2007). For example, if the CRU model is a four-state Markov model, then there are four advection-reaction equations, one for each CRU state, and the reaction terms are read off directly from the ODEs for the Markov model. Similarly, if the fluxes depend on both the Ca^{2+} concentration in the diadic cleft (c_d) and in the junctional SR (c_{js}), then each probability density is a function of t, c_d, and c_{js}. In the example above this would give

$$\frac{\partial \rho_C}{dt} = -\frac{\partial}{\partial c_d}\left[J_C^d(c_d, c_{js})\rho_C\right] - \frac{\partial}{\partial c_{js}}\left[J_C^{js}(c_d, c_{js})\rho_C\right] + k_1\rho_O - k_{-1}\rho_C,$$

$$\frac{\partial \rho_O}{dt} = -\frac{\partial}{\partial c_d}\left[J_O^d(c_d, c_{js})\rho_O\right] - \frac{\partial}{\partial c_{js}}\left[J_O^{js}(c_d, c_{js})\rho_O\right] - k_1\rho_O + k_{-1}\rho_C,$$

where J_C^d contains all the fluxes in or out of the diadic cleft when the channel is closed, with analogous definitions for the other Js. Although the notation gets a little baroque, as it is necessary to keep careful track of how the flux terms change as the channel state changes, the principle is the same as in the simpler case.

Williams et al (2007) present a detailed comparison of the probability density method to a direct Monte-Carlo simulation. As expected, the Monte-Carlo simulation converges to the solution of the probability density method, but once the number of CRUs gets greater than a few thousand, the probability density method is substantially more computationally efficient.

The major disadvantage of the probability density method is that it becomes inefficient once the number of states in the CRU Markov model gets large; if the CRU is modelled by an n-state Markov model, one would need n (possibly two-dimensional) advection-reaction equations, a method which quickly becomes unwieldy as the CRU model becomes more complex. To avoid this problem, Williams et al (2008) derived a further simplification by, firstly, assuming that Ca^{2+} in the diadic cleft rapidly reaches its equilibrium concentration, and, secondly, by reducing the model to a system of ordinary differential equations for the moments of the distributions. The details are too complex to be presented here, but simulation of the moment closure model shows excellent agreement with solutions of the full probability density model, while being about 10,000 times faster than the corresponding Monte-Carlo simulations. This approach thus provides an intriguing and potentially powerful new way of approaching whole-cell computations.

7.2.7 Atrial Myocytes

We have seen in Section 3.5.1 that a particularly simple and attractive model for Ca^{2+} wave propagation is the fire-diffuse-fire model, where discrete release sites instantaneously fire (i.e., release Ca^{2+} from the ER) when $[Ca^{2+}]$ reaches a threshold value. Thul et al (2012a,b) have used this formalism to develop a model for the subcellular Ca^{2+} dynamics in atrial myocytes. Atria are the cardiac chambers that receive blood, while ventricles push the blood out of the heart. However, when necessary, atrial contraction occurs, which increases the amount of blood within the ventricles. This results in increased pumping by the whole heart. In pathological situations, atrial fibrillation (i.e., disorganised electrical activity) greatly increases the risk of stroke. Atrial contraction is triggered by elevations in $[Ca^{2+}]$, and atrial fibrillation is linked to a dysregulation of Ca^{2+} signalling, which gives us two good reasons to be interested in the Ca^{2+} dynamics of atrial myocytes.

In contrast to the types of myocytes described above, T-tubules are virtually absent in the atria of small mammals. Ca^{2+} release from the SR is initiated at the cell periphery, where RyR and voltage-gated Ca^{2+} channels are very close. Waves then propagate into the interior. In these cells, Ca^{2+}-

releasing sites made of clusters of RyR are spatially organised as shown in Fig. 7.8.

Besides the typical organisation in z planes, an important characteristic is the gap (approximately 1 μm) between the junctional and nonjunctional RyR. The regular geometry is plausibly modelled by a fire-diffuse-fire formalism, avoiding the time-consuming resolution of the steep $[\mathrm{Ca}^{2+}]$ gradients that would be required by a more detailed model of the RyR. Similarly to (3.126), the evolution of cytosolic $[\mathrm{Ca}^{2+}]$ is given by

$$\frac{\partial c}{\partial t} = D\nabla^2 c - \frac{c}{\tau} + \sum_n \sum_m \delta(r - r_n)\eta(t - T_n^m), \tag{7.15}$$

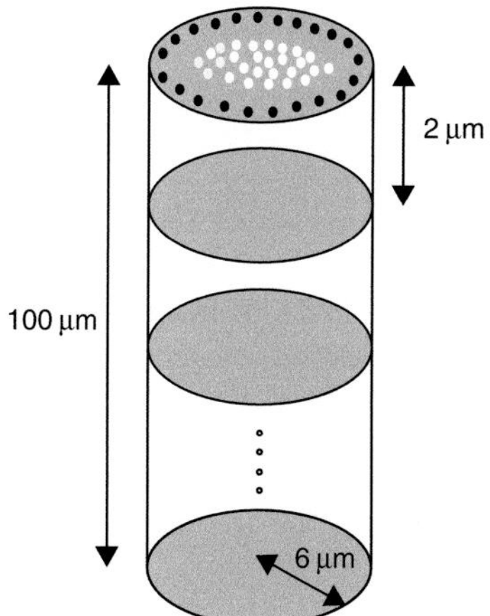

Fig. 7.8 Schematic representation of the cylindrical atrial myocyte geometry (Thul et al, 2012b). The myocyte is modelled as a stack of z planes on which the RyR are expressed. RyR are drawn in the upper plane only, but their distribution is identical in each plane. Black dots represent clusters of junctional RyR (facing the VGCC in diadic clefts), whereas white dots represent clusters of inner, nonjunctional RyR.

where r_n is the position of the nth release site, and T_n^m is the time when the nth site releases Ca^{2+} for the mth time. In the model, junctional and nonjunctional RyR are modelled the same way. The wave is triggered by opening a set of junctional RyR (via appropriate initial conditions), which simulates their activation in response to the opening of plasma membrane VGCC.

There are two main differences with the basic model described by (3.126). First, three spatial dimensions are considered. Second, release is no longer considered to be instantaneous. Instead,

$$\eta(t - T_n^m) = \sigma H(t)H(t_{\mathrm{rel}} - t), \tag{7.16}$$

where H is the Heaviside function. Thus, from time T_n^m on, Ca^{2+} is released for the duration $t_{\rm rel}$ with a constant conductance σ.

As in the basic fire-diffuse-fire model, a release site (i, j, k) opens when the local $[Ca^{2+}]$ reaches a threshold $c_{\rm th}$, but only if this occurs some time, $t_{\rm ref}$, after the previous opening. Hence $t_{\rm ref}$ represents an absolute refractory period. Thus

$$T_n^m = \inf\left\{t \,|\, c(r_n, t) > c_{\rm th},\, T_n^m > T_n^{m-1} + t_{\rm ref}\right\}. \tag{7.17}$$

To model the plasma membrane, Thul et al imposed fluxes on the boundaries:

$$\frac{\partial c}{\partial r} + h_r c = 0, \qquad r = R, \tag{7.18}$$

$$\frac{\partial c}{\partial z} - h_0 c = 0, \qquad z = 0, \tag{7.19}$$

$$\frac{\partial c}{\partial z} + h_l c = 0, \qquad z = l. \tag{7.20}$$

This allows Ca^{2+} extrusion to have different properties on the top, bottom, or sides of the cell.

Solving (7.15) analytically in the geometry of Fig. 7.8, and with boundary conditions defined by (7.18)–(7.20), is a complex task. To do so, it was assumed that the release events occur at multiples of some dt. It then follows that the concentration at $(p+1)dt$ is given by

$$c(r, (p+1)dt) = \sigma \sum_n a_n(p) K(r, r_n, dt) + J_p(r), \tag{7.21}$$

with

$$K(r, r', t) = \int_0^t G(r, r', s)\, ds, \tag{7.22}$$

$$J_p(r) = \int_V G(r, r', dt) c_p(r')\, dr', \tag{7.23}$$

where $G(r, r', dt)$ is a modified Green's function whose specific form depends on the boundary conditions. The function $a_n(p)$ is a recursively-defined indicator function that takes into account the release duration and the refractory period. The detailed expressions for these functions can be found in Thul et al (2012a). Although the calculations are complex, the main advantage is that the G's and K's need be computed only once for each set of parameter values. The whole system of equations thus provides a computationally cheap algorithm that reproduces a realistic three-dimensional geometry and allows an extensive scanning of the parameter space.

Simulations reveal that the gap between the junctional and nonjunctional SR prevents inward propagation of the Ca^{2+} wave for small release strengths σ. This suggests that the gap plays a primary role in limiting atrial contraction under resting conditions. As an unexpected result, simulations also predict the existence of a previously unknown arrhythmic pattern of Ca^{2+} signalling. For sufficiently low release strengths, a wave initiated at the central z plane induces RyR activity that continues indefinitely. Ca^{2+} waves rotate not only in the stimulated plane, but also in neighbouring ones. The parameter values required for getting this spatiotemporal pattern of Ca^{2+} release – which would presumably correspond to fibrillation – are pertinent to pathological conditions related to heart failure, namely a weak Ca^{2+} pumping into the SR and a reduced expression of RyR. These waves, called "ping waves", cannot be resolved with current experimental techniques.

7.3 Skeletal Myocytes

Ca^{2+} dynamics in skeletal muscle is inherently simpler than in cardiac muscle, as skeletal muscle has no diadic cleft. Instead, the LCC activate the RyR directly to release Ca^{2+} from the SR. Thus, the question of how Ca^{2+} behaves inside the very small region of the diad is mostly bypassed, and models of Ca^{2+} in skeletal muscle have mostly focused on quantitative descriptions of how Ca^{2+} diffuses from the release site, often with detailed inclusion of Ca^{2+} buffers.

The earliest model of this type is Cannell and Allen (1984), who assumed a cylindrically symmetric myofibril with a compartmental structure. One of the principal results to come from this model was how dramatically the Ca^{2+} buffers, including binding to troponin, could slow down the Ca^{2+} response in the centre of the myofibril. Although the peak of the Ca^{2+} response close to the RyR occurred at 3–4 ms, the response further away could peak as late as 20 ms. Clearly, Ca^{2+} buffering causes substantial response delays. Although such a result does not appear surprising to us today, we must remember that, at the time this model was constructed, our understanding of Ca^{2+} buffers was far less advanced than it is now.

A similar model structure was used in a series of papers by Baylor and Hollingworth (Baylor and Hollingworth, 1998; Baylor et al, 2002; Baylor and Hollingworth, 2012) whose results differ from the earlier work of Cannell principally in the quantitative details. A more recent model by Liu and Olson (2015) dispenses almost entirely with Ca^{2+} spatial gradients to focus instead on modelling the membrane ion channels, and is thus of less relevance here.

7.4 Smooth Muscle

Smooth muscle, in all its many types, forms layers surrounding the hollow organs of the body, such as blood vessels, the gastrointestinal tract, the bladder, the uterus, and the airways of the lung. It is, in its biochemistry and function, quite different from skeletal and cardiac muscle. Smooth muscle, for example, is nonstriated. Although it contains the actin and myosin proteins necessary for contraction, they are not arranged in regular arrays. In addition, although contraction in smooth muscle depends on a rise in $[Ca^{2+}]$, the biochemistry of contraction is quite different from that in striated muscle. In striated muscle the binding of Ca^{2+} to troponin exposes the myosin binding site on the actin filament and allows the crossbridge cycle to generate force. However, smooth muscle contains no troponin. Instead, the crossbridge cycle can occur only when myosin is phosphorylated at a specific site. This phosphorylation is carried out by *myosin light chain kinase* (MLCK), which is activated by Ca^{2+}/calmodulin (Section 7.5.2). Hence, a rise in Ca^{2+} leads to a rise in Ca^{2+}/calmodulin, leading to increased activation of MLCK, phosphorylation of the myosin, and thus contraction.

Dephosphorylation of the myosin by *myosin light chain phosphatase* (MLCP) prevents the crossbridge cycle from occurring. However, this does not necessarily lead to relaxation of the muscle. If dephosphorylation occurs while the myosin is bound to actin, the crossbridge enters the *latch state.* While in this state, myosin can unbind from actin only very slowly; although no active force can be generated (since the crossbridge cycle cannot occur), neither can the crossbridge relax, as the myosin remains bound. Thus, if all the crossbridges were in the latch state, the muscle would be rigid and unable to relax, but would require little energy to stay in this state. For this reason, smooth muscle is able to sustain a load while using little ATP, which allows for the maintenance of organ shape and dimension without undue use of energy. Striated muscle, in contrast, has no latch state and cannot maintain a force without continuous crossbridge cycling.

Although smooth muscle is diverse, it can be usefully classified into two broad groups (Berridge, 2008; Hill-Eubanks et al, 2011). In the first group are smooth muscle types that are electrically excitable and exhibit action potentials that are mediated by the inward flow of Ca^{2+} through voltage-gated Ca^{2+} channels. Examples of this type include smooth muscle in the urinary bladder, the uterus or the vas deferens. In the second group are smooth muscle types where the increase in cytosolic Ca^{2+} comes about, not from Ca^{2+} influx from the outside, but mostly from Ca^{2+} release from the SR, through either IPR or RyR. Airway and vascular smooth muscle are both in this second group.

In general, models of excitation-contraction coupling in smooth muscle have a similar structure to many of the models we see elsewhere in this book (see, for example, Section 3.7, as well as the models of hypothalamic and pituitary cells in Section 8.7). There are two basic model components: firstly, a model of the release and reuptake of cytosolic Ca^{2+}, including fluxes through IPR and/or RyR as well as across the cell membrane, and secondly, a model of the ionic fluxes across the cell membrane, which control the membrane potential. The cytosolic module is coupled to the membrane potential module via Ca^{2+}-dependent K^+ and Cl^- channels, while the membrane module affects the cytosolic module via the entry of Ca^{2+} through ion channels.

The majority of modelling work on the Ca^{2+} dynamics of smooth muscle has been done in vascular and airway smooth muscle, both of which are nonexcitable. Additionally, in both muscle types Ca^{2+} release from the SR appears to play a major role, and thus the dynamics of cytosolic Ca^{2+} can show considerable complexity. For this reason, we focus here on models of these two types of smooth muscle.

There are detailed quantitative models of other types of smooth muscle also, but these all concentrate on the electrical aspects, and pay less attention to the control of Ca^{2+} via release from the SR. For example, in the uterine smooth muscle model of Bursztyn et al (2007), used also in a modified form by Tong et al (2011), Ca^{2+} simply enters the cell through voltage-gated Ca^{2+} channels and is removed by plasma membrane ATPase pumps and a Na^+/Ca^{2+} exchanger. Although, of course, it is still not a trivial matter to fit to experimental data, the underlying Ca^{2+} dynamics is necessarily simple. Ca^{2+} handling in the model of Rihana et al (2009) is even simpler, with removal of Ca^{2+} from the cytoplasm being modelled by a linear removal term. It is still not clear whether or not a more complex model of Ca^{2+} handling by the SR is even required in uterine smooth muscle, as there is evidence that Ca^{2+} release from the SR does not play an important role (Berridge, 2008; Noble et al, 2009).

In the gastrointestinal tract, smooth muscle Ca^{2+} handling tends to be of the simple type seen in uterine smooth muscle. More interesting Ca^{2+} dynamics occurs in the pacemaker cells, the interstitial cells of Cajal, but since the interaction between intracellular Ca^{2+} dynamics and slow waves in those cells is not well understood, we shall not discuss them in detail (but see Lees-Green et al (2014)).

7.4.1 Airway Smooth Muscle

Airway smooth muscle (ASM) is one of the most important cell types in the study of airway hyperresponsiveness, as major breathing difficulties occur when the contraction of ASM is either too strong or too sensitive. Since contraction of ASM is caused by a rise in cytosolic $[Ca^{2+}]$, the understanding of hyperresponsiveness relies fundamentally on our understanding of the Ca^{2+} dynamics of ASM. The most comprehensive single presentation of Ca^{2+} signalling in airway smooth muscle is the book of Wang (2014), which includes discussions of both experimental and theoretical approaches.

In response to agonist stimulation, $[Ca^{2+}]$ in ASM displays a range of complex spatiotemporal behaviours, including periodic waves of varying frequency (Roux et al, 1997, 1998; Bai et al, 2009; Perez and Sanderson, 2005b; Pabelick et al, 1999; Prakash et al, 2000; Ressmeyer et al, 2010; Sanderson et al, 2010), as illustrated in Fig. 1.6. The oscillation period depends both on the agonist and on the species, with, for example, mouse ASM having oscillations of much higher frequency than human. Although Ca^{2+} oscillations in ASM are often presented simply as functions of time, it should always be kept in mind that these are not whole-cell oscillations that are homogeneous in space. Rather, they invariably take the form of periodic waves. These periodic waves continue, at least for a time, in the absence of extracellular Ca^{2+}, and thus the majority of the Ca^{2+} in each transient is being released from internal stores, mostly through IPR.

The major questions that can be addressed by modelling are, firstly, what are the mechanisms that generate the oscillations, secondly, how do the IPR and RyR interact to set the oscillation period, and thirdly, how do the Ca^{2+} oscillations affect the membrane potential and vice versa? An additional question, not directly related to the control of the Ca^{2+} dynamics, is how the contractile force depends on the frequency, shape, and amplitude of the periodic Ca^{2+} waves. However, it is important to note that the answers to these questions depend to some extent on the details of the experimental method; in particular, isolated ASM do not behave the same as ASM in a lung slice preparation, while there are also quantitative differences between species. Thus one must always consider carefully the provenance of the experimental results before drawing conclusions about mechanism. Here, we shall be considering mostly data from lung slices as opposed to isolated cells, and this will affect our conclusions.

7.4.1.1 Stochastic or Deterministic?

The first question to be addressed is whether the periodic waves arise via a stochastic or a deterministic mechanism. As we saw in Chapter 4, one way to determine this is to plot the oscillation standard deviation against the period to determine the coefficient of variation. Typical results are shown in Fig. 7.9, whence it can be seen that there is a linear relationship between

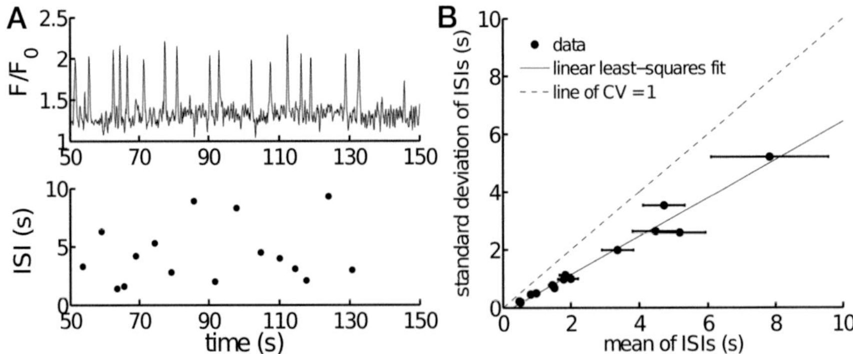

Fig. 7.9 **A:** Calcium spiking in airway smooth muscle cells stimulated with 50 nM methacholine (MCh).The interspike interval (ISI) calculated from the upper panel is shown in the lower panel. **B:** ISI standard deviation plotted against the ISI mean. The relationship is approximately linear with a slope of 0.66, which implies that the spiking is generated by an inhomogeneous Poisson process. The dashed line indicates where the coefficient of variation (CV) is 1 (i.e., a pure Poisson process). Reproduced from Cao et al (2014), Fig. 2.

the mean and the standard deviation, and that the coefficient of variation is less than one. Hence we can infer that the oscillations are generated by an inhomogeneous Poisson process (Cao et al, 2014).

Stochasticity appears to arise principally from stochastic openings and closings of the IPR, which lead to the emergence of larger-scale periodic waves as the effects of multiple IPR are summed. This conclusion is supported by simulation of a stochastic model of a cluster of IPR (Cao et al, 2014). However, it appears that ASM are somewhat unusual, in that the intrinsic stochastic behaviour of the IPR does not seem to result in any Ca^{2+} puffs, which have not been observed in lung slice preparations. Ca^{2+} sparks have been observed in isolated ASM (ZhuGe et al, 2010; Lifshitz et al, 2011; Bao et al, 2008) and in other smooth muscle types (Jaggar et al, 2000; Pozo et al, 2002; Burdyga and Wray, 2005; Essin and Gollasch, 2009) but these are generated by RyR not IPR. The absence of IPR-generated Ca^{2+} puffs in ASM is somewhat puzzling, but it has been hypothesised that this is a result of the tight coupling of IPR in ASM in lung slice preparations (Cao et al, 2014). In this scenario, the IPR are so tightly coupled that every puff results in a global wave, resulting in waves with a period that are governed solely by the stochastic dynamics of a cluster of IPR, rather than by the intercluster coupling. This hypothesis is consistent with the high frequency of Ca^{2+} oscillations at higher agonist concentrations.

Although the periodic waves are generated by a stochastic mechanism, this does not necessarily mean that a stochastic model is required to study them. The most important reason for constructing a model is to make predictions about behaviour that can be tested experimentally. Thus, for parameter values where a deterministic model behaves in a similar way to a stochastic model, the former has the same predictive power as the latter. This is clearly the case in the limit as the number of IPR tends to infinity, but that is not the case in ASM, where the number of IPR is small enough to leave a clear stochastic footprint. However, Cao et al (2014) have shown that a deterministic model of Ca^{2+} dynamics in ASM has almost the same predictive power as a more accurate stochastic model. Although a deterministic model cannot be used to predict the precise parameter values where oscillations begin or end, the dependence of oscillation frequency upon model parameters remains qualitatively the same. Being much easier to construct and solve, a deterministic model is thus an acceptable and convenient simplification.

7.4.1.2 The Cytosolic Oscillator

For simplicity, we begin by describing a spatially homogeneous model of Ca^{2+} oscillations in ASM. The first such models were those of Haberichter et al (2002) and Brumen et al (2005), but our discussion here is based on the later, very similar, models of Wang et al (2010), Croisier et al (2013), and Cao et al (2014).

Before constructing a model, we note two important experimental results that have a major impact on the structure of the model.

- In ASM, agonist-induced Ca^{2+} oscillations occur in the absence of extracellular Ca^{2+} (Perez and Sanderson, 2005b), and thus must necessarily arise as a result of the cycling of Ca^{2+} into and out of internal stores. In this case, the most important internal store is the sarcoplasmic reticulum (SR). Hence the model must be one in which oscillations are modified by Ca^{2+} influx, but do not depend on it.
- In addition, there is strong evidence that Ca^{2+} oscillations occur for a constant $[IP_3]$, and are thus caused by a Class I mechanism (Section 3.2.7). Although $[IP_3]$ and $[Ca^{2+}]$ have not been measured simultaneously in ASM, results from experiments where exogenous pulses of IP_3 were applied indicate that $[IP_3]$ is constant during the oscillations (Sneyd et al, 2006).

The model is based on the following assumptions:

- Calcium is removed from the cytoplasm in two ways. It is pumped out of the cells by ATPase pumps on the plasma membrane, and it is pumped into the SR by SERCA pumps. Thus we are ignoring the influence of mitochondria on Ca^{2+} removal. Although it has been shown that mitochondria probably play an important role in Ca^{2+} removal (Roux and Marhl, 2004), it is not clear how much this will affect the predictions of this model. Here, we focus more on the interplay between the Ca^{2+} entry pathways, and, for an initial approximation, we believe a simpler model of Ca^{2+} removal to be sufficient. The PM pumps and the SERCA pumps are both modelled by Hill functions, with Hill coefficients of 4 and 2, respectively, as in (2.37).
- Calcium entry into the cell is via three different pathways:
  - Receptor-operated Ca^{2+} channels (ROCC) are opened indirectly by agonists, and allow Ca^{2+} to enter the cell. Here, the ROCC current is assumed to be dependent on $[IP_3]$, as this is a surrogate for the agonist concentration. Note that this is only true if Ca^{2+} oscillations occur at a constant $[IP_3]$, as they do in ASM. Since ROCC appear to play only a minor role (Croisier et al, 2013), a simple ROCC model such as (2.99) is sufficient.
  - Store-operated Ca^{2+} channels (SOCC) open in response to a decrease in SR $[Ca^{2+}]$, but have an intrinsic delay, in a manner similar to that of (2.100)–(2.101). Thus, the steady-state SOCC flux is proportional to $1/K_s^4 + c_s^4$, where c_s is the $[Ca^{2+}]$ in the SR, and the steady state is reached with a time constant of τ_s. In the model of Croisier et al (2013), $\tau_s = 30$ seconds.

  - The L-type channel is modelled very simply as a voltage-gated Ca^{2+} channel, with the reversal potential given by the Goldman-Hodgkin-Katz equation, and a single activation gating variable that is a time-independent increasing function of the membrane potential. The more complex models of Section 2.5 are not used.

  However, although Ca^{2+} influx is an important model component, it remains true that the Ca^{2+} oscillations and waves are caused by release and reuptake of Ca^{2+} from the SR, not by Ca^{2+} transport across the plasma membrane.
- Calcium moves from the SR to the cytoplasm through RyR and IPR. The Ca^{2+} flux through the IPR is the most difficult and complex part of the model. Most recent models of Ca^{2+} dynamics in ASM have either used the simple IPR model of De Young and Keizer (1992) or the more recent model of Cao et al (2013, 2014). Both of these models have a similar structure, although quite different derivations and parameter values, and both are discussed in detail in Section 2.7. For ASM, the choice of IPR model is dictated by the objectives of the modelling. For a detailed understanding of the stochastic properties of Ca^{2+} waves in ASM it is necessary to use the model of Cao et al (2013, 2014), but for a qualitative understanding of how the IPR and RyR are interacting with the other Ca^{2+} fluxes to control oscillations, it suffices to use either model.
- There are two principal components of the RyR model. Firstly, it exhibits CICR, as the flux through the receptor is an increasing function of c. Wang et al (2010) used the model of Friel (1995), which is an excellent quantitative description of CICR through RyR, although in a different cell type. Secondly, the RyR Ca^{2+} flux is dependent on $[Ca^{2+}]$ in the SR, c_s. As c_s increases, so does the RyR flux, while, conversely, as the SR gets depleted, the RyR flux decreases rapidly. This feature of the RyR is borrowed from similar models in cardiac cells, where it is well known that overloading of the SR leads to spontaneous release of Ca^{2+} through the RyR (Cheng et al, 1993; Shannon et al, 2004).
- Ca^{2+} buffering is assumed to be fast and linear, so that all fluxes may be interpreted as effective fluxes (Section 2.9).

Full details of the model equations, and of variants, can be found in Wang et al (2010), Croisier et al (2013) and Cao et al (2013, 2014).

7.4.1.3 The Interplay Between IP_3R and RyR

Typical model solutions, showing Ca^{2+} oscillations and waves are shown in Fig. 7.10. In response to agonist, there is an initial release of Ca^{2+} through the IPR. This initial release causes a larger release of Ca^{2+} through the RyR, via CICR, which decreases the SR $[Ca^{2+}]$, at least partially. This decrease in

SR [Ca^{2+}] inactivates the RyR, but has less effect on the IPR, which initiate a cycle of Ca^{2+} release and reuptake to and from the SR, leading to Ca^{2+} oscillations and periodic waves. The Ca^{2+} flux through the RyR continues to oscillate, but only with small amplitude.

The cycles of Ca^{2+} release through the IPR occur by the same mechanism as in the generic models of Ca^{2+} oscillations discussed in Chapter 3. An increase in [Ca^{2+}] increases the open probability of the IPR, leading to a positive feedback cycle that operates on a subsecond time scale. The IPR also inactivates in response to increased [Ca^{2+}], but on a slower time scale. The interaction between the fast activation and the slower inactivation leads to cycles of IPR opening and closing. The small-amplitude oscillations in the RyR flux are purely a response to the IPR-based oscillations, and play little role in the control of oscillation period.

The model also reproduces the response to KCl (Fig. 7.10 C and D), which depolarises the cell, leading to an increased influx of Ca^{2+}. In this case, the SR overloads with Ca^{2+}, leading to spontaneous release of Ca^{2+} through the RyR, subsequent depletion of the SR, inhibition of the RyR, and termination of the Ca^{2+} transient. Once the Ca^{2+} transient ends, the SR can refill and the cycle can repeat, leading to long-period oscillations, quite different in nature from those caused by agonists that activate the IPR. In this case, because [IP_3] is zero (or at least very low), the IPR are not able to sustain high-frequency Ca^{2+} oscillations.

There is thus a close interplay between the RyR and the IPR. In some cases oscillations are governed by one receptor type, in other cases the other receptor type is more important, and during the initial response to agonist both RyR and IPR play a role. The dominant receptor type is essentially determined by the [Ca^{2+}] in the SR. When the SR is overfilled, the RyR predominate, when the SR [Ca^{2+}] is lower, the IPR predominate.

By applying agonist, KCl, and ryanodine in various orders, we can explore these interactions further. If KCl is applied first, followed by methacholine (MCh), the model predicts that long-period oscillations (mediated by the RyR) will be replaced by short-period oscillations (mediated by the IPR). Conversely, if MCh is applied first, followed by KCl, the KCl has no significant effect on the oscillatory pattern. Model simulations and the experimental verification are shown in Fig. 7.11. The intuitive explanation of these results is as follows:

- Upon application of KCl, Ca^{2+} entry increases, which overloads the SR, leading to spontaneous activation of the RyR and long-period spiking. Although there is some temporary depletion of the SR at the peak of the Ca^{2+} spike, the SR, on average, remains overloaded.

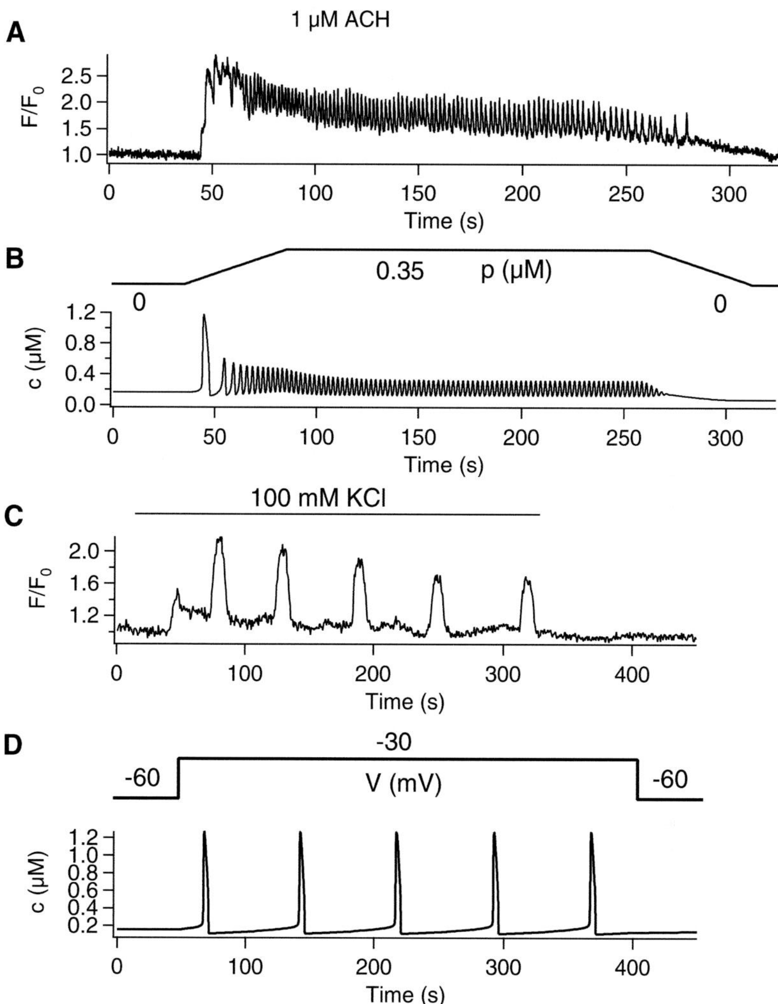

Fig. 7.10 Experimental results and model simulations. Basic oscillations under control conditions. **A:** application of ACh causes sustained oscillations in $[Ca^{2+}]$ in mouse ASM. These oscillations occur on a raised baseline. **B:** model simulation of ACh-induced Ca^{2+} oscillations. ACh is simulated by an increase in $[IP_3]$. Although the model shows good qualitative agreement with the experimental results in panel A, the quantitative agreement is poor, with a large initial peak, and with the oscillations not occurring on a raised baseline. **C:** application of KCl to the ASM cell causes depolarisation of the cell membrane, entry of Ca^{2+} through voltage-gated channels, and cycles of Ca^{2+} release and reuptake. These Ca^{2+} oscillations are quite different from ACh-stimulated oscillations (panel A), being much wider, with a larger period. **D:** model simulation of the response to depolarisation. Ca^{2+} oscillations occur with a much larger period. Reproduced from Wang et al (2010), Figs. 2 and 3, with permission from Elsevier.

- When MCh is added, the IPR also open, leading to increased Ca^{2+} flux from the SR into the cytoplasm. Because of this, the SR becomes, on average, depleted, which inactivates the RyR, leading to oscillations mediated almost solely by the IPR.
- Conversely, if MCh is added first, the SR becomes depleted. Subsequent addition of KCl, despite increasing the Ca^{2+} influx, is unable to cause an overload of the SR, and thus the RyR remain inactivated.

We can also use the model to predict the effects of adding ryanodine. Ryanodine acts in a slightly unusual manner, in that it locks the RyR open, but only once it has already been opened – ryanodine has no effect on a closed RyR. If we include that effect in the model, we see that pre-treatment with ryanodine will eliminate agonist-induced Ca^{2+} oscillations. This is because the initial agonist-induced release of Ca^{2+} through the IPR causes the RyR to open (via Ca^{2+}-induced Ca^{2+} release), but these RyR are then locked in the open state by ryanodine, resulting in a large leak of Ca^{2+} out of the ER, which cannot then support oscillations through the IPR. Again, this model prediction is confirmed experimentally.

Conversely, addition of ryanodine has little effect on MCh-induced oscillations. This is because addition of MCh results in depletion of the ER, and subsequent closure of most of the RyR. Thus, when ryanodine is added to a cell exhibiting MCh-induced oscillations, there are few open RyR upon which to act, and so the effect of the ryanodine is minimal. When ryanodine is added *before* MCh this is not the case; the initial Ca^{2+} spike opens up a substantial fraction of the RyR, which are then locked open by the ryanodine, leading to SR depletion and cessation of the oscillations. Hence the order of application of MCh and RyR is of crucial importance.

The interactions between the IPR and the RyR are thus an important factor in the control of Ca^{2+} oscillations, and this interplay occurs since both channel types are releasing Ca^{2+} from the same internal store. The amount of Ca^{2+} in the SR is controlled by both IPR and RyR, and, in turn, the SR $[Ca^{2+}]$ modulates release through both these Ca^{2+} channels. In this way, although MCh-induced oscillations are mediated primarily by the IPR, they are significantly affected by the RyR, via the control of how much Ca^{2+} is in the SR. In the same way, Ca^{2+} release through the RyR is indirectly affected by the IPR.

7.4.1.4 Periodic Waves in the Model

When Ca^{2+} diffusion is included in the model, the oscillations turn into periodic waves (Fig. 7.12). In this case, each part of the cytoplasm is an autonomous oscillator, weakly linked to its neighbours by diffusion of Ca^{2+}.

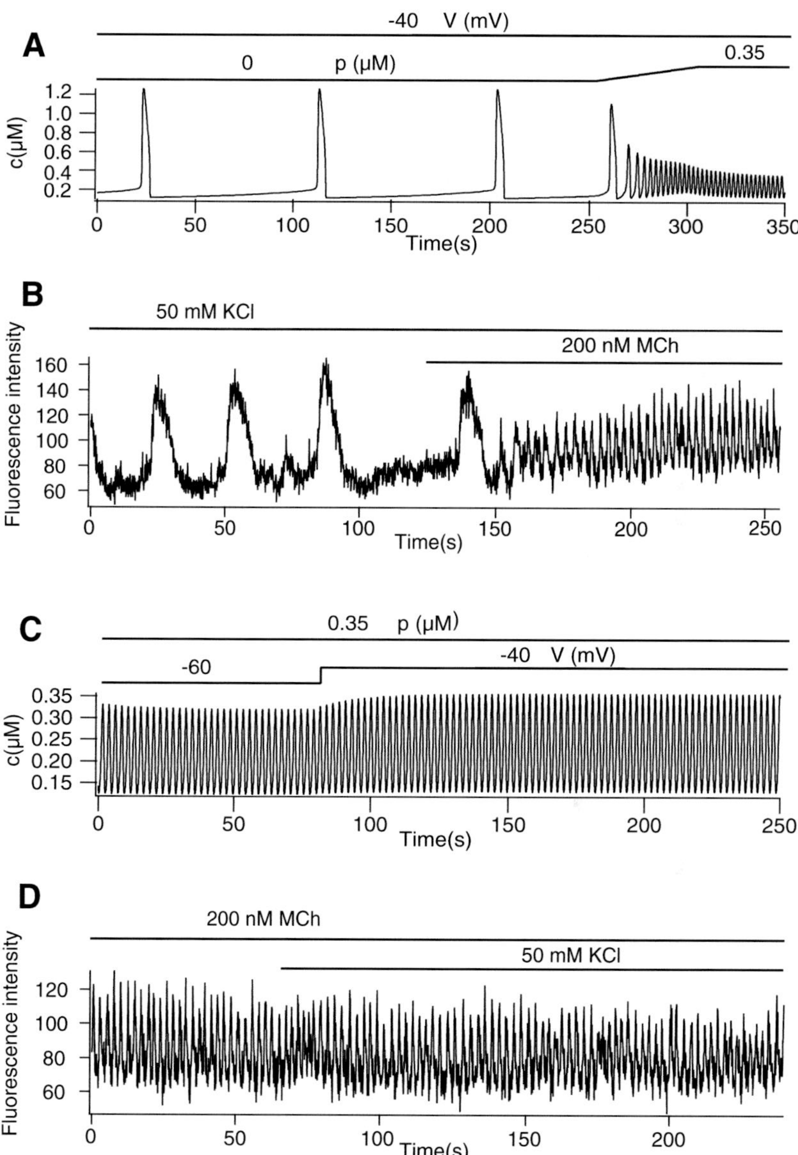

Fig. 7.11 **A:** Model simulations, predicting the response when KCl is applied first, followed by MCh (simulated by an increase in $[IP_3]$). The model predicts that the Ca^{2+} oscillations speed up and decrease in magnitude. **B:** Experimental confirmation of the model prediction. Note that the model is predicting $[Ca^{2+}]$, while the experiments measure the fluorescence ratio. This difference is the cause of some of the quantitative discrepancy. **C:** Model simulations, predicting the response of the application of MCh first, followed by KCl. The model predicts that the oscillations remain practically unchanged. **D:** Experimental confirmation of the model prediction. Reproduced from Wang et al (2010), Fig. 4, with permission from Elsevier.

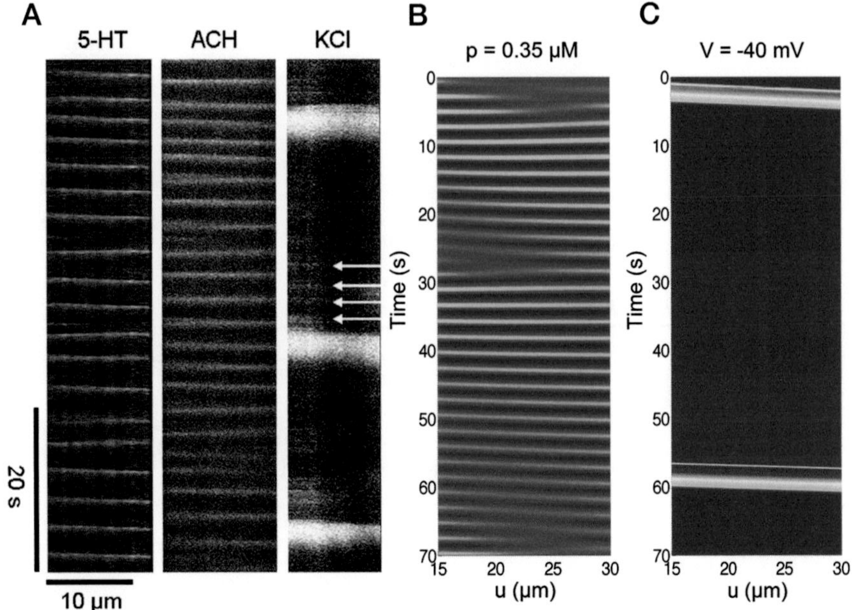

Fig. 7.12 Periodic waves in ASM, in experiment and simulations. Lighter, or warmer (i.e., red or orange), colours denote higher $[Ca^{2+}]$. **A:** Periodic waves of Ca^{2+} in ASM, in response to three agonists. In the response to 5-HT one can see that the wave reverses direction. Often, the waves in different directions are separated by abortive waves that do not travel the full length of the cell. The arrows mark abortive waves that result from localised stochastic release of Ca^{2+} through RyR. **B:** Model simulation of MCh-induced periodic Ca^{2+} waves. The model waves also reverse direction periodically. **C:** Model simulation of KCl-induced periodic Ca^{2+} waves. Reproduced from Wang et al (2010), Fig. 5, with permission from Elsevier.

However, there is one feature here of particular interest; in the cell, the periodic waves are often observed changing direction, as can be seen in the leftmost panel of Fig. 7.12 A. Waves in different directions are separated by abortive waves that do not travel the full length of the cell. It is not possible (without additional intervention) to reproduce this qualitative feature in the model if the density of IPR is the same throughout the cell. Instead, the model predicts that the density of IPR is higher at one end of the cell than the other. With such an IPR density gradient, one end of the cell is trying to oscillate at a slightly higher frequency than the other end. Although the higher frequency oscillation can drive the lower frequency oscillation for a short time, eventually it gets so far ahead that the wave travelling from the high-frequency end encounters a refractory region propagating from the low-frequency end, which aborts the wave (Fig. 7.12 B). Then, before the high-frequency end can initiate the next wave, the low-frequency end does so, resulting in a wave which moves in the opposite direction. However, this wave direction cannot long be maintained, because of the intrinsic frequency difference between the ends.

Given the apparent symmetry of the ASM cells, this prediction is somewhat surprising, as there is no obvious reason to suppose that the IPR density is asymmetrical in this way. Only additional experimental work will suffice to answer this question.

Other points of interest in Fig. 7.12 A are the abortive Ca^{2+} waves in response to KCl, as marked by the arrows. Recall that the period of KCl-induced waves is determined by the time taken to refill the SR after it has been depleted by a wave. When the SR is almost refilled, stochastic openings of the RyR can lead to abortive waves that fail to generate a whole-cell response; the SR $[Ca^{2+}]$ is high enough to support local release of Ca^{2+}, but not quite high enough to support wave propagation. The more the SR refills, the more these abortive waves occur, leading to a clustering of localised responses that occur just before the larger whole-cell response.

7.4.1.5 More Detailed Treatment of the Membrane Currents

In the model discussed above there was only the most cursory treatment of the membrane currents and the membrane potential. The actual situation is far more complex, as ASM contain a variety of ion channels, including Ca^{2+}-dependent Cl^- and K^+ channels. The potential effects of this are two fold.

Firstly, oscillations in $[Ca^{2+}]$ will cause analogous oscillations of the membrane potential, which will in turn affect the driving force for Ca^{2+} entry into the cell. Since Ca^{2+} entry is an important factor in the control of oscillation period, the membrane potential could possibly exert a fine control over the properties of the oscillations. The most detailed theoretical study of this possibility is that of Roux et al (2006), who combined a model of the cytosolic Ca^{2+} oscillator with models of the various ion channels (Fig. 7.13). From their modelling work they predicted how the membrane potential would vary during changes in $[Ca^{2+}]$, but as yet there has been no direct experimental confirmation. Their most important conclusion is that a rise in Ca^{2+} causes a depolarisation of the cell membrane, but not one of sufficient magnitude to open voltage-gated channels. Thus, they predict that cytosolic Ca^{2+} oscillations will occur without any necessary large changes in flux through membrane Ca^{2+} channels, such as the L-type channel, in which case the cytosolic oscillator would be mostly independent of the membrane potential. This offers a possible explanation for why the cytosolic oscillator persists in the absence of extracellular Ca^{2+}, but is also consistent with the observation that depolarisation of the membrane by KCl will lead to a much larger Ca^{2+} influx through the L-type channels, overloading of the SR, and consequent release of Ca^{2+} through RyR, as described above. In this scenario, the membrane potential is playing only a minor modulatory role in the cytosolic oscillations.

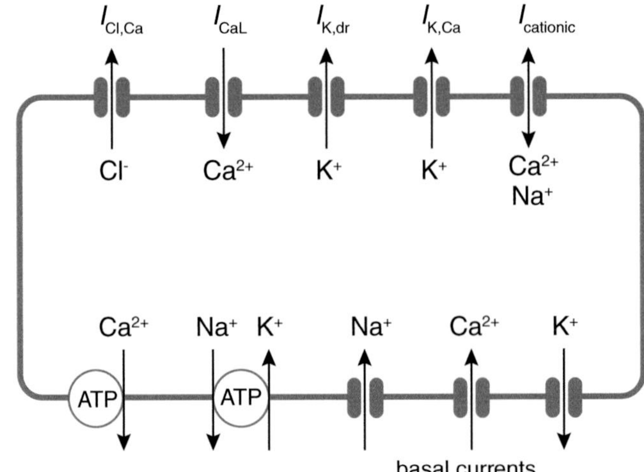

Fig. 7.13 Schematic diagram of the ionic currents in the model of Roux et al (2006). The model includes a Ca^{2+}-activated Cl^- current ($I_{Cl,Ca}$), an L-type Ca^{2+} current ($I_{Ca,L}$), a delayed rectifier K^+ channel ($I_{K,dr}$), a Ca^{2+}-activated K^+ channel ($I_{K,Ca}$), a nonspecific cationic current ($I_{cationic}$), and three generic background currents for Ca^{2+}, Na^+, and K^+. Ca^{2+} is removed from the cell by a PMCA pump, and Na^+/K^+ balance is maintained by the usual Na^+/K^+ ATPase.

Secondly, local aspects could be important. It is possible that a significant fraction of Ca^{2+}-sensitive BK channels are located close to RyR (Lifshitz et al, 2011), and thus local elevations of Ca^{2+} at the mouth of the RyR could activate significant K^+ currents that would serve to hyperpolarise the cell, thus decreasing the driving force for Ca^{2+} influx and causing relaxation (ZhuGe et al, 2010; Janssen, 2012). However, there is as yet no quantitative description that supports this hypothesis; what modelling evidence there is suggests that it is inconsistent with other results (Roux et al, 2006).

7.4.2 Vascular Smooth Muscle

The diameter of many arteries and arterioles exhibits rhythmic contractions and dilations, a phenomenon known as *vasomotion* (Haddock and Hill, 2005; Aalkjaer and Nilsson, 2005). Although vasomotion is widely observed, and has been extensively studied, its physiological role remains unclear. Vasomotion is a different kind of movement from the kind of peristaltic movement that is exhibited by smooth muscle in, for example, the gastrointestinal tract. There, contraction of gastrointestinal smooth muscle is not intrinsic but is driven by interstitial cells of Cajal (Hirst, 2003), and takes the form of slowly-moving waves that aid in the movement of food along the gut. However, vasomotion appears, firstly, to be intrinsic to the vascular smooth muscle, and, secondly, is often found to be synchronous over considerable lengths of artery, causing a more uniform constriction or dilation of the diameter rather than a peristaltic motion. Vasomotion can also propagate along arteries in a more peristaltic fashion (Seppey et al, 2010), but the physiological function of this behaviour is even less clear.

Although not all the details are known, or agreed on, there is a great deal of evidence showing that vasomotion is the result of spontaneous Ca^{2+} oscillations caused by the release of Ca^{2+} through IPR and RyR. These Ca^{2+} oscillations are modulated by the membrane potential, and modulate the potential in their turn, and thus models of vascular smooth muscle necessarily include both cytosolic and membrane components. Furthermore, one of the major questions in the theoretical study of vasomotion is that of synchronisation; can models help us understand how Ca^{2+} oscillations and contraction are synchronised over regions much larger than that of a single cell? It is highly plausible that electrical connectivity will be important for interactions over long distances, but it is still not known exactly how the cytosolic Ca^{2+} oscillator interacts with the membrane potential to control such behaviour.

In response to agonists such as 5-HT, vascular smooth muscle exhibits Ca^{2+} oscillations that range in period (depending on the species and the agonist concentration) from about a second to over a minute (Fig. 7.14 A, B, and C). Each Ca^{2+} transient is associated with a membrane depolarisation and an increase in contractile force (Fig. 7.14 D).

Models of the Ca^{2+} dynamics of vascular smooth muscle have the same basic structure as for any cell where a cytosolic Ca^{2+} oscillator interacts with the membrane (Section 3.7), the same structure as seen in the model of airway smooth muscle discussed above. The cytosolic oscillator model typically includes equations for cytosolic $[Ca^{2+}]$ and SR $[Ca^{2+}]$, and, if a Class II model is used (Section 3.2.7), may also involve an equation for the dynamic control of $[IP_3]$. Cytosolic Ca^{2+} typically modulates the conductance of Cl^- and K^+ channels in the membrane, thus changing the membrane potential, while the membrane potential itself controls the current through voltage-gated Ca^{2+} channels (Section 2.5). In this way the cytosolic oscillator is connected to the membrane potential and vice versa. The membrane potential model is usually based on the traditional circuit model of the cell membrane (Hodgkin and Huxley, 1952; Keener and Sneyd, 2008). Thus, the model equations have the same general form as (3.150) and (3.153), differing only in the details of exactly which ionic currents are included and how the various Ca^{2+} fluxes are modelled.

An excellent and thorough review of computational models of Ca^{2+} dynamics in vascular smooth muscle is Tsoukias (2011). Here we give only a very brief overview of the different models.

The earliest model of Ca^{2+} dynamics in vascular smooth muscle was that of Wong and Klassen (1993). However, since the various model components

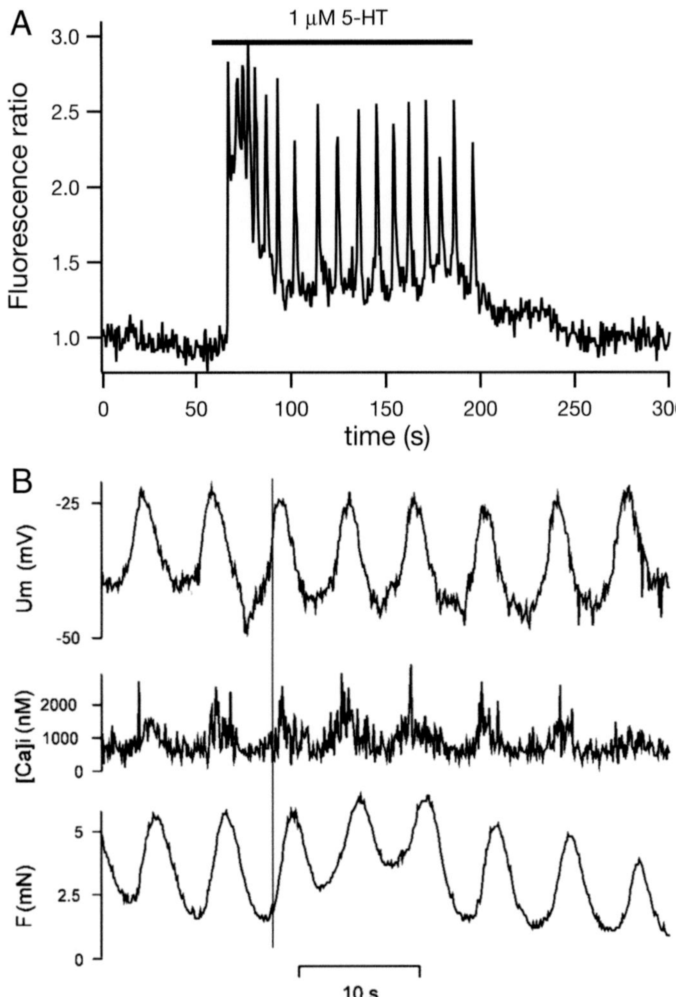

Fig. 7.14 **A**: Calcium oscillations in mouse pulmonary arteriole smooth muscle (measured in a lung slice) in response to 5-HT (Perez and Sanderson, 2005a). Original data provided by Michael Sanderson (University of Massachusetts Medical School). **B:** Membrane potential (U_m), [Ca^{2+}] and isometric force (F) in an isolated rat mesenteric small artery. Note the temporal sequence, as indicated by the vertical line; the depolarisation occurs before the Ca^{2+} transient, which itself occurs before the increase in force. Reproduced from Aalkjaer and Nilsson (2005), Fig. 2, with permission from Wiley.

are now showing their age, the Wong/Klassen model has been largely superseded by more recent work from a number of groups. Another early model is Gonzalez-Fernandez and Ermentrout (1994), which incorporated the mechanics of the cell walls, but, like the Wong/Klassen model, has been superseded by more recent models that use more realistic descriptions of the Ca^{2+} dynamics.

A typical model (Kapela et al, 2008) tracks four ionic concentrations (K^+, Na^+, Cl^-, and Ca^{2+}), with a total of 10 different currents across the plasma membrane, including three types of K^+ channels, one of which is Ca^{2+}-sensitive, a Ca^{2+}-sensitive Cl^- channel, various exchangers and cotransporters, and two inward Ca^{2+} currents (a voltage-gated Ca^{2+} current and a store-operated current). Since each of these channels, exchangers, or cotransporters comes with its own detailed model, each with multiple parameters, it is clear that such models quickly become highly complex. Another complex model of note is that of Yang et al (2003) who coupled their Ca^{2+} and voltage model to a simplified model of the muscle mechanics and force generation.

More recent work from the Tsoukias group (Kapela et al, 2010, 2012) has extended model complexity yet further by coupling two smooth muscle cells, to investigate how each of the model components affects the ability of vascular smooth muscle cells to synchronise their oscillatory behaviour. Although these results are still entirely theoretical, they do indicate which of the potential coupling mechanisms are most important, and thus which should be preferentially investigated experimentally.

A somewhat simpler model, although still with the same basic structure, is that of Parthimos et al (1999) (with an extended analysis in Parthimos et al (2007)), and this was also used as the basis of a series of investigations by Koenigsberger et al (2004, 2005, 2008, 2010). One of the major goals of this series of models is to understand how intercellular Ca^{2+} waves in vascular smooth muscle (Seppey et al, 2010) could control the propagation of vasomotion along the artery. Synchronisation, with a particular emphasis on the role of a cGMP-dependent Ca^{2+}-activated Cl^- channel, was also studied in a series of papers by Jacobsen, Holstein-Rathlou, and their colleagues (Jacobsen et al, 2007b,a; Postnov et al, 2011).

7.5 Calcium and the Generation of Force in Smooth Muscle

In muscle we are often interested, not in the $[Ca^{2+}]$ per se, but rather in the force that concentration generates. In striated muscle there is a relatively straightforward relationship between $[Ca^{2+}]$ and force. An action potential causes an increase in $[Ca^{2+}]$, Ca^{2+} binds to troponin which exposes the myosin binding site on the actin, and the crossbridge cycle generates force. Quantitative models of this process, mostly based on the original crossbridge model of Huxley (1957), are common (Keener and Sneyd, 2008). The part played by Ca^{2+} in these models is mostly permissive; there is some feedback from the crossbridges to the $[Ca^{2+}]$ – the crossbridges act as a Ca^{2+} buffer, for example – but, particularly in cardiac muscle, there is often an implicit assumption that the contraction can be simply deduced from the Ca^{2+} transient (Bers, 2002). Although, of course, this is not true in detail, it is not entirely unreasonable as a first approximation for cardiac muscle.

However, in smooth muscle the role played by Ca^{2+} is considerably more complicated, since, as discussed above, the crossbridge cycle can occur only when myosin is phosphorylated by myosin light chain kinase. In addition, the crossbridge cycle of smooth muscle is more complex than that of striated muscle, because of the possibility that attached myosin can be dephosphorylated, putting the crossbridge into the latch state.

7.5.1 The Hai–Murphy Model

Most quantitative studies of smooth muscle are based on the work of Hai and Murphy (1988a,b) and Murphy et al (1990) who assumed that a crossbridge could exist in one of four forms: myosin either unattached or attached, either phosphorylated or not (Fig. 7.15). The rate of crossbridge phosphorylation is denoted by k_1, and is a function of $[Ca^{2+}]$, denoted, as usual, by c. The rate of dephosphorylation was assumed by Hai and Murphy to be constant, although

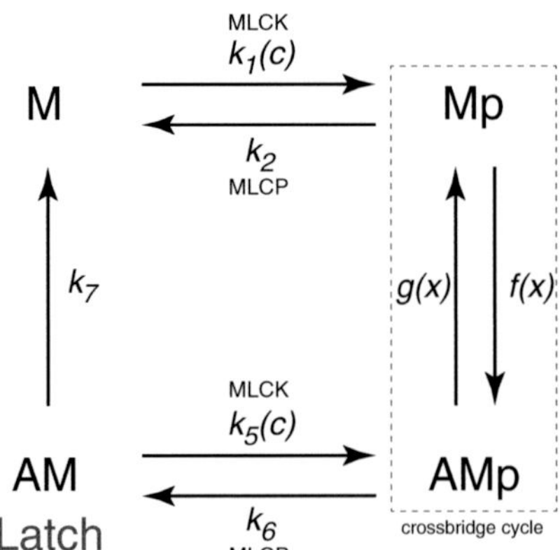

Fig. 7.15 Schematic diagram of the Hai–Murphy model of smooth muscle (Hai and Murphy, 1989a,b). M – myosin; Mp – phosphorylated myosin; AMp – attached phosphorylated myosin; AM – attached myosin. MLCK is activated myosin light chain kinase and its activity is a function of $[Ca^{2+}]$ (denoted by c), via Ca^{2+}/calmodulin. MLCP is myosin light chain phosphatase, and its activity may well be a function of c, time and other variables, although this is not shown here. As in the Huxley model, x is crossbridge displacement. The cycle between Mp and AMp corresponds to the usual crossbridge cycle of striated muscle.

it now seems likely that, in some cell types, it also depends on c as well as the level of agonist stimulation. In general, unattached and attached myosin (M and AM, respectively) could be phosphorylated at different rates ($k_1(c)$ and $k_5(c)$, respectively), although it is often assumed that these two rates are the same. Mijailovich et al (2000) extended the original Hai–Murphy model by including dependence on the crossbridge displacement, x, in the same way as in the Huxley model (Huxley, 1957), thus obtaining a set of partial differential equations to solve for the crossbridge states. However, unphosphorylated myosin cannot attach to actin, and detaches only slowly. Hence, k_7 is smaller than $g(x)$, and there is no reverse reaction.

The Hai–Murphy crossbridge model is connected to models of intracellular Ca^{2+} dynamics via $k_1(c)$ and $k_5(c)$ (Koenigsberger et al, 2004, 2005, 2006; Fajmut et al, 2005; Payne and Stephens, 2005; Bursztyn et al, 2007; Wang et al, 2010).

7.5.2 Calcium, Calmodulin, and MLCK

In the Hai–Murphy model, $k_1(c)$ is proportional to the concentration of activated MLCK, which depends on $[Ca^{2+}]$, but not in any simple way, as Ca^{2+} must first bind the protein calmodulin (CaM; calmodulin is an abbreviation of "calcium-modulated protein"), which must then activate MLCK.

The simplest (and first) model of this process is that of Kato et al (1984). Calcium activation of CaM is assumed to be a one-step process, whereby four Ca^{2+} ions bind to CaM, which then, in another one-step process, activates MLCK. Thus

$$\mathrm{CaM} + 4\mathrm{Ca}^{2+} \underset{k_{-1}}{\overset{k_1}{\rightleftharpoons}} \mathrm{CaM}^*, \tag{7.24}$$

$$\mathrm{CaM}^* + \mathrm{MLCK} \underset{k_{-2}}{\overset{k_2}{\rightleftharpoons}} \mathrm{MLCK}^*, \tag{7.25}$$

where a superscript $*$ denotes an activated form. If we assume that both equations are at equilibrium, and use the conservation conditions

$$[\mathrm{CaM}] + [\mathrm{CaM}^*] + [\mathrm{MLCK}^*] = M_0, \tag{7.26}$$

$$[\mathrm{MLCK}] + [\mathrm{MLCK}^*] = L_0, \tag{7.27}$$

for some total concentrations M_0 and L_0, then one finds that

$$c^4 = \frac{K_1 K_2 L^*}{(1-\alpha L^*)(1-L^*) - K_2 L^*}, \tag{7.28}$$

where $K_i = k_{-i}/k_i$, for $i = 1, 2$, $L^* = [\mathrm{MLCK}^*]$, and $\alpha = L_0/M_0$. From this it follows that L^* is a sigmoidally increasing function of c, with a maximum of

λ, where λ is the smaller root of $(1-\alpha\lambda)(1-\lambda) = K_2\lambda$, and thus $0 < \lambda < 1$. When α is small this reduces to

$$L^* = \frac{c^4}{(1+K_2)c^4 + K_1K_2}, \tag{7.29}$$

which makes the sigmoidal nature of the curve even more evident.

A more complicated model was constructed by Fajmut et al (2005), which takes into account the fact that Ca^{2+} binding to CaM is not simply a one-step process in which four Ca^{2+} bind simultaneously. This model assumes that CaM can bind two Ca^{2+} in each of two different binding sites (two Ca^{2+} ions can bind to binding sites in the C terminus, while two can bind to binding sites in the N terminus). The model then assumes that CaM can bind to MLCK when it has no Ca^{2+} bound, or two Ca^{2+} bound to the C or N terminus binding sites, or when all the binding sites have Ca^{2+} bound.

This leads to the 8-state model shown in Fig. 7.16. CaM is assumed to have three binding regions – the N and C termini, and the binding region for MLCK – each of which can be occupied or unoccupied, giving a total of $2^3 = 8$ states, which can be distributed at the vertices of a three-dimensional cube. The differential equations for this model follow simply from mass action, in combination with the conservation laws which assume that the total amounts of CaM and MLCK are constant. Parameters of the model can be estimated from a somewhat bewildering array of inconsistent and diverse experimental data; the full details of this process are given in Fajmut et al (2005). For example,

- The affinity of CaM for MLCK is higher when CaM has four Ca^{2+} bound (with a K_d of around 1 nM) than when it has only two Ca^{2+} bound ($K_d \approx$ 5 nM. When CaM has no Ca^{2+} bound, its affinity for MLCK is very low.
- In the absence of MLCK, the affinity for Ca^{2+} of the C terminal binding site is around three times higher than the affinity of the N terminal binding site. In the presence of MLCK, the affinity for Ca^{2+} of the C terminal binding site is decreased by a factor of approximately 10.
- Ca^{2+} binds approximately 70 times faster to the N terminal than to the C terminal.
- The presence of MLCK slows down unbinding of Ca^{2+} from the C terminal by a factor of approximately 40, and slows down the unbinding of Ca^{2+} from the N terminus by a factor of approximately 140–225. Thus, binding of MLCK stabilises Ca^{2+} binding.

The model equations were solved to give each CaM state as a function of $[Ca^{2+}]$, and the results compared to additional experimental data obtained by Geguchadze (2004) who used a FRET-based biosensor to measure Ca^{2+}/CaM binding to MLCK, and thus MLCK activity. We are interested particularly in the ratio

$$A = \frac{[\mathrm{CaM_{NCM}}]}{[\mathrm{MLCK}]_{\mathrm{tot}}}, \tag{7.30}$$

as this denotes the fraction of MCLK that has been activated by Ca^{2+}/CaM.

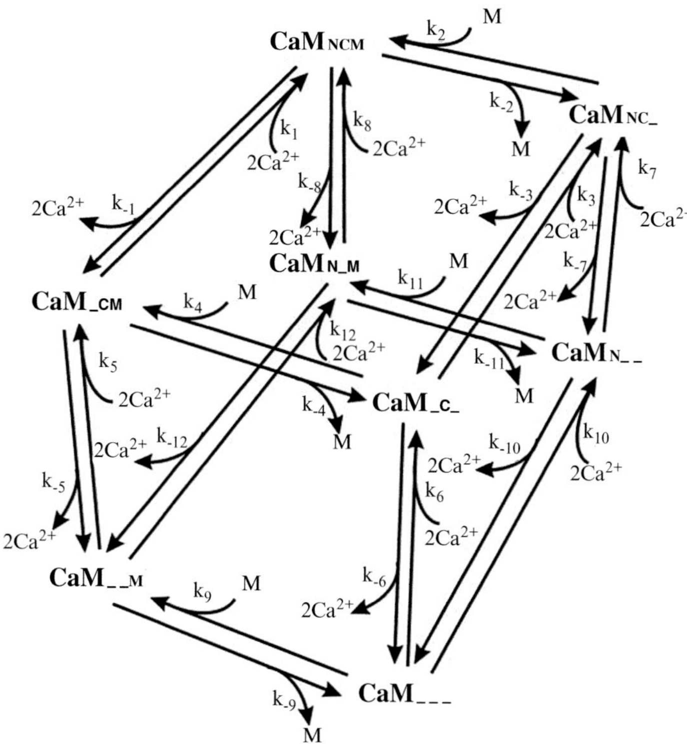

Fig. 7.16 Schematic diagram of the Ca^{2+}/calmodulin model of Fajmut et al (2005). Calmodulin (CaM) can bind two Ca^{2+} ions at the N terminus (denoted by a subscript N), two Ca^{2+} ions at the C terminus (subscript C) and can bind to MLCK (subscript M). Thus, for example, CaM_{NCM} is a calmodulin molecule that has two Ca^{2+} bound to the N and C termini, and is bound to MLCK. A subscript _ denotes an empty binding site. Reproduced from Fajmut et al (2005), Fig. 1, with permission from Elsevier.

It turns out, not unexpectedly, that $[CaM_{NCM}]/[MLCK]_{tot}$ is a sigmoidally increasing function of $[Ca^{2+}]$, as shown in Fig. 7.17. However, what is particularly noteworthy is that, although the model was not fitted to these data, the agreement is nevertheless extremely good. Thus, a complex process of parameter estimation from a variety of sources has resulted in a model that has the ability to predict a wider range of results.

7.5.3 The Frequency Response of Airway Smooth Muscle

In smooth muscle, the crossbridges are, in general, responding to an oscillating $[Ca^{2+}]$, and thus an oscillating concentration of activated MLCK. It is thus of interest to determine how the contractile response depends on the

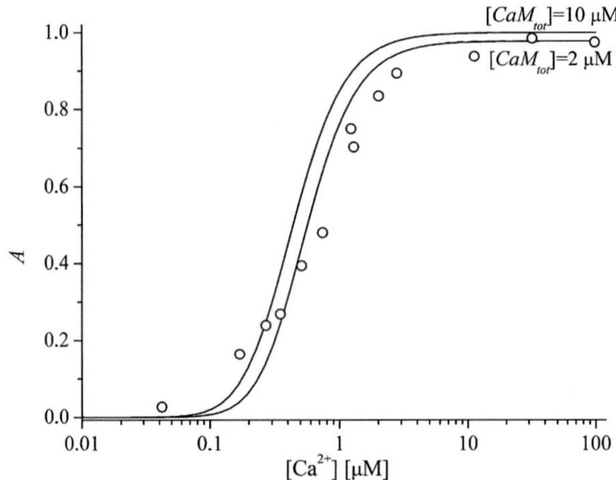

Fig. 7.17 Activation of MLCK (A) as a function of $[Ca^{2+}]$. Symbols are experimental data from Geguchadze (2004), smooth curves are from the model of Fajmut et al (2005), for two different values of the total amount of CaM. Reproduced from Fajmut et al (2005), Fig. 2A, with permission from Elsevier.

frequency of the Ca^{2+} oscillation. This has been measured experimentally in airway smooth muscle (Perez and Sanderson, 2005b), as shown in Fig. 7.18. As the oscillation frequency increases, so does the contraction (measured by the change of lumen area of the bronchiole in a lung slice). Interestingly, for the same frequency of oscillation, different agonists (which, in general, cause oscillations of different shapes) cause different amounts of contraction, a result which shows clearly that the shape of the oscillation is important also, not only the frequency.

These data are somewhat difficult to interpret, as it is impossible to control for the average $[Ca^{2+}]$. In this experiment, as the frequency increases so does the average $[Ca^{2+}]$, and thus it is not immediately clear how much of the change in contractile force is due solely to the change in frequency.

To investigate this in more detail, Wang et al (2008, 2010) computed the frequency response of a model of Ca^{2+} oscillations in airway smooth muscle, and showed that, for a fixed average $[Ca^{2+}]$, the force generated by the muscle is an increasing function of the Ca^{2+} oscillation frequency (Fig. 7.19 A and B), and that Ca^{2+} oscillations always give a greater force than does the same average constant $[Ca^{2+}]$. Once the oscillations become faster than about 10 per minute there is little change in force. Thus, the changes in force that are observed experimentally (see Fig. 7.18) at higher frequencies are due to an increase in average $[Ca^{2+}]$, rather than an increase in frequency. These results are essentially a concrete application of the general results of Salazar et al (2008), who studied the interaction of Ca^{2+}-dependent kinases and phosphatases in a more general setting. However, it is important to note that the frequency response of ASM is dependent also on the species, as, for

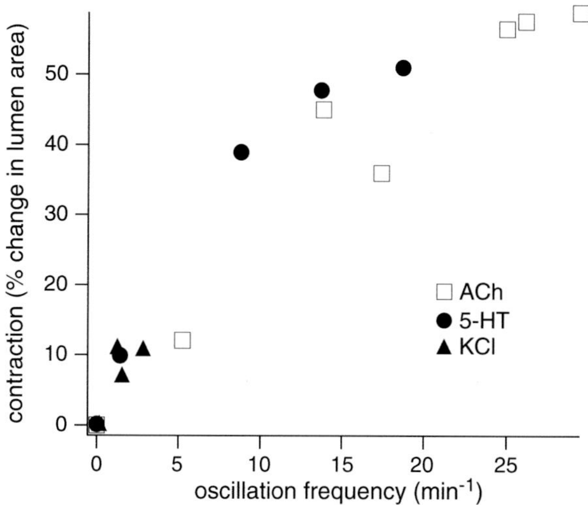

Fig. 7.18 The force-frequency relationship in airway smooth muscle (Perez and Sanderson, 2005b). The vertical axis shows the airway contraction (plotted as a percent change in lumen area) induced by 5-HT, ACH, and KCl; the contraction is directly related to the force exerted by the ASM, although there is no simple relationship. As the oscillation frequency increases, so does the force. Original data provided by Michael Sanderson (University of Massachusetts Medical School).

example, MLCP in mice has a qualitatively different dependence on $[Ca^{2+}]$ than does MLCP in rats or humans. As yet, the only theoretical frequency response studies in ASM have been based on data from mice.

It is interesting to note that periodic Ca^{2+} waves result in considerably less force than whole-cell Ca^{2+} oscillations (Fig. 7.19 C). This is not due to changes in average $[Ca^{2+}]$ (as is shown by Wang et al (2010)), but is instead caused by changes in the shape of the Ca^{2+} transients. It is a rather sobering thought that the accurate computation of force in airway smooth muscle (and possibly other smooth muscle types as well) is so sensitively dependent on the wave shape, and thus on the detailed spatial structures within each cell. If accurate forces can be generated only by spatial models, the problem of how to construct useful multiscale models becomes even more difficult.

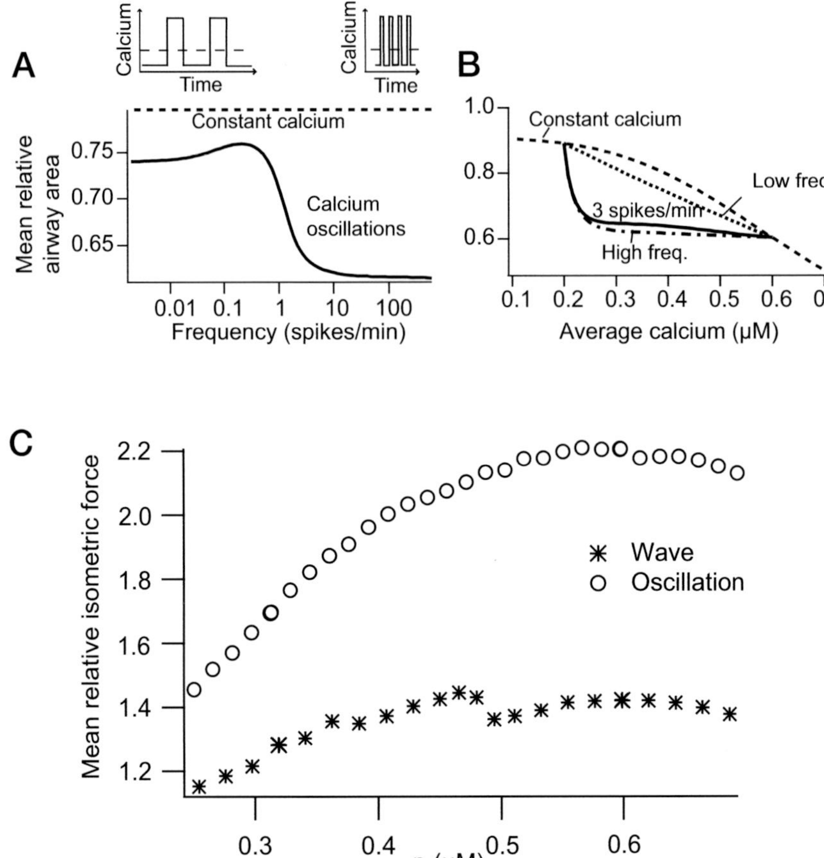

Fig. 7.19 **A:** Airway area as a function of Ca^{2+} oscillation frequency. Note that a smaller airway area corresponds to a greater force exerted by the muscle, as the muscle is causing contraction of the airway. The oscillations have a constant average $[Ca^{2+}]$ and a caricature shape, as shown in the upper insets. The mean average $[Ca^{2+}]$ is maintained at $0.4\,\mu$M by varying the spike width together with the period so that their ratio is 0.5. Calcium oscillations always result in a greater force than the same maintained mean $[Ca^{2+}]$. **B:** Airway area as a function of average $[Ca^{2+}]$ for different oscillation frequencies. For a given average $[Ca^{2+}]$, the force increases (i.e., the airway area decreases) as the oscillation frequency increases. Similarly, for a given oscillation frequency, the force increases as the average $[Ca^{2+}]$ increases. **C:** force as a function of $[IP_3]$ (p) for waves and oscillations. For each value of p the waves and oscillations have similar frequencies, but the waves consistently exert less force than whole-cell oscillations. This is because the Ca^{2+} spike has a different shape in waves and oscillations, and these differences are sufficient to cause significant changes in force. Reproduced from Wang et al (2008), Fig. 8 and Wang et al (2010), Fig. 6, with permission from Elsevier.

Chapter 8
Neurons and Other Excitable Cells

8.1 Introduction

Calcium plays a major role in every neuron. Not only does it play a part in controlling the membrane potential, but it is also a crucial ingredient of both the pre-synaptic and post-synaptic terminals. At the pre-synaptic terminal the secretion of neurotransmitter is controlled by Ca^{2+}, as is short-term plasticity, while in the post-synaptic terminal Ca^{2+} is both necessary and sufficient for long-term synaptic plasticity, i.e., long-term depression, long-term potentiation, and spike-timing-dependent plasticity. In neuroendocrine cells from the hypothalamus and pituitary, as well as in the endocrine pancreas, Ca^{2+} controls the membrane potential, the shape of the bursts of electrical spiking, and the secretion of hormones. In photoreceptors and olfactory receptors, Ca^{2+} is at the centre of the biochemical networks that control adaptation to a maintained stimulus, while in interstitial cells of Cajal Ca^{2+} controls the membrane potential in ways that we do not yet fully understand. There are many other examples of the importance of Ca^{2+} in neurons and excitable cells.

Here we present only a selection of topics, focusing in each case on the dynamics of Ca^{2+}, and thus a brief description of the physiological context must suffice.

8.2 Pre-synaptic Calcium Dynamics

Synaptic transmission begins with the arrival of an action potential at the pre-synaptic terminal. This action potential opens voltage-gated Ca^{2+} channels (VGCC), increasing Ca^{2+} influx into the pre-synaptic neuron, and the resultant increase in $[Ca^{2+}]$ causes exocytosis of neurotransmitter (Neher, 1998a,b; Meinrenken et al, 2003; Neher and Sakaba, 2008; Eggermann et al,

G. Dupont et al., *Models of Calcium Signalling*, Interdisciplinary Applied Mathematics 43, DOI 10.1007/978-3-319-29647-0_8

2012; Neher and Taschenberger, 2013). Exocytosis requires a high $[Ca^{2+}]$ and there is thus a close physical association between the VGCC and the vesicles. The highly localised microdomains of high $[Ca^{2+}]$ around the mouth of the VGCC suffice to cause fast neurotransmitter release. However, although this basic scheme is relatively simple and robust, changes in $[Ca^{2+}]$ over longer time scales also modulate the amount of neurotransmitter released by an action potential. For example, in some synapses, sustained action potential firing can cause depletion of the vesicle pool and short-term depression of the response. However, high $[Ca^{2+}]$ appears also to lead to increased vesicle recruitment, and thus there is a complex set of Ca^{2+}-mediated processes, some fast, some slow, which interact to control the synaptic strength. In other synapses, prolonged bursts of action potentials lead to enhanced synaptic strength.

There have been many models of pre-synaptic Ca^{2+} dynamics. One of the earliest models is that of Fogelson and Zucker (1985), and most subsequent models have been similar in structure, although differing considerably in detail. In essence, Ca^{2+} is assumed to enter through VGCC into a three-dimensional domain, is buffered by a variety of fast and slow buffers, binds to synaptic vesicles at some specified distance from the VGCC, and is removed from the cytoplasm by SERCA and PMCA pumps. For example, a recent model (Pan and Zucker, 2009) assumes two different Ca^{2+} buffers, one fast and one slow, assumes that the vesicles can exist in multiple states, with Ca^{2+}-dependent transitions, and solves the full time-dependent problem in a three-dimensional domain using the CalC software developed by Victor Matveev (`http://web.njit.edu/~matveev/calc.html`). Nevertheless, despite the plethora of complicating details, the model structure remains essentially the same as that of Fogelson and Zucker (1985).

Matveev and his colleagues have constructed a series of models (Matveev et al, 2002, 2004, 2006, 2009, 2011), again similar in basic structure to the original work of Fogelson and Zucker (1985), while Neher and his colleagues have constructed a family of models paying particular attention to the properties and consequences of Ca^{2+} buffering (Neher, 1998b; Sakaba and Neher, 2001; Wadel et al, 2007; Hosoi et al, 2007; Neher and Sakaba, 2008; Neher and Taschenberger, 2013).

These models are generally too detailed to present at length, and so here we present only a brief selection of simpler models. However, these simpler models still capture the basic structure of most pre-synaptic Ca^{2+} models, and thus remain important.

8.2.1 Facilitation

One of the major challenges in the study of pre-synaptic Ca^{2+} dynamics is the phenomenon of *facilitation*, which occurs when the amount of neurotransmitter release caused by an action potential is increased by an earlier action potential, provided that the time interval between the action potentials is not too great. There is a lot of evidence that facilitation is primarily a pre-synaptic mechanism, and thus quantitative explanations have focused on the pre-synaptic terminal.

One of the earliest hypotheses (Katz and Miledi, 1968) was that facilitation is caused by the buildup of Ca^{2+} in the pre-synaptic terminal, since the Ca^{2+} introduced by one action potential is not completely removed before the next one comes along, and this hypothesis, in its general form, is still widely accepted.

However, although there is general agreement that buildup of Ca^{2+} is, one way or another, responsible for facilitation, the devil is in the details; there is still no agreement as to where, or in what form, the Ca^{2+} is building up. The *residual free calcium* hypothesis claims that it is an increase in free pre-synaptic Ca^{2+} that underlies facilitation, and there is considerable evidence in support of this claim (Zucker and Regehr, 2002). For example, Kamiya and Zucker (1994) tested how facilitation responds to a rapid increase of the concentration of Ca^{2+} buffer in the cytoplasm. This rapid increase was accomplished by the photorelease of a Ca^{2+} buffer, diazo-2. They found that the fast release of diazo-2 into the pre-synaptic terminal causes a decrease of the synaptic response within milliseconds. Since, on this time scale, the main effect of the added buffers should be to decrease the free $[Ca^{2+}]$, it implies that it is the resting free $[Ca^{2+}]$ that is responsible for facilitation.

Mathematical models of the residual free Ca^{2+} hypothesis (Tang et al, 2000; Matveev et al, 2002; Bennett et al, 2004, 2007) have shown that this hypothesis is capable of explaining many experimental results. These models require that the Ca^{2+} facilitation binding site (to the vesicle release machinery) be situated at least 150 nm away from the release binding site, for if this were not so, then every action potential would saturate the facilitation binding site, leading to no increase in neurotransmitter release for subsequent action potentials. In addition, these models require low diffusion coefficients of Ca^{2+} close to the Ca^{2+} channel, and low diffusion coefficients of the exogenous buffer.

A similar hypothesis is that facilitation is caused by the local saturation of Ca^{2+} buffers close to the mouth of the Ca^{2+} channel, leading to the accumulation of cytoplasmic Ca^{2+} during a train of action potentials, and thus facilitation (Klingauf and Neher, 1997). Although a mathematical model of this *buffer saturation* hypothesis (Matveev et al, 2004) can explain many experimental results (at least from the crayfish neuromuscular junction), some questions remain.

The third hypothesis discussed here was, historically, the first to be proposed. In 1968, Katz and Miledi proposed that facilitation was the result of residual Ca^{2+} bound to the vesicle release site; i.e., if the action potentials followed upon one another too closely there would be not enough time for Ca^{2+} to unbind, making it easier for the subsequent action potential to release neurotransmitter. Although this *residual bound calcium* hypothesis has fallen somewhat out of favour in recent years, due principally to the experiments described briefly above, in which the application of an exogenous buffer was shown to decrease the synaptic response quickly, more recent detailed models have shown that it is, after all, consistent with this experiment (Matveev et al, 2006). Indeed, the residual bound Ca^{2+} hypothesis is superior in some ways, as it does not suffer from the necessity of somewhat artificial requirements, such as the 150 nm separation between the release and the facilitation binding sites, or the small diffusion coefficients around the mouth of the Ca^{2+} channel.

Nevertheless, the question of what causes facilitation is still far from completely resolved. As is often the case, the most likely explanation is a combination of all the above three hypotheses, each acting with different strengths depending on the exact situation.

8.2.2 A Model of the Residual Bound Calcium Hypothesis

A simple model of the residual bound Ca^{2+} hypothesis does not include any spatial information, but is based entirely on the kinetics of Ca^{2+} binding to the vesicle binding sites (Bertram et al, 1996). This simplification is justified by experimental results showing that the minimum latency between Ca^{2+} influx and the onset of transmitter release can be as short as 200 μs. Since the Ca^{2+} binding site must thus be close to the Ca^{2+} channel, in a simple model we can neglect Ca^{2+} diffusion and assume that Ca^{2+} entering through the channel is immediately available for binding to the vesicle binding site.

It is also assumed that transmitter release is the result of Ca^{2+} entering through a single channel, the so-called Ca^{2+}-domain hypothesis. If the Ca^{2+} channels are far enough apart, or if only few open during each action potential, the Ca^{2+} domains of individual channels are independent.

Our principal goal here is to provide a plausible explanation for the intriguing experimental observation that facilitation increases in a steplike fashion as a function of the frequency of the conditioning action potential train, as illustrated in Fig. 8.1.

We assume that Ca^{2+} entering through the VGCC is immediately available to bind to the transmitter release site, which itself consists of four independent, but not identical, gates, denoted by S_1 through S_4. Gate S_j can either be closed (state C_j, with probability C_j) or open (state O_j, with probability O_j), with the transition governed by

$$\mathrm{Ca}^{2+} + \mathrm{C}_j \underset{k_{-j}}{\overset{k_j}{\rightleftharpoons}} \mathrm{O}_j. \tag{8.1}$$

It follows that

$$\frac{dO_j}{dt} = k_j c - \frac{O_j}{\tau_j(c)}, \tag{8.2}$$

where $\tau_j(c) = 1/(k_j c + k_{-j})$, and c is the $[Ca^{2+}]$. Finally, the probability, P, that the release site is activated is

$$P = O_1 O_2 O_3 O_4. \tag{8.3}$$

The rates of closure of S_3 and S_4 are much greater than for S_1 and S_2, and thus Ca^{2+} remains bound to S_1 and S_2 for a relatively long time, providing the possibility of facilitation.

To demonstrate how facilitation works in this model, we suppose that a train of square pulses of Ca^{2+} (each of width t_p and amplitude c_p) arrives at the synapse. We want to calculate the level of activation at the end of each pulse and show that this is an increasing function of time. The reason for this increase is clear from the governing differential equation, (8.2). If a population of gates is initially closed, then a Ca^{2+} pulse begins to open them, but when Ca^{2+} is absent, the gates close. If the interval between pulses is sufficiently short and the decay time constant sufficiently large, then when the next pulse arrives, some gates are already open, so the new pulse activates a larger fraction of transmitter release sites than the first, and so on.

To quantify this observation we define t_n to be the time at the end of the nth pulse,

$$t_n = t_p + (n-1)T, \tag{8.4}$$

where $T = t_p + t_I$ is the period and t_I is the interpulse interval. For any gate (temporarily omitting the subscript j) with $O(0) = 0$, the open probability at the end of the first pulse is

$$O(t_1) = O_\infty (1 - e^{-t_p/\tau_p}), \tag{8.5}$$

where $O_\infty = k c_p \tau_p$ is the steady-state probability corresponding to a steady $[Ca^{2+}]$, c_p, and $\tau_p = \tau(c_p) = 1/(k c_p + k_-)$.

Suppose that $O(t_{n-1})$ is the open probability at the end of the $(n-1)$st Ca^{2+} pulse. During the interpulse period, O decays with rate constant $\tau(0)$. Thus, at the start of the nth pulse,

$$O(t_{n-1} + t_I) = O(t_{n-1}) e^{-t_I/\tau(0)}, \tag{8.6}$$

and so at the end of the nth pulse,

$$\begin{aligned} O(t_n) &= O(t_{n-1}) e^{-t_I/\tau(0)} e^{-t_p/\tau_p} + O_\infty (1 - e^{-t_p/\tau_p}) \\ &= \alpha O(t_{n-1}) + O(t_1), \end{aligned} \tag{8.7}$$

where $\alpha = \exp(-(t_I/\tau(0) + t_p/\tau_p)) = \exp(-k_-(t_p c_p/K))$ and $K = k_-/k_+$. This is a difference equation for $O(t_n)$, which can be solved by setting $O(t_n) = A\alpha^n + B$ and substituting into (8.7), from which we find that

$$\frac{O(t_n)}{O(t_1)} = \frac{1-\alpha^n}{1-\alpha}. \tag{8.8}$$

Note that $\alpha \to 0$ as the interpulse interval gets large (i.e., as $t_I \to \infty$), so that $O(t_n)$ is independent of n. On the other hand, α increases if the Ca^{2+} pulses are shortened (t_p is decreased).

Now we define facilitation as the ratio

$$F_n = \frac{P(t_n)}{P(t_1)}, \tag{8.9}$$

and find that

$$F_n = \left(\frac{1-\alpha_1^n}{1-\alpha_1}\right)\left(\frac{1-\alpha_2^n}{1-\alpha_2}\right)\left(\frac{1-\alpha_3^n}{1-\alpha_3}\right)\left(\frac{1-\alpha_4^n}{1-\alpha_4}\right), \tag{8.10}$$

where α_j is the α corresponding to gate j. Since α_4 is nearly zero in the physiologically relevant range of frequencies, it can be safely ignored. A plot of F_n against the pulse train frequency shows a steplike function, as is observed experimentally. In Fig. 8.1 is shown the maximal facilitation,

$$F_{\max} = \lim_{n\to\infty} F_n = \left(\frac{1}{1-\alpha_1}\right)\left(\frac{1}{1-\alpha_2}\right)\left(\frac{1}{1-\alpha_3}\right), \tag{8.11}$$

which also has a steplike appearance.

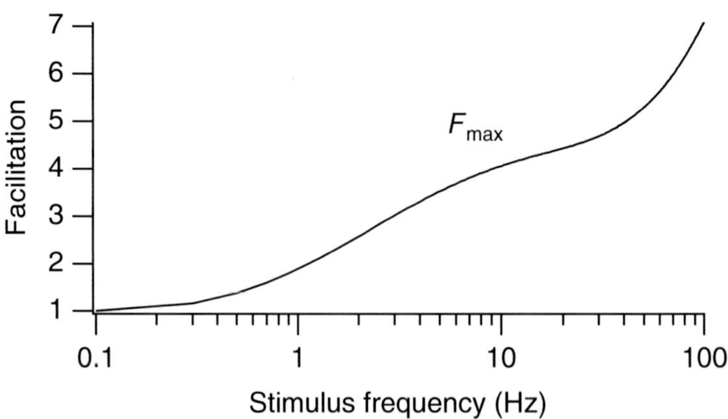

Fig. 8.1 Facilitation as a function of stimulus frequency in the binding model of synaptic facilitation, calculated using $t_p c_p = 200\,\mu$M ms. Here $F_{\max} = \lim_{n\to\infty} F_n$ is the maximal facilitation produced by a pulse train.

8.2.3 A More Complex Version

A more complex model of the residual bound Ca^{2+} hypothesis takes into account diffusion of Ca^{2+} from the channel to the binding site and the presence of Ca^{2+} buffers (Matveev et al, 2006).

We assume that each of n Ca^{2+} channels is located at a discrete position r_j, $j = 1, \ldots, n$, and that there are two Ca^{2+} buffers, one with slower kinetics. Thus, following the discussion of buffers and discrete Ca^{2+} release in Sections 2.9 and 3.5.1, we get

$$\frac{\partial c}{\partial t} = D_c \nabla^2 c + \sum_{i=1}^{2} [k_{-,i} b_i - k_{+,i} c(b_{t,i} - b_i)] + \frac{1}{2F} I_{\mathrm{Ca}} \sum_{j=1}^{n} \delta(r - r_j), \tag{8.12}$$

$$\frac{\partial b_i}{\partial t} = D_b \nabla^2 b_i + \sum_{i=1}^{2} [k_{-,i} b_i - k_{+,i} c(b_{t,i} - b_i)]. \tag{8.13}$$

These equations are solved numerically on the domain illustrated in Fig. 8.2. Calcium channels are clustered into active zones, with 16 channels per zone, and the symmetry of the domain around each active zone is used to reduce the problem to one on the quarter box bordered by the dashed lines. In each quarter box there is a single Ca^{2+} binding domain, denoted by the filled circle, which contains multiple Ca^{2+} binding sites, all colocalised. Calcium is pumped out only on the top and bottom surfaces (shaded), while the other surfaces have no-flux boundary conditions, which assumes a regular array of active sites. Calcium pumping is assumed to be a simple saturating mechanism, with Hill coefficient 1, and is balanced by a constant leak into the cell. Thus, on the shaded boundaries,

$$\nabla c \cdot \mathbf{n} = J_{\mathrm{leak}} - \frac{V_p c}{K_p + c}, \tag{8.14}$$

where $\mathbf{n}$ is the normal to the boundary.

The binding domain is assumed to contain two different types of binding sites, one faster (type X), one slower (type Y). The X binding site can bind two Ca^{2+} ions, and thus

$$\mathrm{X} \underset{k_{-x}}{\overset{2ck_x}{\rightleftharpoons}} \mathrm{CaX} \underset{2k_{-x}}{\overset{ck_x}{\rightleftharpoons}} \mathrm{Ca_2X}, \tag{8.15}$$

while binding sites Y_1 and Y_2 can bind one Ca^{2+} ion each, and thus

$$\mathrm{Y_1} \underset{k_{-1}}{\overset{ck_1}{\rightleftharpoons}} \mathrm{CaY_1}, \tag{8.16}$$

$$\mathrm{Y_2} \underset{k_{-2}}{\overset{ck_2}{\rightleftharpoons}} \mathrm{CaY_2}. \tag{8.17}$$

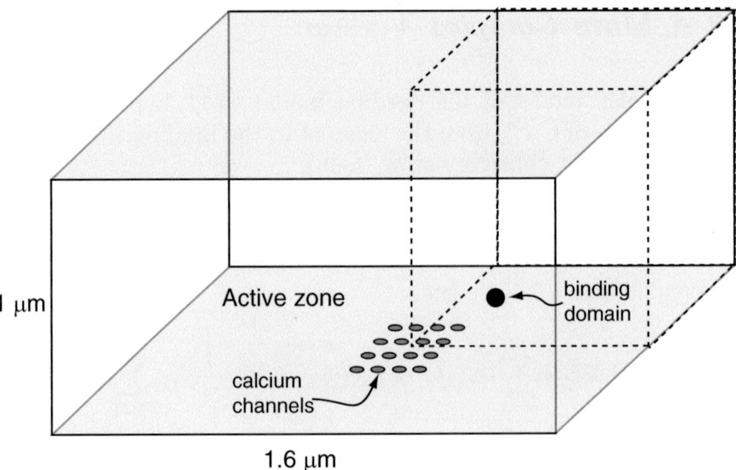

Fig. 8.2 Schematic diagram of a pre-synaptic active zone, the Ca^{2+} channels, and the Ca^{2+} binding site (Matveev et al, 2006). The binding domain is situated 130 nm away from the centre of the active zone, and about 55 nm from the nearest Ca^{2+} channel.

If R denotes the rate of release of neurotransmitter, then the rate of increase of R is assumed to be proportional to $[\mathrm{Ca_2X}][\mathrm{CaY_1}][\mathrm{CaY_2}]$:

$$\frac{dR}{dt} = k_R[\mathrm{Ca_2X}][\mathrm{CaY_1}][\mathrm{CaY_2}] - k_I R. \tag{8.18}$$

Significantly, model results show that the release of an exogenous buffer by a UV flash does indeed decrease the synaptic response within milliseconds (Matveev et al, 2006). Thus, the experimental results of Kamiya and Zucker (1994), which have often been taken as evidence that the residual bound Ca^{2+} hypothesis is incorrect, show no such thing. Although we cannot conclude that the residual bound Ca^{2+} hypothesis is correct, neither can it be rejected on the basis of such experiments. However, what is clear is that there is a complex interplay between free and bound Ca^{2+}, with both playing important roles in the control of facilitation.

8.3 Post-Synaptic Plasticity

In the post-synaptic neuron Ca^{2+} plays a major role in the control of synaptic plasticity, including long-term potentiation (LTP), long-term depression (LTD), and spike-timing-dependent plasticity (STDP) (Abbott and Nelson, 2000; Graupner and Brunel, 2010; Bush and Jin, 2012; Graupner and Brunel, 2012).

Post-synaptic plasticity is the change in synaptic strength caused by near-simultaneous arrival at a dendritic spine of a stimulus from the pre-synaptic neuron (an EPSP, or excitatory post-synaptic potential), and a back-propagating action potential in the post-synaptic neuron. In the classical view, if the EPSP occurs before the back-propagating action potential, the strength of the synapse is increased (LTP), while if the post-synaptic neuron spikes just before the EPSP occurs, then the synaptic strength is decreased (LTD). If the EPSP and back-propagating action potential do not occur close to one another, no changes in synaptic strength occur. Such changes in synaptic strength are believed to be the neural mechanisms that underlie memory, and can be maintained for very long times.

However, although this classical view of STDP captures a great deal of behaviour, there is a wide range of different patterns of STDP. In some neurons the effect of ordering is reversed, while in other neurons the temporal sequence has no effect, only the time between the two stimuli (Abbott and Nelson, 2000).

Studies on individual synapses show that changes in synaptic strength are all-or-nothing, and that gradations in strength only appear when the activity of a large population of neurons is averaged. There is thus a great deal of interest in identifying the biochemical mechanisms that underlie such bistable behaviour, whereby a single synapse can be switched from a stable, low-strength, state to a stable, high-strength, state. In almost every synapse, activation of the synapse leads to an increase in $[Ca^{2+}]$ in the post-synaptic neuron. This can be via Ca^{2+} influx through NMDA receptors, or via Ca^{2+} influx through voltage-gated channels, which open in response to depolarisation. These changes in post-synaptic $[Ca^{2+}]$ are both necessary and sufficient for the bistable nature of synaptic strength (Graupner and Brunel, 2010), and there is considerable evidence that Ca^{2+} is working via the activation of multiple pathways, including Ca^{2+}/calmodulin-dependent protein kinase II (CaM kinase II, or CaMKII), calcineurin, and protein kinase A (PKA). This network controls the number of activated NMDA receptors, both via direct phosphorylation and by modifying the rate of protein synthesis (see Fig. 4 of Graupner and Brunel (2010)). Here we focus on a single component of this network, CaMKII.

8.3.1 Calcium/Calmodulin-Dependent Protein Kinase II as a Bistable Switch

An important component of the network that controls synaptic strength is the bistable behaviour of CaMKII. The initial step leading to phosphorylation (and thus activation) of CaMKII is the binding of Ca^{2+} to the protein calmodulin. Calmodulin is activated by four Ca^{2+} ions, and is one of the common pathways for Ca^{2+} to act as a second messenger, as many other cellular proteins can interact with activated calmodulin, but not with Ca^{2+} directly. (We saw another example of this in Section 7.5.2, where Ca^{2+}/calmodulin activates MLCK, leading to myosin phosphorylation and the development of force in smooth muscle.) Activated calmodulin binds to CaMKII, which allows autophosphorylation to begin, a process in which one subunit of CaMKII catalyses the phosphorylation of neighbouring subunits. This leads to a positive feedback process whereby a small initial stimulus can lead to a large and maintained activation of CaMKII. In principle, this is similar to any generic excitable system with a threshold. However, there are additional complexities associated with the autophosphorylation process that lead to greater dynamic complexity, with significant implications for the interpretation of experimental data.

There have been a number of models of CaMKII, with particular reference to its importance for post-synaptic plasticity. The earliest model is that of Lisman (1985), who considered the case of an autophosphorylating protein, where the autophosphorylation is not intramolecular (as it is with CaMKII), but intermolecular.

To show bistability in this case, we begin by deriving the rate at which autophosphorylation proceeds. Consider a substrate, S, that reacts with an enzyme, P, to form the product, P. Thus P is both the product and the catalysing enzyme. Following the usual Michaelis-Menten approach, and assuming the existence of an intermediate complex, we can write this reaction as

$$\mathrm{S} + \mathrm{P} \underset{k_{-1}}{\overset{k_1}{\rightleftharpoons}} \mathrm{C} \overset{k_2}{\longrightarrow} 2\mathrm{P}. \tag{8.19}$$

Assuming, as usual, that C is at quasi-steady state, and thus $dC/dt = 0$, gives

$$C = \frac{k_1 SP}{k_{-1} + k_2}, \tag{8.20}$$

where C, S, and P denote [C], [S], and [P], respectively. This, together with the conservation law

$$S + P + 2C = T, \tag{8.21}$$

where T is the total concentration of the autophosphorylating protein, gives

$$C = \frac{(T - P)P}{K_m + 2P}, \tag{8.22}$$

where $K_m = (k_{-1} + k_2)/k_1$. Hence, the rate of autophosphorylation, V_{phos}, is given by

$$V_{\mathrm{phos}} = 2k_2 C = \frac{2k_2(T - P)P}{K_m + 2P}. \tag{8.23}$$

Note how $V_{\mathrm{phos}} = 0$ when $P = 0$ (as there is no enzyme), but also when $P = T$ (as there is no substrate).

We now combine an autophosphorylation step with a dephosphorylation step to obtain a bistable system. Suppose there is another enzyme that dephosphorylates P back to S. Again according to standard Michaelis-Menten theory, the rate, V_{dephos} of this dephosphorylation will be

$$V_{\text{dephos}} = \frac{V_{\max} P}{K_{\text{dephos}} + P}. \tag{8.24}$$

Finally, the net rate at which S is converted to P is

$$\frac{dP}{dt} = V_{\text{phos}} - V_{\text{dephos}} = \frac{2k_2(T-P)P}{K_m + 2P} - \frac{V_{\max} P}{K_{\text{dephos}} + P}. \tag{8.25}$$

A simple qualitative argument now suffices to demonstrate bistability in this model. Steady states occur when $V_{\text{phos}} = V_{\text{dephos}}$. These rates are sketched as functions of P in Fig. 8.3 A. Since V_{phos} is a quadratic and V_{dephos} is a monotonically increasing hyperbolic curve, it follows that, when $V_{\max}$ is large there is a single intersection point at $P = 0$, when $V_{\max}$ is small there are two intersection points (one at $P = 0$ the other close to $P = T$), and when $V_{\max}$ has intermediate values there can be three intersection points (for some, but not all, parameters). The middle intersection point is unstable, but the upper and lower ones are stable. Thus, this model can have a region of bistability. These intersection points are sketched as functions of $V_{\max}$ (Fig. 8.3 B).

This simple model of autophosphorylation is not directly applicable to CaMKII, as it models a reaction where the autophosphorylation is intermolecular, rather than intramolecular, as is the case for CaMKII. Thus a

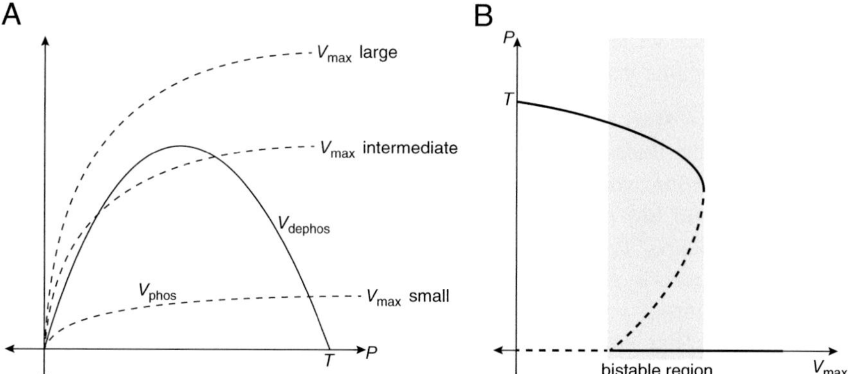

Fig. 8.3 Schematic diagram of the steady states of the model of Lisman (1985). **A:** plots of V_{phos} (solid curve) and V_{dephos} (dashed curves). Three different V_{dephos} curves are shown, for three different values of $V_{\max}$. Intersections of these curves are steady states of the model. **B:** bifurcation diagram showing the steady states of the model as functions of $V_{\max}$. Stable steady states are solid lines, unstable steady states are dashed lines. For intermediate values of $V_{\max}$ there is a region of bistability. This does not occur for all parameter values in the model, only for some.

number of groups have constructed more detailed models of CaMKII autophosphorylation (Okamoto and Ichikawa, 2000; Zhabotinsky, 2000; Lisman and Zhabotinsky, 2001; Kubota and Bower, 2001; Graupner and Brunel, 2007; Earnshaw and Bressloff, 2010; Graupner and Brunel, 2012; Michalski, 2013). Discussion of these models is rather outside the scope of this book, as they have less to do with Ca^{2+} dynamics, and more to do with complex intramolecular autophosphorylation. Nevertheless, a brief discussion is useful.

CaMKII consists of 12 subunits grouped into two groups of six functionally connected subunits. Each subunit can bind Ca^{2+}/calmodulin and can be phosphorylated, and can thus (in the simplest version of the model) exist in one of four states. Additional possible complexities are discussed in Section 8.3.3. For convenience we label these states as

- S: the basic subunit, neither phosphorylated, nor with Ca^{2+}/CaM bound.
- S-CaM: the subunit with Ca^{2+}/CaM bound.
- S-CaM-P: the phosphorylated subunit with Ca^{2+}/CaM bound.
- S-P: the phosphorylated subunit.

Autophosphorylation and bistability result from the following two facts. Firstly, the only subunit state that can be phosphorylated is S-CaM. That is, a subunit can be phosphorylated only when it is bound to CaM. Secondly, phosphorylation can be catalysed by any of the states S-P, S-CaM, or S-CaM-P. However, phosphorylation cannot be catalysed by a subunit in the state S. Due to spatial constraints, a subunit (in one of the states S-P, S-CaM, or S-CaM-P) can only phosphorylate a neighbouring subunit (in state S-CaM), and thus there are immediate combinatorial complexities, as, for a complete model, one must keep track of each subunit individually. That is, one cannot calculate overall reaction rates merely by counting the number of subunits in any particular state; their position on the ring is also crucial.

Even without writing down a mathematical model we can draw some qualitative conclusions from the above properties.

- Suppose we start with every subunit in the state S, with no catalytic activity, and unable to be phosphorylated. Suppose now that $[Ca^{2+}]$ is raised to a low value. Then the concentration of CaM is also low, and there will be few subunits in the state S-CaM, with most remaining in state S. Hence, typically, any subunit in state S-CaM will not have a neighbour also in state S-CaM. Since S-CaM is the only state that can be phosphorylated, and since the plain subunit S has no catalytic activity, this means that none of the S-CaM states can be phosphorylated (as they do not have a neighbour with catalytic activity). Thus, small increases in $[Ca^{2+}]$ have no effect.
- As $[Ca^{2+}]$ is raised further, and CaM increases, there will eventually be two neighbouring subunits both in state S-CaM. When this happens, one subunit can now phosphorylate its neighbour, which moves to state S-CaM-P.
- At high $[Ca^{2+}]$, all the subunits will have been phosphorylated by their neighbours, and all the subunits will be in state S-CaM-P.

- If $[Ca^{2+}]$ is now decreased again, many of the subunits will lose their CaM, and move to state S-P. However, S-P retains catalytic activity, and thus even at low $[Ca^{2+}]$ the CaMKII remains fully phosphorylated and active.

One can therefore see, in a qualitative fashion, how bistability arises. Small increases in $[Ca^{2+}]$ have no effect if the CaMKII is inactive. However, once a threshold is crossed, and there are enough neighbouring subunits both with CaM bound, then they start catalysing the phosphorylation of their neighbours. Once phosphorylated, a subunit retains its catalytic activity even when CaM unbinds, and thus the CaMKII remains in an activated state even after Ca^{2+} decreases.

A model of CaMKII autophosphorylation was proposed by Okamoto and Ichikawa (2000). This model keeps track only of which subunits are phosphorylated, and incorporates the effects of CaM indirectly via the modification of the phosphorylation rate. Since it does not track the positions of each subunit, it cannot take into account that fact that only neighbouring subunits can interact. However, despite these simplifying assumptions, it still gives a good example of the basic structure of CaMKII autophosphorylating models.

Let S_n denote the CaMKII enzyme with n phosphorylated subunits and $N - n$ unphosphorylated subunits. In other words, an enzyme in state S_n has n subunits in state S-P and $N - n$ subunits in state S. Then the basic stepwise autophosphorylation scheme can be described as

$$S_n \xrightarrow{G(n)} S_{n+1}, \qquad n = 0, \cdots, N-1. \tag{8.26}$$

There are three ways in which S_n can be converted to S_{n+1}.

(a) A subunit in state S-CaM can phosphorylate another subunit (which must be in state S-CaM). Call this rate $G_a(n)$.
(b) A subunit in state S-P-CaM can phosphorylate another subunit (which must be in state S-CaM). Call this rate $G_b(n)$.
(c) A subunit in state S-P can phosphorylate another subunit (which must be in state S-CaM). Call this rate $G_c(n)$.

Thus

$$G(n) = G_a(n) + G_b(n) + G_c(n). \tag{8.27}$$

We now derive simplistic models for these rates. First, for simplicity assume that all pairwise subunit interactions proceed at the same rate, k. Also, suppose that γ is the probability per unit time that CaM binds to a subunit in state S, and that γ^* is the probability per unit time that CaM binds to a subunit in state S-P. Note that γ and γ^* need not be different, although in general they will be.

As an example, consider case (b) of (8.26). For any given two subunits, one in state S, the other in state S-P, the rate at which this reaction proceeds is $k\gamma\gamma^*$, as each subunit must first bind CaM and then the two subunits react at the rate k. However, given a population of N subunits, n of which are in state S-P and $N-n$ of which are in state S, there are $n(N-n)$ ways to select two subunits, one in state S, the other in state S-P. Thus,

$$G_b(n) = k\gamma\gamma^* n(N-n). \tag{8.28}$$

Using similar arguments it follows that

$$G_a(n) = k\gamma^2(N-n)(N-n-1), \tag{8.29}$$

$$G_c(n) = k\gamma(1-\gamma^*)n(N-n), \tag{8.30}$$

and thus

$$G(n) = k\big[\gamma^2(N-n)(N-n-1) + \gamma n(N-n)\big]. \tag{8.31}$$

The terms involving γ^* cancel out, as a result of the assumption that all the reactions proceed at the same rate.

To model the Ca^{2+}-dependent activation of CaMKII we need an expression for γ in terms of $[Ca^{2+}]$. For this, we simply assume that the rate at which a subunit in state S reacts with activated CaM is a saturating increasing function of the concentration of activated CaM, which we denote by M. Thus

$$\gamma = \frac{M}{M+K_1}, \tag{8.32}$$

for some K_1. Note how γ has been scaled to have a maximum value of 1, and thus can be interpreted as a probability of reaction per unit time. If we further assume that the reaction of CaM with Ca^{2+} is at equilibrium, then we know that

$$M = \frac{M_{\text{total}}c^4}{c^4+K_2}, \tag{8.33}$$

for some K_2, whereupon we now have γ as a function of c (which, as usual, denotes $[Ca^{2+}]$).

To complete the model construction we now need to model the dephosphorylation steps, which can be written as

$$\mathrm{S}_n \xrightarrow{R(n)} \mathrm{S}_{n-1}, \qquad n = 1, \cdots, N, \tag{8.34}$$

for some rate $R(n)$. The rate of dephosphorylation of a single subunit is not immediately obvious, as all the other phosphorylated subunits act as competitive inhibitors. To calculate this rate, we first remind ourselves of the kinetics of competitive inhibitors.

Suppose we have an enzyme with two substrates; substrate 1 (with concentration s_1 and Michaelis constant K_1) and substrate 2 (with concentration s_2 and Michaelis constant K_2). Then it is a standard result that the rate, V_1, of the reaction of substrate 1 is given by

$$V_1 = \frac{V_{\max,1} s_1}{s_1 + K_1 \left(1 + \frac{s_2}{K_2}\right)}. \tag{8.35}$$

Thus, a competitive inhibitor works by shifting the effective K_m of the reaction while leaving the maximum velocity unchanged. In the special case that $K_1 = K_2$, then

$$V_1 = \frac{V_{\max,1} s_1}{K_1 + s_1 + s_2}. \tag{8.36}$$

If there are multiple competitive substrates, s_1 through s_N, all with the same Michaelis constant, this result generalises simply to

$$V_1 = \frac{V_{\max,1} s_1}{K_1 + \sum_{i=1}^N s_i}. \tag{8.37}$$

A useful interpretation of this equation is that the rate of reaction per unit of substrate 1 is $\frac{V_{\max,1}}{K_1 + \sum_{i=1}^N s_i}$. For a fixed amount of dephosphorylating enzyme, the rate of dephosphorylation per unit of substrate 1 decreases as the total amount of available substrate increases, and thus the competition increases.

Returning now to the dephosphorylation of CaMKII we see that, for any phosphorylated subunit, the competitive inhibitors are all the other phosphorylated subunits on all the other CaMKII enzymes, and thus the total concentration of phosphorylated subunits is $\sum_{n=0}^N n[\mathrm{S}_n]$. Hence, if we let k_D denote the rate of dephosphorylation per subunit, it follows that

$$k_D = \frac{V_D}{K_D + \sum_{n=0}^N n S_n}, \tag{8.38}$$

for some constants V_D and K_D, and where $S_n = [\mathrm{S}_n]$.

From this it follows that $R(n)$ in (8.34) is given by

$$R(n) = n k_D = \frac{n V_D}{K_D + \sum_{n=0}^N n S_n}. \tag{8.39}$$

This follows since S_n has n subunits in state S-P, and is thus dephosphorylated at n times the rate of an individual (phosphorylated) subunit.

In summary, the model of CaMKII autophosphorylation consists of the following equations:

$$\frac{dS_0}{dt} = R(1)S_1 - [R(0) + G(0)]S_0, \tag{8.40}$$

$$\frac{dS_n}{dt} = R(n+1)S_{n+1} - [R(n) + G(n)]S_n + G(n-1)S_{n-1}, \quad n = 1, \ldots, N-1, \tag{8.41}$$

$$\frac{dS_N}{dt} = G(N-1)S_{N-1} - [R(N) + G(N)]S_N, \tag{8.42}$$

where R and G are defined by (8.39) and (8.31).

Qualitatively, the model results are the same as shown in Fig. 8.3 B. There is a region of bistability where, for a given physiologically reasonable $[\mathrm{Ca}^{2+}]$, CaMKII can exist in either a stable inactive, or a stable active state. A small increase in $[\mathrm{Ca}^{2+}]$ from the resting state will lead to little change in the phosphorylation state, but a large enough change will cause CaMKII to switch to the active high phosphorylation state and remain there, even when $[\mathrm{Ca}^{2+}]$ is decreased back to the resting level.

This basic model was modified by Graupner and Brunel (2007) in order to take better into account the fact that autophosphorylation can occur only between neighbouring subunits. However, although the complexities are thus increased, the basic model behaviour remains unchanged. Similarly, the models of Zhabotinsky (2000) and Michalski (2013) have increased complexity, but little qualitative difference in behaviour.

One major difficulty with the model is that it is unable to reproduce some important experimental results. It is well established that high-frequency Ca^{2+} pulses cause LTP (which is reproduced well by the bifurcation diagram in Fig. 8.3), but also that low-frequency pulses of Ca^{2+} cause LTD (which is not reproduced at all by the above model). Graupner and Brunel (2007) included Ca^{2+}-dependent phosphatase activity, via Ca^{2+}-dependent activation of calcineurin and PKA, to add additional complexity to the bifurcation diagram in such a way as to explain a wider range of experimental results. However, many questions remain. It is most likely that the full range of LTP, LTD, and STDP will only be explained by post-synaptic models that incorporate the full range of Ca^{2+}-dependent pathways. The thought of such a model is daunting.

8.3.2 Phenomenological Models

There are other models of Ca^{2+}-dependent post-synaptic plasticity that do not include explicit models of CaMKII. Typically, such models include a phenomenological model of a Ca^{2+} detector that translates post-synaptic $[\mathrm{Ca}^{2+}]$

into synaptic strength. Models of this type include Rubin et al (2005), Rackham et al (2010) and Bush and Jin (2012). Although these are all fascinating dynamic models, there is not enough space here to discuss them in any detail, and so interested readers are referred to the original papers. The review by Graupner and Brunel (2010) covers many such models, and explains well their different approaches and assumptions.

8.3.3 CaMKII as a Frequency Decoder in the Absence of Dephosphorylation

Besides exhibiting a switch-like behaviour, CaMKII activity is also sensitive to the frequency of the Ca^{2+} spikes, even in the absence of phosphatases. This was demonstrated by De Koninck and Schulman (1998) who developed an experimental setup to investigate how the activity of CaMKII depends on the characteristics of high-frequency Ca^{2+} oscillations. The enzyme was immobilised in PVC tubing, and a pulsatile flow allowed rapid changes of the perfusion solution. Square-wave pulses of Ca^{2+}, simulating oscillations, were generated by alternate perfusion with either Ca^{2+}/CaM or EGTA. Autonomous kinase activity (i.e., in the absence of Ca^{2+}/CaM) was assayed after a variable number of pulses at frequencies ranging from 0.1 to 1 Hz. In other words, the Ca^{2+} spikes pushed the CaMKII molecule into a state where it remains active even once the Ca^{2+} is removed, and the amount of remaining activity was measured as a function of the Ca^{2+} spike frequency.

The main result of this study is that autonomous CaMKII activity increases in a roughly exponential manner with the frequency of Ca^{2+} spikes when the calmodulin concentration is around 100 nM, a relatively low value which, when applied constantly, yields 20% of the maximum autonomous activity. When the duration of each Ca^{2+} pulse is fixed at 200 ms, the frequency leading to half-maximal autonomous activity after 100 pulses is about 2.5 Hz. If CaMKII is exposed to lower pulse frequencies it fails to integrate them, and its level of autonomous activity remains low (i.e., little activity remains when Ca^{2+} is removed).

In the framework of the bistable behaviour of CaMKII, one could imagine that successive spikes could activate the kinase in an incremental manner and turn on the switch if the timing of stimulation is appropriate. However, bistability cannot be the correct explanation, because frequency sensitivity occurs in the absence of phosphatase. Furthermore, the CaMKII activities that are measured experimentally at different frequencies of stimulation range over the whole possible scale, thus excluding a switch-like behaviour.

To account for such frequency sensitivity, one needs to describe the mode of activation of CaMKII in more detail. In particular, states S-CAM, S-CAM-P, and S-P described above have different levels of kinase activity. Not only this, but an S-CaM-P subunit even has different behaviour depending on whether, in this complex, Ca^{2+} is bound to CaM or not. In the following, we will thus distinguish CaM* (with concentration M^*) that represents CaM bound to four Ca^{2+} ions, from CaM (with concentration M) that has no Ca^{2+} bound. The transitions between the different possible states for each subunit are sketched in Fig. 8.4 (Dupont et al, 2003).

Dividing all subunit concentrations by the total concentration of subunits (i.e., 12 $[\text{CaMKII}]_{\text{tot}}$), one can describe the model by four equations for the concentrations of states S, S_α, S_β, and S_γ. The concentration of state S_δ is then given by the conservation relation. Reversible binding of CaM* to a basic subunit is described by the mass-action law

$$\frac{dS}{dt} = -k_{+1}SM^* + k_{-1}S_\alpha, \tag{8.43}$$

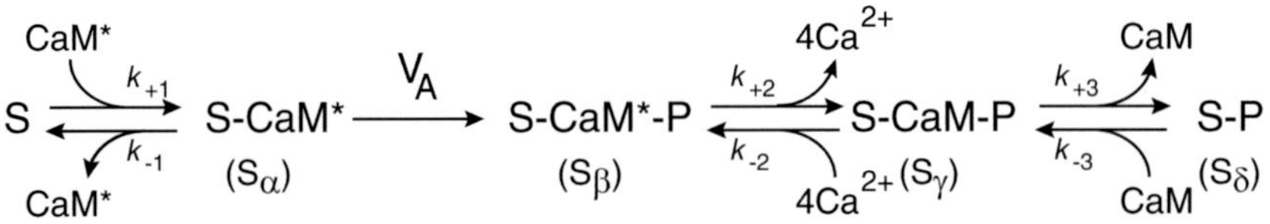

Fig. 8.4 Schematic representation of the mode of activation of a CaMKII subunit (S) by the Ca^{2+}/CaM complex. CaM denotes calmodulin and CaM* denotes the Ca^{2+}/CaM complex. The S_i are the names of the subunits in each state. V_A is the autophosphorylation rate, so called because this reaction occurs between two adjacent subunits of the ring-shaped CaMKII. This requirement is included phenomenologically in the model via (8.47). To mimic experiments in a phosphatase-free medium, as in the experiments by De Koninck and Schulman (1998), autophosphorylation is assumed to be irreversible.

where

$$M^* = M_{\text{tot}}\frac{c^4}{K^4 + c^4} \tag{8.44}$$

with K the dissociation constant between Ca^{2+} and CaM.

For a subunit bound to CaM*,

$$\frac{dS_\alpha}{dt} = k_{+1}SM^* - k_{-1}S_\alpha - V_A. \tag{8.45}$$

S_α can be phosphorylated either by another S_α, or by any phosphorylated subunit. Thus,

$$V_A = k_A\left[2(\alpha S_\alpha)^2 + (\alpha S_\alpha)(\beta S_\beta) + (\alpha S_\alpha)(\gamma S_\gamma) + (\alpha S_\alpha)(\delta S_\delta)\right]. \tag{8.46}$$

α, β, γ, and δ represent the coefficients of kinase activity, with values 0.75, 1, 0.8, and 0.8, respectively (Hudmon and Schulman, 2002). The factor 2 in the first term accounts for the fact that autophosphorylation is bidirectional. Thus, if two adjacent subunits are in state S_α, both become phosphorylated.

As such, (8.46) does not consider that subunits have to be neighbours within a holoenzyme for phosphorylation to occur. To incorporate this requirement, k_A is made dependent on the total fraction of active subunits ($T = S_\alpha + S_\beta + S_\gamma + S_\delta$): if T is low, the probability of an arbitrarily selected CaM*-bound subunit to be adjacent to an active subunit is very low, and thus k_A must be small. The probability increases with T in a nonlinear manner as the active subunits progressively fill the ring-shaped CaMKII. Among other possibilities, such conditions are qualitatively satisfied by a cubic function of T such as

$$k_A = k'_A(aT + bT^2 + cT^3). \tag{8.47}$$

This empirical function greatly simplifies the model, as the location of the subunits does not need to be considered, which would increase the number of variables by a factor of 12. Parameters k'_A, a, b, and c are determined by fitting to experimental data.

Once phosphorylated, a subunit can release Ca^{2+} but keep CaM, as autophosphorylation greatly increases the affinity of a CaMKII subunit for CaM. This phenomenon, characteristic of CaMKII, is known as "CaM trapping". Thus,

$$\frac{dS_\beta}{dt} = V_A - k_{+2}S_\beta + k_{-2}c^4 S_\gamma. \tag{8.48}$$

The exponent 4 accounts for the binding of four Ca^{2+} ions to a CaM molecule.

Finally, CaM can dissociate, and thus

$$\frac{dS_\gamma}{dt} = k_{+2}S_\beta - k_{-2}c^4 S_\gamma + k_{-3}MS_\delta - k_{+3}S_\gamma. \tag{8.49}$$

Because of trapping, k_{+3} is 1000 times smaller than k_{-1}.

The activating protocol corresponds to square-wave pulses of CaM*, each protocol being characterised by the number, amplitude, and frequency of the pulses. These Ca^{2+} pulses push a certain fraction of the CaMKII subunits through to states S_β, S_γ, and S_δ. When the Ca^{2+} pulses are removed, the subunit states S_α or S_β both disappear. This is because in the absence of Ca^{2+} there can be no CaM*, and no reverse reaction from S_γ to S_β, and thus, at steady state, there can be no CaMKII subunits in states S_α or S_β. However, S_δ and S_γ represent CaMKII subunit states that remain active in the absence of Ca^{2+}, and thus represent the autonomous activity of the CaMKII.

Results of the model are compared to experimental results in Fig. 8.5. Fig. 8.5 A shows the evolution of autonomous activity when CaMKII is stimulated by Ca^{2+}-CaM spikes at frequencies of 4, 2.5, or 1 Hz. How does the model reproduce this strong dependence of autonomy on Ca^{2+} spike frequency? At each Ca^{2+} spike, a significant portion of subunits becomes activated as the concentration of Ca^{2+}-saturated CaM (i.e., CaM*) is above its K_D for S. At high stimulation frequency (4Hz), dissociation of CaM is very limited between two spikes. In consequence, S_α accumulates and, after a few seconds, autophosphorylation can be initiated because the number of CaM*-bound subunits in close proximity becomes significant. The rate of autophosphorylation then keeps on increasing for two reasons. First, the probability that two active subunits are neighbours increases in a nonlinear way as the number of activated subunits increases. Second, once phosphorylated, a subunit remains active even between spikes (as there is no dephosphorylation). In consequence, the fraction of phosphorylated subunits increases in an autocatalytic manner. In contrast, at low frequency (1 Hz), nearly all CaM* dissociates from CaMKII between two spikes, in which case the number of bound subunits S_α always remains below the value necessary for significant autophosphorylation.

The shape of the frequency-activity response largely depends on the spike amplitude and duration, and this response becomes less steep if the width of the spike is increased (Fig. 8.5 B). For example, a spike of width 1000 ms provides enough CaM* to generate some significant autophosphorylation. In contrast, when the spikes are very brief (200 or 80 ms), shorter interspike intervals are required to allow for accumulation of bound subunits sufficient for autophosphorylation. Therefore, the probability of simultaneous binding of CaM to neighbouring subunits determines the threshold frequency required

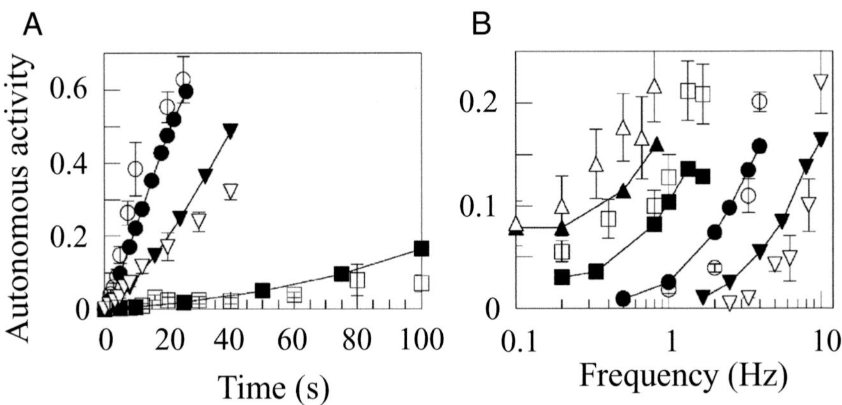

Fig. 8.5 Comparison of the model to experimental results (De Koninck and Schulman, 1998). **A:** activity of CaMKII after repetitive pulses of Ca^{2+} (500 μM) and CaM (100 nM), with duration 200 ms and frequency equal to 1 (squares), 2.5 (triangles), and 4 Hz (circles). **B:** the effect of changing the duration of the CaM* pulse (triangles: 1000 ms, squares: 500 ms, circles: 200 ms, inverted triangles: 80 ms). In each case, the total exposure time of the kinase to the activators was 6 s (500 μM Ca^{2+}, 100 nM CaM). In both panels, filled symbols have been obtained by numerical simulation of the model with parameters from Dupont et al (2003), while open symbols represent the experimental results reproduced from De Koninck and Schulman (1998). Reproduced from Dupont et al (2003), Fig. 3C and E, with permission from Elsevier.

for significant autophosphorylation: the lower the probability, the higher the threshold.

In agreement with experimental data, the model thus indicates that CaMKII can act as a decoder in a range of frequencies of 1–5 Hz. In the context of synaptic plasticity, Ca^{2+} spikes following an action potential are very brief (a few ms in width), yet can locally reach very high amplitudes, such as in dendritic spines. Simulations indicate that, in these ranges, CaMKII is still able to decode frequencies larger than 5 Hz. However, CaMKII is not able to decode low-frequency Ca^{2+} oscillations (< 0.5 Hz), whatever the durations and amplitudes of the spikes. Thus, it cannot serve as a frequency decoder for Ca^{2+} oscillations in electrically nonexcitable cells, where the periods are most often on the time scale of tens of seconds, or even minutes. However, cellular responses in these cells are often still sensitive to the frequency of Ca^{2+} oscillations. There is no evidence for a molecular decoder in such cases. Instead, the interplay between Ca^{2+}-activated kinases and phosphatases has been suggested to provide a plausible and robust frequency decoder (Goldbeter et al, 1990; Pahle et al, 2008; Salazar et al, 2008), as it does in the case of airway smooth muscle (Section 7.5.3).

8.4 Pancreatic Beta Cells

In response to glucose, β cells of the pancreatic islet secrete insulin, which causes the increased use or uptake of glucose in target tissues such as muscle, liver, and adipose tissue. When blood levels of glucose decline, insulin secretion stops, and the tissues begin to use their energy stores instead. Interruption of this control system results in diabetes, a disease that, if left uncontrolled, can result in kidney failure, heart disease, and death. The immediate controller of insulin secretion from pancreatic β cells is the intracellular $[Ca^{2+}]$, which activates exocytosis. $[Ca^{2+}]$ is itself raised principally by the influx of Ca^{2+} through voltage-gated Ca^{2+} channels, which occurs during periods when the β cell membrane potential is spiking. Such electrical spikes occur in bursts, separated by quiescent periods, a behaviour which is called *electrical bursting*, or simply bursting. A typical example is shown in Fig. 8.6, whence it can be seen that, on average, $[Ca^{2+}]$ rises during the bursting phase but decreases during the quiescent phase.

Although the secretion of insulin occurs periodically, with a range of different periods, and there is yet no consensus on the mechanisms underlying these oscillatory processes, it is believed that electrical bursting plays an important (but not exclusive) role. The review by Félix-Martínez and Godínez-Fernández (2014) is a comprehensive and readable introduction to the array of β cell models in the literature, while Fridlyand et al (2010) is a more physiologically based review with some quantitative aspects.

Pancreatic β cells are one of the earliest cell types for which the interaction of a membrane oscillator with cytosolic Ca^{2+} dynamics was investigated in detail (Chay and Keizer, 1983; Rinzel, 1985; Rinzel and Lee, 1987). Although many of the details of this early work have been revised in the light of further experimental work, the theory of these early models still provides an excellent example of how such models can be analysed, and the types of behaviour that can be expected.

8.4.1 Bursting in the Pancreatic Beta Cell

Although bursting has been studied extensively for many years, most mathematical studies are based on the pioneering work of Rinzel (Rinzel, 1985; Rinzel and Lee, 1987), which was in turn based on one of the first biophysical models of a pancreatic β cell (Chay and Keizer, 1983). Rinzel's interpretation of bursting in terms of nonlinear dynamics provides an excellent example of how mathematics can be used to understand the interaction between membrane oscillators and their cytosolic controllers.

Models of bursting in pancreatic β cells can be divided into two major groups. Earlier models were generally based on the assumption that bursting was caused by an underlying slow oscillation in the intracellular$[Ca^{2+}]$

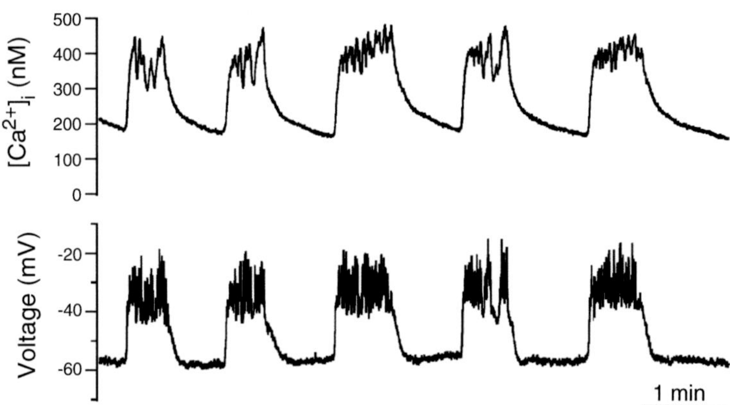

Fig. 8.6 Simultaneous Ca^{2+} and voltage measurements from a bursting pancreatic β cell. Reproduced from Zhang et al (2003) Fig. 4, with permission from Elsevier.

(Chay, 1986, 1987; Chay and Cook, 1988; Himmel and Chay, 1987; Keizer and Magnus, 1989). However, in light of experimental evidence showing that Ca^{2+} is not the slow variable underlying bursting, more recent models have modified this assumption, relying on alternative mechanisms to produce the underlying slow oscillation (Keizer and Smolen, 1991; Smolen and Keizer, 1992; Wierschem and Bertram, 2004; Bertram and Sherman, 2004; Bertram et al, 2004; Nunemaker, 2006; Tsaneva-Atanasova et al, 2006; Bertram et al, 2007a,b, 2010; Ren et al, 2013).

One of the first models of bursting was proposed by Atwater et al (1980), and was later developed into a mathematical model (Chay and Keizer, 1983) which was able to reproduce many of the basic properties of bursting.

The basic form of the model is that described in general terms in Section 3.7. More specifically, the ionic currents are

1. A Ca^{2+}-activated K^+ channel with conductance an increasing function of $c = [Ca^{2+}]$ of the form

$$g_{K,Ca} = \bar{g}_{K,Ca}\frac{c}{K_d + c}, \tag{8.50}$$

   for some constant $\bar{g}_{K,Ca}$.
2. A voltage-gated K^+ channel modelled in the same way as in the Hodgkin-Huxley equations, with

$$g_K = \bar{g}_K n^4, \tag{8.51}$$

   where n obeys the same differential equation as in the Hodgkin-Huxley equations (Hodgkin and Huxley, 1952; Keener and Sneyd, 2008), except that the voltage is shifted by V^*.
3. A voltage-gated Ca^{2+} channel, with conductance

$$g_{Ca} = \bar{g}_{Ca} m^3 h, \tag{8.52}$$

   where m and h satisfy Hodgkin-Huxley differential equations for Na^+ gating, shifted along the voltage axis by V'.

Combining these ionic currents and adding a leak current, with conductance g_L and Nernst potential V_L, gives

$$C_m\frac{dV}{dt} = -(g_{K,Ca} + g_K)(V - V_K) - 2g_{Ca}(V - V_{Ca}) - g_L(V - V_L), \tag{8.53}$$

where C_m is the membrane capacitance.

To complete the model, the conservation equation for Ca^{2+} is

$$\frac{dc}{dt} = f(-k_1 I_{Ca} - k_c c), \tag{8.54}$$

where the Ca^{2+} current is $I_{Ca} = \bar{g}_{Ca} m^3 h(V - V_{Ca})$ and where k_1 and k_c are constants. The constant f is a buffering scale factor relating total changes in $[Ca^{2+}]$ to the changes in free $[Ca^{2+}]$ (as discussed in Section 2.9) and is usually a small number, while k_c is the rate at which Ca^{2+} is removed from the cytoplasm by the membrane ATPase pump.

For this model it is assumed that glucose regulates the rate of removal of Ca^{2+} from the cytoplasm. Thus, k_c is assumed to be an (unspecified) increasing function of glucose concentration. However, the concentration of glucose is not a dynamic variable of the model, so that k_c can be regarded as fixed, and the behaviour of the model can be studied for a range of values of k_c.

A numerically computed solution of this model, shown in Fig. 8.7, exhibits bursts that bear a qualitative resemblance to those seen experimentally. There is a slow oscillation in c underlying the bursts, with fast oscillations in V occurring during the rising phase of the Ca^{2+} oscillation. The fact that Ca^{2+} oscillations occur on a slower time scale than the voltage is built into the Ca^{2+} equation (8.54) explicitly by means of the parameter f; as f becomes smaller, the Ca^{2+} equation evolves more slowly relative to the voltage. This is the basis of the phase-plane analysis that we describe next.

8.4.1.1 Phase-Plane Analysis

The β cell model can be simplified by ignoring the dynamics of m and h, thus removing the time dependence (but not the voltage dependence) of the Ca^{2+} current (Rinzel and Lee, 1986). The simplified model equations are

$$C_m \frac{dV}{dt} = -I_{\mathrm{Ca}}(V) - \left(\bar{g}_{\mathrm{K}} n^4 + \frac{\bar{g}_{\mathrm{K,Ca}} c}{K_d + c} \right)(V - V_{\mathrm{K}}) - \bar{g}_{\mathrm{L}}(V - V_{\mathrm{L}}), \quad (8.55)$$

$$\tau_n(V) \frac{dn}{dt} = n_\infty(V) - n, \quad (8.56)$$

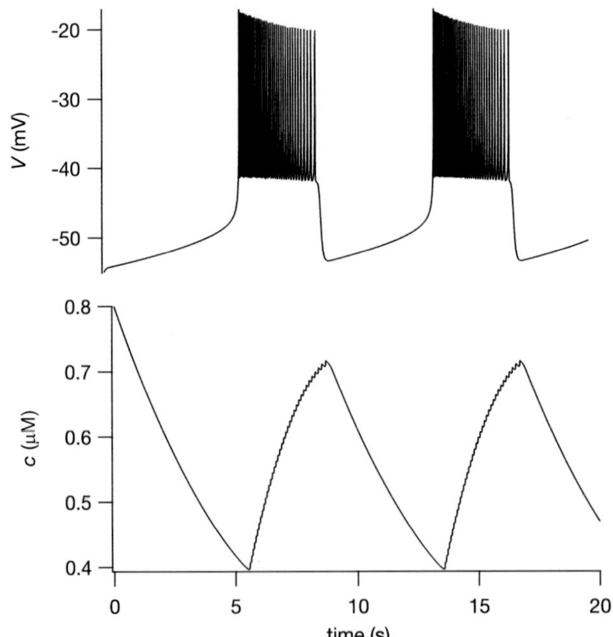

Fig. 8.7 Bursting oscillations in the Chay-Keizer β cell model.

$$\frac{dc}{dt} = f(-k_1 I_{\text{Ca}}(V) - k_c c), \tag{8.57}$$

where $I_{\text{Ca}} = \bar{g}_{\text{Ca}} m_\infty^3(V) h_\infty(V)(V - V_{\text{Ca}})$.

Since f is small, it is common to assume that V and n evolve much faster than c, and hence to separate the model into a fast subsystem (the V and n equations) and a slow equation for c. The fast subsystem is then studied using phase-plane methods, and the behaviour of the full system is explained as a slow variation of the fast subsystem.

This general approach, of separating the model into fast and slow subsystems, is based on ideas from geometric singular perturbation theory, and is discussed at length in Section 5.4. The approach of "freezing" a slow variable, as described in more detail below, is not without some dangers, as the behaviour of complex models can change in unexpected ways when time scales are omitted, but it is still convenient, powerful and widely used.

We first consider the structure of the fast subsystem as a function of c, treating c as a fixed parameter. The sequence of bifurcations can be summarised in a bifurcation diagram, with V plotted against the control parameter c (Fig. 8.8 A). The Z-shaped curve, labelled V_{ss}, is the curve of fixed points, and as usual, the stable limit cycle (see below) around the upper steady state is depicted by the maximum and minimum of V through one cycle.

When c is low, the Ca^{2+}-activated K^+ channel is not activated, and the fast subsystem has a unique fixed point with V high. Conversely, when c is high, the Ca^{2+}-activated K^+ channel is fully activated, and the fast subsystem has a unique fixed point with V low, as the high conductance of the Ca^{2+}-activated K^+ channels pulls the membrane potential closer to the K^+ Nernst potential of about -75 mV. However, for intermediate values of c there are three fixed points, and the phase plane is more intricate.

As c increases from low values, the upper fixed point becomes unstable in a Hopf bifurcation (c_{hb}) and a branch of stable oscillations appears, as illustrated in Fig. 8.8 A. These oscillations eventually collide with the fixed points on the middle branch and disappear via a homoclinic bifurcation (c_{hc}). Thus, for a range of values of c, the fast subsystem is bistable, with a lower stable fixed point and an upper stable periodic orbit. This bistability is crucial to the appearance of bursting.

We now couple the dynamics of the fast subsystem to the slower dynamics of c. Included in Fig. 8.8 A is the curve defined by $dc/dt = 0$, i.e., the c nullcline. When V is above the c nullcline, $dc/dt > 0$, and so c increases, but when V is below the c nullcline, c decreases. Now suppose V starts on the lower fixed point for a value of c that is greater than c_{hc}. Since V is below the c nullcline, c starts to decrease, and V follows the lower branch of fixed points. However, when c becomes too small, this lower branch of fixed points disappears in a saddle-node bifurcation (SN), and so V must switch to another attractor. Since the upper and middle branches are not attracting

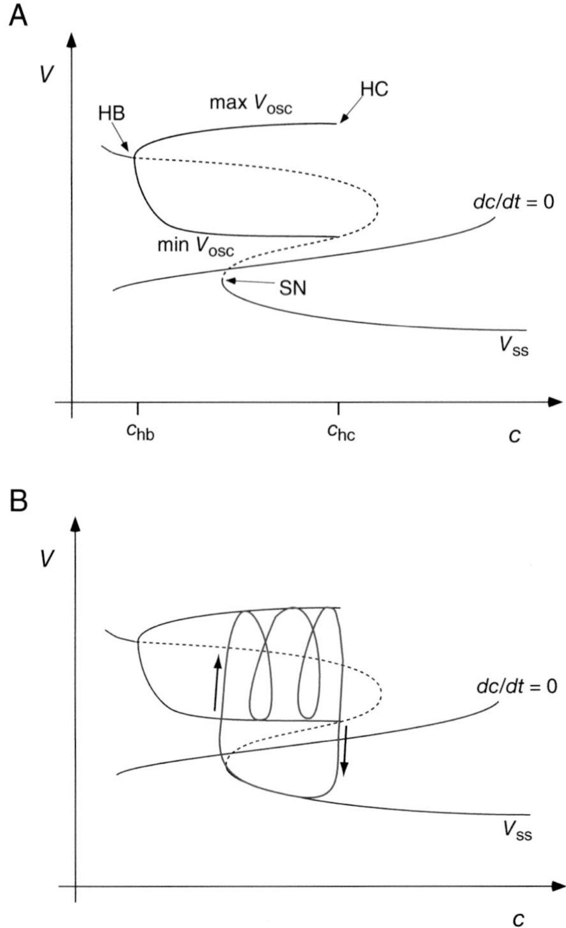

Fig. 8.8 **A:** Sketch (not drawn to scale) of the bifurcation diagram of the simplified Chay-Keizer β cell model, with c as the bifurcation parameter. V_{ss} denotes the curve of steady states of V as a function of c. A solid line indicates a stable steady state; a dashed line indicates an unstable steady state. The two branches of V_{osc} denote the maximum and minimum of V over one oscillatory cycle. HB denotes a Hopf bifurcation, HC denotes a homoclinic bifurcation, and SN denotes a saddle-node bifurcation. **B:** A burst cycle (red curve) projected on the (V, c) plane. Not drawn to scale. When the burst cycle lies below the curve $dc/dt = 0$ then c is decreasing. When the burst cycle lies above the curve $dc/dt = 0$ then c is increasing; in this case, the loops in the burst cycle denote changes in the third coordinate (i.e., in n).

for c values near this saddle-node bifurcation, V follows the stable limit cycle and starts to oscillate.

However, since V now lies entirely above the c nullcline, c begins to increase. Eventually, c increases enough to cross the homoclinic bifurcation at c_{hc}, the stable limit cycles disappear, and V switches back to the lower branch, completing the cycle. Repetition of this process causes bursting. The quiescent phase of the bursting cycle is when V is on the lower branch of the Z-shaped curve, and during this phase V increases slowly. A burst of oscillations occurs when V switches to the branch of stable oscillations, and disappears again after passage through the homoclinic bifurcation.

Clearly, in this scenario, bursting relies on the coexistence of a stable fixed point and a stable limit cycle, and the bursting cycle is a hysteresis loop that switches between branches of the Z-shaped curve. Bursting also relies on the c nullcline intersecting the Z-shaped curve in the right location. For example, if the c nullcline intersects the Z-shaped curve on its lower branch, there is a unique stable fixed point for the whole system, and bursting does not occur. A projection of the bursting cycle on the (V, c) phase plane is shown in Fig. 8.8 B. The periods of the oscillations in the burst increase through the burst, as the limit cycles get closer to the homoclinic trajectory, which has infinite period.

The relationship between bursting patterns and glucose concentration can also be deduced from Fig. 8.8. Notice that the $dc/dt = 0$ nullcline, given by $c = -\frac{k_1}{k_c} I_{\text{Ca}}(V)$, is inversely proportional to k_c. Increasing k_c moves the $dc/dt = 0$ nullcline to the left while decreasing k_c moves it to the right. Thus, when k_c is sufficiently small, the nullcline intersects the lower branch of the V nullcline (i.e., the Z-shaped curve where $dV/dt = 0$). On the other hand, if k_c is extremely large, the c nullcline intersects the upper branch of the V nullcline, possibly to the left of c_{hb}. At intermediate values of c, the c nullclines intersects the middle branch of the V nullcline.

Under the assumption that k_c is monotonically related to the glucose concentration, when the glucose concentration is low, the system is at a stable rest point on the lower V nullcline; there is no bursting. If glucose is increased so that the c nullcline intersects the middle V nullcline with $c < c_{\text{hc}}$, there is bursting. However, the length of the bursting phase increases and the length of the resting phase decreases with increasing glucose, because $[\text{Ca}^{2+}]$ increases at a slower rate and decreases at a faster rate when k_c is increased (recall that k_c is the rate at which Ca^{2+} is removed from the cell). For large enough k_c the bursting is sustained with no rest phase, as c becomes stalled below c_{hc}. Finally, at extremely high k_c values, bursting is replaced by a permanent high membrane potential, with $c < c_{\text{hb}}$. This dependence of the bursting phase on glucose has been seen in experiments, and thus the model is able to provide an explanation for these observations.

8.4.2 ER Calcium as a Slow Controlling Variable

Although it is undeniably elegant, there are two major problems with the above model. Firstly, it does not reproduce the wide variety of periods and bursting patterns actually seen in bursting pancreatic β cells, being limited to a narrow range of fast bursting frequencies. Secondly, experiments have shown that $[Ca^{2+}]$ oscillates too fast to be viewed as a slow control variable. This is illustrated in Fig. 8.6, which shows simultaneous $[Ca^{2+}]$ and voltage measurements. The bursting oscillations in the voltage are mirrored by bursting oscillations in the cytoplasmic $[Ca^{2+}]$, and the rise in $[Ca^{2+}]$ is almost as fast as the rise in voltage.

So the question arises of what controls the length of the bursting active phase. One possibility is that ER $[Ca^{2+}]$ varies much more slowly than cytoplasmic $[Ca^{2+}]$ and could provide the necessary control mechanism. This would be the case if most of the Ca^{2+} entering the cell during the active phase was coming from the extracellular space through transmembrane ion channels, with only a small amount of Ca^{2+} flowing between the cytoplasm and the ER. If this were the case, ER Ca^{2+} would act like a low-pass filter for cytoplasmic Ca^{2+}, and therefore could be used to detect and regulate the length of bursting activity. The possible usefulness of a low-pass filter is seen in the Ca^{2+} traces shown in Fig. 8.6, where, during a burst, $[Ca^{2+}]$ oscillates rapidly around a raised baseline. A low-pass filter would measure the length of time that the $[Ca^{2+}]$ baseline is elevated, but filter out the rapid oscillations.

Thus, it was proposed by Chay (1996a,b, 1997) that slow variations in ER Ca^{2+} could be an important control mechanism, and could, in addition, generate a wider range of bursting periods. This proposal was analysed in detail by Bertram and Sherman (2004) who showed that the interaction of a (not very slow) cytoplasmic Ca^{2+} variable, with a (much slower) ER Ca^{2+} variable, could lead to bursting with a wide range of periods. They named this a *phantom bursting* model, as it can lead to bursting with a period intermediate between that of the slow variables.

8.4.3 Other Models

Although we are more concerned here with Ca^{2+} dynamics per se rather than the myriad details of pancreatic β cell models, for completeness we note that there are a number of models in the literature. Wierschem and Bertram (2004), Bertram et al (2004), and Bertram et al (2007b) have proposed that the coupling of glycolytic oscillations to Ca^{2+} dynamics and the membrane potential can result in bursting with a wide range of periods, while Pedersen (2010) proposed a model for the human β cell, but one that almost entirely omits consideration of the Ca^{2+} dynamics (although this was modified in

Riz et al (2014)). Models have also appeared from Fridlyand et al (2003, 2009, 2013). At a more detailed level, the dynamics of Ca^{2+} removal from β cells was studied quantitatively by Chen et al (2003), while the most complex model of Ca^{2+} handling by β cell mitochondria is that of Magnus and Keizer (1997, 1998a,b).

As we mentioned above, the review of Félix-Martínez and Godínez-Fernández (2014) is an excellent place to obtain a broad overview of the many models in the literature.

8.5 Pancreatic Alpha Cells

Pancreatic β cells are not the only cells responsible for glucose homeostasis. In fact, three types of exocytotic pancreatic cells (α, β, and δ) coordinate the control of this vital process. While the role of δ cells still remains to be clarified, the role of α cells is to secrete glucagon when blood glucose levels are low. Glucagon in turn promotes the breakdown of glycogen into glucose in the liver, and thus has an effect opposite to that of the insulin secreted by β cells.

As is the case for many secretory processes, secretion of insulin and glucagon is mediated by Ca^{2+}. In α cells, as in β cells, the increase in $[Ca^{2+}]$ relies mainly on Ca^{2+} entry through voltage-gated channels, with ER Ca^{2+} behaving more like a modulator of the electrical activity of the plasma membrane. There is, however, an important difference between these two cell types; electrical activity is promoted by high extracellular glucose in β cells, but by low glucose in α cells. These different responses occur although both cell types express nearly all the same channels. As another difference, the electrical activity of α cells occurs as repetitive action potentials and not in the form of bursting as we have just seen for β cells.

There are fewer models for α cells than for β cells. Diderichsen and Göpel (2006) proposed a model for the repetitive action potentials occurring in α cells when external glucose is low. The strongest aspect of their model is that the gating functions of all model currents were determined from experimental data. This model has been modified and extended by Watts and Sherman (2014) to incorporate the effect of store-operated Ca^{2+} channels (SOCC), that open in response to a decrease in ER $[Ca^{2+}]$ (Section 2.6). The pathway linking Ca^{2+} and secretion has also been investigated theoretically. A model, applicable to most Ca^{2+}-mediated secretory processes, has been proposed by Voets (2000), and can be extended to take into account the specific properties of β and α cells, as explained later in this section (Chen et al, 2008a; González-Vélez et al, 2012).

8.5.1 Electrical Activity of Pancreatic Alpha Cells

The electrical activity of α cells can be described by the interplay between seven types of ionic currents (Diderichsen and Göpel, 2006), each described using the Hodgkin-Huxley formalism. Both T (transient) and L (long lasting) Ca^{2+} currents are included. Using standard notation, they can be written as

$$I_{\mathrm{Ca,T}} = g_{\mathrm{Ca,T}} m_{\mathrm{Ca,T}}^3 h_{\mathrm{Ca,T}} (V - V_{\mathrm{Ca}}), \tag{8.58}$$

$$I_{\mathrm{Ca,L}} = g_{\mathrm{Ca,L}} m_{\mathrm{Ca,L}}^2 h_{\mathrm{Ca,L}} (V - V_{\mathrm{Ca}}). \tag{8.59}$$

In (8.59), it is assumed for simplicity that L-type Ca^{2+} channels inactivate in a voltage-dependent manner; inhibition by high $[\mathrm{Ca}^{2+}]$ is not included.

The inward Na^+ current is given by

$$I_{\mathrm{Na}} = g_{\mathrm{Na}} m_{\mathrm{Na}}^3 h_{\mathrm{Na}} (V - V_{\mathrm{Na}}). \tag{8.60}$$

Two types of outward K^+ currents are included; A-type K^+ channels that inactivate rapidly, and delayed-rectifier K^+ channels that do not inactivate. They are described by

$$I_{\mathrm{K,A}} = g_{\mathrm{K,A}} m_{\mathrm{K,A}}^3 h_{\mathrm{K,A}} (V - V_{\mathrm{K}}), \tag{8.61}$$

$$I_{\mathrm{K,dr}} = g_{\mathrm{K,dr}} m_{\mathrm{K,dr}}^4 (V - V_{\mathrm{K}}). \tag{8.62}$$

Pancreatic α (and β) cells express a specific type of ATP-sensitive K^+ channel, with its activity determined by the ATP/ADP ratio. Under normal conditions, these channels allow K^+ to flow out of the cell, but close when the concentration of ATP increases, thus depolarising the cell. This rise in ATP concentration occurs when glucose metabolism is very active. Thus, in the model, $g_{\mathrm{K,ATP}}$ is the parameter that represents the concentration of external glucose. The higher the glucose concentration, the lower $g_{\mathrm{K,ATP}}$. The activity of K,ATP channels can be modelled as

$$I_{\mathrm{K,ATP}} = g_{\mathrm{K,ATP}} (V - V_{\mathrm{K}}). \tag{8.63}$$

The final current to be included in the model is the usual small-amplitude leak

$$I_{\mathrm{leak}} = g_{\mathrm{leak}} (V - V_{\mathrm{leak}}). \tag{8.64}$$

The currents defined by (8.58)–(8.64) are combined in the usual way (see Section 3.7), to give

$$C_m \frac{dV}{dt} = -(I_{\mathrm{Ca,T}} + I_{\mathrm{Ca,L}} + I_{\mathrm{Na}} + I_{\mathrm{K,A}} + I_{\mathrm{K,dr}} + I_{\mathrm{K,ATP}} + I_{\mathrm{leak}}). \tag{8.65}$$

Note that here, in contrast to (8.53), the factor of 2 that accounts for the charge of the Ca^{2+} ions is included in the constants $g_{\mathrm{Ca,T}}$ and $g_{\mathrm{Ca,L}}$.

The other equations of the model are those for the activation and inactivation functions appearing in the currents, and these were determined by fitting to voltage-clamp data (Diderichsen and Göpel, 2006).

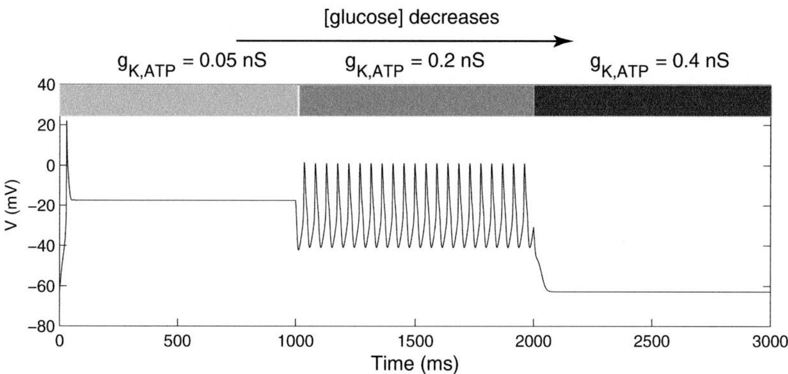

Fig. 8.9 Electrical activity of α cells computed from the model of Diderichsen and Göpel (2006), for different values of the conductance of the ATP-dependent K^+ channel, $g_{K,ATP}$. This channel acts as a sensor of external glucose as its conductance increases when glucose decreases, which hyperpolarises the cell.

The model can reproduce the repetitive action potentials seen in α cells (Fig. 8.9). At low values of $g_{K,ATP}$ (corresponding to high glucose concentrations), α cells are highly depolarised and electrically inactive because of the reduced K^+ efflux. Repetitive electrical activity occurs for intermediate values of $g_{K,ATP}$, while the cell remains in a hyperpolarised, electrically inactive, state at high values of $g_{K,ATP}$. What is specific to this model, as compared to models for β cells, is the transition from an inactive to an active state when lowering glucose (as shown by the increase of $g_{K,ATP}$ from 0.05 to 0.2 nS in Fig. 8.9).

What is the specific property of α cells that is responsible for this behaviour? At intermediate glucose concentrations, spontaneous action potentials are driven by activation and inactivation of the voltage-gated channels. In particular, depolarisation triggers the activation of I_{Na} that causes the upstroke of the action potential. Voltage-clamp experiments reveal that in α cells, I_{Na} is already half inactivated at around -45 mV. Thus, at high glucose (small K,ATP current), α cells are depolarised (because K^+ efflux is not sufficient) and Na^+ channels cannot initiate the action potentials. In principle, T-type Ca^{2+} channels that are present in α cells (but not in β cells) could take over, but they are inactivated in the same voltage range as Na^+ channels. Another possibility would be that $I_{K,A}$ could compensate for the closure of K,ATP channels but, again, this is not the case as this current is nearly totally inactivated at about -45 mV in α cells. Diderichsen and Göpel (2006) thus predict that glucose could have small effects on inactivation of the voltage-gated currents in α cells. An alternative possibility could be that α cells express voltage-gated channels with slightly different properties, either because they express other isoforms, or because they are regulated differently by accessory proteins.

8.5.2 Calcium Dynamics in Pancreatic Alpha Cells

As it is not electrical activity per se, but Ca^{2+} entering through voltage-gated channels that triggers secretion, glucagon release by α cells is best modelled by taking Ca^{2+} explicitly into account (Watts and Sherman, 2014). The model is sketched in Fig. 8.10.

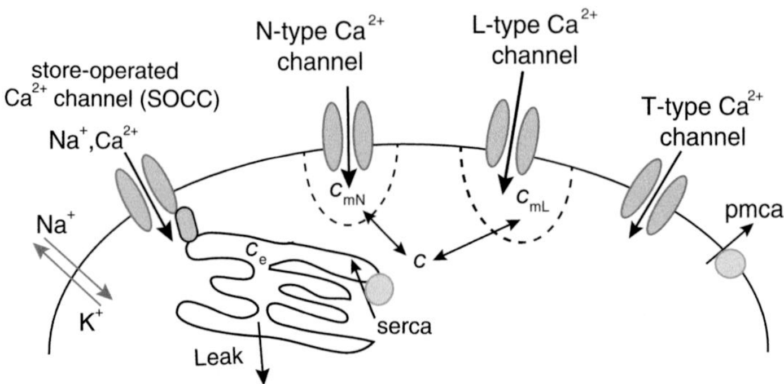

Fig. 8.10 Schematic representation of the modifications by Watts and Sherman (2014) to the original model of Diderichsen and Göpel (2006) for α cells. These changes pertain to the inclusion of Ca^{2+} dynamics by considering microdomains around N- and L-type Ca^{2+} channels, as well as store-operated Ca^{2+} channels, which are also permeable to Na^+.

As in the previous model, $[Ca^{2+}]$ increases in the cell when T- and L-type Ca^{2+} channels are active. In addition, two other types of Ca^{2+} channel are included:

- N-type Ca^{2+} channels, which need strong depolarisation to open, and are known to play an important role in glucagon secretion,

$$I_{\mathrm{Ca,N}} = g_{\mathrm{Ca,T}} m_{\mathrm{Ca,T}} h_{\mathrm{Ca,T}} (V - V_{\mathrm{Ca}}), \tag{8.66}$$

- Store-operated Ca^{2+} channels (SOCC) that open upon a decrease in ER $[Ca^{2+}]$,

$$I_{\mathrm{SOCC}} = \frac{g_{\mathrm{SOCC}}}{1 + \exp\left(\frac{c_e - \bar{c}_e}{20}\right)} (V - V_{\mathrm{SOCC}}), \tag{8.67}$$

where $\bar{c}_e$ is the value of c_e that gives half the maximal store-operated current for any given voltage. This expression is similar to the model described in Section 2.6, but also includes a dependence on voltage as it has been observed that SERCA inhibition depolarises α cells (Liu et al, 2004). In addition, SOCC are assumed to be permeable also to Na^+, as this has been observed to be the case in pancreatic β cells (Worley et al, 1994; Mears and Zimliki, 2004).

Microdomains around L-type and N-type Ca^{2+} channels are included, as Ca^{2+} entering through these channels is particularly effective at triggering secretion, probably because the channels often colocalise with docked secretory granules. These microdomains are modelled in the same way as those around the IPR, as described in Section 3.2.3. For example, the evolution of cytosolic $[Ca^{2+}]$ around the N-type channels is given by

$$\frac{dc_{mN}}{dt} = f_m\big(J_N - k_{\text{diff}}(c_{mN} - c)\big), \tag{8.68}$$

where f_m denotes the buffering capacity of the microdomain, i.e., f_m converts a Ca^{2+} flux into a change in the concentration of free Ca^{2+} in the microdomain (Section 2.9). J_N is directly computed from $I_{\text{Ca,N}}$ through

$$J_N = -\frac{1}{2F}\frac{I_{\text{Ca,N}}}{V_{mN}}, \tag{8.69}$$

where F is Faraday's constant, and V_{mN} is the volume of the microdomain considered (on the order of femtolitres).

The equation for c_{mL}, the $[Ca^{2+}]$ in the microdomain surrounding the L-type Ca^{2+} channels, is written in the same way. From these microdomains, Ca^{2+} diffuses passively to the bulk cytoplasmic space. The equation for cytoplasmic $[Ca^{2+}]$ is thus

$$\begin{aligned}\frac{1}{f_c}\frac{dc}{dt} &= J_T + \frac{V_{mL}}{V_{\text{cell}}}k_{\text{diff}}(c_{mL} - c) + \frac{V_{mN}}{V_{\text{cell}}}k_{\text{diff}}(c_{mN} - c) \\ &\quad + p_{\text{leak}}(c_e - c) - k_c c - k_{\text{pmca}} c,\end{aligned} \tag{8.70}$$

where f_c is the cytosolic buffering factor, analogous to f_m for the microdomain. The flux through the T-type channel is computed from the corresponding current in the same way as in (8.69). Ca^{2+} exchanges between the cytosol and the microdomains are scaled by the volume ratios of the respective compartments. Surprisingly, SOCC currents do not appear in (8.70) as it is assumed that in α cells, as in β cells, SOCC are non-specific channels and mainly carry Na^+ (Mears and Zimliki, 2004). Ca^{2+} exchanges with the ER are modelled by a linear SERCA pump (k_c) and a passive leak, thus neglecting the possible involvement of IPR and RyR.

The final equation of the model is that for ER $[Ca^{2+}]$, and is immediately deduced from (8.70) to be

$$\frac{dc_e}{dt} = -f_{\rm ER} \frac{V_{\rm cell}}{V_{\rm ER}} \left[p_{\rm leak}(c_e - c) - k_c c \right], \tag{8.71}$$

where $f_{\rm ER}$ is the ER buffering factor. When combined with the expressions for the voltage-gated currents proposed by Diderichsen and Göpel (2006), this model accounts for repetitive action potentials, and reproduces the cessation of electrical activity upon membrane depolarisation following a decrease in $g_{\rm K,ATP}$. Because cytosolic $[{\rm Ca}^{2+}]$ is a slow variable (the buffering factors f_c and $f_{\rm ER}$ are small, and so the cytosolic and ER Ca^{2+} concentrations have much slower dynamics than the membrane potential), it remains at a quasi-steady level that increases with the membrane electrical activity. This is not the case for c_{mN} and c_{mL}, the $[{\rm Ca}^{2+}]$ in the microdomains where secretion is triggered. These variables follow the rapid action potentials.

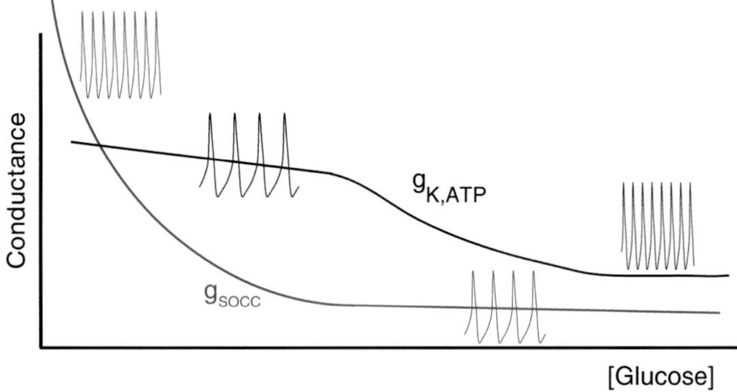

Fig. 8.11 Origin of the biphasic response of α cells to changes in external glucose concentration. As the glucose concentration increases, the ATP concentration also increases, which increases the rate of SERCA pumping, thus decreasing the SOCC conductance. Since a decrease in SOCC current leads to a decrease in the α cell electrical activity, upon an increase in glucose, secretion first decreases (the red traces). In contrast, $I_{\rm K,ATP}$ becomes inhibited at higher levels of glucose. As the electrical activity of α cells increases when $g_{\rm K,ATP}$ decreases, this restores secretion (the black traces).

The store-operated currents in the model have an interesting effect, as they allow glucose to affect the dynamics through a parameter other than $g_{\rm K,ATP}$. For example, the rate of Ca^{2+} pumping through the SERCA pumps (k_c) can be modelled as an increasing function of glucose concentration, as pumping is an ATP-driven process. This dependency is also taken into account in models for electrical bursting in β cells (Section 8.4.1). Hence, a change in glucose modulates k_c thereby modifying c_e, and, through (8.67), also modifying the amplitude of $I_{\rm SOCC}$. Bifurcation analysis shows that increasing $g_{\rm SOCC}$ leads first to an increase, and then to a decrease in electrical activity and $[{\rm Ca}^{2+}]$. As the density of these channels in α cells is very low, only the increasing part is physiologically relevant. As g_{SOCC} is modulated by c_e, increasing ATP will decrease the store-operated current and thereby decrease electrical activity.

In the whole model, a rise in glucose concentration thus has two opposite effects, which are schematically represented in Fig. 8.11. By lowering $g_{\mathrm{K,ATP}}$, it tends to boost electrical activity. On the other hand, by increasing k_c, it decreases SOCC activity (because the ER has a higher $[\mathrm{Ca}^{2+}]$), which diminishes the electrical activity. This dual effect of glucose in the model provides an explanation for the experimental observation that in the range of glucose concentrations leading to electrical activity, secretion first decreases and then increases when raising external glucose. If k_c is sensitive to lower glucose concentrations than $g_{\mathrm{K,ATP}}$, the inhibiting effect occurs first, as glucose is increased. Upon further glucose increase, SERCA are already saturated and the predominant effect is the decrease in $g_{\mathrm{K,ATP}}$, which boosts electrical activity. This U-shaped dose-response relationship for glucose-regulated glucagon secretion has been observed experimentally by Salehi et al (2006) and may explain why diabetic patients exhibiting pronounced hyperglycaemia sometimes exhibit anomalously high levels of glucagon.

8.6 Calcium-Mediated Secretion

Secretion is a complex process, involving a coordinated sequence of steps, many of which are sensitive to Ca^{2+}. Calcium entry through voltage-gated Ca^{2+} channels during repetitive action potential firing is the predominant trigger for secretion of insulin and glucagon by β and α cells, respectively. Watts and Sherman (2014) thus define the rate of glucagon secretion as

$$J_S = N_c m_{\mathrm{Ca,N}} h_{\mathrm{Ca,N}} \left(\frac{c_{mN}}{K_N + c_{mN}} \right)^4 \tag{8.72}$$

assuming that Ca^{2+} entry through the N-type Ca^{2+} channels plays a predominant role in triggering secretion. N_c denotes the number of channels, and other symbols have been defined in the previous section. Although (8.72) provides a valuable first approximation, secretion depends on many other factors than just the activity of N-type Ca^{2+} channels. A clear presentation of the biological mechanism of exocytosis in α and β cells can be found in the review by Barg (2003).

Exocytosis occurs when vesicles, full of the peptide to be secreted, fuse with the plasma membrane and thereby release their contents to the extracellular space. Fusion rapidly follows $[\mathrm{Ca}^{2+}]$ increase, but this requires the vesicles to be located in the immediate vicinity of the plasma membrane. Thus, the vesicles that are formed in the bulk of the cytoplasm (in the ER and Golgi) need to be translocated to the plasma membrane where they dock. When they are attached to the plasma membrane they become primed for release by an ATP-dependent process, and can be triggered by a Ca^{2+} influx to undergo fusion with the membrane and release their contents outside the cell. Because fusion of the vesicles to the plasma membrane is inherent to secretion, exocytosis can be measured as a rise in capacitance resulting from an increase of the cell membrane area.

8.6.1 Prototypic Model for Calcium-Mediated Secretion

Most models for Ca^{2+}-mediated secretion in pancreatic α and β cells are based on the work of Voets et al (1999) and Voets (2000), who modelled secretion in chromaffin cells. The experimental basis of this model is the observation of two phases of exocytosis in response to depolarisation; a fast transient component synchronised with channel opening, followed by a slower sustained phase. By precisely controlling $[Ca^{2+}]$, the authors found that Ca^{2+} not only triggers the final stage of exocytosis (fusion) but also regulates the initial transfer of the vesicles to the membrane. In contrast, priming of the vesicles depends on ATP but not on Ca^{2+}. The model assumes that docked, but unprimed, vesicles can also fuse and release their contents, but at a much slower rate than primed ones.

The model (Fig. 8.12) thus considers three types of vesicles: reserve, slowly releasable, and ready to be released. Transitions between these three states are reversible. The filling of the pool of docked vesicles from the reserve is stimulated by Ca^{2+}, and there is an unlimited number of reserve vesicles. Thus, the corresponding rate is given by

$$r_1 = r_{\max} \frac{c}{K_D + c}, \tag{8.73}$$

where K_D is the apparent Ca^{2+} affinity of the putative regulatory site. Both docked and primed granules need to bind Ca^{2+} to fuse with the membrane. For priming, the concentration of ATP is assumed to be high and constant, and thus does not appear explicitly in the equations. All reaction rates (except r_1 in (8.73)) are described by the mass-action law. To obtain satisfactory agreement with measured rates of exocytosis, the model takes into account that fusion requires three Ca^{2+} binding steps. The rate constants $k_{\text{on,r}}$ and $k_{\text{off,r}}$ are approximately 10 times larger than $k_{\text{on,s}}$ and $k_{\text{off,s}}$. Furthermore,

$$K_{D,r} = \frac{k_{\text{off,r}}}{k_{\text{on,r}}} > K_{D,s} = \frac{k_{\text{off,s}}}{k_{\text{on,s}}}, \tag{8.74}$$

meaning that docked and unprimed granules are released only at higher concentrations of Ca^{2+}.

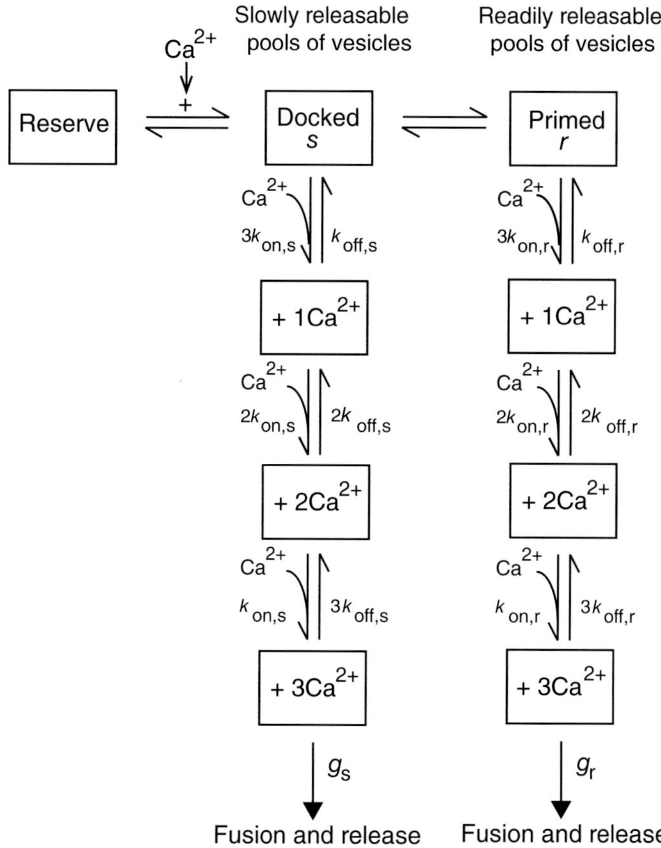

Fig. 8.12 Model proposed by Voets (2000) for Ca^{2+}-mediated secretion. There are three different kinds of vesicles; a reserve pool, docked vesicles, and primed vesicles. Both docked and primed vesicles can bind Ca^{2+} and fuse with the membrane to release their contents, but the primed vesicles do this much faster.

8.6.2 Secretion of Insulin by Pancreatic Beta Cells

The specific case of insulin secretion by pancreatic β cells has been investigated by Chen et al (2008b) and Pedersen and Sherman (2009). Similarly to chromaffin cells, pancreatic β cells secrete insulin in a biphasic pattern. In vivo, a pancreatic islet stimulated by a step increase in glucose first displays a burst of high secretion for a few minutes, followed by a sustained slow secretion lasting for more than an hour. Again, this is physiologically important as it has been suggested that diabetes is correlated with the loss of the first phase (Gerish, 2002). In their model, the authors considered that docking and priming occur in a single step (thus keeping only one column of the model shown in Fig. 8.12). They postulated that the biphasic kinetics of secretion can be accounted for by including Ca^{2+} microdomains around the L-type channels, i.e.,

$$\frac{dc_{mL}}{dt} = f_{mL}\left[J(t) - k_{\text{diff}}(c_{mL} - c)\right]. \tag{8.75}$$

Variables and parameters are defined as in Section 8.5.2. This microdomain $[Ca^{2+}]$ regulates the three Ca^{2+} binding steps leading to fusion. In contrast, the evolution towards the docked state is regulated by bulk cytosolic $[Ca^{2+}]$ through a Michaelis-Menten term identical to (8.73). The evolution of c is given by

$$\frac{dc}{dt} = f_c\left[J(t) + \frac{V_m L}{V_{\text{cell}}} k_{\text{diff}}(c_{mL} - c) - J_{\text{serca}} - J_{\text{pmca}} - J_{\text{NCX}} - J_{\text{leak}}\right], \tag{8.76}$$

where $J(t)$ represents the Ca^{2+} flux through the voltage-gated Ca^{2+} channels other than the L-type.

In the framework of this model, the transient first phase of secretion is due to exocytosis of docked granules. The second sustained phase is mainly due to newcomer granules that are supplied at a moderate rate because they are sensitive to the slow variable c.

As an alternative hypothesis, the biphasic secretion pattern of β cells can be accounted for by keeping the two types of releasable granules (as initially suggested by Voets (2000)). According to this hypothesis, the two vesicle pools differ because they are regulated by distinct Ca^{2+} concentrations, not because they have (artificially imposed) different values for the kinetic constants. In this case, the rapid phase is due to the release of vesicles that are sensitive to the $[Ca^{2+}]$ in the microdomains, while the sustained phase results from the activity of another group of vesicles which have a high affinity for Ca^{2+} and which are regulated by the cytoplasmic concentration (Pedersen and Sherman, 2009).

8.6.3 Secretion of Glucagon by Pancreatic Alpha Cells

To focus on the mechanism of glucagon secretion by pancreatic α cells in vivo, a somewhat different approach was adopted by González-Vélez et al (2012). Ca^{2+} dynamics is indeed not well understood in α cells. In the intact islet, the changes in $[Ca^{2+}]$ occurring at low external glucose occur in the form of irregular, long lasting and very slow Ca^{2+} transients. Each increase corresponds to a transient phase of electrical activity (Quoix et al, 2009). The reason why these cells show bursts of activity remains unknown. Thus, to bypass this problem, González-Vélez et al (2012) developed a model of glucagon secretion using, as input, experimentally obtained records of intracellular $[Ca^{2+}]$ in α cells in different glucose conditions.

The model includes only one type of releasable granule because this was found to be sufficient to reproduce experimental observations. As another difference with the model of Voets (2000), a slow rate of granule formation was assumed, instead of an unlimited pool of reserve granules. This is justified by the limited number of granules present in α cells. Besides that, the secretory process is described in a similar way as that sketched in Fig. 8.12, and involves three steps of Ca^{2+} binding with similar affinities. As for β cells, granule supply is sensitive to bulk cytosolic $[Ca^{2+}]$, while the rates of binding preceding granule fusion are controlled by the $[Ca^{2+}]$ below the plasma membrane (called c_{sub}). This is somewhat different from the microdomain approach, as c_{sub} is supposed to be the concentration in a whole shell located just below the plasma membrane. To extract the values of c and c_{sub} from the experimental time series of average whole-cell $[Ca^{2+}]$, electrical activity was simulated using the equations proposed by Diderichsen and Göpel (2006) (Section 8.5.1). Assuming an average cell radius of $5.3\,\mu$m and a shell radius of 100 nm (corresponding to the region where granules are seen at a high density), simulations indicate that

$$\frac{c_{\mathrm{sub}}}{c} \approx 1.3. \tag{8.77}$$

Such a low value is related to the limited volume of the cytosol in α cells where the nucleus occupies approximately 60% of the cell volume. As an

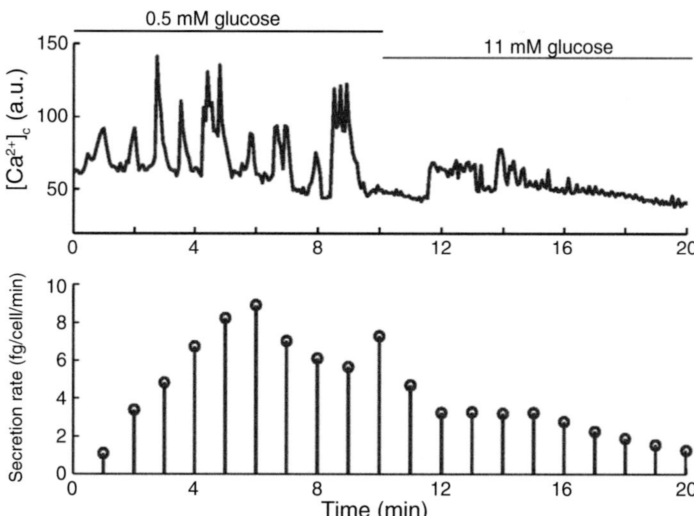

Fig. 8.13 Ca^{2+}-mediated secretion by pancreatic α cells. The upper trace shows cytosolic $[Ca^{2+}]$ in an α cell (in an intact islet) for two different levels of external glucose. The lower panel shows the average secretion rate in the model of González-Vélez et al (2012), computing by averaging the results of 14 different simulations, using 14 experimental time series of Ca^{2+} concentrations as inputs. Reproduced from González-Vélez et al (2012), Fig. 3.

important prediction of the model, agreement with experimental observations on secretion rates could only be achieved when assuming that the rate of granule resupply increases linearly with glucose. Thus,

$$r_1 = r_1^*[\text{Glu}]\frac{c}{K_D + c}. \tag{8.78}$$

This suggests that α cells have their own glucose-sensing mechanisms, as reported in some studies (Vieira et al, 2007). Assuming mass-action kinetics for the other reaction steps, the model quantitatively reproduces the dynamics of secretion in response to a decrease in glucose. In contrast to insulin secretion by β cells, α cells start to secrete glucagon slowly after a decrease in external glucose (Fig. 8.13). Also, a basal level of secretion still occurs even at high glucose. In Fig. 8.13 the lower panel shows the secretion rate, computed by averaging the results obtained from 14 individual cells (i.e., 14 recordings of Ca^{2+} dynamics such as those shown in the upper panel). In vivo, α cells display asynchronous Ca^{2+} oscillations, in contrast to β cells that are synchronised. Secretion is computed as $\int F(t)\,dt$ over 1 minute, where F is the number of fused granules. One granule corresponds to 2 femtograms of glucagon.

The model reproduces the delayed maximal response to low glucose, as reported by Gustavsson et al (2009). This delay corresponds to the time taken by granules to become releasable, as about half the granules are in the reserve pool at resting $[Ca^{2+}]$. As expected, steady-state secretion (attained after 5 min) at low glucose is greater than at high glucose.

That half the granules are in the reserve pool at high glucose is in agreement with experimental reports (Gustavsson et al, 2009). In contrast, at low glucose approximately 70% of the granules are in the reserve pool, both in the model and in the experiments. This paradoxical result stems from the dependence of the resupply rate on glucose concentration, as seen in (8.78). On the one hand, a decrease in external glucose initiates electrical activity that triggers secretion, but on the other hand, this decrease in external glucose also slows down the resupply of reserve granules into subplasmalemmal granules, ready to be released. A possible interpretation of this dual effect of glucose could be that it avoids the exhaustion of granules and thereby allows for α cells to secrete glucagon on very long time scales to prevent possible hypoglycaemia.

The model can also be used to assess the possible physiological advantage of Ca^{2+} oscillations as compared to an increased steady state. A plot of the rate of secretion versus the frequency of oscillations obtained with the model indicates that there is no correlation between these two quantities. This suggests the absence of frequency encoding, which is not surprising given the high level of irregularity of Ca^{2+} oscillations in α cells. Interestingly, curves of the secretion rate as a function of average $[Ca^{2+}]$ are very steep. This provides a clue to understand why oscillations favour secretion. Because of the highly nonlinear relationship between mean $[Ca^{2+}]$ and secre-

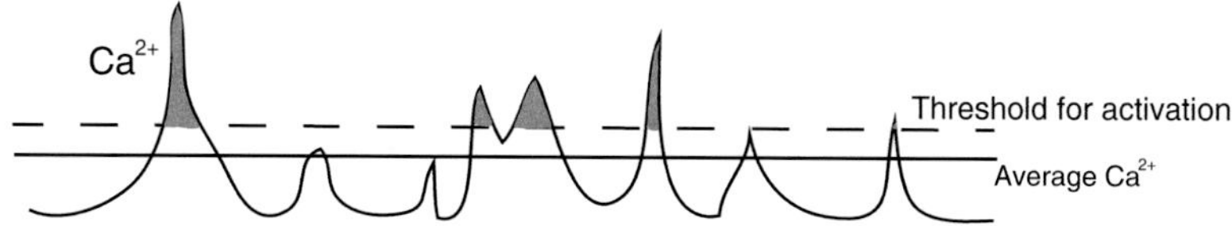

Fig. 8.14 Schematic representation of the mechanism by which Ca^{2+} oscillations can potentiate a Ca^{2+}-activated cellular response, while a constant $[Ca^{2+}]$, of the same average value, cannot. If the threshold for a response lies above the average concentration, then a response will only occur if Ca^{2+} oscillations push the concentration over the threshold.

tion rate, a small increase in $[Ca^{2+}]$ produces a large increase in secretion. Thus, secretion during oscillations is larger than the secretion obtained over the same period of time with $[Ca^{2+}]$ kept constant at a value corresponding to its average during oscillations. Such a potentiating effect of oscillations on Ca^{2+}-activated responses has been demonstrated experimentally for gene expression (Dolmetsch et al, 1998) and is illustrated in Fig. 8.14. This figure represents the hypothetical situation of Ca^{2+} oscillations with the average value indicated by the solid line, which activate a physiological response, whose effective K_D for Ca^{2+} is represented by the dashed line. Because the K_D is larger than the average $[Ca^{2+}]$, no response would be triggered at a constant value corresponding to the average. In contrast, during oscillations, the response is activated each time $[Ca^{2+}]$ exceeds the threshold, which is represented by the grey regions in Fig. 8.14.

8.7 Hypothalamic and Pituitary Cells

In pancreatic β cells, the interaction of membrane-based electrical excitability with ER-based Ca^{2+} dynamics is crucial for the generation of fast bursting oscillations (Section 8.4). Similar interactions are also known to be important in many cell types, including many types of hypothalamic neurons and pituitary cells (Stojilkovic, 2012). Hypothalamic neurons secrete hormones which typically travel a short distance to the pituitary along the inferior hypophyseal artery and the portal vein, to act there on specific target cells. For example, gonadotropin-releasing hormone (GnRH) is secreted from GnRH neurons in the hypothalamus; after travelling to the pituitary it stimulates the release of gonadotropin from pituitary gonadotrophs.

There are already a number of detailed models of electrical and Ca^{2+} excitability in hypothalamic and pituitary cells. The release of corticotropin from pituitary corticotrophs in response to corticotropin-releasing hormone has been studied in detail by LeBeau et al (1997), LeBeau et al (1998), Shorten et al (2000), and Shorten and Wall (2000), the interaction of Ca^{2+} and membrane potential in vasopressin hypothalamic neurons has been studied by Roper et al (2003, 2004), while somatotrophs and lactotrophs were studied by Tsaneva-Atanasova et al (2007), Tomaiuolo et al (2010) and Tabak et al (2011). More theoretical approaches are those of Vo et al (2010, 2014), and Teka et al (2011a,b, 2012).

8.7.1 The Gonadotroph

An early model that shows clearly the complex connections between membrane potential, cytosolic Ca^{2+} and ER Ca^{2+} is that of Li et al (1995b, 1997). In gonadotrophs, Ca^{2+} spiking occurs in two different ways. First, spontaneous action potentials lead to the influx of Ca^{2+} through voltage-gated channels, and thus Ca^{2+} spikes. Second, in response to the agonist GnRH, gonadotrophs produce IP_3 which releases Ca^{2+} from the ER and leads to oscillations in the cytoplasmic $[Ca^{2+}]$. Both the electrophysiological and agonist-induced Ca^{2+} signalling pathways are well-characterised experimentally (Kukuljan et al, 1994; Stojilkovic et al, 1994).

Similarly to the model of the pancreatic β cell, the gonadotroph model consists of two pieces: a model of the membrane electrical potential, and a model of the intracellular Ca^{2+} dynamics, including IP_3-induced Ca^{2+} release from the ER. It is, however, more complicated in application, as Li et al modelled the cell as a spherically symmetric region, with a gradient of $[Ca^{2+}]$ from the plasma membrane to the interior of the cell.

8.7.1.1 The Membrane Model

The membrane potential is modelled in the usual way:

$$C\frac{dV}{dt} = -I_{\mathrm{Ca,T}} - I_{\mathrm{Ca,L}} - I_{\mathrm{K}} - I_{\mathrm{K,Ca}} - I_{\mathrm{L}}, \tag{8.79}$$

where the Ca^{2+} current has been divided into T-type and L-type currents, I_{K} is the current through K^+ channels, $I_{\mathrm{K,Ca}}$ is the current through Ca^{2+}-sensitive K^+ channels, and I_{L} is a leak current.

Each of these currents is modelled in Hodgkin-Huxley fashion, in much the same way as in the model of the pancreatic β cell. Most importantly, the conductance of the Ca^{2+}-sensitive K^+ channel is assumed to be an increasing function of c_R, the $[Ca^{2+}]$ at the plasma membrane. Thus,

$$I_{\mathrm{K,Ca}} = g_{\mathrm{K,Ca}}\frac{c_R^4}{c_R^4 + K_c^4}\phi_K(V), \tag{8.80}$$

where $\phi_K(V)$ denotes the Goldman–Hodgkin–Katz current–voltage relationship (Keener and Sneyd, 2008) used here instead of the linear one used in the pancreatic β cell model.

8.7.1.2 The Calcium Model

Let c and c_e denote, respectively, the intracellular and ER [Ca^{2+}]. Note that both c and c_e are functions of r, the radial distance from the centre of the spherically symmetric cell. In the body of the cell, Ca^{2+} can enter the cytoplasm only from the ER (through IP_3 receptors), and can leave the cytoplasm only via the action of SERCA pumps in the ER membrane. However, on the boundary of the cell, Ca^{2+} can enter or leave via Ca^{2+} currents or plasma membrane Ca^{2+} ATPase pumps. Calcium buffering is assumed to be fast and linear (Section 2.9).

The equations for c and c_e are thus

$$\frac{\partial c}{\partial t} = D\nabla^2 c + J_{\mathrm{IPR}} - J_{\mathrm{serca}}, \tag{8.81}$$

$$\frac{\partial c_e}{\partial t} = D_e\nabla^2 c_e - \gamma(J_{\mathrm{IPR}} - J_{\mathrm{serca}}), \tag{8.82}$$

where, as usual, D and D_e are the effective diffusion coefficients of c and c_e, respectively, and γ is the scale factor relating ER volume to cytoplasmic volume. On the boundary of the cell, i.e., at $r = R$, we have

$$D\left.\frac{\partial c}{\partial r}\right|_{r=R} = -\alpha(I_{\mathrm{Ca,T}} + I_{\mathrm{Ca,L}}) - J_{\mathrm{pm}}, \tag{8.83}$$

$$\left.\frac{\partial c_e}{\partial r}\right|_{r=R} = 0, \tag{8.84}$$

where J_{pm} denotes the flux out of the cell due to the action of plasma membrane Ca^{2+} ATPase pumps. The scale factor $\alpha = 1/(2FA_{\mathrm{cell}})$, where A_{cell} is the surface area of the cell and F is Faraday's constant, converts the current (in coulombs per second) to a mole flux density (moles per area per second).

The flux through the IP_3 receptor, J_{IPR}, is modelled by the Li-Rinzel simplification of the De Young-Keizer model (Section 2.7), while the ATPase pump fluxes, J_{serca} and J_{pm}, are modelled by Hill functions (Section 2.2).

8.7.1.3 Results

When this model is solved for a range of [IP_3] (corresponding to a range of GnRH concentrations), the agreement with experimental data is impressive (Fig. 8.15). Before the addition of agonist the cells exhibit continuous spiking; once agonist is applied the frequency of spiking drops dramatically, changing to a more complex burst pattern. At the highest agonist concentrations, bursting is initially suppressed by the agonist, although spiking eventually reappears.

In all cases, the spiking frequency is initially greatly decreased by the agonist, but recovers gradually. This recovery of the spike frequency is an interesting demonstration of the importance of the amount of Ca^{2+} in the ER. Before the addition of agonist, the cell exhibits tonic spiking. Upon addition of IP_3, c starts to oscillate and the voltage spikes more slowly. The slowdown of the voltage spiking is caused by the greater release of Ca^{2+} from the ER, which activates the Ca^{2+}-sensitive K^+ current to a greater extent,

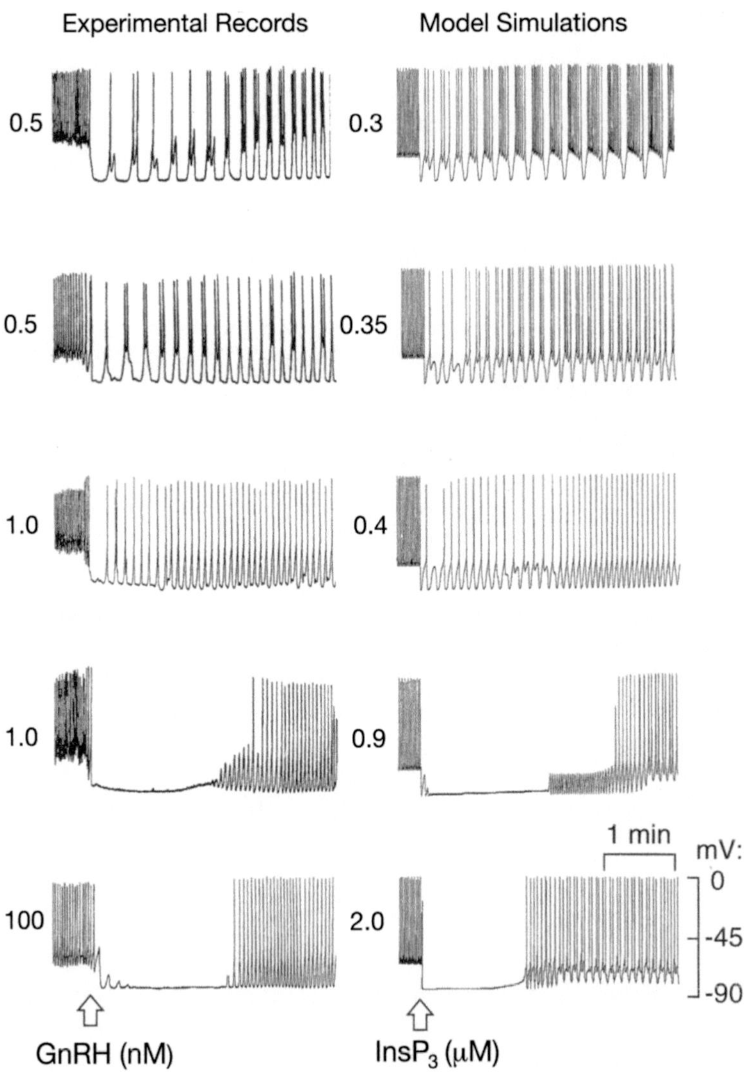

Fig. 8.15 The response of gonadotrophs, in experiment and theory, to increasing concentrations of agonist. In the experiments (left panels), the indicated concentration of GnRH was added at the arrow, while this was simulated in the model (right panels) by addition of the indicated [IP_3]. Reproduced from Li et al (1997), Fig 2, with permission from Elsevier.

thus making it more difficult for a voltage spike to occur. However, as time goes on, Ca^{2+} is lost from the cell by the action of the plasma membrane ATPase pumps. This overall decline in $[Ca^{2+}]$ manifests itself in two principal ways: firstly, the concentration in the ER declines slowly, and secondly, the baseline of the agonist-induced Ca^{2+} spikes gradually decreases. It is this decrease that causes the slow increase in the frequency of the voltage spikes, as $I_{\mathrm{K,Ca}}$ is gradually decreased.

Eventually, c_e decreases to such an extent that the ER can no longer support large Ca^{2+} spikes. The oscillations in c become smaller and faster, thus decreasing the extent of activation of the K^+ current, allowing the ER to refill, and allowing fast electrical spiking to occur again.

8.7.2 GnRH Neurons

The gonadotroph model discussed in the previous section gives excellent insight into how the membrane voltage, the cytosolic $[Ca^{2+}]$, and the ER $[Ca^{2+}]$ can interact, over a range of different time scales. A more recent model of gonadotropin-releasing hormone (GnRH) neurons shows how the different time scales in the interaction of cytosolic Ca^{2+} with the membrane ion channels can shape bursting oscillations.

GnRH neurons are hypothalamic neurons that secrete GnRH, which travels to the pituitary and stimulates the secretion of gonadotropin from gonadotrophs. There are comparatively few GnRH neurons in a mammalian brain, and their cell bodies are scattered throughout the hypothalamus (Constantin et al, 2012). Since they are therefore difficult to find and study in a brain slice, many earlier studies of GnRH neurons were performed on GT1 immortalised cells (Krsmanović et al, 1992; Van Goor et al, 1999). However, the development of genetically encoded fluorescent Ca^{2+} dyes that could be targeted to GnRH neurons allowed for the simultaneous measurement of both membrane potential and $[Ca^{2+}]$ in GnRH cells in situ in mouse brain slices (Jasoni et al, 2007, 2010; Lee et al, 2010; Constantin et al, 2012).

Mirroring the experimental methods, much of the previous modelling work on GnRH neurons has been based, not on data from GnRH neurons themselves, but on data from GT1 cells. Although these early models thus have only limited direct applicability to GnRH neurons in slice preparations, they still remain the basis for all subsequent modelling work.

The earliest models of electrical spiking in GT1 cells were those of Van Goor et al (2000) and LeBeau et al (2000), and were based on the usual structure of a membrane oscillator connected to the cytosolic $[Ca^{2+}]$. Thus,

$$C\frac{dV}{dt} = -(I_{\mathrm{Na}} + I_{\mathrm{Ca,L}} + I_{\mathrm{Ca,T}} + I_{\mathrm{K,dr}} + I_{\mathrm{M}} + I_{\mathrm{K,ir}} + I_{\mathrm{SK}} + I_{\mathrm{SOCC}} + I_{\mathrm{d}}).$$

$I_{\rm Na}$ is a TTX-sensitive Na^+ current, $I_{\rm Ca,L}$ and $I_{\rm Ca,T}$ are L-type and T-type Ca^{2+} currents, and $I_{\rm K,dr}$, $I_{\rm M}$ and $I_{\rm K,ir}$ are delayed-rectifier, M-type and inward-rectifier K^+ currents. $I_{\rm SK}$ is a Ca^{2+}-activated K^+ current, while $I_{\rm SOCC}$ is a store-operated Ca^{2+} channel, i.e., a Ca^{2+} channel that opens when the endoplasmic reticulum is depleted of Ca^{2+} (see Section 2.6). Finally, $I_{\rm d}$ is a nonselective inward cation current.

In the model of Van Goor et al (2000), there was no detailed model of the intracellular Ca^{2+} dynamics, but this was substantially improved in the subsequent version of LeBeau et al (2000). In the LeBeau model, spiking was essentially caused by the Na^+ and K^+ currents, with the frequency modulated by the three pacemaker currents, $I_{\rm SK}$, $I_{\rm SOCC}$, and $I_{\rm d}$, all three currents being coupled to the intracellular $[Ca^{2+}]$, and thus to the Ca^{2+} influx through $I_{\rm Ca,L}$ and $I_{\rm Ca,T}$.

One of the most interesting results of the LeBeau model was their prediction of the existence of a nonselective cation current, $I_{\rm d}$, activated by cAMP and inhibited by Ca^{2+}. This current was necessary, principally, to explain the response to forskolin, which activates adenylate cyclase, leading to the production of cAMP and an increase in the firing rate of GT1 neurons. Although they were unable to confirm the existence of this current experimentally, the model still demonstrated the components needed to explain the experimental results.

The Ca^{2+} handling in this model was further improved by Fletcher and Li (2009), who constructed a spatial model of the intracellular $[Ca^{2+}]$, as well as simplifying the currents in the model.

In a series of papers, Roberts, Suter and others (Roberts et al, 2006, 2008b; Roberts and Suter, 2008; Roberts et al, 2008a, 2009; Hemond et al, 2012) constructed a GnRH model of a rather different type. Although their model was also based to an extent on the earlier work of LeBeau and van Goor, Roberts and Suter constructed a more complex compartmental model using the software package GENESIS, and determined some of the parameters (the membrane resistance, the membrane capacitance and the axial resistance) by fitting to experimental data, with the other parameters taken from previous work, mostly from the LeBeau and van Goor models. The model gave excellent agreement with the shape of the individual action potentials in each burst. They also performed an elementary bifurcation analysis on a single-compartment version of the model, although their analysis did not focus on the mechanisms of bursting.

Possibly the most comprehensive model to date of GnRH neurons in vivo is that of Csercsik et al (2012). They incorporated the earlier models of van Goor and LeBeau with the later model of Lee et al (2010) to provide a single integrated model that reproduced the shape of the action potentials in detail, as well as incorporating the mechanisms that controlled burst length and interburst interval. Finally, the Csercsik model incorporated a simple form of soma-dendrite interaction, thus including, in a simplified form, the approach of Roberts and Suter.

For illustration, we consider the model of Lee et al (2010) in slightly more detail. Typical experimental results are shown in Fig. 8.16. A number of important points stand out. Firstly, as is usual in neuroendocrine cells, each burst of action potentials is tightly correlated with an increase in intracellular [Ca^{2+}]. However, in GnRH neurons, the Ca^{2+} transients are long lasting, and continue to increase after the action potentials have stopped. This suggests strongly that the Ca^{2+} in this transient is coming not only from voltage-gated Ca^{2+} channels, but also from internal stores. This conclusion is consistent with other evidence showing that depletion of the ER by thapsigargin, or blockage of the IPR, diminishes or prevents Ca^{2+} transients (Lee et al, 2010). Furthermore, experiments also show that the apamin-sensitive, Ca^{2+}-sensitive SK channel is responsible for termination of the electrical bursting, and that bursting is intrinsic to the GnRH neuron, not merely a response to external stimuli.

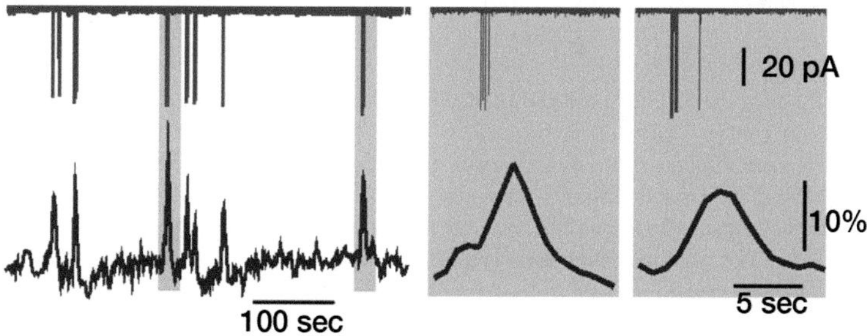

Fig. 8.16 Simultaneous recordings of membrane currents (upper trace) and intracellular [Ca^{2+}] (lower trace) from GnRH neurons in a GnRH-Pericam transgenic mouse brain slice. Each group of current spikes is accompanied by a spike in [Ca^{2+}]. The panels on the right show two close-up views of a single burst of action currents (upper trace) and the associated Ca^{2+} spike (lower trace), showing how [Ca^{2+}] continues to increase after the current spikes have stopped. Reproduced from Duan et al (2011), Fig. 1A, with permission from Elsevier.

However, this raises the question of what sets the interburst interval. For much of the period between bursts [Ca^{2+}] is sitting at its low resting level, and thus the SK channel cannot be the sole mechanism that determines the interburst interval.

To address this particular question (and others), Lee et al (2010) constructed a model of the GnRH neuron, closely based on the earlier work of LeBeau et al (2000). A schematic diagram of the model is shown in Fig. 8.17. The model includes ten different ionic currents: two types of Na^+ current ($I_{\rm Naf}$ and $I_{\rm Nap}$), three types of K^+ current ($I_{\rm K,dr}$, $I_{\rm K,ir}$ and $I_{\rm Km}$), two types of Ca^{2+} current ($I_{\rm Ca,L}$ and $I_{\rm Ca,T}$), a passive membrane leakage current ($I_{\rm leak}$), and two pacemaker K^+ currents ($I_{\rm AHP-SK}$ and $I_{\rm AHP-UCL}$). (The terminology here is a little baroque. AHP stands for After-HyperPolarisation, SK stands for Small-conductance K^+ channel, while the AHP-UCL K^+ channel is blocked by the compound 3-(triphenylmethylaminomethyl)pyridine, or UCL2077 for short; this compound was first developed by researchers at University College London.)

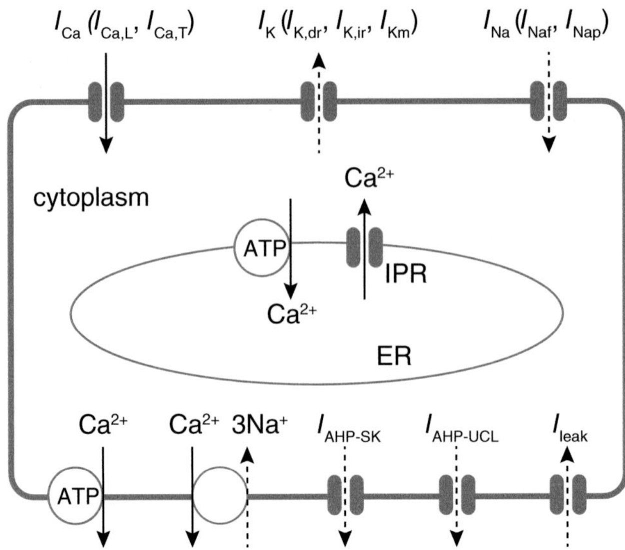

Fig. 8.17 Schematic diagram of the model of a GnRH neuron. I_{Na} amalgamates two different Na^+ currents, while I_K amalgamates three different K^+ currents. The I_K currents combine with the I_{Na} currents to generate oscillatory action current spiking, in the manner of the Hodgkin-Huxley model. I_{AHP-SK} is a Ca^{2+}-sensitive and apamin-sensitive K^+ current that is activated when $[Ca^{2+}]$ is raised, while $I_{AHP-UCL}$ is a Ca^{2+}-sensitive and time-dependent K^+ current that is activated by raised $[Ca^{2+}]$ and turns off only gradually, in a Ca^{2+}-independent manner. An inward leakage current (I_{leak}) is also included. There are five influx or efflux pathways for Ca^{2+}; the Na^+/Ca^{2+} exchanger, the plasma membrane ATPase pump, the IP_3 receptor, membrane Ca^{2+} channels, and the SERCA pump. The dashed lines represent the direction of net ionic fluxes and the Ca^{2+} fluxes are denoted by solid lines.

The Ca^{2+} submodel consists of two spatially homogeneous compartments, the cytosol and the ER, with fluxes described by standard equations as described in Chapter 2. Five Ca^{2+} fluxes are included in the Ca^{2+} submodel: influx through plasma-membrane channels (J_{in}), efflux through the Ca^{2+} ATPase (J_{pm}) and NCX (J_{NCX}), and release (J_{IPR}) and uptake (J_{SERCA}) of Ca^{2+} from the ER.

The most important Ca^{2+} flux is J_{IPR}, which models release of Ca^{2+} from the ER through the IPR, and thus mediates the Ca^{2+}-induced Ca^{2+} release that is a central feature of the model. The model assumes that the flux through the IPR is an algebraic function of $[Ca^{2+}]$, and thus all dynamic aspects of IPR modulation by Ca^{2+} are ignored. This simplification is not based on experimental evidence, but is made purely for convenience. It is justified in a purely post hoc manner, by the fact that the model makes predictions that have been confirmed experimentally.

Rather than show detailed simulations of the model, it is more useful instead to gain a qualitative understanding of how the model behaves. Each of these channels and Ca^{2+} fluxes has a specific role to play. The Na^+ and K^+ currents generate spontaneous electrical spiking, and bring in Ca^{2+} through the voltage-gated Ca^{2+} channels. This Ca^{2+} entry causes the release of more Ca^{2+} through the IPR (CICR modelled by J_{IPR}), which activates the pacemaker currents, $I_{\mathrm{AHP-SK}}$ and $I_{\mathrm{AHP-UCL}}$. $I_{\mathrm{AHP-SK}}$ acts relatively quickly to increase K^+ current, hyperpolarise the cell and stop the spiking, while $I_{\mathrm{AHP-UCL}}$ is a slower and longer-lasting K^+ current that prevents bursting for a longer time. Thus, to a good approximation, $I_{\mathrm{AHP-SK}}$ determines the length of the burst, while $I_{\mathrm{AHP-UCL}}$ sets the interburst interval. It is the different time scales of Ca^{2+} interaction with the two different Ca^{2+}-sensitive K^+ channels that modulates both the intraburst and the interburst properties.

Interestingly, the existence of the $I_{\mathrm{AHP-UCL}}$ current was predicted by the model before it was discovered experimentally. As the modelling work showed, without a slow Ca^{2+}-sensitive K^+ current (i.e., a K^+ current that is activated by Ca^{2+} but only slowly inactivates) it is difficult to reproduce the correct interburst interval. Thus motivated, additional experiments were performed, and such a current was found by Lee et al (2010).

The model of Lee et al (2010) was analysed and simplified by Duan et al (2011), and was extended into a spatial model of the GnRH dendron by Chen et al (2013) and Chen and Sneyd (2015). However, since these are concerned largely with electrical aspects of the GnRH neuron, they are beyond the scope of this book.

References

Aalkjaer C, Nilsson H (2005) Vasomotion: cellular background for the oscillator and for the synchronization of smooth muscle cells. Br J Pharmacol 144(5):605–16, DOI: 10.1038/sj.bjp.0706084 326, 328

Abbott LF, Nelson SB (2000) Synaptic plasticity: taming the beast. Nat Neurosci 3 Suppl:1178–83, DOI: 10.1038/81453 344, 345

Akopian A, Witkovsky P (2002) Calcium and retinal function. Mol Neurobiol 25(2):113–32, DOI: 10.1385/MN:25:2:113 16

Alberts B, Johnson A, Lewis J, Raff M, Roberts K, Walter P (2002) Molecular Biology of the Cell. Garland Science, New York, London viii

Almassy J, Won JH, Begenisich TB, Yule DI (2012) Apical Ca^{2+}-activated potassium channels in mouse parotid acinar cells. J Gen Physiol 139(2):121–33, DOI: 10.1085/jgp.201110718 283

Alzayady KJ, Wagner LE 2nd, Chandrasekhar R, Monteagudo A, Godiska R, Tall GG, Joseph SK, Yule DI (2013) Functional inositol 1,4,5-trisphosphate receptors assembled from concatenated homo- and heteromeric subunits. J Biol Chem 288(41):29,772–84, DOI: 10.1074/jbc.M113.502203 11

Ambudkar IS (2014) Ca^{2+} signaling and regulation of fluid secretion in salivary gland acinar cells. Cell Calcium 55(6):297–305, DOI: 10.1016/j.ceca.2014.02.009 274

Arizono M, Bannai H, Nakamura K, Niwa F, Enomoto M, Matsu-Ura T, Miyamoto A, Sherwood MW, Nakamura T, Mikoshiba K (2012) Receptor-selective diffusion barrier enhances sensitivity of astrocytic processes to metabotropic glutamate receptor stimulation. Sci Signal 5(218):ra27, DOI: 10.1126/scisignal.2002498 294

Ashby MC, Tepikin AV (2002) Polarized calcium and calmodulin signaling in secretory epithelia. Physiol Rev 82(3):701–34, DOI: 10.1152/physrev.00006.2002 274

Atkinson PJ, Young KW, Ennion SJ, Kew JNC, Nahorski SR, Challiss RAJ (2006) Altered expression of $G_{q/11}\alpha$ protein shapes mGlu1 and mGlu5 receptor-mediated single cell inositol 1,4,5-trisphosphate and Ca^{2+} signaling. Mol Pharmacol 69(1):174–84, DOI: 10.1124/mol.105.014258 287

Atri A, Amundson J, Clapham D, Sneyd J (1993) A single-pool model for intracellular calcium oscillations and waves in the *Xenopus laevis* oocyte. Biophys J 65(4):1727–39, DOI: 10.1016/S0006-3495(93)81191-3 25, 76, 81, 117, 153, 208, 246, 247, 248, 248

Atwater I, Dawson CM, Scott A, Eddlestone G, Rojas E (1980) The nature of the oscillatory behavior in electrical activity for pancreatic β-cell. J Horm Metabol Res 10 (suppl.):100–107 359

Backx PH, de Tombe P, Van Deen J, Mulder BJM, ter Keurs H (1989) A model of propagating calcium-induced calcium release mediated by calcium diffusion. J Gen Physiol 93:963–77, DOI: 10.1085/jgp.93.5.963 160

G. Dupont et al., *Models of Calcium Signalling*, Interdisciplinary Applied Mathematics 43, DOI 10.1007/978-3-319-29647-0

Bai Y, Edelmann M, Sanderson MJ (2009) The contribution of inositol 1,4,5-trisphosphate and ryanodine receptors to agonist-induced Ca^{2+} signaling of airway smooth muscle cells. Am J Physiol Lung Cell Mol Physiol 297(2):L347–61, DOI: 10.1152/ajplung.90559.2008 315

Bao R, Lifshitz LM, Tuft RA, Bellvé K, Fogarty KE, ZhuGe R (2008) A close association of RyRs with highly dense clusters of Ca^{2+}-activated Cl^- channels underlies the activation of STICs by Ca^{2+} sparks in mouse airway smooth muscle. J Gen Physiol 132(1):145–60, DOI: 10.1085/jgp.200709933 317

Barg S (2003) Mechanisms of exocytosis in insulin-secreting β-cells and glucagon-secreting α-cells. Pharmacol Toxicol 92(1):3–13, DOI: 10.1034/j.1600-0773.2003.920102.x 371

Bassani JW, Yuan W, Bers DM (1995) Fractional SR Ca release is regulated by trigger Ca and SR Ca content in cardiac myocytes. Am J Physiol 268(5):C1313–19 85

Bauer PJ (2001) The local Ca concentration profile in the vicinity of a Ca channel. Cell Biochem Biophys 35(1):49–61, DOI: 10.1385/CBB:35:1:49 128

Baughman JM, Perocchi F, Girgis HS, Plovanich M, Belcher-Timme CA, Sancak Y, Bao XR, Strittmatter L, Goldberger O, Bogorad RL, Koteliansky V, Mootha VK (2011) Integrative genomics identifies MCU as an essential component of the mitochondrial calcium uniporter. Nature 476(7360):341–5, DOI: 10.1038/nature10234 55

Baylor SM, Hollingworth S (1998) Model of sarcomeric Ca^{2+} movements, including ATP Ca^{2+} binding and diffusion, during activation of frog skeletal muscle. J Gen Physiol 112(3):297–316, DOI: 10.1085/jgp.112.3.297 313

Baylor SM, Hollingworth S (2012) Intracellular calcium movements during excitation-contraction coupling in mammalian slow-twitch and fast-twitch muscle fibers. J Gen Physiol 139(4):261–72, DOI: 10.1085/jgp.201210773 313

Baylor SM, Hollingworth S, Chandler WK (2002) Comparison of simulated and measured calcium sparks in intact skeletal muscle fibers of the frog. J Gen Physiol 120(3):349–68, DOI: 10.1085/jgp.20028620 313

Beeler G, Reuter H (1977) Reconstruction of the action potential of ventricular myocardial fibers. J Physiol 268:177–210, DOI: 10.1113/jphysiol.1977.sp011853 296

Bell D, Deng B (2002) Singular perturbation of N-front travelling waves in the FitzHugh–Nagumo equations. Nonlinear analysis: Real world applications 3(4):515–41, DOI: 10.1016/s1468-1218(01)00046-3 238

Bennett MR, Farnell L, Gibson WG (2004) The facilitated probability of quantal secretion within an array of calcium channels of an active zone at the amphibian neuromuscular junction. Biophys J 86(5):2674–90, DOI: 10.1016/S0006-3495(04)74323-4 339

Bennett MR, Farnell L, Gibson WG (2005) A quantitative model of purinergic junctional transmission of calcium waves in astrocyte networks. Biophys J 89(4):2235–50, DOI: 10.1529/biophysj.105.062968 156, 156, 156

Bennett MR, Farnell L, Gibson WG, Dickens P (2007) Mechanisms of calcium sequestration during facilitation at active zones of an amphibian neuromuscular junction. J Theor Biol 247(2):230–41, DOI: 10.1016/j.jtbi.2007.03.022 339

Bentele K, Falcke M (2007) Quasi-steady approximation for ion channel currents. Biophys J 93(8):2597–608, DOI: 10.1529/biophysj.107.104299 134, 134, 134, 135, 173, 184

Berridge MJ (1990) Calcium oscillations. J Biol Chem 265:9583–86 174

Berridge MJ (2008) Smooth muscle cell calcium activation mechanisms. J Physiol 586(21):5047–61, DOI: 10.1113/jphysiol.2008.160440 314, 315

Berridge MJ, Irvine R (1989) Inositol phosphates and cell signalling. Nature 341:197–205, DOI: 10.1038/341197a0 13

Berridge MJ, Lipp P, Bootman MD (2000) The versatility and universality of calcium signalling. Nat Rev Mol Cell Biol 1(1):11–21, DOI: 10.1038/35036035 6

Bers DM (2001) Excitation-Contraction Coupling and Cardiac Contractile Force., 2nd edn. Kluwer, New York, DOI: 10.1007/978-94-010-0658-3 295, 297

Bers DM (2002) Cardiac excitation–contraction coupling. Nature 415(6868):198–205, DOI: 10.1038/415198a 296, 297, 330

Bertram R, Sherman A (2000) Dynamical complexity and temporal plasticity in pancreatic beta-cells. J Biosci 25(2):197–209 232

Bertram R, Sherman A (2004) A calcium-based phantom bursting model for pancreatic islets. Bull Math Biol 66(5):1313–44, DOI: 10.1016/j.bulm.2003.12.005 358, 364

Bertram R, Sherman A, Stanley EF (1996) Single-domain/bound calcium hypothesis of transmitter release and facilitation. J Neurophysiol 75(5):1919–31 340

Bertram R, Smith GD, Sherman A (1999) Modeling study of the effects of overlapping Ca^{2+} microdomains on neurotransmitter release. Biophys J 76(2):735–50, DOI: 10.1016/S0006-3495(99)77240-1 308

Bertram R, Satin L, Zhang M, Smolen P, Sherman A (2004) Calcium and glycolysis mediate multiple bursting modes in pancreatic islets. Biophys J 87(5):3074–87, DOI: 10.1529/biophysj.104.049262 358, 364

Bertram R, Egli M, Toporikova N, Freeman ME (2006) A mathematical model for the mating-induced prolactin rhythm of female rats. Am J Physiol Endocrinol Metab 290(3):E573–82, DOI: 10.1152/ajpendo.00428.2005 232

Bertram R, Satin LS, Pedersen MG, Luciani DS, Sherman A (2007a) Interaction of glycolysis and mitochondrial respiration in metabolic oscillations of pancreatic islets. Biophys J 92(5):1544–55, DOI: 10.1529/biophysj.106.097154 358

Bertram R, Sherman A, Satin LS (2007b) Metabolic and electrical oscillations: partners in controlling pulsatile insulin secretion. Am J Physiol Endocrinol Metab 293(4):E890–900, DOI: 10.1152/ajpendo.00359.2007 358, 364

Bertram R, Sherman A, Satin LS (2010) Electrical bursting, calcium oscillations, and synchronization of pancreatic islets. Adv Exp Med Biol 654:261–79, DOI: 10.1007/978-90-481-3271-3_12 232, 358

Beuter A, Glass L, Mackey M, Titcombe M (2003) Nonlinear Dynamics in Physiology and Medicine. Springer-Verlag, New York viii

Bezprozvanny I, Watras J, Ehrlich BE (1991) Bell-shaped calcium-response curves of Ins(1,4,5)P_3- and calcium-gated channels from endoplasmic reticulum of cerebellum. Nature 351(6329):751–4, DOI: 10.1038/351751a0 73

Bird GS, Obie JF, Putney JW (1997) Effect of cytoplasmic Ca^{2+} on (1,4,5)IP_3 formation in vasopressin-activated hepatocytes. Cell Calcium 21(3):253–6, DOI: 10.1016/s0143-4160(97)90049-x 12, 94, 259

Blank JL, Ross AH, Exton JH (1991) Purification and characterization of two G-proteins that activate the $\beta 1$ isozyme of phosphoinositide-specific phospholipase C. Identification as members of the G_q class. J Biol Chem 266(27):18,206–16 94

Blaustein MP, Lederer WJ (1999) Sodium/calcium exchange: its physiological implications. Physiol Rev 79(3):763–854 44

Boie S, Kirk V, Sneyd J, Wechselberger M (2016) Effects of quasi-steady state reduction on biophysical models with oscillations. J Theor Biol 393:16–31, DOI: 10.1016/j.jtbi.2015.12.011 216, 216, 218, 219, 219

Boitano S, Dirksen ER, Sanderson MJ (1992) Intercellular propagation of calcium waves mediated by inositol trisphosphate. Science 258(5080):292–5, DOI: 10.1126/science.1411526 26, 151

Bootman M, Niggli E, Berridge M, Lipp P (1997a) Imaging the hierarchical Ca^{2+} signalling system in HeLa cells. J Physiol 499(2):307–14, DOI: 10.1113/jphysiol.1997.sp021928 20

Bootman MD, Berridge MJ, Lipp P (1997b) Cooking with calcium: the recipes for composing global signals from elementary events. Cell 91(3):367–73, DOI: 10.1016/s0092-8674(00)80420-1 20, 176, 176

Bradley SJ, Challiss RAJ (2011) Defining protein kinase/phosphatase isoenzymic regulation of $mGlu_5$ receptor-stimulated phospholipase C and Ca^{2+} responses in astrocytes. Br J Pharmacol 164(2b):755–71, DOI: 10.1111/j.1476-5381.2011.01421.x 289, 291, 291

Bradley SJ, Challiss RAJ (2012) G protein-coupled receptor signalling in astrocytes in health and disease: a focus on metabotropic glutamate receptors. Biochem Pharmacol 84(3):249–59, DOI: 10.1016/j.bcp.2012.04.009 285

van Breemen C, Fameli N, Evans AM (2013) Pan-junctional sarcoplasmic reticulum in vascular smooth muscle: nanospace Ca^{2+} transport for site- and function-specific Ca^{2+} signalling. J Physiol 591(8):2043–54, DOI: 10.1113/jphysiol.2012.246348 135

Britton N (1986) Reaction-Diffusion Equations and their Applications to Biology. Academic Press, London 144

Brown SA, Morgan F, Watras J, Loew LM (2008) Analysis of phosphatidylinositol-4,5-bisphosphate signaling in cerebellar purkinje spines. Biophys J 95(4):1795–812, DOI: 10.1529/biophysj.108.130195 95

Brumen M, Fajmut A, Dobovišek A, Roux E (2005) Mathematical modelling of Ca^{2+} oscillations in airway smooth muscle cells. J Biol Phys 31(3–4):515–24, DOI: 10.1007/s10867-005-2409-4 317

Bugrim A, Fontanilla R, Eutenier BB, Keizer J, Nuccitelli R (2003) Sperm initiate a Ca^{2+} wave in frog eggs that is more similar to Ca^{2+} waves initiated by IP_3 than by Ca^{2+}. Biophys J 84(3):1580–90, DOI: 10.1016/S0006-3495(03)74968-6 251, 253, 254

Burdyga T, Wray S (2005) Action potential refractory period in ureter smooth muscle is set by Ca sparks and BK channels. Nature 436(7050):559–62, DOI: 10.1038/nature03834 317

Burnashev N (1998) Calcium permeability of ligand-gated channels. Cell Calcium 24(5–6):325–32, DOI: 10.1016/s0143-4160(98)90056-2 16

Bursztyn L, Eytan O, Jaffa AJ, Elad D (2007) Mathematical model of excitation-contraction in a uterine smooth muscle cell. Am J Physiol Cell Physiol 292(5):C1816–29, DOI: 10.1152/ajpcell.00478.2006 315, 331

Bush D, Jin Y (2012) Calcium control of triphasic hippocampal STDP. J Comput Neurosci 33(3):495–514, DOI: 10.1007/s10827-012-0397-5 344, 353

Callamaras N, Marchant JS, Sun XP, Parker I (1998) Activation and co-ordination of $InsP_3$-mediated elementary Ca^{2+} events during global Ca^{2+} signals in *Xenopus* oocytes. J Physiol 509(1):81–91, DOI: 10.1111/j.1469-7793.1998.081bo.x 20

Cannell MB, Allen DG (1984) Model of calcium movements during activation in the sarcomere of frog skeletal muscle. Biophys J 45:913–25, DOI: 10.1016/s0006-3495(84)84238-1 313

Cannell MB, Soeller C (1997) Numerical analysis of ryanodine receptor activation by L-type channel activity in the cardiac muscle diad. Biophys J 73(1):112–22, DOI: 10.1016/S0006-3495(97)78052-4 301

Cannell MB, Soeller C (1999) Mechanisms underlying calcium sparks in cardiac muscle. J Gen Physiol 113(3):373–6, DOI: 10.1085/jgp.113.3.373 20, 299

Cannell MB, Berlin JR, Lederer WJ (1987) Effect of membrane potential changes on the calcium transient in single rat cardiac muscle cells. Science 238(4832):1419–23, DOI: 10.1126/science.2446391 297

Cannell MB, Kong CHT, Imtiaz MS, Laver DR (2013) Control of sarcoplasmic reticulum Ca^{2+} release by stochastic RyR gating within a 3D model of the cardiac dyad and importance of induction decay for CICR termination. Biophys J 104(10):2149–59, DOI: 10.1016/j.bpj.2013.03.058 89, 89, 302, 302, 302, 303, 303

Cao P, Donovan G, Falcke M, Sneyd J (2013) A stochastic model of calcium puffs based on single-channel data. Biophys J 105(5):1133–42, DOI: 10.1016/j.bpj.2013.07.034 71, 78, 185, 189, 319, 319, 319

Cao P, Tan X, Donovan G, Sanderson MJ, Sneyd J (2014) A deterministic model predicts the properties of stochastic calcium oscillations in airway smooth muscle cells. PLoS Comput Biol 10(8):e1003,783, DOI: 10.1371/journal.pcbi.1003783 78, 79, 81, 81, 176, 316, 317, 317, 317, 317, 317, 319, 319, 319

Capite JD, Ng SW, Parekh AB (2009) Decoding of cytoplasmic Ca^{2+} oscillations through the spatial signature drives gene expression. Curr Biol 19(10):853–8, DOI: 10.1016/j.cub.2009.03.063 174

Carafoli E (1987) Intracellular calcium homeostasis. Annu Rev Biochem 56:395–433, DOI: 10.1146/annurev.bi.56.070187.002143 5

Carafoli E (2003) Historical review: Mitochondria and calcium: ups and downs of an unusual relationship. Trends Biochem Sci 28(4):175–81, DOI: 10.1016/S0968-0004(03)00053-7 8, 51

Carafoli E, Gamble RL, Lehninger AL (1966) Rebounds and oscillations in respiration-linked movements of Ca^{++} and H^+ in rat liver mitochondria. J Biol Chem 241(11):2644–52 52

Carter P, Sandstede B (2014) Fast pulses with oscillatory tails in the FitzHugh–Nagumo system 238, 238

Catterall WA (2000) Structure and regulation of voltage-gated Ca^{2+} channels. Annu Rev Cell Dev Biol 16:521–55, DOI: 10.1146/annurev.cellbio.16.1.521 14, 56

Catterall WA (2011) Voltage-gated calcium channels. Cold Spring Harb Perspect Biol 3(8):a003,947, DOI: 10.1101/cshperspect.a003947 14, 56, 57

Chad J, Eckert R, Ewald D (1984) Kinetics of calcium-dependent inactivation of calcium current in voltage-clamped neurones of *Aplysia californica*. J Physiol 347:279–300, DOI: 10.1113/jphysiol.1984.sp015066 59

Chad JE, Eckert R (1986) An enzymatic mechanism for calcium current inactivation in dialysed Helix neurones. J Physiol 378:31–51, DOI: 10.1113/jphysiol.1986.sp016206 59

Champneys A, Kirk V, Knobloch E, Oldeman B, Sneyd J (2007) When Shil'nikov meets Hopf in excitable systems. SIAM J Appl Dyn Syst 6:663–93, DOI: 10.1137/070682654 236, 237, 237, 238, 241

Charles AC, Merrill JE, Dirksen ER, Sanderson MJ (1991) Intercellular signaling in glial cells: calcium waves and oscillations in response to mechanical stimulation and glutamate. Neuron 6(6):983–92, DOI: 10.1016/0896-6273(91)90238-u 26

Charles AC, Naus CC, Zhu D, Kidder GM, Dirksen ER, Sanderson MJ (1992) Intercellular calcium signaling via gap junctions in glioma cells. J Cell Biol 118(1):195–201, DOI: 10.1083/jcb.118.1.195 26

Chatton JY, Cao Y, Stucki JW (1998) Perturbation of myo-inositol-1,4,5-trisphosphate levels during agonist-induced Ca^{2+} oscillations. Biophys J 74(1):523–31, DOI: 10.1016/S0006-3495(98)77809-9 259

Chay TR (1986) On the effect of the intracellular calcium-sensitive K^+ channel in the bursting pancreatic β-cell. Biophys J 50:765–77, DOI: 10.1016/s0006-3495(86)83517-2 358

Chay TR (1987) The effect of inactivation of calcium channels by intracellular Ca^{2+} ions in the bursting pancreatic β-cell. Cell Biophysics 11:77–90, DOI: 10.1007/bf02797114 358

Chay TR (1996a) Electrical bursting and luminal calcium oscillation in excitable cell models. Biol Cybern 75(5):419–31, DOI: 10.1007/s004220050307 364

Chay TR (1996b) Modeling slowly bursting neurons via calcium store and voltage-independent calcium current. Neural Comput 8(5):951–78, DOI: 10.1162/neco.1996.8.5.951 364

Chay TR (1997) Effects of extracellular calcium on electrical bursting and intracellular and luminal calcium oscillations in insulin secreting pancreatic beta-cells. Biophys J 73(3):1673–88, DOI: 10.1016/S0006-3495(97)78199-2 364

Chay TR, Cook DL (1988) Endogenous bursting patterns in excitable cells. Math Biosci 90:139–53, DOI: 10.1016/0025-5564(88)90062-4 358

Chay TR, Keizer J (1983) Minimal model for membrane oscillations in the pancreatic β-cell. Biophys J 42:181–90, DOI: 10.1016/s0006-3495(83)84384-7 57, 358, 358, 359

Chen D, Eisenberg R (1993a) Charges, currents, and potentials in ionic channels of one conformation. Biophys J 64:1405–21, DOI: 10.1016/s0006-3495(93)81507-8 59

Chen DP, Eisenberg RS (1993b) Flux, coupling, and selectivity in ionic channels of one conformation. Biophys J 65(2):727–46, DOI: 10.1016/s0006-3495(93)81099-3 59

Chen DP, Barcilon V, Eisenberg RS (1992) Constant fields and constant gradients in open ionic channels. Biophys J 61(5):1372–93, DOI: 10.1016/s0006-3495(92)81944-6 59

Chen L, Koh DS, Hille B (2003) Dynamics of calcium clearance in mouse pancreatic beta-cells. Diabetes 52(7):1723–31, DOI: 10.2337/diabetes.52.7.1723 365

Chen X, Sneyd J (2015) A computational model of the dendron of the GnRH neuron. Bull Math Biol 76(6):904–26, DOI: 10.1007/s11538-014-0052-6 385

Chen X, Iremonger K, Herbison A, Kirk V, Sneyd J (2013) Regulation of electrical bursting in a spatiotemporal model of a GnRH neuron. Bull Math Biol 75(10):1941–60, DOI: 10.1007/s11538-013-9877-7 385

Chen XF, Li CX, Wang PY, Li M, Wang WC (2008a) Dynamic simulation of the effect of calcium-release activated calcium channel on cytoplasmic Ca^{2+} oscillation. Biophys Chem 136(2–3):87–95, DOI: 10.1016/j.bpc.2008.04.010 365

Chen Yd, Wang S, Sherman A (2008b) Identifying the targets of the amplifying pathway for insulin secretion in pancreatic beta-cells by kinetic modeling of granule exocytosis. Biophys J 95(5):2226–41, DOI: 10.1529/biophysj.107.124990 373

Cheng H, Lederer WJ, Cannell MB (1993) Calcium sparks: elementary events underlying excitation-contraction coupling in heart muscle. Science 262(5134):740–4, DOI: 10.1126/science.8235594 20, 301, 319

Cheng H, Lederer MR, Lederer WJ, Cannell MB (1996) Calcium sparks and $[Ca^{2+}]_i$ waves in cardiac myocytes. Am J Physiol 270(1):C148–59 20

Christel C, Lee A (2012) Ca^{2+}-dependent modulation of voltage-gated Ca^{2+} channels. Biochim Biophys Acta 1820(8):1243–52, DOI: 10.1016/j.bbagen.2011.12.012 14

Cinlar E (1969) Markov renewal theory. Advances in Applied Probability 1(2):123–87, DOI: 10.2307/1426216 185

Cinlar E (1975) Markov renewal theory: a survey. Management Science 21(7):727–52, DOI: 10.1287/mnsc.21.7.727 185

Clair C, Chalumeau C, Tordjmann T, Poggioli J, Erneux C, Dupont G, Combettes L (2001) Investigation of the roles of Ca^{2+} and $InsP_3$ diffusion in the coordination of Ca^{2+} signals between connected hepatocytes. J Cell Sci 114(11):1999–2007 271, 272

Colegrove S, Albrecht M, Friel D (2000a) Quantitative analysis of mitochondrial Ca^{2+} uptake and release pathways in sympathetic neurons. Reconstruction of the recovery after depolarization-evoked $[Ca^{2+}]_i$ elevations. J Gen Physiol 115:371–88, DOI: 10.1085/jgp.115.3.371 56

Colegrove SL, Albrecht MA, Friel D (2000b) Dissection of mitochondrial Ca^{2+} uptake and release fluxes in situ after depolarization-evoked $[Ca^{2+}]_i$ elevations in sympathetic neurons. J Gen Physiol 115(3):351–70 56, 280

Colquhoun D (2006) The quantitative analysis of drug–receptor interactions: a short history. Trends Pharmacol Sci 27(3):149–57, DOI: 10.1016/j.tips.2006.01.008 30

Constantin S, Jasoni C, Romanò N, Lee K, Herbison AE (2012) Understanding calcium homeostasis in postnatal gonadotropin-releasing hormone neurons using cell-specific pericam transgenics. Cell Calcium 51(3–4):267–76, DOI: 10.1016/j.ceca.2011.11.005 381, 381

Coombes S (2001) The effect of ion pumps on the speed of travelling waves in the fire-diffuse-fire model of Ca^{2+} release. Bull Math Biol 63(1):1–20, DOI: 10.1006/bulm.2000.0193 140, 142

Coombes S, Bressloff PC (2003) Saltatory waves in the spike-diffuse-spike model of active dendritic spines. Phys Rev Lett 91(2):028,102, DOI: 10.1103/PhysRevLett.91.028102 140

Coombes S, Timofeeva Y (2003) Sparks and waves in a stochastic fire-diffuse-fire model of Ca^{2+} release. Phys Rev E Stat Nonlin Soft Matter Phys 68(2):021,915, DOI: 10.1103/PhysRevE.68.021915 140

Coombes S, Hinch R, Timofeeva Y (2004) Receptors, sparks and waves in a fire-diffuse-fire framework for calcium release. Prog Biophys Mol Biol 85(2–3):197–216, DOI: 10.1016/j.pbiomolbio.2004.01.015 140

Cornell-Bell AH, Finkbeiner SM, Cooper MS, Smith SJ (1990) Glutamate induces calcium waves in cultured astrocytes: long-range glial signaling. Science 247(4941):470–3, DOI: 10.1126/science.1967852 27

Corry B, Allen TW, Kuyucak S, Chung SH (2000) Mechanisms of permeation and selectivity in calcium channels. Biophys J 80(1):195–214, DOI: 10.1016/s0006-3495(01)76007-9 59

Cortassa S, Aon MA, Marbán E, Winslow RL, O'Rourke B (2003) An integrated model of cardiac mitochondrial energy metabolism and calcium dynamics. Biophys J 84(4):2734–55, DOI: 10.1016/S0006-3495(03)75079-6 52, 56, 56

Cox D (1970) Renewal Theory. Methuen & Co, London. 185

Croisier H, Tan X, Perez-Zoghbi JF, Sanderson MJ, Sneyd J, Brook BS (2013) Activation of store-operated calcium entry in airway smooth muscle cells: insight from a mathematical model. PLoS One 8(7):e69,598, DOI: 10.1371/journal.pone.0069598 65, 317, 318, 318, 319

Csercsik D, Farkas I, Hrabovszky E, Liposits Z (2012) A simple integrative electrophysiological model of bursting GnRH neurons. J Comput Neurosci 32(1):119–36, DOI: 10.1007/s10827-011-0343-y 382

Csordás G, Hajnóczky G (2003) Plasticity of mitochondrial calcium signaling. J Biol Chem 278(43):42,273–82, DOI: 10.1074/jbc.M305248200 53

Csordás G, Hajnóczky G (2009) SR/ER-mitochondrial local communication: calcium and ROS. Biochim Biophys Acta 1787(11):1352–62, DOI: 10.1016/j.bbabio.2009.06.004 9, 17, 52

Csordás G, Thomas AP, Hajnóczky G (2001) Calcium signal transmission between ryanodine receptors and mitochondria in cardiac muscle. Trends Cardiovasc Med 11(7):269–75, DOI: 10.1016/S1050-1738(01)00123-2 9, 52

Csordás G, Golenár T, Seifert EL, Kamer KJ, Sancak Y, Perocchi F, Moffat C, Weaver D, de la Fuente Perez S, Bogorad R, Koteliansky V, Adijanto J, Mootha VK, Hajnóczky G (2013) MICU1 controls both the threshold and cooperative activation of the mitochondrial Ca^{2+} uniporter. Cell Metab 17(6):976–87, DOI: 10.1016/j.cmet.2013.04.020 17, 53, 55

Cuthbertson KSR, Chay T (1991) Modelling receptor-controlled intracellular calcium oscillators. Cell Calcium 12:97–109, DOI: 10.1016/0143-4160(91)90012-4 30, 259

Dakin K, Li WH (2007) Cell membrane permeable esters of D-myo-inositol 1,4,5-trisphosphate. Cell Calcium 42(3):291–301, DOI: 10.1016/j.ceca.2006.12.003 119

Dargan SL, Parker I (2003) Buffer kinetics shape the spatiotemporal patterns of IP_3-evoked Ca^{2+} signals. J Physiol 553(3):775–88, DOI: 10.1113/jphysiol.2003.054247 159, 161

Dargan SL, Schwaller B, Parker I (2004) Spatiotemporal patterning of IP_3-mediated Ca^{2+} signals in *Xenopus* oocytes by Ca^{2+}-binding proteins. J Physiol 556(2):447–61, DOI: 10.1113/jphysiol.2003.059204 161

Dash RK, Beard DA (2008) Analysis of cardiac mitochondrial Na^+-Ca^{2+} exchanger kinetics with a biophysical model of mitochondrial Ca^{2+} handling suggests a 3:1 stoichiometry. J Physiol 586(13):3267–85, DOI: 10.1113/jphysiol.2008.151977 56

Dash RK, Qi F, Beard DA (2009) A biophysically based mathematical model for the kinetics of mitochondrial calcium uniporter. Biophys J 96(4):1318–32, DOI: 10.1016/j.bpj.2008.11.005 54, 55

De Koninck P, Schulman H (1998) Sensitivity of CaM kinase II to the frequency of Ca^{2+} oscillations. Science 279(5348):227–30, DOI: 10.1126/science.279.5348.227 353, 354, 356, 356

De Lean A, Stadel JM, Lefkowitz RJ (1980) A ternary complex model explains the agonist-specific binding properties of the adenylate cyclase-coupled beta-adrenergic receptor. J Biol Chem 255(15):7108–17 33, 33

De Pittà M, Goldberg M, Volman V, Berry H, Ben-Jacob E (2009) Glutamate regulation of calcium and IP_3 oscillating and pulsating dynamics in astrocytes. J Biol Phys 35(4):383–411, DOI: 10.1007/s10867-009-9155-y 294

De Pittà M, Volman V, Levine H, Ben-Jacob E (2009) Multimodal encoding in a simplified model of intracellular calcium signaling. Cogn Process 10 Suppl 1:S55–70, DOI: 10.1007/s10339-008-0242-y 273

De Stefani D, Raffaello A, Teardo E, Szabò I, Rizzuto R (2011) A forty-kilodalton protein of the inner membrane is the mitochondrial calcium uniporter. Nature 476(7360):336–40, DOI: 10.1038/nature10230 55

De Young GW, Keizer J (1992) A single-pool inositol 1,4,5-trisphosphate-receptor-based model for agonist-stimulated oscillations in Ca^{2+} concentration. Proc Natl Acad Sci USA 89(20):9895–9, DOI: 10.1073/pnas.89.20.9895 72, 73, 81, 101, 189, 319

Destexhe A, Huguenard JR (2000) Nonlinear thermodynamic models of voltage-dependent currents. J Comp Neurosci 9(3):259–70 58

Dickinson GD, Parker I (2013) Factors determining the recruitment of inositol trisphosphate receptor channels during calcium puffs. Biophys J 105(11):2474–84, DOI: http://dx.doi.org/10.1016/j.bpj.2013.10.028 166, 166, 167, 167, 169, 169, 169, 169, 186, 187

Dickinson GD, Swaminathan D, Parker I (2012) The probability of triggering calcium puffs is linearly related to the number of inositol trisphosphate receptors in a cluster. Biophys J 102(8):1826–36, DOI: 10.1016/j.bpj.2012.03.029 169, 173, 173, 173, 189, 200

Dickson EJ, Falkenburger BH, Hille B (2013) Quantitative properties and receptor reserve of the IP_3 and calcium branch of G_q-coupled receptor signaling. J Gen Physiol 141(5):521–35, DOI: 10.1085/jgp.201210886 34

Diderichsen PM, Göpel SO (2006) Modelling the electrical activity of pancreatic α-cells based on experimental data from intact mouse islets. J Biol Phys 32(3–4):209–29, DOI: 10.1007/s10867-006-9013-0 365, 366, 367, 367, 367, 368, 370, 375

Doedel EJ (1981) AUTO: A program for the automatic bifurcation analysis of autonomous systems. Congr Numer 30:265–84 209

Doedel EJ, Champneys AR, Fairgrieve TF, Kuznetsov YA, Sandstede B, Wang X (1998) Auto97: Continuation and bifurcation software for ordinary differential equations. Available for download from http://indy.cs.concordia.ca/auto 209

Dolmetsch RE, Xu K, Lewis RS (1998) Calcium oscillations increase the efficiency and specificity of gene expression. Nature 392(6679):933–6, DOI: 10.1038/31960 377

Dolphin AC (2006) A short history of voltage-gated calcium channels. Br J Pharmacol 147 Suppl 1:S56–62, DOI: 10.1038/sj.bjp.0706442 14, 56

Domijan M, Murray R, Sneyd J (2006) Dynamical probing of the mechanisms underlying calcium oscillations. J Nonlin Sci 16(5):483–506, DOI: 10.1007/s00332-005-0744-z 115, 117, 208, 209, 228

Duan W, Lee K, Herbison AE, Sneyd J (2011) A mathematical model of adult GnRH neurons in mouse brain and its bifurcation analysis. J Theor Biol 276(1):22–34, DOI: 10.1016/j.jtbi.2011.01.035 219, 383, 385

Dufour JF, Arias I, Turner T (1997) Inositol 1,4,5-trisphosphate and calcium regulate the calcium channel function of the hepatic inositol 1,4,5-trisphosphate receptor. J Biol Chem 272:2675–81, DOI: 10.1074/jbc.272.5.2675 11, 71, 76

Duman JG, Chen L, Hille B (2008) Calcium transport mechanisms of PC12 cells. J Gen Physiol 131(4):307–23, DOI: 10.1085/jgp.200709915 17

Dupont G (1998) Theoretical insights into the mechanism of spiral Ca^{2+} wave initiation in *Xenopus* oocytes. Am J Physiol 275(1 Pt 1):C317–C322 248

Dupont G, Combettes L (2006) Modelling the effect of specific inositol 1,4,5-trisphosphate receptor isoforms on cellular Ca^{2+} signals. Biol Cell 98(3):171–82, DOI: 10.1042/BC20050032 70, 76

Dupont G, Combettes L (2009) What can we learn from the irregularity of Ca^{2+} oscillations? Chaos 19(3):037,112, DOI: 10.1063/1.3160569 267, 267

Dupont G, Erneux C (1997) Simulations of the effects of inositol 1,4,5-trisphosphate 3-kinase and 5-phosphatase activities on Ca^{2+} oscillations. Cell Calcium 22(5):321–31, DOI: 10.1016/s0143-4160(97)90017-8 75, 94, 96, 115, 259

Dupont G, Goldbeter A (1993) One-pool model for Ca^{2+} oscillations involving Ca^{2+} and inositol 1,4,5-trisphosphate as co-agonists for Ca^{2+} release. Cell Calcium 14:311–22, DOI: 10.1016/0143-4160(93)90052-8 65, 104

Dupont G, Swillens S (1996) Quantal release, incremental detection, and long-period Ca^{2+} oscillations in a model based on regulatory Ca^{2+}-binding sites along the permeation pathway. Biophys J 71(4):1714–22, DOI: 10.1016/s0006-3495(96)79373-6 82, 105, 105

Dupont G, Swillens S, Clair C, Tordjmann T, Combettes L (2000a) Hierarchical organization of calcium signals in hepatocytes: from experiments to models. Biochim Biophys Acta 1498(2–3):134–52, DOI: 10.1016/S0167-4889(00)00090-2 258, 269

Dupont G, Tordjmann T, Clair C, Swillens S, Claret M, Combettes L (2000b) Mechanism of receptor-oriented intercellular calcium wave propagation in hepatocytes. FASEB J 14(2):279–89 271, 272

Dupont G, Koukoui O, Clair C, Erneux C, Swillens S, Combettes L (2003) Ca^{2+} oscillations in hepatocytes do not require the modulation of $InsP_3$ 3-kinase activity by Ca^{2+}. FEBS Lett 534(1–3):101–5, DOI: 10.1016/s0014-5793(02)03789-4 115, 260, 262, 263, 354, 356, 356

Dupont G, Combettes L, Leybaert L (2007) Calcium dynamics: spatio-temporal organization from the subcellular to the organ level. Int Rev Cytol 261:193–245, DOI: 10.1016/S0074-7696(07)61005-5 146, 147

Dupont G, Abou-Lovergne A, Combettes L (2008) Stochastic aspects of oscillatory Ca^{2+} dynamics in hepatocytes. Biophys J 95(5):2193–202, DOI: 10.1529/biophysj.108.133777 264, 265, 268

Dupont G, Combettes L, Bird GS, Putney JW (2011a) Calcium oscillations. Cold Spring Harb Perspect Biol 3(3):a004,226, DOI: 10.1101/cshperspect.a004226 174

Dupont G, Lokenye EFL, Challiss RAJ (2011b) A model for Ca^{2+} oscillations stimulated by the type 5 metabotropic glutamate receptor: an unusual mechanism based on repetitive, reversible phosphorylation of the receptor. Biochimie 93(12):2132–38, DOI: 10.1016/j.biochi.2011.09.010 36, 287, 287, 290, 292, 293

Earnshaw BA, Bressloff PC (2010) A diffusion-activation model of CaMKII translocation waves in dendrites. J Comput Neurosci 28(1):77–89, DOI: 10.1007/s10827-009-0188-9 348

Edelstein-Keshet L (1988) Mathematical Models in Biology. Random House, New York viii

Edwards JR, Gibson WG (2010) A model for Ca^{2+} waves in networks of glial cells incorporating both intercellular and extracellular communication pathways. J Theor Biol 263(1):45–58, DOI: 10.1016/j.jtbi.2009.12.002 156

Eggermann E, Bucurenciu I, Goswami SP, Jonas P (2012) Nanodomain coupling between Ca^{2+} channels and sensors of exocytosis at fast mammalian synapses. Nat Rev Neurosci 13(1):7–21, DOI: 10.1038/nrn3125 338

Endo M, Tanaka M, Ogawa Y (1970) Calcium-induced release of calcium from the sarcoplasmic reticulum of skinned skeletal muscle fibres. Nature 228:34–6, DOI: 10.1038/228034a0 10, 83, 108

Ermentrout B (1996) Type I membranes, phase resetting curves, and synchrony. Neural Comput 8(5):979–1001, DOI: 10.1162/neco.1996.8.5.979 118

Ermentrout B (2002) Simulating, analyzing, and animating dynamical systems: a guide to XPPAUT for researchers and students. SIAM 209

Ermentrout G, Kopell N (1984) Frequency plateaus in a chain of weakly coupled oscillators. SIAM J Math Anal 15:215–37, DOI: 10.1137/0515019 149

Erneux T, Goldbeter A (2006) Rescue of the quasi-steady-state approximation in a model for oscillations in an enzymatic cascade. SIAM J Appl Math 67(2):305–20, DOI: 10.1137/060654359 216

Essin K, Gollasch M (2009) Role of ryanodine receptor subtypes in initiation and formation of calcium sparks in arterial smooth muscle: comparison with striated muscle. J Biomed Biotechnol 2009:135,249, DOI: 10.1155/2009/135249 317

Eva R, Bouyoucef-Cherchalli D, Patel K, Cullen PJ, Banting G (2012) IP_3 3-kinase opposes NGF driven neurite outgrowth. PLoS One 7(2):e32,386, DOI: 10.1371/journal.pone.0032386 13

Eyring H, Lumry R, Woodbury JW (1949) Some applications of modern rate theory to physiological systems. Record Chem Prog 10:100–14 58

Fabiato A, Fabiato F (1975) Contractions induced by a calcium-triggered release of calcium from the sarcoplasmic reticulum of single skinned cardiac cells. J Physiol 249(3):469–95, DOI: 10.1113/jphysiol.1975.sp011026 108

Fabiato A, Fabiato F (1978) Calcium-induced release of calcium from the sarcoplasmic reticulum of skinned cells from adult human, dog, cat, rabbit, rat, and frog hearts and from fetal and new-born rat ventricles. Ann N Y Acad Sci 307:491–522, DOI: 10.1111/j.1749-6632.1978.tb41979.x 108

Fajmut A, Brumen M, Schuster S (2005) Theoretical model of the interactions between Ca, calmodulin and myosin light chain kinase. FEBS Letters 579(20):4361–66, DOI: 10.1016/j.febslet.2005.06.076 331, 332, 332, 333, 333, 334, 334

Falcke M (2003a) Buffers and oscillations in intracellular Ca^{2+} dynamics. Biophys J 84(1):28–41, DOI: 10.1016/s0006-3495(03)74830-9 160, 190

Falcke M (2003b) On the role of stochastic channel behavior in intracellular Ca^{2+} dynamics. Biophys J 84(1):42–56, DOI: 10.1016/s0006-3495(03)74831-0 164, 176, 190

Falcke M (2004) Reading the patterns in living cells – the physics of Ca^{2+} signaling. Advances in Physics 53(3):255–440, DOI: 10.1080/00018730410001703159 173, 174

Falcke M, Hudson JL, Camacho P, Lechleiter JD (1999) Impact of mitochondrial Ca^{2+} cycling on pattern formation and stability. Biophys J 77(1):37–44, DOI: 10.1016/s0006-3495(99)76870-0 52, 248, 249, 249, 250

Falcke M, Or-Guil M, Bär M (2000) Dispersion gap and localized spiral waves in a model for intracellular Ca^{2+} dynamics. Phys Rev Lett 84(20):4753–6, DOI: 10.1103/physrevlett.84.4753 251

Falkenburger BH, Jensen JB, Hille B (2010a) Kinetics of M_1 muscarinic receptor and G protein signaling to phospholipase C in living cells. J Gen Physiol 135(2):81–97, DOI: 10.1085/jgp.200910344 34, 34, 35, 36

Falkenburger BH, Jensen JB, Hille B (2010b) Kinetics of PIP_2 metabolism and KCNQ2/3 channel regulation studied with a voltage-sensitive phosphatase in living cells. J Gen Physiol 135(2):99–114, DOI: 10.1085/jgp.200910345 34

Falkenburger BH, Dickson EJ, Hille B (2013) Quantitative properties and receptor reserve of the DAG and PKC branch of G_q-coupled receptor signaling. J Gen Physiol 141(5):537–55, DOI: 10.1085/jgp.201210887 34

Fall CP, Marland ES, Wagner JM, Tyson JJ (2002) Computational cell biology. Springer Verlag, New York viii

Fall CP, Wagner JM, Loew LM, Nuccitelli R (2004) Cortically restricted production of IP_3 leads to propagation of the fertilization Ca^{2+} wave along the cell surface in a model of the *Xenopus* egg. J Theor Biol 231(4):487–96, DOI: 10.1016/j.jtbi.2004.06.019 251, 254, 256

Fameli N, van Breemen C, Kuo KH (2007) A quantitative model for linking Na^+/Ca^{2+} exchanger to SERCA during refilling of the sarcoplasmic reticulum to sustain $[Ca^{2+}]$ oscillations in vascular smooth muscle. Cell Calcium 42(6):565–75, DOI: 10.1016/j.ceca.2007.02.001 135

Félix-Martínez GJ, Godínez-Fernández JR (2014) Mathematical models of electrical activity of the pancreatic β-cell: a physiological review. Islets 6(3):e949,195, DOI: 10.4161/19382014.2014.949195 357, 365

Fenichel N (1979) Geometric singular perturbation theory for ordinary differential equations. J Diff Eq 31:53–98, DOI: 10.1016/0022-0396(79)90152-9 222, 227

Ferrell JE Jr, Ha SH (2014) Ultrasensitivity part III: cascades, bistable switches, and oscillators. Trends Biochem Sci 39(12):612–8, DOI: 10.1016/j.tibs.2014.10.002 294

Fife P (1979) Mathematical aspects of reacting and diffusing systems, Springer Lecture Notes in Biomathematics, vol 28. Springer-Verlag, Berlin, DOI: 10.1007/978-3-642-93111-6 144

Fill M, Copello JA (2002) Ryanodine receptor calcium release channels. Physiol Rev 82(4):893–922, DOI: 10.1152/physrev.00013.2002 10, 87, 89

Finch E, Goldin S (1994) Calcium and inositol 1,4,5-trisphosphate-induced Ca^{2+} release. Science 265:813–5, DOI: 10.1126/science.8047889 11, 76

Finch EA, Turner TJ, Goldin SM (1991) Calcium as a coagonist of inositol 1,4,5-trisphosphate-induced calcium release. Science 252(5004):443–6, DOI: 10.1126/science.2017683 11, 71, 76

FitzHugh R (1960) Thresholds and plateaus in the Hodgkin-Huxley nerve equations. J Gen Physiol 43:867–96, DOI: 10.1085/jgp.43.5.867 108

FitzHugh R (1961) Impulses and physiological states in theoretical models of nerve membrane. Biophys J 1(6):445–66, DOI: 10.1016/s0006-3495(61)86902-6 108

FitzHugh R (1969a) Impulses and physiological states in theoretical models of nerve membrane. In: Schwan HP (ed) Biological Engineering, McGraw-Hill, New York, pp 445–66 108

FitzHugh R (1969b) Mathematical models of excitation and propagation in nerve. In: Schwan H (ed) Biological Engineering, McGraw-Hill, New York, pp 1–85 108

Flach EH, Schnell S (2006) Use and abuse of the quasi-steady-state approximation. IEE Proceedings-Systems Biology 153(4):187–91, DOI: 10.1049/ip-syb:20050104 216

Fletcher PA, Li YX (2009) An integrated model of electrical spiking, bursting, and calcium oscillations in GnRH neurons. Biophys J 96(11):4514–24, DOI: 10.1016/j.bpj.2009.03.037 232, 382

Fogarty KE, Kidd JF, Tuft DA, Thorn P (2000) Mechanisms underlying $InsP_3$-evoked global Ca^{2+} signals in mouse pancreatic acinar cells. J Physiol 526:515–26, DOI: 10.1111/j.1469-7793.2000.t01-1-00515.x 274

Fogelson A, Zucker R (1985) Presynaptic calcium diffusion from various arrays of single channels. Biophys J 48:1003–17, DOI: 10.1016/s0006-3495(85)83863-7 338, 338, 338

Fontanilla RA, Nuccitelli R (1998) Characterization of the sperm-induced calcium wave in *Xenopus* eggs using confocal microscopy. Biophys J 75(4):2079–87, DOI: 10.1016/S0006-3495(98)77650-7 251

Foskett JK, Melvin JE (1989) Activation of salivary secretion: coupling of cell volume and $[Ca^{2+}]_i$ in single cells. Science 244(4912):1582–5, DOI: 10.1126/science.2500708 283

Foskett JK, White C, Cheung KH, Mak DOD (2007) Inositol trisphosphate receptor Ca^{2+} release channels. Physiol Rev 87(2):593–658, DOI: 10.1152/physrev.00035.2006 10, 70, 71, 71

Fridlyand LE, Tamarina N, Philipson LH (2003) Modeling of Ca^{2+} flux in pancreatic beta-cells: role of the plasma membrane and intracellular stores. Am J Physiol Endocrinol Metab 285(1):E138–54, DOI: 10.1152/ajpendo.00194.2002 365

Fridlyand LE, Jacobson DA, Kuznetsov A, Philipson LH (2009) A model of action potentials and fast Ca^{2+} dynamics in pancreatic beta-cells. Biophys J 96(8):3126–39, DOI: 10.1016/j.bpj.2009.01.029 365

Fridlyand LE, Tamarina N, Philipson LH (2010) Bursting and calcium oscillations in pancreatic beta-cells: specific pacemakers for specific mechanisms. Am J Physiol Endocrinol Metab 299(4):E517–32, DOI: 10.1152/ajpendo.00177.2010 357

Fridlyand LE, Jacobson DA, Philipson LH (2013) Ion channels and regulation of insulin secretion in human β-cells: a computational systems analysis. Islets 5(1):1–15, DOI: 10.4161/isl.24166 365

Friel DD (1995) $[Ca^{2+}]_i$ oscillations in sympathetic neurons: an experimental test of a theoretical model. Biophys J 68(5):1752–66, DOI: 10.1016/s0006-3495(95)80352-8 84, 319

Galione A (2011) NAADP receptors. Cold Spring Harb Perspect Biol 3(1):a004,036, DOI: 10.1101/cshperspect.a004036 9

Galione A, Ruas M (2005) NAADP receptors. Cell Calcium 38(3–4):273–80, DOI: 10.1016/j.ceca.2005.06.031 9

Gallegos LL, Newton AC (2008) Spatiotemporal dynamics of lipid signaling: protein kinase C as a paradigm. IUBMB Life 60(12):782–9, DOI: 10.1002/iub.122 287

Gaspers LD, Thomas AP (2005) Calcium signaling in liver. Cell Calcium 38(3–4):329–42, DOI: 10.1016/j.ceca.2005.06.009 259, 269

Gaspers LD, Bartlett PJ, Politi A, Burnett P, Metzger W, Johnston J, Joseph SK, Höfer T, Thomas AP (2014) Hormone-induced calcium oscillations depend on cross-coupling with inositol 1,4,5-trisphosphate oscillations. Cell Rep 9(4):1209–18, DOI: 10.1016/j.celrep.2014.10.033 94, 117

Gauthier LD, Greenstein JL, Winslow RL (2011) Toward an integrative computational model of the guinea pig cardiac myocyte. Front Physiol 3:1–19, DOI: 10.3389/fphys.2012.00244 64

Geguchadze R (2004) Quantitative measurements of Ca^{2+}/calmodulin binding and activation of myosin light chain kinase in cells. FEBS Lett 557(1–3):121–4, DOI: 10.1016/S0014-5793(03)01456-X 332, 334

Gerish J (2002) Is reduced first phase insulin release the earliest detectable abnormality in individuals destined to develop type 2 diabetes? Diabetes 51(Suppl. 1):S117–21 373

Gillespie D (1976) A general method for numerically simulating the stochastic time evolution of coupled chemical reactions. J Comput Phys 22(4):403–34, DOI: 10.1016/0021-9991(76)90041-3 264

Gin E, Kirk V, Sneyd J (2006) A bifurcation analysis of calcium buffering. J Theor Biol 242(1):1–15, DOI: 10.1016/j.jtbi.2006.01.030 91, 161

Gin E, Crampin EJ, Brown DA, Shuttleworth TJ, Yule DI, Sneyd J (2007) A mathematical model of fluid secretion from a parotid acinar cell. J Theor Biol 248(1):64–80, DOI: 10.1016/j.jtbi.2007.04.021 276

Gin E, Falcke M, Wagner LE, Yule DI, Sneyd J (2009) Markov chain Monte Carlo fitting of single-channel data from inositol trisphosphate receptors. J Theor Biol 257(3):460–74, DOI: 10.1016/j.jtbi.2008.12.020 76

Giovannucci DR, Groblewski GE, Sneyd J, Yule DI (2000) Targeted phosphorylation of inositol 1,4,5-trisphosphate receptors selectively inhibits localized Ca^{2+} release and shapes oscillatory Ca^{2+} signals. J Biol Chem 275(43):33,704–11, DOI: 10.1074/jbc.M004278200 277

Giovannucci DR, Bruce JIE, Straub SV, Arreola J, Sneyd J, Shuttleworth TJ, Yule DI (2002) Cytosolic Ca^{2+} and Ca^{2+}-activated Cl^- current dynamics: insights from two functionally distinct mouse exocrine cells. J Physiol 540(2):469–84, DOI: 10.1113/jphysiol.2001.013453 278, 279, 279

Girard S, Clapham D (1993) Acceleration of intracellular calcium waves in *Xenopus* oocytes by calcium influx. Science 260:229–32, DOI: 10.1126/science.8385801 113, 247

Girard S, Lückhoff A, Lechleiter J, Sneyd J, Clapham D (1992) Two-dimensional model of calcium waves reproduces the patterns observed in *Xenopus* oocytes. Biophys J 61:509–517, DOI: 10.1016/s0006-3495(92)81855-6 247

Glass L, Mackey M (1988) From Clocks to Chaos. Princeton University Press, Princeton 118

Goel P, Sneyd J, Friedman A (2006) Homogenization of the cell cytoplasm: the calcium bidomain equations. SIAM J Multiscale Modeling and Simulation 5:1045–62, DOI: 10.1137/060660783 123, 124

Goldbeter A, Koshland DE Jr (1981) An amplified sensitivity arising from covalent modification in biological systems. Proc Natl Acad Sci USA 78(11):6840–4, DOI: 10.1073/pnas.78.11.6840 293

Goldbeter A, Decroly O, Li Y, Martiel JL, Moran F (1988) Finding complex oscillatory phenomena in biochemical systems. An empirical approach. Biophys Chem 29(1–2):211–7, DOI: 10.1016/0301-4622(88)87040-6 292

Goldbeter A, Dupont G, Berridge M (1990) Minimal model for signal-induced Ca^{2+} oscillations and for their frequency encoding through protein phosphorylation. Proc Natl Acad Sci USA 87:1461–5, DOI: 10.1073/pnas.87.4.1461 104, 273, 357

Gonzalez-Fernandez JM, Ermentrout B (1994) On the origin and dynamics of the vasomotion of small arteries. Math Biosci 119(2):127–67, DOI: 10.1016/0025-5564(94)90074-4 329

González-Vélez V, Dupont G, Gil A, González A, Quesada I (2012) Model for glucagon secretion by pancreatic α-cells. PLoS One 7(3):e32,282, DOI: 10.1371/journal.pone.0032282 365, 374, 374, 375, 375

Gonze D, Halloy J, Goldbeter A (2002) Robustness of circadian rhythms with respect to molecular noise. Proc Natl Acad Sci USA 99(2):673–8, DOI: 10.1073/pnas.022628299 265, 267

Gould GW, East JM, Froud RJ, McWhirter JM, Stefanova HI, Lee AG (1986) A kinetic model for the Ca^{2+} + Mg^{2+}-activated ATPase of sarcoplasmic reticulum. Biochem J 237(1):217–27, DOI: 10.1042/bj2370217 43

Graupner M, Brunel N (2007) STDP in a bistable synapse model based on CaMKII and associated signaling pathways. PLoS Comput Biol 3(11):e221, DOI: 10.1371/journal.pcbi.0030221 348, 352, 352

Graupner M, Brunel N (2010) Mechanisms of induction and maintenance of spike-timing dependent plasticity in biophysical synapse models. Front Comput Neurosci 4, DOI: 10.3389/fncom.2010.00136 344, 345, 345, 353

Graupner M, Brunel N (2012) Calcium-based plasticity model explains sensitivity of synaptic changes to spike pattern, rate, and dendritic location. Proc Natl Acad Sci USA 109(10):3991–6, DOI: 10.1073/pnas.1109359109 344, 348

Greenstein JL, Winslow RL (2002) An integrative model of the cardiac ventricular myocyte incorporating local control of Ca^{2+} release. Biophys J 83(6):2918–45, DOI: 10.1016/s0006-3495(02)75301-0 87, 88, 306, 306

Greenstein JL, Winslow RL (2011) Integrative systems models of cardiac excitation-contraction coupling. Circ Res 108(1):70–84, DOI: 10.1161/CIRCRESAHA.110.223578 87, 297

Greenstein JL, Hinch R, Winslow RL (2006) Mechanisms of excitation-contraction coupling in an integrative model of the cardiac ventricular myocyte. Biophys J 90(1):77–91, DOI: 10.1529/biophysj.105.065169 64, 87, 136

Grindrod P (1991) Patterns and Waves: the Theory and Application of Reaction-Diffusion Equations. Clarendon Press, Oxford 144, 144

Grynkiewicz G, Poenie M, Tsien R (1985) A new generation of Ca^{2+} indicators with greatly improved fluorescent properties. J Biol Chem 260:3440–50 18

Guevara MR, Shrier A, Glass L (1986) Phase resetting of spontaneously beating embryonic ventricular heart cell aggregates. Am J Physiol 251(6):H1298–305 118

Gupta RD, Kundu D (2007) Generalized exponential distribution: existing results and some recent developments. J Statist Plann Inference 137(11):3537–47, DOI: 10.1016/j.jspi.2007.03.030 188, 188

Gustavsson N, Wei SH, Hoang DN, Lao Y, Zhang Q, Radda GK, Rorsman P, Südhof TC, Han W (2009) Synaptotagmin-7 is a principal Ca^{2+} sensor for Ca^{2+} -induced glucagon exocytosis in pancreas. J Physiol 587(6):1169–78, DOI: 10.1113/jphysiol.2008.168005 376, 376

Györke I, Hester N, Jones LR, Györke S (2004) The role of calsequestrin, triadin, and junctin in conferring cardiac ryanodine receptor responsiveness to luminal calcium. Biophys J 86(4):2121–8, DOI: 10.1016/S0006-3495(04)74271-X 85

Györke S, Fill M (1993) Ryanodine receptor adaptation: control mechanism of Ca^{2+}-induced Ca^{2+} release in heart. Science 260(5109):807–9, DOI: 10.1126/science.8387229 87, 87

Haberichter T, Roux E, Marhl M, Mazat JP (2002) The influence of different $InsP_3$ receptor isoforms on Ca^{2+} signaling in tracheal smooth muscle cells. Bioelectrochemistry 57(2):129–38, DOI: 10.1016/s1567-5394(02)00063-4 317

Haddock RE, Hill CE (2005) Rhythmicity in arterial smooth muscle. J Physiol 566(3):645–56, DOI: 10.1113/jphysiol.2005.086405 326

Hagar RE, Burgstahler AD, Nathanson MH, Ehrlich BE (1998) Type III $InsP_3$ receptor channel stays open in the presence of increased calcium. Nature 396(6706):81–4, DOI: 10.1038/23954 71

Hai CM, Murphy RA (1988a) Cross-bridge phosphorylation and regulation of latch state in smooth muscle. Am J Physiol 254(1):C99–106 330

Hai CM, Murphy RA (1988b) Regulation of shortening velocity by cross-bridge phosphorylation in smooth muscle. Am J Physiol 255(1):C86–94 330

Hake J, Lines GT (2008) Stochastic binding of Ca^{2+} ions in the dyadic cleft; continuous versus random walk description of diffusion. Biophys J 94(11):4184–201, DOI: 10.1529/biophysj.106.103523 139, 139, 304, 305, 305

Hake J, Edwards AG, Yu Z, Kekenes-Huskey PM, Michailova AP, McCammon JA, Holst MJ, Hoshijima M, McCulloch AD (2012) Modelling cardiac calcium sparks in a three-dimensional reconstruction of a calcium release unit. J Physiol 590(18):4403–22, DOI: 10.1113/jphysiol.2012.227926 302, 304

Hamer RD (2000) Computational analysis of vertebrate phototransduction: combined quantitative and qualitative modeling of dark- and light-adapted responses in amphibian rods. Vis Neurosci 17(5):679–99, DOI: 10.1017/s0952523800175030 16

Hamilton NB, Attwell D (2010) Do astrocytes really exocytose neurotransmitters? Nat Rev Neurosci 11(4):227–38, DOI: 10.1038/nrn2803 285

Hänggi P, Talkner P, Borkovec M (1990) Reaction-rate theory: fifty years after Kramers. Reviews of Modern Physics 62(2):251, DOI: 10.1103/revmodphys.62.251 58

Harootunian AT, Kao JP, Tsien RY (1988) Agonist-induced calcium oscillations in depolarized fibroblasts and their manipulation by photoreleased Ins(1,4,5)P_3, Ca^{2+}, and Ca^{2+} buffer. Cold Spring Harb Symp Quant Biol 53:935–43, DOI: 10.1101/sqb.1988.053.01.108 117

Harvey E, Kirk V, Osinga HM, Sneyd J, Wechselberger M (2010) Understanding anomalous delays in a model of intracellular calcium dynamics. Chaos 20(4):045,104, DOI: 10.1063/1.3523264 117, 208, 222, 228, 230, 231, 232, 232

Harvey E, Kirk V, Wechselberger M, Sneyd J (2011) Multiple timescales, mixed mode oscillations and canards in models of intracellular calcium dynamics. J Nonlin Sci 21(5):639–83, DOI: 10.1007/s00332-011-9096-z 117, 117, 212, 214, 222, 226

Hassinger T, Guthrie P, Atkinson P, Bennett M, Kater S (1996) An extracellular signaling component in propagation of astrocytic calcium waves. Proc Natl Acad Sci USA 93:13,268–73, DOI: 10.1073/pnas.93.23.13268 155, 156

Hastings S (1976) On the existence of homoclinic and periodic orbits for the FitzHugh-Nagumo equations. Quart J Math 27(1):123–34, DOI: 10.1093/qmath/27.1.123 108

Hastings S (1982) Single and multiple pulse waves for the FitzHugh-Nagumo equations. SIAM J Appl Math 42:247–60, DOI: 10.1137/0142018 108

Haynes DH, Mandveno A (1987) Computer modeling of Ca^{2+} pump function of Ca^{2+}-Mg^{2+}-ATPase of sarcoplasmic reticulum. Physiol Rev 67(1):244–84 43

Hemond PJ, O'Boyle MP, Roberts CB, Delgado-Reyes A, Hemond Z, Suter KJ (2012) Simulated GABA synaptic input and L-type calcium channels form functional microdomains in hypothalamic gonadotropin-releasing hormone neurons. J Neurosci 32(26):8756–66, DOI: 10.1523/JNEUROSCI.4188-11.2012 382

Herbison AE, Pape JR, Simonian SX, Skynner MJ, Sim JA (2001) Molecular and cellular properties of GnRH neurons revealed through transgenics in the mouse. Mol Cell Endocrinol 185(1–2):185–94, DOI: 10.1016/s0303-7207(01)00618-9 18

Hess P, Tsien RW (1984) Mechanism of ion permeation through calcium channels. Nature 309(5967):453–6, DOI: 10.1038/309453a0 58

Higgins ER, Goel P, Puglisi JL, Bers DM, Cannell M, Sneyd J (2007) Modelling calcium microdomains using homogenisation. J Theor Biol 247(4):623–44, DOI: 10.1016/j.jtbi.2007.03.019 137, 138, 301

Higgins ER, Schmidle H, Falcke M (2009) Waiting time distributions for clusters of IP_3 receptors. J Theor Biol 259(2):338–49, DOI: DOI: 10.1016/j.jtbi.2009.03.018 185, 185, 189, 189

Hilgemann DW (2004) New insights into the molecular and cellular workings of the cardiac Na^+/Ca^{2+} exchanger. Am J Physiol 287(5):C1167–72, DOI: 10.1152/ajpcell.00288.2004 44

Hilgemann DW, Nicoll DA, Philipson KD (1991) Charge movement during Na^+ translocation by native and cloned cardiac Na^+/Ca^{2+} exchanger. Nature 352(6337):715–8, DOI: 10.1038/352715a0 44, 51

Hilgemann DW, Matsuoka S, Nagel GA, Collins A (1992) Steady-state and dynamic properties of cardiac sodium-calcium exchange. Sodium-dependent inactivation. J Gen Physiol 100(6):905–32, DOI: 10.1085/jgp.100.6.905 51

Hill-Eubanks DC, Werner ME, Heppner TJ, Nelson MT (2011) Calcium signaling in smooth muscle. Cold Spring Harb Perspect Biol 3(9):a004,549, DOI: 10.1101/cshperspect.a004549 314

Hille B (2001) Ionic Channels of Excitable Membranes, 3rd edn. Sinauer, Sunderland, MA 54, 58

Himmel DM, Chay TR (1987) Theoretical studies on the electrical activity of pancreatic β-cells as a function of glucose. Biophys J 51:89–107, DOI: 10.1016/s0006-3495(87)83314-3 358

Hinch R (2004) A mathematical analysis of the generation and termination of calcium sparks. Biophys J 86(3):1293–307, DOI: 10.1016/S0006-3495(04)74203-4 301, 306, 306, 307

Hinch R, Greenstein J, Winslow R (2006) Multi-scale models of local control of calcium induced calcium release. Prog Biophys Mol Biol 90(1–3):136–50, DOI: 10.1016/j.pbiomolbio.2005.05.014 63, 297

Hindmarsh JL, Rose RM (1984) A model of neuronal bursting using three coupled first order differential equations. Proc R Soc Lond B 221:87–102, DOI: 10.1098/rspb.1984.0024 233

Hirose K, Kadowaki S, Tanabe M, Takeshima H, Iino M (1999) Spatiotemporal dynamics of inositol 1,4,5-trisphosphate that underlies complex Ca^{2+} mobilization patterns. Science 284(5419):1527–30 114

Hirst GDS (2003) Interstitial cells: involvement in rhythmicity and neural control of gut smooth muscle. J Physiol 550(2):337–46, DOI: 10.1113/jphysiol.2003.043299 326

Hodgkin AL, Huxley AF (1952) A quantitative description of membrane current and its application to conduction and excitation in nerve. J Physiol 117(4):500–44, DOI: 10.1113/jphysiol.1952.sp004764 53, 75, 157, 327, 359

Höfer T, Politi A, Heinrich R (2001) Intercellular Ca^{2+} wave propagation through gap-junctional Ca^{2+} diffusion: a theoretical study. Biophys J 80(1):75–87, DOI: 10.1016/S0006-3495(01)75996-6 148

Höfer T, Venance L, Giaume C (2002) Control and plasticity of intercellular calcium waves in astrocytes: a modeling approach. J Neurosci 22(12):4850–9 154, 154

Hoover PJ, Lewis RS (2011) Stoichiometric requirements for trapping and gating of Ca^{2+} release-activated Ca^{2+} (CRAC) channels by stromal interaction molecule 1 (STIM1). Proc Natl Acad Sci USA 108(32):13,299–304, DOI: 10.1073/pnas.1101664108 67, 68, 68, 68

Hoppensteadt F, Peskin C (2002) Modeling and Simulation in Medicine and the Life Sciences. Springer-Verlag, New York viii

Horikawa K (2015) Recent progress in the development of genetically encoded Ca^{2+} indicators. J Med Invest 62(1–2):24–8, DOI: 10.2152/jmi.62.24 18

Horng TL, Lin TC, Liu C, Eisenberg R (2012) PNP equations with steric effects: a model of ion flow through channels. J Phys Chem B 116(37):11,422–41, DOI: 10.1021/jp305273n 59

Horowitz LF, Hirdes W, Suh BC, Hilgemann DW, Mackie K, Hille B (2005) Phospholipase C in living cells: activation, inhibition, Ca^{2+} requirement, and regulation of M current. J Gen Physiol 126(3):243–62, DOI: 10.1085/jgp.200509309 12

Hosoi N, Sakaba T, Neher E (2007) Quantitative analysis of calcium-dependent vesicle recruitment and its functional role at the calyx of Held synapse. J Neurosci 27(52):14,286–98, DOI: 10.1523/JNEUROSCI.4122-07.2007 338

Hoth M, Penner R (1992) Depletion of intracellular calcium stores activates a calcium current in mast cells. Nature 355(6358):353–6, DOI: 10.1038/355353a0 14

Howard AD, McAllister G, Feighner SD, Liu Q, Nargund RP, Van der Ploeg LH, Patchett AA (2001) Orphan G-protein-coupled receptors and natural ligand discovery. Trends Pharmacol Sci 22(3):132–40, DOI: 10.1016/s0165-6147(00)01636-9 7

Hudmon A, Schulman H (2002) Neuronal Ca^{2+}/calmodulin-dependent protein kinase II: the role of structure and autoregulation in cellular function. Annu Rev Biochem 71:473–510, DOI: 10.1146/annurev.biochem.71.110601.135410 354

Huertas MA, Smith GD (2007) The dynamics of luminal depletion and the stochastic gating of Ca^{2+}-activated Ca^{2+} channels and release sites. J Theor Biol 246(2):332–54, DOI: 10.1016/j.jtbi.2007.01.003 306

Huguenard JR, Prince DA (1992) A novel T-type current underlies prolonged Ca^{2+}-dependent burst firing in GABAergic neurons of rat thalamic reticular nucleus. J Neurosci 12(10):3804–17 58

Huxley A (1957) Muscle structure and theories of contraction. Progress in Biophysics 7:255–318 295, 329, 331

Imredy JP, Yue DT (1994) Mechanism of Ca^{2+}-sensitive inactivation of L-type Ca^{2+} channels. Neuron 12(6):1301–18, DOI: 10.1016/0896-6273(94)90446-4 62

Ionescu L, White C, Cheung KH, Shuai J, Parker I, Pearson JE, Foskett JK, Mak DO (2007) Mode switching is the major mechanism of ligand regulation of $InsP_3$ receptor calcium release channels. J Gen Physiol 130(6):631–45, DOI: 10.1085/jgp.200709859 76, 76

Irving M, Maylie J, Sizto N, Chandler W (1990) Intracellular diffusion in the presence of mobile buffers: application to proton movement in muscle. Biophys J 57:717–21, DOI: 10.1016/s0006-3495(90)82592-3 91

Ishii K, Hirose K, Iino M (2006) Ca^{2+} shuttling between endoplasmic reticulum and mitochondria underlying Ca^{2+} oscillations. EMBO Rep 7(4):390–6, DOI: 10.1038/sj.embor.7400620 52

Izhikevich E (2000) Neural excitability, spiking and bursting. Int J Bif Chaos 10(6):1171–266, DOI: 10.1142/s0218127400000840 220, 233

Izu LT, Wier WG, Balke CW (2001) Evolution of cardiac calcium waves from stochastic calcium sparks. Biophys J 80(1):103–20, DOI: 10.1016/s0006-3495(01)75998-x 306

Jacobsen JCB, Aalkjaer C, Nilsson H, Matchkov VV, Freiberg J, Holstein-Rathlou NH (2007a) Activation of a cGMP-sensitive calcium-dependent chloride channel may cause transition from calcium waves to whole cell oscillations in smooth muscle cells. Am J Physiol Heart Circ Physiol 293(1):H215–28, DOI: 10.1152/ajpheart.00726.2006 329

Jacobsen JCB, Aalkjaer C, Nilsson H, Matchkov VV, Freiberg J, Holstein-Rathlou NH (2007b) A model of smooth muscle cell synchronization in the arterial wall. Am J Physiol Heart Circ Physiol 293(1):H229–37, DOI: 10.1152/ajpheart.00727.2006 329

Jafri M, Keizer J (1994) Diffusion of inositol 1,4,5-trisphosphate, but not Ca^{2+}, is necessary for a class of inositol 1,4,5-trisphosphate-induced Ca^{2+} waves. Proc Natl Acad Sci USA 91:9485–89, DOI: 10.1073/pnas.91.20.9485 145, 146, 146

Jafri MS, Keizer J (1995) On the roles of Ca^{2+} diffusion, Ca^{2+} buffers, and the endoplasmic reticulum in IP_3-induced Ca^{2+} waves. Biophys J 69(5):2139–53, DOI: 10.1016/S0006-3495(95)80088-3 160, 160

Jafri MS, Rice JJ, Winslow RL (1998) Cardiac Ca^{2+} dynamics: the roles of ryanodine receptor adaptation and sarcoplasmic reticulum load. Biophys J 74(3):1149–68, DOI: 10.1016/s0006-3495(98)77832-4 62, 62, 136, 299, 300

Jaggar JH, Porter VA, Lederer WJ, Nelson MT (2000) Calcium sparks in smooth muscle. Am J Physiol Cell Physiol 278(2):C235–56 20, 317

Jaiswal J (2001) Calcium – how and why? Journal of Biosciences 26(3):357–363, DOI: 10.1007/bf02703745 5

Jakobsson E (1980) Interactions of cell volume, membrane potential, and membrane transport parameters. Am J Physiol Cell Physiol 238:C196–C206 282

Janssen LJ (2012) Airway smooth muscle electrophysiology in a state of flux? Am J Physiol Lung Cell Mol Physiol 302(8):L730–2, DOI: 10.1152/ajplung.00032.2012 326

Jasoni CL, Todman MG, Strumia MM, Herbison AE (2007) Cell type-specific expression of a genetically encoded calcium indicator reveals intrinsic calcium oscillations in adult gonadotropin-releasing hormone neurons. J Neurosci 27(4):860–7, DOI: 10.1523/JNEUROSCI.3579-06.2007 381

Jasoni CL, Romano N, Constantin S, Lee K, Herbison AE (2010) Calcium dynamics in gonadotropin-releasing hormone neurons. Front Neuroendocrinol 31(3):259–69, DOI: 10.1016/j.yfrne.2010.05.005 381

Jensen JB, Lyssand JS, Hague C, Hille B (2009) Fluorescence changes reveal kinetic steps of muscarinic receptor-mediated modulation of phosphoinositides and Kv7.2/7.3 K^+ channels. J Gen Physiol 133(4):347–59, DOI: 10.1085/jgp.200810075 34

Ji Y, Loukianov E, Loukianova T, Jones LR, Periasamy M (1999) SERCA1a can functionally substitute for SERCA2a in the heart. Am J Physiol 276(1):H89–97 43, 44

Jones C (1984) Stability of the traveling wave solutions of the FitzHugh-Nagumo system. Trans Amer Math Soc 286:431–69 238

Jones CKRT (1995) Geometric singular perturbation theory. In: Dynamical Systems, Springer Verlag, pp 44–118 222

Joseph SK (1996) The inositol trisphosphate receptor family. Cell Signal 8(1):1–7, DOI: 10.1016/0898-6568(95)02012-8 10

Jouaville LS, Ichas F, Holmuhamedov EL, Camacho P, Lechleiter JD (1995) Synchronization of calcium waves by mitochondrial substrates in *Xenopus laevis* oocytes. Nature 377(6548):438–41, DOI: 10.1038/377438a0 52, 246, 248, 249

Kaftan EJ, Ehrlich BE, Watras J (1997) Inositol 1,4,5-trisphosphate ($InsP_3$) and calcium interact to increase the dynamic range of $InsP_3$ receptor-dependent calcium signaling. J Gen Physiol 110(5):529–38, DOI: 10.1085/jgp.110.5.529 71

Kamiya H, Zucker RS (1994) Residual Ca^{2+} and short-term synaptic plasticity. Nature 371(6498):603–6, DOI: 10.1038/371603a0 339, 344

van Kampen N (2001) Stochastic Processes in Physics and Chemistry. North-Holland, Amsterdam 190, 191, 191

Kang M, Othmer HG (2007) The variety of cytosolic calcium responses and possible roles of PLC and PKC. Phys Biol 4(4):325–43, DOI: 10.1088/1478-3975/4/4/009 30, 294

Kang TM, Hilgemann DW (2004) Multiple transport modes of the cardiac Na^+/Ca^{2+} exchanger. Nature 427(6974):544–8, DOI: 10.1038/nature02271 44, 51

Kapela A, Bezerianos A, Tsoukias NM (2008) A mathematical model of Ca^{2+} dynamics in rat mesenteric smooth muscle cell: agonist and NO stimulation. J Theor Biol 253(2):238–60, DOI: 10.1016/j.jtbi.2008.03.004 329

Kapela A, Nagaraja S, Tsoukias NM (2010) A mathematical model of vasoreactivity in rat mesenteric arterioles. II. Conducted vasoreactivity. Am J Physiol Heart Circ Physiol 298(1):H52–65, DOI: 10.1152/ajpheart.00546.2009 329

Kapela A, Parikh J, Tsoukias NM (2012) Multiple factors influence calcium synchronization in arterial vasomotion. Biophys J 102(2):211–20, DOI: 10.1016/j.bpj.2011.12.032 329

Kaplan W (1981) Advanced Engineering Mathematics. Addison-Wesley, Reading, MA viii

Karlstad J, Sun Y, Singh BB (2012) Ca^{2+} signaling: an outlook on the characterization of Ca^{2+} channels and their importance in cellular functions. Adv Exp Med Biol 740:143–57, DOI: 10.1007/978-94-007-2888-2_6 16

Kasai H, Li YX, Miyashita Y (1993) Subcellular distribution of Ca^{2+} release channels underlying Ca^{2+} waves and oscillations in exocrine pancreas. Cell 74:669–77, DOI: 10.1016/0092-8674(93)90514-q 274, 277

Kato S, Osa T, Ogasawara T (1984) Kinetic model for isometric contraction in smooth muscle on the basis of myosin phosphorylation hypothesis. Biophys J 46(1):35–44, DOI: 10.1016/s0006-3495(84)83996-x 331

Katz B, Miledi R (1968) The role of calcium in neuromuscular facilitation. J Physiol 195:481–92, DOI: 10.1113/jphysiol.1968.sp008469 339

Kawabata S, Tsutsumi R, Kohara A, Yamaguchi T, Nakanishi S, Okada M (1996) Control of calcium oscillations by phosphorylation of metabotropic glutamate receptors. Nature 383(6595):89–92, DOI: 10.1038/383089a0 285, 286

Kaźmierczak B, Volpert V (2007) Calcium waves in systems with immobile buffers as a limit of waves for systems with nonzero diffusion. Nonlinearity 21(1):71–96, DOI: 10.1088/0951-7715/21/1/004 159

Keener J, Sneyd J (2008) Mathematical Physiology, 2nd edn. Springer-Verlag, New York, DOI: 10.1007/978-0-387-79388-7 viii, 38, 38, 48, 54, 58, 75, 108, 122, 124, 126, 142, 143, 144, 157, 233, 282, 309, 327, 329, 359, 378

Keener JP (1980) Waves in excitable media. SIAM J Appl Math 39(3):528–48, DOI: 10.1137/0139043 108

Keener JP (2000) Principles of Applied Mathematics: Transformation and Approximation, 2nd edn. Perseus Books, Cambridge, Mass. 141

Keizer J, De Young GW (1992) Two roles of Ca^{2+} in agonist stimulated Ca^{2+} oscillations. Biophys J 61(3):649–60, DOI: 10.1016/S0006-3495(92)81870-2 94, 115

Keizer J, Levine L (1996) Ryanodine receptor adaptation and Ca^{2+}-induced Ca^{2+} release-dependent Ca^{2+} oscillations. Biophys J 71(6):3477–87, DOI: 10.1016/S0006-3495(96)79543-7 87, 87, 280

Keizer J, Magnus G (1989) ATP-sensitive potassium channel and bursting in the pancreatic beta cell. Biophys J 56:229–42, DOI: 10.1016/s0006-3495(89)82669-4 358

Keizer J, Smith GD (1998) Spark-to-wave transition: saltatory transmission of calcium waves in cardiac myocytes. Biophys Chem 72(1–2):87–100, DOI: 10.1016/s0301-4622(98)00125-2 87, 87, 89

Keizer J, Smolen P (1991) Bursting electrical activity in pancreatic β-cells caused by Ca^{2+} and voltage-inactivated Ca^{2+} channels. Proc Natl Acad Sci USA 88:3897–901, DOI: 10.1073/pnas.88.9.3897 358

Keizer J, Smith GD, Ponce-Dawson S, Pearson JE (1998) Saltatory propagation of Ca^{2+} waves by Ca^{2+} sparks. Biophys J 75(2):595–600, DOI: 10.1016/s0006-3495(98)77550-2 140, 306

Kevorkian J (2000) Partial Differential Equations: Analytical Solution Techniques, 2nd edn. Springer, New York, DOI: 10.1007/978-1-4757-3266-5 141

Kidd JF, Fogarty KE, Tuft RA, Thorn P (1999) The role of Ca^{2+} feedback in shaping $InsP_3$-evoked Ca^{2+} signals in mouse pancreatic acinar cells. J Physiol 520 (1):187–201, DOI: 10.1111/j.1469-7793.1999.00187.x 159, 161

Kim CH, Braud S, Isaac JTR, Roche KW (2005) Protein kinase C phosphorylation of the metabotropic glutamate receptor $mGluR_5$ on Serine 839 regulates Ca^{2+} oscillations. J Biol Chem 280(27):25,409–15, DOI: 10.1074/jbc.M502644200 286, 288

Kim WT, Rioult MG, Cornell-Bell AH (1994) Glutamate-induced calcium signaling in astrocytes. Glia 11(2):173–84, DOI: 10.1002/glia.440110211 27

Klingauf J, Neher E (1997) Modeling buffered Ca^{2+} diffusion near the membrane: implications for secretion in neuroendocrine cells. Biophys J 72(2):674–90, DOI: 10.1016/s0006-3495(97)78704-6 339

Koch KW, Dell'Orco D (2013) A calcium-relay mechanism in vertebrate phototransduction. ACS Chem Neurosci 4(6):909–17, DOI: 10.1021/cn400027z 16

Koenigsberger M, Sauser R, Lamboley M, Bény JL, Meister JJ (2004) Ca^{2+} dynamics in a population of smooth muscle cells: modeling the recruitment and synchronization. Biophys J 87(1):92–104, DOI: 10.1529/biophysj.103.037853 329, 331

Koenigsberger M, Sauser R, Meister JJ (2005) Emergent properties of electrically coupled smooth muscle cells. Bull Math Biol 67(6):1253–72, DOI: 10.1016/j.bulm.2005.02.001 329, 331

Koenigsberger M, Sauser R, Bény JL, Meister JJ (2006) Effects of arterial wall stress on vasomotion. Biophys J 91(5):1663–74, DOI: 10.1529/biophysj.106.083311 331

Koenigsberger M, Sauser R, Seppey D, Bény JL, Meister JJ (2008) Calcium dynamics and vasomotion in arteries subject to isometric, isobaric, and isotonic conditions. Biophys J 95(6):2728–38, DOI: 10.1529/biophysj.108.131136 329

Koenigsberger M, Seppey D, Bény JL, Meister JJ (2010) Mechanisms of propagation of intercellular calcium waves in arterial smooth muscle cells. Biophys J 99(2):333–43, DOI: 10.1016/j.bpj.2010.04.031 329

Koizumi S, Lipp P, Berridge MJ, Bootman MD (1999) Regulation of ryanodine receptor opening by lumenal Ca^{2+} underlies quantal Ca^{2+} release in PC12 cells. J Biol Chem 274(47):33,327–33, DOI: 10.1074/jbc.274.47.33327 85

Kopell N, Ermentrout G (1986) Symmetry and phaselocking in chains of weakly coupled oscillators. Comm Pure Appl Math 39(5):623–60, DOI: 10.1002/cpa.3160390504 149

Kopell N, Ermentrout G (1990) Phase transitions and other phenomena in chains of coupled oscillators. SIAM J Appl Math 50(4):1014–52, DOI: 10.1137/0150062 149

Kopell N, Howard L (1973a) Plane wave solutions to reaction-diffusion equations. Stud App Math 52:291–328, DOI: 10.1002/sapm1973524291 144

Kopell N, Howard L (1981) Target pattern and spiral solutions to reaction-diffusion equations with more than one space dimension. Adv App Math 2(4):417–49, DOI: 10.1016/0196-8858(81)90043-9 144

Kopell N, Howard LN (1973b) Horizontal bands in the Belousov reaction. Science 180(4091):1171–73, DOI: 10.1126/science.180.4091.1171 145

Korenbrot JI (2012) Speed, sensitivity, and stability of the light response in rod and cone photoreceptors: facts and models. Prog Retin Eye Res 31(5):442–66, DOI: 10.1016/j.preteyeres.2012.05.002 16

Kotlikoff MI (2007) Genetically encoded Ca^{2+} indicators: using genetics and molecular design to understand complex physiology. J Physiol 578(Pt 1):55–67, DOI: 10.1113/jphysiol.2006.120212 18

Kreyszig E (1994) Advanced Engineering Mathematics. John Wiley and Sons, New York viii

Kroeber S, Schomerus C, Korf HW (1997) Calcium oscillations in a subpopulation of S-antigen-immunoreactive pinealocytes of the rainbow trout (*Oncorhynchus mykiss*). Brain Res 744(1):68–76, DOI: 10.1016/s0006-8993(96)01084-0 273

Krsmanović LZ, Stojilković SS, Merelli F, Dufour SM, Virmani MA, Catt KJ (1992) Calcium signaling and episodic secretion of gonadotropin-releasing hormone in hypothalamic neurons. Proc Natl Acad Sci USA 89(18):8462–6, DOI: 10.1073/pnas.89.18.8462 381

Krupa M, Sandstede B, Szmolyan P (1997) Fast and slow waves in the FitzHugh-Nagumo equation. J Diff Eq 133(1):49–97, DOI: 10.1006/jdeq.1996.3198 238

Krupa M, Popovic N, Kopell N (2008) Mixed-mode oscillations in three time-scale systems: a prototypical example. SIAM J Appl Dyn Syst 7(2):361–420, DOI: 10.1137/070688912 226

Krupa M, Vidal A, Desroches M, Clément F (2012) Mixed-mode oscillations in a multiple time scale phantom bursting system. SIAM J Appl Dyn Syst 11:1458–98, DOI: 10.1137/110860136 226, 233

Kuba K, Takeshita S (1981) Simulation of intracellular Ca^{2+} oscillation in a sympathetic neurone. J Theor Biol 93:1009–31, DOI: 10.1016/0022-5193(81)90352-0 104

Kubota Y, Bower JM (2001) Transient versus asymptotic dynamics of CaM kinase II: possible roles of phosphatase. J Comput Neurosci 11(3):263–79, DOI: 10.1023/a:1013727331979 348

Kukuljan M, Rojas E, Catt KJ, Stojilkovic SS (1994) Membrane potential regulates inositol 1,4,5-trisphosphate-controlled cytoplasmic Ca^{2+} oscillations in pituitary gonadotrophs. J Biol Chem 269(7):4860–5 378

Kummer U, Olsen L, Dixon C, Green A, Bornberg-Bauer E, Baier G (2000) Switching from simple to complex oscillations in calcium signaling. Biophys J 79:1188–95, DOI: 10.1016/s0006-3495(00)76373-9 30, 65, 259

Kuyucak S, Chung SH (2002) Permeation models and structure-function relationships in ion channels. J Biol Phys 28(2):289–308, DOI: 10.1023/A:1019939900568 59

Lallouette J, De Pittà M, Ben-Jacob E, Berry H (2014) Sparse short-distance connections enhance calcium wave propagation in a 3D model of astrocyte networks. Front Comput Neurosci 8:45, DOI: 10.3389/fncom.2014.00045 149

Langer GA, Peskoff A (1996) Calcium concentration and movement in the diadic cleft space of the cardiac ventricular cell. Biophys J 70(3):1169–82, DOI: 10.1016/S0006-3495(96)79677-7 301

Larina O, Thorn P (2005) Ca^{2+} dynamics in salivary acinar cells: distinct morphology of the acinar lumen underlies near-synchronous global Ca^{2+} responses. J Cell Sci 118(18):4131–9, DOI: 10.1242/jcs.02533 279, 281

Larsen AZ, Olsen LF, Kummer U (2004) On the encoding and decoding of calcium signals in hepatocytes. Biophys Chem 107(1):83–99, DOI: 10.1016/j.bpc.2003.08.010 259

Lauffenburger DA, Linderman JJ (1993) Receptors: Models for Binding, Trafficking, and Signaling. Oxford University Press, New York 29

Lazzari C, Kipanyula MJ, Agostini M, Pozzan T, Fasolato C (2014) Aβ42 oligomers selectively disrupt neuronal calcium release. Neurobiol Aging DOI: 10.1016/j.neurobiolaging.2014.10.020 286

LeBeau AP, Robson AB, McKinnon AE, Donald RA, Sneyd J (1997) Generation of action potentials in a mathematical model of corticotrophs. Biophys J 73(3):1263–75, DOI: 10.1016/s0006-3495(97)78159-1 232, 377

LeBeau AP, Robson AB, McKinnon AE, Sneyd J (1998) Analysis of a reduced model of corticotroph action potentials. J Theor Biol 192(3):319–39, DOI: 10.1006/jtbi.1998.0656 377

LeBeau AP, Yule DI, Groblewski GE, Sneyd J (1999) Agonist-dependent phosphorylation of the inositol 1,4,5-trisphosphate receptor: a possible mechanism for agonist-specific calcium oscillations in pancreatic acinar cells. J Gen Physiol 113(6):851–72, DOI: 10.1085/jgp.113.6.851 276, 276, 277

LeBeau AP, Van Goor F, Stojilkovic SS, Sherman A (2000) Modeling of membrane excitability in gonadotropin-releasing hormone-secreting hypothalamic neurons regulated by Ca^{2+}-mobilizing and adenylyl cyclase-coupled receptors. J Neurosci 20(24):9290–7 58, 381, 382, 383

Lechleiter J, Clapham D (1992) Molecular mechanisms of intracellular calcium excitability in *X. laevis* oocytes. Cell 69:283–94, DOI: 10.1016/0092-8674(92)90409-6 145, 159

Lechleiter J, Girard S, Peralta E, Clapham D (1991) Spiral calcium wave propagation and annihilation in *Xenopus laevis* oocytes. Science 252:123–6, DOI: 10.1126/science.2011747 23, 145

Lechleiter JD, John LM, Camacho P (1998) Ca^{2+} wave dispersion and spiral wave entrainment in *Xenopus laevis* oocytes overexpressing Ca^{2+} ATPases. Biophys Chem 72(1–2):123–9, DOI: 10.1016/s0301-4622(98)00128-8 251

Lee K, Duan W, Sneyd J, Herbison AE (2010) Two slow calcium-activated afterhyperpolarization currents control burst firing dynamics in gonadotropin-releasing hormone neurons. J Neurosci 30(18):6214–24, DOI: 10.1523/JNEUROSCI.6156-09.2010 219, 232, 233, 381, 382, 383, 383, 383, 385, 385

Lees-Green R, Gibbons SJ, Farrugia G, Sneyd J, Cheng LK (2014) Computational modeling of anoctamin 1 calcium-activated chloride channels as pacemaker channels in interstitial cells of Cajal. Am J Physiol Gastrointest Liver Physiol 306(8):G711–27, DOI: 10.1152/ajpgi.00449.2013 315

Lemon G, Gibson WG, Bennett MR (2003) Metabotropic receptor activation, desensitization and sequestration—I: modelling calcium and inositol 1,4,5-trisphosphate dynamics following receptor activation. J Theor Biol 223(1):93–111, DOI: 10.1016/s0022-5193(03)00079-1 36

Levy D, Seigneuret M, Bluzat A, Rigaud JL (1990) Evidence for proton countertransport by the sarcoplasmic reticulum Ca^{2+}-ATPase during calcium transport in reconstituted proteoliposomes with low ionic permeability. J Biol Chem 265(32):19,524–34 43

Levy MN, Berne RM, Koeppen BM, Stanton BA (2006) Principles of Physiology. Mosby viii

Lewis RS (2007) The molecular choreography of a store-operated calcium channel. Nature 446(7133):284–7, DOI: 10.1038/nature05637 14

Leybaert L, Sanderson MJ (2012) Intercellular Ca^{2+} waves: mechanisms and function. Physiol Rev 92(3):1359–92, DOI: 10.1152/physrev.00029.2011 26, 146

Li YX, Rinzel J (1994) Equations for $InsP_3$ receptor-mediated Ca^{2+} oscillations derived from a detailed kinetic model: a Hodgkin-Huxley-like formalism. J Theor Biol 166:461–73, DOI: 10.1006/jtbi.1994.1041 73, 80, 81

Li YX, Rinzel J, Keizer J, Stojilković S (1994) Calcium oscillations in pituitary gonadotrophs: comparison of experiment and theory. Proc Natl Acad Sci USA 91:58–62, DOI: 10.1073/pnas.91.1.58 232

Li YX, Keizer J, Stojilković SS, Rinzel J (1995a) Ca^{2+} excitability of the ER membrane: an explanation for IP_3-induced Ca^{2+} oscillations. Am J Physiol 269(5):C1079–92 108

Li YX, Rinzel J, Vergara L, Stojilković SS (1995b) Spontaneous electrical and calcium oscillations in unstimulated pituitary gonadotrophs. Biophys J 69(3):785–95, DOI: 10.1016/s0006-3495(95)79952-0 378

Li YX, Stojilković S, Keizer J, Rinzel J (1997) Sensing and refilling calcium stores in an excitable cell. Biophys J 72:1080–91, DOI: 10.1016/s0006-3495(97)78758-7 378, 380

Lifshitz LM, Carmichael JD, Lai FA, Sorrentino V, Bellvé K, Fogarty KE, ZhuGe R (2011) Spatial organization of RYRs and BK channels underlying the activation of STOCs by Ca^{2+} sparks in airway myocytes. J Gen Physiol 138(2):195–209, DOI: 10.1085/jgp.201110626 317, 326

Linderman JJ (2008) Modeling of G-protein-coupled receptor signaling pathways. J Biol Chem 284(9):5427–31, DOI: 10.1074/jbc.R800028200 30

Lipp P, Niggli E (1996) Submicroscopic calcium signals as fundamental events of excitation-contraction coupling in guinea-pig cardiac myocytes. J Physiol 492 (1):31–8, DOI: 10.1113/jphysiol.1996.sp021286 20

Lisman JE (1985) A mechanism for memory storage insensitive to molecular turnover: a bistable autophosphorylating kinase. Proc Natl Acad Sci USA 82(9):3055–7, DOI: 10.1073/pnas.82.9.3055 346, 347

Lisman JE, Zhabotinsky AM (2001) A model of synaptic memory: a CaMKII/PP1 switch that potentiates transmission by organizing an AMPA receptor anchoring assembly. Neuron 31(2):191–201, DOI: 10.1016/s0896-6273(01)00364-6 348

Liu JL, Eisenberg B (2013) Correlated ions in a calcium channel model: a Poisson-Fermi theory. J Phys Chem B 117(40):12,051–8, DOI: 10.1021/jp408330f 59

Liu JL, Eisenberg B (2014a) Analytical models of calcium binding in a calcium channel. J Chem Phys 141(7):075,102, DOI: 10.1063/1.4892839 59

Liu JL, Eisenberg B (2014b) Poisson-Nernst-Planck-Fermi theory for modeling biological ion channels. J Chem Phys 141(22):22D532, DOI: 10.1063/1.4902973 59

Liu W, Olson SD (2015) Compartment calcium model of frog skeletal muscle during activation. J Theor Biol 364:139–53, DOI: 10.1016/j.jtbi.2014.08.050 313

Liu W, Tang F, Chen J (2010) Designing dynamical output feedback controllers for store-operated Ca^{2+} entry. Math Biosci 228(1):110–8, DOI: 10.1016/j.mbs.2010.08.013 66

Liu YJ, Vieira E, Gylfe E (2004) A store-operated mechanism determines the activity of the electrically excitable glucagon-secreting pancreatic α-cell. Cell Calcium 35(4):357–65, DOI: 10.1016/j.ceca.2003.10.002 369

Lock JT, Parker I, Smith IF (2015) A comparison of fluorescent Ca^{2+} indicators for imaging local Ca^{2+} signals in cultured cells. Cell Calcium 58(6):638–48, DOI: 10.1016/j.ceca.2015.10.003 18

Low JT, Shukla A, Behrendorff N, Thorn P (2010) Exocytosis, dependent on Ca^{2+} release from Ca^{2+} stores, is regulated by Ca^{2+} microdomains. J Cell Sci 123(18):3201–8, DOI: 10.1242/jcs.071225 275

Luik RM, Wang B, Prakriya M, Wu MM, Lewis RS (2008) Oligomerization of STIM1 couples ER calcium depletion to CRAC channel activation. Nature 454(7203):538–42, DOI: 10.1038/nature07065 66, 67

Lukyanenko V, Chikando A, Lederer WJ (2009) Mitochondria in cardiomyocyte Ca^{2+} signaling. Int J Biochem Cell Biol 41(10):1957–71, DOI: 10.1016/j.biocel.2009.03.011 52, 52

Luo CH, Rudy Y (1991) A model of the ventricular cardiac action potential; depolarization, repolarization and their interaction. Circ Res 68:1501–26, DOI: 10.1161/01.res.68.6.1501 296

Luttrell LM (2006) Transmembrane signaling by G protein-coupled receptors. Methods Mol Biol 332:3–49, DOI: 10.1385/1-59745-048-0:1 7, 29

Lytton J, Westlin M, Burk S, Shull G, MacLennan D (1992) Functional comparisons between isoforms of the sarcoplasmic or endoplasmic reticulum family of calcium pumps. J Biol Chem 267:14,483–9 40

Machaca K (2004) Increased sensitivity and clustering of elementary Ca^{2+} release events during oocyte maturation. Dev Biol 275(1):170–82, DOI: 10.1016/j.ydbio.2004.08.004 255

MacLennan DH, Rice WJ, Green NM (1997) The mechanism of Ca^{2+} transport by sarco(endo)plasmic reticulum Ca^{2+}-ATPases. J Biol Chem 272(46):28,815–8, DOI: 10.1074/jbc.272.46.28815 37, 41, 41

Maginu K (1985) Geometrical characteristics associated with stability and bifurcations of periodic travelling waves in reaction-diffusion equations. SIAM J Appl Math 45:750–74, DOI: 10.1137/0145044 144, 236

Magnus G, Keizer J (1997) Minimal model of beta-cell mitochondrial Ca^{2+} handling. Am J Physiol 273(2):C717–33 52, 54, 56, 365

Magnus G, Keizer J (1998a) Model of beta-cell mitochondrial calcium handling and electrical activity. I. Cytoplasmic variables. Am J Physiol 274(4):C1158–73 52, 365

Magnus G, Keizer J (1998b) Model of beta-cell mitochondrial calcium handling and electrical activity. II. Mitochondrial variables. Am J Physiol 274(4):C1174–84 52, 365

Mahama PA, Linderman JJ (1994) Calcium signaling in individual BC3H1 cells: speed of calcium mobilization and heterogeneity. Biotech Prog 10(1):45–54, DOI: 10.1021/bp00025a005 30

Mak DOD, Foskett JK (2015) Inositol 1,4,5-trisphosphate receptors in the endoplasmic reticulum: a single-channel point of view. Cell Calcium 58(1):67–78, DOI: 10.1016/j.ceca.2014.12.008 10, 69

Mak DOD, McBride S, Foskett J (2001) Regulation by Ca^{2+} and inositol 1,4,5-trisphosphate ($InsP_3$) of single recombinant type 3 $InsP_3$ receptor channels: Ca^{2+} activation uniquely distinguishes types 1 and 3 $InsP_3$ receptors. J Gen Physiol 117:435–46, DOI: 10.1085/jgp.117.5.435 135

Mak DOD, Pearson JE, Loong KPC, Datta S, Fernández-Mongil M, Foskett JK (2007) Rapid ligand-regulated gating kinetics of single inositol 1,4,5-trisphosphate receptor Ca^{2+} release channels. EMBO Rep 8(11):1044–51, DOI: 10.1038/sj.embor.7401087 11, 71, 76, 78

Marchant J, Callamaras N, Parker I (1999) Initiation of IP_3-mediated Ca^{2+} waves in *Xenopus* oocytes. EMBO J 18(19):5285–99, DOI: 10.1093/emboj/18.19.5285 20, 175, 176, 176, 176, 190

Marchant JS, Parker I (2001) Role of elementary Ca^{2+} puffs in generating repetitive Ca^{2+} oscillations. EMBO J 20(1–2):65–76, DOI: 10.1093/emboj/20.1.65 20, 172, 175

Marchant JS, Taylor CW (1998) Rapid activation and partial inactivation of inositol trisphosphate receptors by inositol trisphosphate. Biochemistry 37(33):11,524–33, DOI: 10.1021/bi980808k 71, 76

Marhl M, Schuster S, Brumen M (1998) Mitochondria as an important factor in the maintenance of constant amplitudes of cytosolic calcium oscillations. Biophys Chem 2(3):125–32, DOI: 10.1016/s0301-4622(97)00139-7 52

Marhl M, Haberichter T, Brumen M, Heinrich R (2000) Complex calcium oscillations and the role of mitochondria and cytosolic proteins. Biosystems 57:75–86, DOI: 10.1016/s0303-2647(00)00090-3 52

Marino MJ, Conn PJ (2006) Glutamate-based therapeutic approaches: allosteric modulators of metabotropic glutamate receptors. Curr Opin Pharmacol 6(1):98–102, DOI: 10.1016/j.coph.2005.09.006 286

Matsu-ura T, Michikawa T, Inoue T, Miyawaki A, Yoshida M, Mikoshiba K (2006) Cytosolic inositol 1,4,5-trisphosphate dynamics during intracellular calcium oscillations in living cells. J Cell Biol 173(5):755–65, DOI: 10.1083/jcb.200512141 291

Matsuoka SS (2002) Stoichiometry of the Na^+/Ca^{2+} exchanger: models and implications. Ann N Y Acad Sci 976:121–132, DOI: 10.1111/j.1749-6632.2002.tb04730.x 44, 51

Matsuoka SS, Hilgemann DWD (1992) Steady-state and dynamic properties of cardiac sodium-calcium exchange. Ion and voltage dependencies of the transport cycle. J Gen Physiol 100(6):963–1001, DOI: 10.1085/jgp.100.6.963 51

Matthews EK, Petersen OH (1973) Pancreatic acinar cells: ionic dependence of the membrane potential and acetycholine-induced depolarization. J Physiol 231(2):283–95, DOI: 10.1113/jphysiol.1973.sp010233 274

Matveev V, Sherman A, Zucker RS (2002) New and corrected simulations of synaptic facilitation. Biophys J 83(3):1368–73, DOI: 10.1016/S0006-3495(02)73907-6 338, 339

Matveev V, Zucker RS, Sherman A (2004) Facilitation through buffer saturation: constraints on endogenous buffering properties. Biophys J 86(5):2691–709, DOI: 10.1016/S0006-3495(04)74324-6 338, 339

Matveev V, Bertram R, Sherman A (2006) Residual bound Ca^{2+} can account for the effects of Ca^{2+} buffers on synaptic facilitation. J Neurophysiol 96(6):3389–97, DOI: 10.1152/jn.00101.2006 338, 340, 343, 344, 344

Matveev V, Bertram R, Sherman A (2009) Ca^{2+} current versus Ca^{2+} channel cooperativity of exocytosis. J Neurosci 29(39):12,196–209, DOI: 10.1523/JNEUROSCI.0263-09.2009 338

Matveev V, Bertram R, Sherman A (2011) Calcium cooperativity of exocytosis as a measure of Ca^{2+} channel domain overlap. Brain Res 1398:126–38, DOI: 10.1016/j.brainres.2011.05.011 338

Mazzag B, Tignanelli CJ, Smith GD (2005) The effect of residual Ca^{2+} on the stochastic gating of Ca^{2+}-regulated Ca^{2+} channel models. J Theor Biol 235(1):121–50, DOI: 10.1016/j.jtbi.2004.12.024 306, 308

McKean H (1970) Nagumo's equation. Adv Math 4:209–23, DOI: 10.1016/0001-8708(70)90023-x 108

Means S, Smith AJ, Shepherd J, Shadid J, Fowler J, Wojcikiewicz RJH, Mazel T, Smith GD, Wilson BS (2006) Reaction diffusion modeling of calcium dynamics with realistic ER geometry. Biophys J 91(2):537–57, DOI: 10.1529/biophysj.105.075036 135

Means SA, Sneyd J (2010) Spatio-temporal calcium dynamics in pacemaking units of the interstitial cells of Cajal. J Theor Biol 267(2):137–52, DOI: 10.1016/j.jtbi.2010.08.008 135

Mears D, Zimliki CL (2004) Muscarinic agonists activate Ca^{2+} store-operated and -independent ionic currents in insulin-secreting HIT-T15 cells and mouse pancreatic β-cells. J Membr Biol 197(1):59–70, DOI: 10.1007/s00232-003-0642-y 369, 369

Meinrenken CJ, Borst JG, Sakmann B (2003) Local routes revisited: the space and time dependence of the Ca^{2+} signal for phasic transmitter release at the rat calyx of Held. J Physiol 547(3):665–89, DOI: 10.1113/jphysiol.2002.032714 337

Messenger SW, Falkowski MA, Groblewski GE (2014) Ca^{2+}-regulated secretory granule exocytosis in pancreatic and parotid acinar cells. Cell Calcium 55(6):369–75, DOI: 10.1016/j.ceca.2014.03.003 275

Meyer T, Stryer L (1988) Molecular model for receptor-stimulated calcium spiking. Proc Natl Acad Sci USA 85(14):5051–5, DOI: 10.1073/pnas.85.14.5051 94, 115, 259

Michalski PJ (2013) The delicate bistability of CaMKII. Biophys J 105(3):794–806, DOI: 10.1016/j.bpj.2013.06.038 348, 352

Mijailovich SM, Butler JP, Fredberg JJ (2000) Perturbed equilibria of myosin binding in airway smooth muscle: bond-length distributions, mechanics, and ATP metabolism. Biophys J 79(5):2667–81, DOI: 10.1016/S0006-3495(00)76505-2 331

Missiaen L, Van Acker K, Van Baelen K, Raeymaekers L, Wuytack F, Parys JB, De Smedt H, Vanoevelen J, Dode L, Rizzuto R, Callewaert G (2004) Calcium release from the Golgi apparatus and the endoplasmic reticulum in HeLa cells stably expressing targeted aequorin to these compartments. Cell Calcium 36(6):479–87, DOI: 10.1016/j.ceca.2004.04.007 9

Moraru I, Kaftan EJ, Ehrlich BE, Watras J (1999) Regulation of type 1 inositol 1,4,5-trisphosphate-gated calcium channels by $InsP_3$ and calcium: simulation of single channel kinetics based on ligand binding and electrophysiological analysis. J Gen Physiol 113(6):837–49, DOI: 10.1085/jgp.113.6.837 71

Muallem S, Schoeffield M, Pandol S, Sachs G (1985) Inositol trisphosphate modification of ion transport in rough endoplasmic reticulum. Proc Natl Acad Sci USA 82(13):4433–7, DOI: 10.1073/pnas.82.13.4433 274

Mullins LJ (1977) A mechanism for Na/Ca transport. J Gen Physiol 70(6):681–95, DOI: 10.1085/jgp.70.6.681 44, 51

Murphy RA, Rembold CM, Hai CM (1990) Contraction in smooth muscle: what is latch? Prog Clin Biol Res 327:39–50 330

Murray J (2003) Mathematical Biology, 2nd edn. Springer Verlag, Berlin, Heidelberg, New York, DOI: 10.1007/978-3-662-08542-4 viii, 144

Nagumo J, Arimoto S, Yoshizawa S (1964) An active pulse transmission line simulating nerve axon. Proc IRE 50:2061–70, DOI: 10.1109/jrproc.1962.288235 108

Nan P, Wang Y, Vivien K, Rubin JE (2015) Understanding and distinguishing three time scale oscillations: case study in a coupled Morris–Lecar system. SIAM J Appl Dyn Syst, in press 225, 226, 233

Naraghi M, Neher E (1997) Linearized buffered Ca^{2+} diffusion in microdomains and its implications for calculation of $[Ca^{2+}]$ at the mouth of a calcium channel. J Neurosci 17(18):6961–73 128, 130

Nash MS, Young KW, Challiss RA, Nahorski SR (2001) Intracellular signalling. Receptor-specific messenger oscillations. Nature 413(6854):381–2, DOI: 10.1038/35096643 114, 285

Nash MS, Schell MJ, Atkinson PJ, Johnston NR, Nahorski SR, Challiss RAJ (2002) Determinants of metabotropic glutamate receptor-5-mediated Ca^{2+} and inositol 1,4,5-trisphosphate oscillation frequency. Receptor density versus agonist concentration. J Biol Chem 277(39):35,947–60, DOI: 10.1074/jbc.M205622200 286, 286, 286, 291, 291, 292

Nathanson MH, Mariwalla K (1996) Characterization and function of ATP receptors on hepatocytes from the little skate *Raja erinacea*. Am J Physiol 270(3):R561–70 273

Nathanson MH, Padfield PJ, O'Sullivan AJ, Burgstahler AD, Jamieson JD (1992) Mechanism of Ca^{2+} wave propagation in pancreatic acinar cells. J Biol Chem 267(25):18,118–21 274

Nathanson MH, Fallon MB, Padfield PJ, Maranto AR (1994) Localization of the type 3 inositol 1,4,5-trisphosphate receptor in the Ca^{2+} wave trigger zone of pancreatic acinar cells. J Biol Chem 269(7):4693–6 277

Navarrete M, Perea G, Maglio L, Pastor J, García de Sola R, Araque A (2013) Astrocyte calcium signal and gliotransmission in human brain tissue. Cereb Cortex 23(5):1240–6, DOI: 10.1093/cercor/bhs122 285

Neher E (1995) The use of fura-2 for estimating Ca buffers and Ca fluxes. Neuropharmacology 34(11):1423–42, DOI: 10.1016/0028-3908(95)00144-u 92

Neher E (1998a) Usefulness and limitations of linear approximations to the understanding of Ca^{++} signals. Cell Calcium 24(5–6):345–57, DOI: 10.1016/s0143-4160(98)90058-6 128, 337

Neher E (1998b) Vesicle pools and Ca^{2+} microdomains: new tools for understanding their roles in neurotransmitter release. Neuron 20(3):389–99, DOI: 10.1016/s0896-6273(00)80983-6 337, 338

Neher E, Sakaba T (2008) Multiple roles of calcium ions in the regulation of neurotransmitter release. Neuron 59(6):861–72, DOI: 10.1016/j.neuron.2008.08.019 337, 338

Neher E, Taschenberger H (2013) Transients in global Ca^{2+} concentration induced by electrical activity in a giant nerve terminal. J Physiol 591(13):3189–95, DOI: 10.1113/jphysiol.2012.248617 338, 338

Nguyen MHT, Jafri MS (2005) Mitochondrial calcium signaling and energy metabolism. Ann N Y Acad Sci 1047:127–37, DOI: 10.1196/annals.1341.012 52, 54, 56

Nilius B (2003) From TRPs to SOCs, CCEs, and CRACs: consensus and controversies. Cell Calcium 33(5–6):293–8, DOI: 10.1016/s0143-4160(03)00042-3 15

Noble D (1962) A modification of the Hodgkin-Huxley equations applicable to Purkinje fiber action and pacemaker potential. J Physiol 160:317–52, DOI: 10.1113/jphysiol.1962.sp006849 296

Noble D (2008) Computational models of the heart and their use in assessing the actions of drugs. J Pharmacol Sci 107(2):107–17, DOI: 10.1254/jphs.CR0070042 297

Noble K, Matthew A, Burdyga T, Wray S (2009) A review of recent insights into the role of the sarcoplasmic reticulum and Ca entry in uterine smooth muscle. Eur J Obstet Gynecol Reprod Biol 144 Suppl 1:S11–9, DOI: 10.1016/j.ejogrb.2009.02.010 315

Nonner W, Eisenberg B (1998) Ion permeation and glutamate residues linked by Poisson-Nernst-Planck theory in L-type calcium channels. Biophys J 75(3):1287–305, DOI: 10.1016/s0006-3495(98)74048-2 59

Nonner W, Chen DP, Eisenberg B (1999) Progress and prospects in permeation. J Gen Physiol 113(6):773–82, DOI: 10.1085/jgp.113.6.773 59

Nowycky M, Pinter M (1993) Time courses of calcium and calcium-bound buffers following calcium influx in a model cell. Biophys J 64:77–91, DOI: 10.1016/s0006-3495(93)81342-0 160

Nuccitelli R, Yim DL, Smart T (1993) The sperm-induced Ca^{2+} wave following fertilization of the *Xenopus* egg requires the production of Ins(1,4,5)P_3. Dev Biol 158(1):200–12, DOI: 10.1006/dbio.1993.1179 159

Nunemaker CS (2006) Glucose modulates $[Ca^{2+}]_i$ oscillations in pancreatic islets via ionic and glycolytic mechanisms. Biophys J 91(6):2082–96, DOI: 10.1529/biophysj.106.087296 358

Nykamp DQ, Tranchina D (2000) A population density approach that facilitates large-scale modeling of neural networks: analysis and an application to orientation tuning. J Comput Neurosci 8(1):19–50, DOI: 10.1023/a:1008912914816 308

Ochiai EI (1991) Why calcium? Principles and applications in bioorganic chemistry-IV. Journal of Chemical Education 68(1):10, DOI: 10.1021/ed068p10 5

Okamoto H, Ichikawa K (2000) Switching characteristics of a model for biochemical-reaction networks describing autophosphorylation versus dephosphorylation of Ca^{2+}/calmodulin-dependent protein kinase II. Biol Cybern 82(1):35–47, DOI: 10.1007/pl00007960 348, 349

Olson ML, Chalmers S, McCarron JG (2012) Mitochondrial organization and Ca^{2+} uptake. Biochem Soc Trans 40(1):158–67, DOI: 10.1042/BST20110705 8, 52, 52

O'Neill AF, Hagar RE, Zipfel WR, Nathanson MH, Ehrlich BE (2002) Regulation of the type III InsP_3 receptor by InsP_3 and calcium. Biochem Biophys Res Commun 294(3):719–25, DOI: 10.1016/S0006-291X(02)00524-7 71

Ong HL, Liu X, Tsaneva-Atanasova K, Singh BB, Bandyopadhyay BC, Swaim WD, Russell JT, Hegde RS, Sherman A, Ambudkar IS (2006) Relocalization of STIM1 for activation of store-operated Ca^{2+} entry is determined by the depletion of subplasma membrane endoplasmic reticulum Ca^{2+} store. J Biol Chem 282(16):12,176–85, DOI: 10.1074/jbc.M609435200 65, 66

Orrenius S, Zhivotovsky B, Nicotera P (2003) Regulation of cell death: the calcium-apoptosis link. Nat Rev Mol Cell Biol 4(7):552–65, DOI: 10.1038/nrm1150 5

Pabelick CM, Prakash YS, Kannan MS, Sieck GC (1999) Spatial and temporal aspects of calcium sparks in porcine tracheal smooth muscle cells. Am J Physiol 277(5):L1018–25 315

Pacher P, Csordás P, Schneider T, Hajnóczky G (2000) Quantification of calcium signal transmission from sarco-endoplasmic reticulum to the mitochondria. J Physiol 529:553–64, DOI: 10.1111/j.1469-7793.2000.00553.x 53

Pahle J, Green AK, Dixon CJ, Kummer U (2008) Information transfer in signaling pathways: a study using coupled simulated and experimental data. BMC Bioinformatics 9:139, DOI: 10.1186/1471-2105-9-139 357

Palk L, Sneyd J, Shuttleworth TJ, Yule DI, Crampin EJ (2010) A dynamic model of saliva secretion. J Theor Biol 266(4):625–40, DOI: 10.1016/j.jtbi.2010.06.027 25, 275, 276, 282

Palk L, Sneyd J, Patterson K, Shuttleworth TJ, Yule DI, Maclaren O, Crampin EJ (2012) Modelling the effects of calcium waves and oscillations on saliva secretion. J Theor Biol 305:45–53, DOI: 10.1016/j.jtbi.2012.04.009 26, 276, 277, 282

Pan B, Zucker RS (2009) A general model of synaptic transmission and short-term plasticity. Neuron 62(4):539–54, DOI: 10.1016/j.neuron.2009.03.025 338

Paredes RM, Etzler JC, Watts LT, Zheng W, Lechleiter JD (2008) Chemical calcium indicators. Methods 46(3):143–51, DOI: 10.1016/j.ymeth.2008.09.025 18

Parekh AB, Putney JW (2005) Store-operated calcium channels. Physiol Rev 85(2):757–810, DOI: 10.1152/physrev.00057.2003 14

Parker I, Ivorra I (1990) Inhibition by Ca^{2+} of inositol trisphosphate-mediated Ca^{2+} liberation: a possible mechanism for oscillatory release of Ca^{2+}. Proc Natl Acad Sci USA 87:260–4, DOI: 10.1073/pnas.87.1.260 11

Parker I, Yao Y, Ilyin V (1996) Fast kinetics of calcium liberation induced in *Xenopus* oocytes by photoreleased inositol trisphosphate. Biophys J 70:222–37, DOI: 10.1016/s0006-3495(96)79565-6 11

Parri HR, Crunelli V (2003) The role of Ca^{2+} in the generation of spontaneous astrocytic Ca^{2+} oscillations. Neuroscience 120(4):979–92, DOI: 10.1016/S0306-4522(03)00379-8 285

Parthimos D, Edwards DH, Griffith TM (1999) Minimal model of arterial chaos generated by coupled intracellular and membrane Ca^{2+} oscillators. Am J Physiol 277(3):H1119–44 329

Parthimos D, Haddock RE, Hill CE, Griffith TM (2007) Dynamics of a three-variable nonlinear model of vasomotion: comparison of theory and experiment. Biophys J 93(5):1534–56, DOI: 10.1529/biophysj.107.106278 329

Parys J, Sernett S, DeLisle S, Snyder P, Welsh M, Campbell K (1992) Isolation, characterization, and localization of the inositol 1,4,5-trisphosphate receptor protein in *Xenopus laevis* oocytes. J Biol Chem 267:18,776–82 76, 246

Parys JB, Bezprozvanny I (1995) The inositol trisphosphate receptor of *Xenopus* oocytes. Cell Calcium 18(5):353–63, DOI: 10.1016/0143-4160(95)90051-9 264

Parys JB, De Smedt H (2012) Inositol 1,4,5-trisphosphate and its receptors. Adv Exp Med Biol 740:255–79, DOI: 10.1007/978-94-007-2888-2_11 70

Patel S, Muallem S (2011) Acidic Ca^{2+} stores come to the fore. Cell Calcium 50(2):109–12, DOI: 10.1016/j.ceca.2011.03.009 9

Patel S, Joseph SK, Thomas AP (1999) Molecular properties of inositol 1,4,5-trisphosphate receptors. Cell Calcium 25(3):247–64, DOI: 10.1054/ceca.1999.0021 10, 70

Patel S, Ramakrishnan L, Rahman T, Hamdoun A, Marchant JS, Taylor CW, Brailoiu E (2011) The endo-lysosomal system as an NAADP-sensitive acidic Ca^{2+} store: Role for the two-pore channels. Cell Calcium 50(2):157–67, DOI: 10.1016/j.ceca.2011.03.011 9, 9

Patron M, Checchetto V, Raffaello A, Teardo E, Vecellio Reane D, Mantoan M, Granatiero V, Szabò I, De Stefani D, Rizzuto R (2014) MICU1 and MICU2 finely tune the mitochondrial Ca^{2+} uniporter by exerting opposite effects on MCU activity. Mol Cell 53(5):726–37, DOI: 10.1016/j.molcel.2014.01.013 55

Patterson M, Sneyd J, Friel DD (2007) Depolarization-induced calcium responses in sympathetic neurons: relative contributions from Ca^{2+} entry, extrusion, ER/mitochondrial Ca^{2+} uptake and release, and Ca^{2+} buffering. J Gen Physiol 129(1):29–56, DOI: 10.1085/jgp.200609660 52

Payne S, Stephens C (2005) The response of the cross-bridge cycle model to oscillations in intracellular calcium: a mathematical analysis. Proc Annu Int Conf IEEE Eng Med Biol Soc 7:7305–8, DOI: 10.1109/IEMBS.2005.1616198 331

Pearson JE, Ponce-Dawson S (1998) Crisis on skid row. Physica A 257(1):141–8, DOI: 10.1016/s0378-4371(98)00136-8 140

Pedersen MG (2010) A biophysical model of electrical activity in human β-cells. Biophys J 99(10):3200–7, DOI: 10.1016/j.bpj.2010.09.004 364

Pedersen MG, Sherman A (2009) Newcomer insulin secretory granules as a highly calcium-sensitive pool. Proc Natl Acad Sci USA 106(18):7432–6, DOI: 10.1073/pnas.0901202106 373, 374

Pedersen MG, Bersani AM, Bersani E (2008) Quasi steady-state approximations in complex intracellular signal transduction networks–a word of caution. J Math Chem 43(4):1318–44, DOI: 10.1007/s10910-007-9248-4 216

Penny CJ, Kilpatrick BS, Han JM, Sneyd J, Patel S (2014) A computational model of lysosome-ER Ca^{2+} microdomains. J Cell Sci 127(13):2934–43, DOI: 10.1242/jcs.149047 136

Perc M, Green AK, Dixon CJ, Marhl M (2008) Establishing the stochastic nature of intracellular calcium oscillations from experimental data. Biophys Chem 132(1):33–8, DOI: 10.1016/j.bpc.2007.10.002 264

Perez JF, Sanderson MJ (2005a) The contraction of smooth muscle cells of intrapulmonary arterioles is determined by the frequency of Ca^{2+} oscillations induced by 5-HT and KCl. J Gen Physiol 125(6):555–67, DOI: 10.1085/jgp.200409217 328

Perez JF, Sanderson MJ (2005b) The frequency of calcium oscillations induced by 5-HT, ACH, and KCl determine the contraction of smooth muscle cells of intrapulmonary bronchioles. J Gen Physiol 125(6):535–53, DOI: 10.1085/jgp.200409216 23, 24, 176, 315, 318, 334, 335

Peskoff A, Langer GA (1998) Calcium concentration and movement in the ventricular cardiac cell during an excitation-contraction cycle. Biophys J 74(1):153–74, DOI: 10.1016/S0006-3495(98)77776-8 301

Peskoff A, Post JA, Langer GA (1992) Sarcolemmal calcium binding sites in heart: II. Mathematical model for diffusion of calcium released from the sarcoplasmic reticulum into the diadic region. J Membr Biol 129(1):59–69, DOI: 10.1007/bf00232055 301

Petersen OH (2014) Calcium signalling and secretory epithelia. Cell Calcium 55(6):282–9, DOI: 10.1016/j.ceca.2014.01.003 158, 274

Pin JP, Kniazeff J, Goudet C, Bessis AS, Liu J, Galvez T, Acher F, Rondard P, Prézeau L (2004) The activation mechanism of class-C G-protein coupled receptors. Biol Cell 96(5):335–42, DOI: 10.1016/j.biolcel.2004.03.005 286

Pinton P, Pozzan T, Rizzuto R (1998) The Golgi apparatus is an inositol 1,4,5-trisphosphate-sensitive Ca^{2+} store, with functional properties distinct from those of the endoplasmic reticulum. EMBO J 17(18):5298–308, DOI: 10.1093/emboj/17.18.5298 9

Pizzo P, Lissandron V, Capitanio P, Pozzan T (2011) Ca^{2+} signalling in the Golgi apparatus. Cell Calcium 50(2):184–92, DOI: 10.1016/j.ceca.2011.01.006 9

Plant TD (1988) Properties and calcium-dependent inactivation of calcium currents in cultured mouse pancreatic β-cells. J Physiol 404:731–47, DOI: 10.1113/jphysiol.1988.sp017316 61

Politi A, Gaspers LD, Thomas AP, Höfer T (2006) Models of IP_3 and Ca^{2+} oscillations: frequency encoding and identification of underlying feedbacks. Biophys J 90(9):3120–33, DOI: 10.1529/biophysj.105.072249 94, 94, 115, 115, 116, 116, 116, 117

Ponce-Dawson S, Keizer J, Pearson JE (1999) Fire-diffuse-fire model of dynamics of intracellular calcium waves. Proc Natl Acad Sci USA 96(11):6060–3, DOI: 10.1073/pnas.96.11.6060 140

Postnov DE, Jacobsen JCB, Holstein-Rathlou NH, Sosnovtseva OV (2011) Functional modeling of the shift in cellular calcium dynamics at the onset of synchronization in smooth muscle cells. Bull Math Biol 73(10):2507–25, DOI: 10.1007/s11538-011-9636-6 329

Pozo MJ, Pérez GJ, Nelson MT, Mawe GM (2002) Ca^{2+} sparks and BK currents in gallbladder myocytes: role in CCK-induced response. Am J Physiol Gastrointest Liver Physiol 282(1):G165–74, DOI: 10.1152/ajpgi.00326.2001 317

Pradhan RK, Beard DA, Dash RK (2010) A biophysically based mathematical model for the kinetics of mitochondrial Na^{+}-Ca^{2+} antiporter. Biophys J 98(2):218–30, DOI: 10.1016/j.bpj.2009.10.005 50, 56, 56

Prager T, Falcke M, Schimansky-Geier L, Zaks MA (2007) Non-Markovian approach to globally coupled excitable systems. Phys Rev E 76(1):011,118, DOI: 10.1103/PhysRevE.76.011118 192

Prakash YS, Pabelick CM, Kannan MS, Sieck GC (2000) Spatial and temporal aspects of ACh-induced $[Ca^{2+}]_i$ oscillations in porcine tracheal smooth muscle. Cell Calcium 27(3):153–62, DOI: 10.1054/ceca.1999.0106 315

Prins D, Michalak M (2011) Organellar calcium buffers. Cold Spring Harb Perspect Biol 3(3), DOI: 10.1101/cshperspect.a004069 18, 89, 92

Puglisi JL, Bers DM (2001) LabHEART: an interactive computer model of rabbit ventricular myocyte ion channels and Ca transport. Am J Physiol Cell Physiol 281(6):C2049–60 22, 136

Puglisi JL, Wang F, Bers DM (2004) Modeling the isolated cardiac myocyte. Prog Biophys Mol Biol 85(2–3):163–78, DOI: 10.1016/j.pbiomolbio.2003.12.003 296

Putney JW (1986) A model for receptor-regulated calcium entry. Cell Calcium 7(1):1–12, DOI: 10.1016/0143-4160(86)90026-6 14

Qin J, Valle G, Nani A, Nori A, Rizzi N, Priori SG, Volpe P, Fill M (2008) Luminal Ca^{2+} regulation of single cardiac ryanodine receptors: insights provided by calsequestrin and its mutants. J Gen Physiol 131(4):325–34, DOI: 10.1085/jgp.200709907 85

Quoix N, Cheng-Xue R, Mattart L, Zeinoun Z, Guiot Y, Beauvois MC, Henquin JC, Gilon P (2009) Glucose and pharmacological modulators of ATP-sensitive K^+ channels control $[Ca^{2+}]_c$ by different mechanisms in isolated mouse α-cells. Diabetes 58(2):412–21, DOI: 10.2337/db07-1298 374

Rackham OJL, Tsaneva-Atanasova K, Ganesh A, Mellor JR (2010) A Ca-based computational model for NMDA receptor-dependent synaptic plasticity at individual post-synaptic spines in the hippocampus. Front Synaptic Neurosci 2:31, DOI: 10.3389/fnsyn.2010.00031 353

Rapp PE, Berridge MJ (1981) The control of transepithelial potential oscillations in the salivary gland of *Calliphora erythrocephala*. J Exp Biol 93(1):119–32 182, 274

Rauch J, Smoller J (1978) Qualitative theory of the FitzHugh-Nagumo equations. Adv Math 27:12–44, DOI: 10.1016/0001-8708(78)90075-0 108

Ren J, Sherman A, Bertram R, Goforth PB, Nunemaker CS, Waters CD, Satin LS (2013) Slow oscillations of K_{ATP} conductance in mouse pancreatic islets provide support for electrical bursting driven by metabolic oscillations. Am J Physiol Endocrinol Metab 305(7):E805–17, DOI: 10.1152/ajpendo.00046.2013 358

Renard D, Poggioli J, Berthon B, Claret M (1987) How far does phospholipase C activity depend on the cell calcium concentration? A study in intact cells. Biochem J 243(2):391–8, DOI: 10.1042/bj2430391 12, 94, 259

Ressmeyer AR, Bai Y, Delmotte P, Uy KF, Thistlethwaite P, Fraire A, Sato O, Ikebe M, Sanderson MJ (2010) Human airway contraction and formoterol-induced relaxation is determined by Ca^{2+} oscillations and Ca^{2+} sensitivity. Am J Respir Cell Mol Biol 43(2):179–91, DOI: 10.1165/rcmb.2009-0222OC 315

Restrepo JG, Weiss JN, Karma A (2008) Calsequestrin-mediated mechanism for cellular calcium transient alternans. Biophys J 95(8):3767–89, DOI: 10.1529/biophysj.108.130419 85, 86, 86

Ribeiro FM, Hamilton A, Doria JG, Guimaraes IM, Cregan SP, Ferguson SS (2014) Metabotropic glutamate receptor 5 as a potential therapeutic target in Huntington's disease. Expert Opin Ther Targets 18(11):1293–304, DOI: 10.1517/14728222.2014.948419 286

Rice JJ, Jafri MS, Winslow RL (1999) Modeling gain and gradedness of Ca^{2+} release in the functional unit of the cardiac diadic space. Biophys J 77(4):1871–84, DOI: 10.1016/s0006-3495(99)77030-x 87, 300, 301

Rihana S, Terrien J, Germain G, Marque C (2009) Mathematical modeling of electrical activity of uterine muscle cells. Med Biol Eng Comput 47(6):665–75, DOI: 10.1007/s11517-009-0433-4 315

Rinzel J (1977) Repetitive activity and Hopf bifurcation under point-stimulation for a simple FitzHugh-Nagumo nerve conduction model. J Math Biol 5(4):363–82, DOI: 10.1007/bf00276107 108

Rinzel J (1985) Bursting oscillations in an excitable membrane model. In: Sleeman B, Jarvis R (eds) Ordinary and partial differential equations, Springer-Verlag, New York, pp 304–16, DOI: 10.1007/bfb0074739 220, 358, 358

Rinzel J, Lee YS (1986) On different mechanisms for membrane potential bursting. In: Othmer HG (ed) Nonlinear oscillations in biology and chemistry. Lecture Notes in Biomathematics, Vol. 66, Springer-Verlag, New York, pp 19–33, DOI: 10.1007/978-3-642-93318-9_2 360

Rinzel J, Lee YS (1987) Dissection of a model for neuronal parabolic bursting. J Math Biol 25(6):653–75, DOI: 10.1007/bf00275501 358, 358

Riz M, Braun M, Pedersen MG (2014) Mathematical modeling of heterogeneous electrophysiological responses in human β-cells. PLoS Comput Biol 10(1):e1003,389, DOI: 10.1371/journal.pcbi.1003389 365

Rizzuto R, Brini M, Murgia M, Pozzan T (1993) Microdomains with high Ca^{2+} close to IP_3-sensitive channels that are sensed by neighboring mitochondria. Science 262(5134):744–7, DOI: 10.1126/science.8235595 9

Rizzuto R, Bastianutto C, Brini M, Murgia M, Pozzan T (1994) Mitochondrial Ca^{2+} homeostasis in intact cells. J Cell Biol 126(5):1183–94, DOI: 10.1083/jcb.126.5.1183 9

Rizzuto R, Pinton P, Brini M, Chiesa A, Filippin L, Pozzan T (1999) Mitochondria as biosensors of calcium microdomains. Cell Calcium 26(5):193–9, DOI: 10.1054/ceca.1999.0076 9, 53

Rizzuto R, De Stefani D, Raffaello A, Mammucari C (2012) Mitochondria as sensors and regulators of calcium signalling. Nat Rev Mol Cell Biol 13(9):566–78, DOI: 10.1038/nrm3412 8, 51

Robb-Gaspers L, Thomas A (1995) Coordination of Ca^{2+} signaling by intercellular propagation of Ca^{2+} waves in the intact liver. J Biol Chem 270:8102–7, DOI: 10.1074/jbc.270.14.8102 27

Roberts CB, Suter KJ (2008) Emerging methodologies for the study of hypothalamic gonadotropin-releasing-hormone (GnRH) neurons. Integr Comp Biol 48(5):548–59, DOI: 10.1093/icb/icn039 382

Roberts CB, Best JA, Suter KJ (2006) Dendritic processing of excitatory synaptic input in hypothalamic gonadotropin releasing-hormone neurons. Endocrinology 147(3):1545–55, DOI: 10.1210/en.2005-1350 382

Roberts CB, Campbell RE, Herbison AE, Suter KJ (2008a) Dendritic action potential initiation in hypothalamic gonadotropin-releasing hormone neurons. Endocrinology 149(7):3355–60, DOI: 10.1210/en.2008-0152 382

Roberts CB, Hemond P, Suter KJ (2008b) Synaptic integration in hypothalamic gonadotropin releasing hormone (GnRH) neurons. Neuroscience 154(4):1337–51, DOI: 10.1016/j.neuroscience.2008.04.067 382

Roberts CB, O'Boyle MP, Suter KJ (2009) Dendrites determine the contribution of after depolarization potentials (ADPs) to generation of repetitive action potentials in hypothalamic gonadotropin releasing-hormone (GnRH) neurons. J Comput Neurosci 26(1):39–53, DOI: 10.1007/s10827-008-0095-5 382

Romeo MM, Jones CKRT (2003) The stability of traveling calcium pulses in a pancreatic acinar cell. Physica D 177(1):242–58, DOI: 10.1016/S0167-2789(02)00772-8 237

Rooney T, Sass E, Thomas A (1989) Characterization of cytosolic calcium oscillations induced by phenylephrine and vasopressin in single fura-2-loaded hepatocytes. J Biol Chem 264:17,131–41 174, 182

Rooney T, Sass E, Thomas A (1990) Agonist-induced cytosolic calcium oscillations originate from a specific locus in single hepatocytes. J Biol Chem 265(18):10,792–6 176, 176

Roper P, Callaway J, Shevchenko T, Teruyama R, Armstrong W (2003) AHPs, HAPs and DAPs: how potassium currents regulate the excitability of rat supraoptic neurones. J Comput Neurosci 15(3):367–89, DOI: 10.1023/A:1027424128972 377

Roper P, Callaway J, Armstrong W (2004) Burst initiation and termination in phasic vasopressin cells of the rat supraoptic nucleus: a combined mathematical, electrical, and calcium fluorescence study. J Neurosci 24(20):4818–31, DOI: 10.1523/JNEUROSCI.4203-03.2004 232, 377

Rose T, Goltstein PM, Portugues R, Griesbeck O (2014) Putting a finishing touch on GECIs. Front Mol Neurosci 7:88, DOI: 10.3389/fnmol.2014.00088 18

Roux E, Marhl M (2004) Role of sarcoplasmic reticulum and mitochondria in Ca^{2+} removal in airway myocytes. Biophys J 86(4):2583–95, DOI: 10.1016/S0006-3495(04)74313-1 318

Roux E, Guibert C, Savineau JP, Marthan R (1997) $[Ca^{2+}]_i$ oscillations induced by muscarinic stimulation in airway smooth muscle cells: receptor subtypes and correlation with the mechanical activity. Br J Pharmacol 120(7):1294–301, DOI: 10.1038/sj.bjp.0701061 315

Roux E, Molimard M, Savineau JP, Marthan R (1998) Muscarinic stimulation of airway smooth muscle cells. Gen Pharmacol 31(3):349–56, DOI: 10.1016/s0306-3623(98)00007-x 315

Roux E, Noble PJ, Noble D, Marhl M (2006) Modelling of calcium handling in airway myocytes. Prog Biophys Mol Biol 90(1–3):64–87, DOI: 10.1016/j.pbiomolbio.2005.05.004 325, 326, 326

Rubin JE, Gerkin RC, Bi GQ, Chow CC (2005) Calcium time course as a signal for spike-timing-dependent plasticity. J Neurophysiol 93(5):2600–13, DOI: 10.1152/jn.00803.2004 353

Rüdiger S, Shuai JW, Huisinga W, Nagaiah C, Warnecke G, Parker I, Falcke M (2007) Hybrid stochastic and deterministic simulations of calcium blips. Biophys J 93(6):1847–57, DOI: 10.1529/biophysj.106.099879 82, 105

Rüdiger S, Nagaiah C, Warnecke G, Shuai JW (2010a) Calcium domains around single and clustered IP_3 receptors and their modulation by buffers. Biophys J 99(1):3–12, DOI: 10.1016/j.bpj.2010.02.059 82, 160, 160

Rüdiger S, Shuai JW, Sokolov IM (2010b) Law of mass action, detailed balance, and the modeling of calcium puffs. Phys Rev Lett 105(4):048,103, DOI: 10.1103/PhysRevLett.105.048103 82, 105

Sakaba T, Neher E (2001) Calmodulin mediates rapid recruitment of fast-releasing synaptic vesicles at a calyx-type synapse. Neuron 32(6):1119–31, DOI: 10.1016/s0896-6273(01)00543-8 338

Sala F, Hernàndez-Cruz A (1990) Calcium diffusion modeling in a spherical neuron: relevance of buffering properties. Biophys J 57:313–24, DOI: 10.1016/s0006-3495(90)82533-9 160

Salazar C, Politi AZ, Höfer T (2008) Decoding of calcium oscillations by phosphorylation cycles: analytic results. Biophys J 94(4):1203–15, DOI: 10.1529/biophysj.107.113084 334, 357

Salehi A, Vieira E, Gylfe E (2006) Paradoxical stimulation of glucagon secretion by high glucose concentrations. Diabetes 55(8):2318–23, DOI: 10.2337/db06-0080 371

Salido GM, Sage SO, Rosado JA (2009) TRPC channels and store-operated Ca^{2+} entry. Biochim Biophys Acta 1793(2):223–30, DOI: 10.1016/j.bbamcr.2008.11.001 15

Sanderson MJ, Charles AC, Dirksen ER (1990) Mechanical stimulation and intercellular communication increases intracellular Ca^{2+} in epithelial cells. Cell Regul 1(8):585–96, DOI: 10.1091/mbc.1.8.585 26, 27, 151

Sanderson MJ, Charles AC, Boitano S, Dirksen ER (1994) Mechanisms and function of intercellular calcium signaling. Mol Cell Endocrinol 98(2):173–87, DOI: 10.1016/0303-7207(94)90136-8 26, 151, 151

Sanderson MJ, Bai Y, Perez-Zoghbi J (2010) Ca^{2+} oscillations regulate contraction of intrapulmonary smooth muscle cells. Adv Exp Med Biol 661:77–96, DOI: 10.1007/978-1-60761-500-2_5 315

Santo-Domingo J, Demaurex N (2010) Calcium uptake mechanisms of mitochondria. Biochim Biophys Acta 1797(6–7):907–12, DOI: 10.1016/j.bbabio.2010.01.005 53, 53

Sato D, Bers DM (2011) How does stochastic ryanodine receptor-mediated Ca leak fail to initiate a Ca spark? Biophys J 101(10):2370–9, DOI: 10.1016/j.bpj.2011.10.017 86

Schlicht R, Winkler G (2008) A delay stochastic process with applications in molecular biology. J Math Biol 57:613–48, DOI: 10.1007/s00285-008-0178-y 185

Schwaller B (2010) Cytosolic Ca^{2+} buffers. Cold Spring Harb Perspect Biol 2(11):a004,051, DOI: 10.1101/cshperspect.a004051 17, 89, 92

Schweizer N, Kummer U, Hercht H, Braunbeck T (2011) Amplitude-encoded calcium oscillations in fish cells. Biophys Chem 159(2–3):294–302, DOI: 10.1016/j.bpc.2011.08.002 273, 273, 273

Segel IH (1975) Enzyme kinetics: behavior and analysis of rapid equilibrium and steady state enzyme systems. Wiley, New York 38, 50

Seppey D, Sauser R, Koenigsberger M, Bény JL, Meister JJ (2010) Intercellular calcium waves are associated with the propagation of vasomotion along arterial strips. Am J Physiol Heart Circ Physiol 298(2):H488–96, DOI: 10.1152/ajpheart.00281.2009 327, 329

Shannon TR, Ginsburg KS, Bers DM (2000) Potentiation of fractional sarcoplasmic reticulum calcium release by total and free intra-sarcoplasmic reticulum calcium concentration. Biophys J 78(1):334–43, DOI: 10.1016/s0006-3495(00)76596-9 85

Shannon TR, Wang F, Puglisi J, Weber C, Bers DM (2004) A mathematical treatment of integrated Ca dynamics within the ventricular myocyte. Biophys J 87(5):3351–71, DOI: 10.1529/biophysj.104.047449 85, 85, 136, 300, 319

Sherman A, Rinzel J (1991) Model for synchronization of pancreatic beta-cells by gap junction coupling. Biophys J 59(3):547–59, DOI: 10.1016/S0006-3495(91)82271-8 149

Sherman A, Rinzel J, Keizer J (1988) Emergence of organized bursting in clusters of pancreatic beta-cells by channel sharing. Biophys J 54(3):411–25, DOI: 10.1016/S0006-3495(88)82975-8 149

Sherman A, Keizer J, Rinzel J (1990) Domain model for Ca^{2+}-inactivation of Ca^{2+} channels at low channel density. Biophys J 58(4):985–95, DOI: 10.1016/s0006-3495(90)82443-7 59, 59, 61, 61

Sherratt J (1993) The amplitude of periodic plane waves depends on initial conditions in a variety of lambda-omega systems. Nonlinearity 6(6):1055, DOI: 10.1088/0951-7715/6/6/013 144

Sherratt JA (1994) On the evolution of periodic plane waves in reaction-diffusion systems of λ-ω type. SIAM J Appl Math 54(5):1374–85, DOI: 10.1137/s0036139993243746 144

Shorten PR, Wall DJ (2000) A Hodgkin-Huxley model exhibiting bursting oscillations. Bull Math Biol 62(4):695–715, DOI: 10.1006/bulm.2000.0172 377

Shorten PR, Robson AB, McKinnon AE, Wall DJ (2000) CRH-induced electrical activity and calcium signalling in pituitary corticotrophs. J Theor Biol 206(3):395–405, DOI: 10.1006/jtbi.2000.2135 377

Shuai J, Pearson JE, Foskett JK, Mak DO, Parker I (2007) A kinetic model of single and clustered IP_3 receptors in the absence of Ca^{2+} feedback. Biophys J 93(4):1151–62, DOI: 10.1529/biophysj.107.108795 160

Shuai J, Pearson JE, Parker I (2008) Modeling Ca^{2+} feedback on a single inositol 1,4,5-trisphosphate receptor and its modulation by Ca^{2+} buffers. Biophys J 95(8):3738–52, DOI: 10.1529/biophysj.108.137182 160

Shuttleworth TJ (2012) STIM and Orai proteins and the non-capacitative ARC channels. Front Biosci 17:847–60, DOI: 10.2741/3960 15, 15

Shuttleworth TJ, Mignen O (2003) Calcium entry and the control of calcium oscillations. Biochem Soc Trans 31(5):916–9, DOI: 10.1042/ 15, 65

Siekmann I, Wagner LE, Yule D, Fox C, Bryant D, Crampin EJ, Sneyd J (2011) MCMC estimation of Markov models for ion channels. Biophys J 100(8):1919–29, DOI: 10.1016/j.bpj.2011.02.059 76

Siekmann I, Wagner LE, Yule D, Crampin EJ, Sneyd J (2012) A kinetic model for type I and II IP_3R accounting for mode changes. Biophys J 103(4):658–68, DOI: 10.1016/j.bpj.2012.07.016 76, 76, 76, 78

Simpson D, Kirk V, Sneyd J (2005) Complex oscillations and waves of calcium in pancreatic acinar cells. Physica D: Nonlinear Phenomena 200(3–4):303–24, DOI: 10.1016/j.physd.2004.11.006 237

Skupin A, Falcke M (2007) Statistical properties and information content of calcium oscillations. Genome Informatics 18:44–53, DOI: 10.1142/9781860949920_0005 179

Skupin A, Kettenmann H, Winkler U, Wartenberg M, Sauer H, Tovey SC, Taylor CW, Falcke M (2008) How does intracellular Ca^{2+} oscillate: by chance or by the clock? Biophys J 94(6):2404–11, DOI: 10.1529/biophysj.107.119495 171, 172, 174, 177, 178

Smith G, Dai L, Miura R, Sherman A (2001) Asymptotic analysis of buffered calcium diffusion near a point source. SIAM J Appl Math 61:1816–38, DOI: 10.1137/s0036139900368996 92, 128, 129, 130

Smith GD (1996) Analytical steady-state solution to the rapid buffering approximation near an open Ca^{2+} channel. Biophys J 71(6):3064–72, DOI: 10.1016/s0006-3495(96)79500-0 128, 130

Smith GD, Wagner J, Keizer J (1996) Validity of the rapid buffering approximation near a point source of calcium ions. Biophys J 70(6):2527–39, DOI: 10.1016/s0006-3495(96)79824-7 128

Smith IF, Parker I (2009) Imaging the quantal substructure of single IP_3R channel activity during Ca^{2+} puffs in intact mammalian cells. Proc Nat Acad Sci USA 106(15):6404–9, DOI: 10.1073/pnas.0810799106 81, 169, 169, 186, 186, 186, 197

Smith IF, Wiltgen SM, Parker I (2009) Localization of puff sites adjacent to the plasma membrane: functional and spatial characterization of Ca^{2+} signaling in SH-SY5Y cells utilizing membrane-permeant caged IP_3. Cell Calcium 45(1):65–76, DOI: 10.1016/j.ceca.2008.06.001 167, 169, 171, 176, 176

Smolen P, Keizer J (1992) Slow voltage inactivation of Ca^{2+} currents and bursting mechanisms for the mouse pancreatic β-cell. J Membr Biol 127:9–19, DOI: 10.1007/bf00232754 358

Smyth JT, Hwang SY, Tomita T, DeHaven WI, Mercer JC, Putney JW (2010) Activation and regulation of store-operated calcium entry. J Cell Mol Med 14(10):2337–49, DOI: 10.1111/j.1582-4934.2010.01168.x 15

Sneyd J, Dufour JF (2002) A dynamic model of the type-2 inositol trisphosphate receptor. Proc Natl Acad Sci USA 99(4):2398–403, DOI: 10.1073/pnas.032281999 280

Sneyd J, Falcke M (2005) Models of the inositol trisphosphate receptor. Prog Biophys Mol Biol 89(3):207–45, DOI: 10.1016/j.pbiomolbio.2004.11.001 70

Sneyd J, Girard S, Clapham D (1993) Calcium wave propagation by calcium-induced calcium release: an unusual excitable system. Bull Math Biol 55(2):315–44, DOI: 10.1016/s0092-8240(05)80268-x 109, 144

Sneyd J, Charles AC, Sanderson MJ (1994) A model for the propagation of intercellular calcium waves. Am J Physiol 266(1):C293–302 151, 153

Sneyd J, Keizer J, Sanderson MJ (1995a) Mechanisms of calcium oscillations and waves: a quantitative analysis. FASEB J 9(14):1463–72 151

Sneyd J, Wetton BT, Charles AC, Sanderson MJ (1995b) Intercellular calcium waves mediated by diffusion of inositol trisphosphate: a two-dimensional model. Am J Physiol 268(6):C1537–45 151, 153

Sneyd J, Dale P, Duffy A (1998a) Traveling waves in buffered systems: applications to calcium waves. SIAM J Appl Math 58:1178–92, DOI: 10.1137/s0036139996305074 93

Sneyd J, Wilkins M, Strahonja A, Sanderson MJ (1998b) Calcium waves and oscillations driven by an intercellular gradient of inositol (1,4,5)-trisphosphate. Biophys Chem 72(1–2):101–9, DOI: 10.1016/s0301-4622(98)00126-4 154

Sneyd J, LeBeau A, Yule D (2000) Traveling waves of calcium in pancreatic acinar cells: model construction and bifurcation analysis. Physica D 145:158–79, DOI: 10.1016/s0167-2789(00)00108-1 276

Sneyd J, Tsaneva-Atanasova K, Bruce JIE, Straub SV, Giovannucci DR, Yule DI (2003) A model of calcium waves in pancreatic and parotid acinar cells. Biophys J 85(3):1392–405, DOI: 10.1016/S0006-3495(03)74572-X 52, 276, 280, 280, 281

Sneyd J, Tsaneva-Atanasova K, Yule DI, Thompson JL, Shuttleworth TJ (2004) Control of calcium oscillations by membrane fluxes. Proc Natl Acad Sci USA 101(5):1392–6, DOI: 10.1073/pnas.0303472101 113, 215

Sneyd J, Tsaneva-Atanasova K, Reznikov V, Bai Y, Sanderson MJ, Yule DI (2006) A method for determining the dependence of calcium oscillations on inositol trisphosphate oscillations. Proc Natl Acad Sci USA 103(6):1675–80, DOI: 10.1073/pnas.0506135103 94, 117, 118, 119, 275, 318

Sobie EA, Dilly KW, dos Santos Cruz J, Lederer WJ, Jafri MS (2002) Termination of cardiac Ca^{2+} sparks: an investigative mathematical model of calcium-induced calcium release. Biophys J 83(1):59–78, DOI: 10.1016/s0006-3495(02)75149-7 301

Soboloff J, Rothberg BS, Madesh M, Gill DL (2012) STIM proteins: dynamic calcium signal transducers. Nat Rev Mol Cell Biol 13(9):549–65, DOI: 10.1038/nrm3414 15

Soeller C, Cannell MB (1997) Numerical simulation of local calcium movements during L-type calcium channel gating in the cardiac diad. Biophys J 73(1):97–111, DOI: 10.1016/S0006-3495(97)78051-2 301

Soeller C, Cannell MB (2004) Analysing cardiac excitation-contraction coupling with mathematical models of local control. Prog Biophys Mol Biol 85(2–3):141–62, DOI: 10.1016/j.pbiomolbio.2003.12.006 297, 299

Sorensen SD, Conn PJ (2003) G protein-coupled receptor kinases regulate metabotropic glutamate receptor 5 function and expression. Neuropharmacology 44(6):699–706, DOI: 10.1016/S0028-3908(03)00053-4 294

Spät A, Bradford PG, McKinney JS, Rubin RP, Putney JW Jr (1986) A saturable receptor for 32P-inositol-1,4,5-trisphosphate in hepatocytes and neutrophils. Nature 319(6053):514–6, DOI: 10.1038/319514a0 266

Spät A, Szanda G, Csordás G, Hajnóczky G (2008) High- and low-calcium-dependent mechanisms of mitochondrial calcium signalling. Cell Calcium 44(1):51–63, DOI: 10.1016/j.ceca.2007.11.015 17

Spray DC, Bennett MV (1985) Physiology and pharmacology of gap junctions. Annu Rev Physiol 47:281–303, DOI: 10.1146/annurev.ph.47.030185.001433 149

Stamatakis M, Mantzaris NV (2006) Modeling of ATP-mediated signal transduction and wave propagation in astrocytic cellular networks. J Theor Biol 241(3):649–68, DOI: 10.1016/j.jtbi.2006.01.002 156, 156

Stern M, Pizarro G, Ríos E (1997) Local control model of excitation-contraction coupling in skeletal muscle. J Gen Physiol 110:415–40, DOI: 10.1085/jgp.110.4.415 84, 85, 87, 87

Stern MD (1992a) Buffering of calcium in the vicinity of a channel pore. Cell Calcium 13(3):183–92, DOI: 10.1016/0143-4160(92)90046-u 128, 130

Stern MD (1992b) Theory of excitation-contraction coupling in cardiac muscle. Biophys J 63(2):497–517, DOI: 10.1016/S0006-3495(92)81615-6 298, 301

Stojilkovic SS (2012) Molecular mechanisms of pituitary endocrine cell calcium handling. Cell Calcium 51(3–4):212–21, DOI: 10.1016/j.ceca.2011.11.003 377

Stojilkovic SS, Reinhart J, Catt KJ (1994) Gonadotropin-releasing hormone receptors: structure and signal transduction pathways. Endocr Rev 15(4):462–99, DOI: 10.1210/edrv-15-4-462 378

Straub SV, Giovannucci DR, Yule DI (2000) Calcium wave propagation in pancreatic acinar cells: functional interaction of inositol 1,4,5-trisphosphate receptors, ryanodine receptors, and mitochondria. J Gen Physiol 116(4):547–60, DOI: 10.1085/jgp.116.4.547 52, 279

Straub SV, Wagner LE 2nd, Bruce JIE, Yule DI (2004) Modulation of cytosolic calcium signaling by protein kinase A-mediated phosphorylation of inositol 1,4,5-trisphosphate receptors. Biol Res 37(4):593–602, DOI: 10.4067/s0716-97602004000400013 277

Streb H, Irvine RF, Berridge MJ, Schulz I (1983) Release of Ca^{2+} from a nonmitochondrial intracellular store in pancreatic acinar cells by inositol-1,4,5-trisphosphate. Nature 306(5938):67–9, DOI: 10.1038/306067a0 274

Sun XP, Callamaras N, Marchant JS, Parker I (1998) A continuum of $InsP_3$-mediated elementary Ca^{2+} signalling events in *Xenopus* oocytes. J Physiol 509(1):67–80, DOI: 10.1111/j.1469-7793.1998.067bo.x 20

Surroca A, Wolff D (2000) Inositol 1,4,5-trisphosphate but not ryanodine-receptor agonists induces calcium release from rat liver Golgi apparatus membrane vesicles. J Membr Biol 177(3):243–9, DOI: 10.1007/s002320010008 9

Suzuki J, Kanemaru K, Ishii K, Ohkura M, Okubo Y, Iino M (2014) Imaging intraorganellar Ca^{2+} at subcellular resolution using CEPIA. Nat Commun 5:4153, DOI: 10.1038/ncomms5153 18

Swann K, Larman MG, Saunders CM, Lai FA (2004) The cytosolic sperm factor that triggers Ca^{2+} oscillations and egg activation in mammals is a novel phospholipase C: PLCζ. Reproduction 127(4):431–9, DOI: 10.1530/rep.1.00169 254

Swanson CJ, Bures M, Johnson MP, Linden AM, Monn JA, Schoepp DD (2005) Metabotropic glutamate receptors as novel targets for anxiety and stress disorders. Nat Rev Drug Discov 4(2):131–44, DOI: 10.1038/nrd1630 286

Szmolyan P, Wechselberger M (2001) Canards in R^3. J Diff Eq 177(2):419–53, DOI: 10.1006/jdeq.2001.4001 222, 225

Tabak J, Tomaiuolo M, Gonzalez-Iglesias AE, Milescu LS, Bertram R (2011) Fast-activating voltage- and calcium-dependent potassium (BK) conductance promotes bursting in pituitary cells: a dynamic clamp study. J Neurosci 31(46):16,855–63, DOI: 10.1523/JNEUROSCI.3235-11.2011 378

Takazawa K, Vandekerckhove J, Dumont JE, Erneux C (1990) Cloning and expression in *Escherichia coli* of a rat brain cDNA encoding a Ca^{2+}/calmodulin-sensitive inositol 1,4,5-trisphosphate 3-kinase. Biochem J 272(1):107–12, DOI: 10.1042/bj2720107 259

Tang Y, Stephenson JL, Othmer HG (1996) Simplification and analysis of models of calcium dynamics based on IP_3-sensitive calcium channel kinetics. Biophys J 70(1):246–63, DOI: 10.1016/s0006-3495(96)79567-x 70, 249

Tang Y, Schlumpberger T, Kim T, Lueker M, Zucker RS (2000) Effects of mobile buffers on facilitation: experimental and computational studies. Biophys J 78(6):2735–51, DOI: 10.1016/s0006-3495(00)76819-6 339

Tanimura A, Morita T, Nezu A, Tojyo Y (2009) Monitoring of IP_3 dynamics during Ca^{2+} oscillations in HSY human parotid cell line with FRET-based IP_3 biosensors. J Med Invest 56 Suppl:357–61, DOI: 10.2152/jmi.56.357 115

Taylor CW, Laude AJ (2002) IP_3 receptors and their regulation by calmodulin and cytosolic Ca^{2+}. Cell Calcium 32(5–6):321–34, DOI: 10.1016/s0143416002001859 10, 70

Taylor CW, Genazzani AA, Morris SA (1999) Expression of inositol trisphosphate receptors. Cell Calcium 26(6):237–51, DOI: 10.1054/ceca.1999.0090 10

Teka W, Tabak J, Vo T, Wechselberger M, Bertram R (2011a) The dynamics underlying pseudo-plateau bursting in a pituitary cell model. J Math Neurosci 1(12), DOI: 10.1186/2190-8567-1-12 378

Teka W, Tsaneva-Atanasova K, Bertram R, Tabak J (2011b) From plateau to pseudo-plateau bursting: making the transition. Bull Math Biol 73(6):1292–311, DOI: 10.1007/s11538-010-9559-7 378

Teka W, Tabak J, Bertram R (2012) The relationship between two fast/slow analysis techniques for bursting oscillations. Chaos 22(4):043,117, DOI: 10.1063/1.4766943 378

Thiel M, Lis A, Penner R (2013) STIM2 drives Ca^{2+} oscillations through store-operated Ca^{2+} entry caused by mild store depletion. J Physiol 591(6):1433–45, DOI: 10.1113/jphysiol.2012.245399 16

Thomas D, Lipp P, Tovey SC, Berridge MJ, Li W, Tsien RY, Bootman MD (2000) Microscopic properties of elementary Ca^{2+} release sites in non-excitable cells. Curr Biol 10(1):8–15, DOI: 10.1016/s0960-9822(99)00258-4 20

Thorn P, Lawrie AM, Smith PM, Gallacher DV, Petersen OH (1993) Local and global cytosolic Ca^{2+} oscillations in exocrine cells evoked by agonists and inositol trisphosphate. Cell 74(4):661–8, DOI: 10.1016/0092-8674(93)90513-p 274

Thul R, Falcke M (2004a) Release currents of IP_3 receptor channel clusters and concentration profiles. Biophys J 86(5):2660–73, DOI: 10.1016/S0006-3495(04)74322-2 128, 134, 135, 135, 160, 173, 177, 184

Thul R, Falcke M (2004b) Stability of membrane bound reactions. Phys Rev Lett 93:188,103, DOI: 10.1103/PhysRevLett.93.188103 173, 185, 185

Thul R, Falcke M (2006) Frequency of elemental events of intracellular Ca^{2+} dynamics. Phys Rev E 73:061,923, DOI: 10.1103/PhysRevE.73.061923 205

Thul R, Falcke M (2007) Waiting time distributions for clusters of complex molecules. EPL (Europhysics Letters) 79(3):38,003, DOI: 10.1209/0295-5075/79/38003 173, 185, 185, 205

Thul R, Coombes S, Bootman MD (2012a) Persistence of pro-arrhythmic spatio-temporal calcium patterns in atrial myocytes: a computational study of ping waves. Front Physiol 3:279, DOI: 10.3389/fphys.2012.00279 310, 312

Thul R, Coombes S, Roderick HL, Bootman MD (2012b) Subcellular calcium dynamics in a whole-cell model of an atrial myocyte. Proc Natl Acad Sci USA 109(6):2150–5, DOI: 10.1073/pnas.1115855109 310, 311

Thurley K, Falcke M (2011) Derivation of Ca^{2+} signals from puff properties reveals that pathway function is robust against cell variability but sensitive for control. Proc Natl Acad Sci USA 108(1):427–32, DOI: 10.1073/pnas.1008435108 167, 198

Thurley K, Smith IF, Tovey SC, Taylor CW, Parker I, Falcke M (2011) Timescales of IP_3-evoked Ca^{2+} spikes emerge from Ca^{2+} puffs only at the cellular level. Biophys J 101(11):2638–44, DOI: 10.1016/j.bpj.2011.10.030 168, 168, 169, 170, 171, 172, 172, 187, 189, 196, 204

Thurley K, Skupin A, Thul R, Falcke M (2012) Fundamental properties of Ca^{2+} signals. Biochim Biophys Acta 1820(8):1185–94, DOI: 10.1016/j.bbagen.2011.10.007 21

Thurley K, Tovey SC, Moenke G, Prince VL, Meena A, Thomas AP, Skupin A, Taylor CW, Falcke M (2014) Reliable encoding of stimulus intensities within random sequences of intracellular Ca^{2+} spikes. Sci Signal 7(331):ra59, DOI: 10.1126/scisignal.2005237 182, 198

Tinel H, Cancela JM, Mogami H, Gerasimenko JV, Gerasimenko OV, Tepikin AV, Petersen OH (1999) Active mitochondria surrounding the pancreatic acinar granule region prevent spreading of inositol trisphosphate-evoked local cytosolic Ca^{2+} signals. EMBO J 18(18):4999–5008, DOI: 10.1093/emboj/18.18.4999 52, 280

Tippmann SC, Ivanek R, Gaidatzis D, Scholer A, Hoerner L, van Nimwegen E, Stadler PF, Stadler MB, Schubeler D (2012) Chromatin measurements reveal contributions of synthesis and decay to steady-state mRNA levels. Mol Syst Biol 8:593, DOI: 10.1038/msb.2012.23 180

Tomaiuolo M, Bertram R, Gonzalez-Iglesias AE, Tabak J (2010) Investigating heterogeneity of intracellular calcium dynamics in anterior pituitary lactotrophs using a combined modelling/experimental approach. J Neuroendocrinol 22(12):1279–89, DOI: 10.1111/j.1365-2826.2010.02061.x 378

Tong WC, Choi CY, Karche S, Holden AV, Zhang H, Taggart MJ (2011) A computational model of the ionic currents, Ca^{2+} dynamics and action potentials underlying contraction of isolated uterine smooth muscle. PLoS ONE 6(4):e18,685, DOI: 10.1371/journal.pone.0018685 315

Tordjmann T, Berthon B, Claret M, Combettes L (1997) Coordinated intercellular calcium waves induced by noradrenaline in rat hepatocytes: dual control by gap junction permeability and agonist. EMBO J 16(17):5398–407, DOI: 10.1093/emboj/16.17.5398 27, 261, 270

Tordjmann T, Berthon B, Jacquemin E, Clair C, Stelly N, Guillon G, Claret M, Combettes L (1998) Receptor-oriented intercellular calcium waves evoked by vasopressin in rat hepatocytes. EMBO J 17(16):4695–703, DOI: 10.1093/emboj/17.16.4695 27, 261, 270, 270

Tosteson D, Hoffman J (1960) Regulation of cell volume by active cation transport in high and low potassium sheep red cells. J Gen Physiol 44:169–94, DOI: 10.1085/jgp.44.1.169 282

Tran K, Smith NP, Loiselle DS, Crampin EJ (2009) A thermodynamic model of the cardiac sarcoplasmic/endoplasmic Ca^{2+} (SERCA) pump. Biophys J 96(5):2029–42, DOI: 10.1016/j.bpj.2008.11.045 43, 44

Troy W (1976) Bifurcation phenomena in FitzHugh's nerve conduction equations. J Math Anal App 54:678–90, DOI: 10.1016/0022-247x(76)90187-6 108

Tsai J, Sneyd J (2005) Existence and stability of traveling waves in buffered systems. SIAM J Appl Math 66(1):237–65, DOI: 10.1137/040618291 159

Tsai JC (2013) Do calcium buffers always slow down the propagation of calcium waves? J Math Biol 67(6–7):1587–632, DOI: 10.1007/s00285-012-0605-y 159

Tsai JC, Sneyd J (2007) Are buffers boring? Uniqueness and asymptotical stability of traveling wave fronts in the buffered bistable system. J Math Biol 54(4):513–53, DOI: 10.1007/s00285-006-0057-3 159

Tsai JC, Sneyd J (2011) Traveling waves in the buffered FitzHugh-Nagumo model. SIAM J Appl Math 71(5):1606–36, DOI: 10.1137/110820348 159

Tsai JC, Zhang W, Kirk V, Sneyd J (2012) Traveling waves in a simplified model of calcium dynamics. SIAM J Appl Dyn Syst 11(4):1149–99, DOI: 10.1137/120867949 109, 144, 159, 234, 236, 237, 238, 238, 241, 242, 242

Tsaneva-Atanasova K, Yule DI, Sneyd J (2005) Calcium oscillations in a triplet of pancreatic acinar cells. Biophys J 88(3):1535–51, DOI: 10.1529/biophysj.104.047357 276

Tsaneva-Atanasova K, Zimliki CL, Bertram R, Sherman A (2006) Diffusion of calcium and metabolites in pancreatic islets: killing oscillations with a pitchfork. Biophys J 90(10):3434–46, DOI: 10.1529/biophysj.105.078360 149, 358

Tsaneva-Atanasova K, Sherman A, van Goor F, Stojilkovic SS (2007) Mechanism of spontaneous and receptor-controlled electrical activity in pituitary somatotrophs: experiments and theory. J Neurophysiol 98(1):131–44, DOI: 10.1152/jn.00872.2006 378

Tsaneva-Atanasova K, Osinga HM, Riess T, Sherman A (2010a) Full system bifurcation analysis of endocrine bursting models. J Theor Biol 264(4):1133–46, DOI: 10.1016/j.jtbi.2010.03.030 232

Tsaneva-Atanasova K, Osinga HM, Tabak J, Pedersen MG (2010b) Modeling mechanisms of cell secretion. Acta Biotheoretica 58(4):315–27, DOI: 10.1007/s10441-010-9115-8 232

Tsien RW, Tsien RY (1990) Calcium channels, stores, and oscillations. Ann Rev Cell Biol 6:715–60, DOI: 10.1146/annurev.cellbio.6.1.715 14, 56

Tsien RY (1998) The green fluorescent protein. Annu Rev Biochem 67:509–44, DOI: 10.1146/annurev.biochem.67.1.509 18

Tsoukias NM (2011) Calcium dynamics and signaling in vascular regulation: computational models. Wiley Interdiscip Rev Syst Biol Med 3(1):93–106, DOI: 10.1002/wsbm.97 327

Tyson J, Keener J (1988) Singular perturbation theory of traveling waves in excitable media. Physica D 32:327–361, DOI: 10.1016/0167-2789(88)90062-0 108

Ullah G, Jung P, Machaca K (2007) Modeling Ca^{2+} signaling differentiation during oocyte maturation. Cell Calcium 42(6):556–64, DOI: 10.1016/j.ceca.2007.01.010 255

Ullah G, Daniel Mak DO, Pearson JE (2012) A data-driven model of a modal gated ion channel: the inositol 1,4,5-trisphosphate receptor in insect Sf9 cells. J Gen Physiol 140(2):159–73, DOI: 10.1085/jgp.201110753 76, 79

Van Acker K, Kasri NN, Smet PD, Parys JB, Smedt HD, Missiaen L, Callewaert G (2002) IP_3-mediated Ca^{2+} signals in human neuroblastoma SH-SY5Y cells with exogenous overexpression of type 3 IP_3 receptor. Cell Calcium 32(2):71 – 81, DOI: 10.1016/S0143-4160(02)00092-1 171

Van Goor F, Krsmanovic LZ, Catt KJ, Stojilkovic SS (1999) Coordinate regulation of gonadotropin-releasing hormone neuronal firing patterns by cytosolic calcium and store depletion. Proc Natl Acad Sci USA 96(7):4101–6, DOI: 10.1073/pnas.96.7.4101 381

Van Goor F, Krsmanovic LZ, Catt KJ, Stojilkovic SS (2000) Autocrine regulation of calcium influx and gonadotropin-releasing hormone secretion in hypothalamic neurons. Biochem Cell Biol 78(3):359–70, DOI: 10.1139/o00-058 381, 382

Van Petegem F (2012) Ryanodine receptors: structure and function. J Biol Chem 287(38):31,624–32, DOI: 10.1074/jbc.R112.349068 10

Van Petegem F (2015) Ryanodine receptors: allosteric ion channel giants. J Mol Biol 427(1):31–53, DOI: 10.1016/j.jmb.2014.08.004 10

Ventura AC, Sneyd J (2006) Calcium oscillations and waves generated by multiple release mechanisms in pancreatic acinar cells. Bull Math Biol 68(8):2205–31, DOI: 10.1007/s11538-006-9101-0 276

Verechtchaguina T, Sokolov I, Schimansky-Geier L (2007) Interspike interval densities of resonate and fire neurons. Biosystems 89(1–3):63–8, DOI: DOI: 10.1016/j.biosystems.2006.03.014 170

Vieira E, Salehi A, Gylfe E (2007) Glucose inhibits glucagon secretion by a direct effect on mouse pancreatic α cells. Diabetologia 50(2):370–9, DOI: 10.1007/s00125-006-0511-1 376

Vines CM (2012) Phospholipase C. Adv Exp Med Biol 740:235–54, DOI: 10.1007/978-94-007-2888-2_10 12

Vo T, Bertram R, Tabak J, Wechselberger M (2010) Mixed mode oscillations as a mechanism for pseudo-plateau bursting. J Comput Neurosci 28(3):443–58, DOI: 10.1007/s10827-010-0226-7 378

Vo T, Bertram R, Wechselberger M (2013) Multiple geometric viewpoints of mixed mode dynamics associated with pseudo-plateau bursting. SIAM J Appl Dyn Syst 12(2):789–830, DOI: 10.1137/120892842 226, 226, 233

Vo T, Tabak J, Bertram R, Wechselberger M (2014) A geometric understanding of how fast activating potassium channels promote bursting in pituitary cells. J Comput Neurosci 36(2):259–78, DOI: 10.1007/s10827-013-0470-8 378

Voets T (2000) Dissection of three Ca^{2+}-dependent steps leading to secretion in chromaffin cells from mouse adrenal slices. Neuron 28(2):537–45, DOI: 10.1016/s0896-6273(00)00131-8 365, 372, 373, 374, 375

Voets T, Neher E, Moser T (1999) Mechanisms underlying phasic and sustained secretion in chromaffin cells from mouse adrenal slices. Neuron 23(3):607–15, DOI: 10.1016/s0896-6273(00)80812-0 372

Wadel K, Neher E, Sakaba T (2007) The coupling between synaptic vesicles and Ca^{2+} channels determines fast neurotransmitter release. Neuron 53(4):563–75, DOI: 10.1016/j.neuron.2007.01.021 338

Wagner J, Keizer J (1994) Effects of rapid buffers on Ca^{2+} diffusion and Ca^{2+} oscillations. Biophys J 67:447–56, DOI: 10.1016/s0006-3495(94)80500-4 91

Wagner J, Li YX, Pearson J, Keizer J (1998) Simulation of the fertilization Ca^{2+} wave in *Xenopus laevis* eggs. Biophys J 75(4):2088–97, DOI: 10.1016/S0006-3495(98)77651-9 251, 252, 252, 253

Wagner J, Fall CP, Hong F, Sims CE, Allbritton NL, Fontanilla RA, Moraru II, Loew LM, Nuccitelli R (2004) A wave of IP_3 production accompanies the fertilization Ca^{2+} wave in the egg of the frog, *Xenopus laevis*: theoretical and experimental support. Cell Calcium 35(5):433–47, DOI: 10.1016/j.ceca.2003.10.009 251, 253, 254, 255, 257

Wagner LE, Yule DI (2012) Differential regulation of the $InsP_3$ receptor type-1 and -2 single channel properties by $InsP_3$, Ca^{2+} and ATP. J Physiol 590(14):3245–59, DOI: 10.1113/jphysiol.2012.228320 70, 71, 77, 78, 82

Wagner LE, Joseph SK, Yule DI (2008) Regulation of single inositol 1,4,5-trisphosphate receptor channel activity by protein kinase A phosphorylation. J Physiol 586(15):3577–96, DOI: 10.1113/jphysiol.2008.152314 277

Wagner LE 2nd, Groom LA, Dirksen RT, Yule DI (2014) Characterization of ryanodine receptor type 1 single channel activity using "on-nucleus" patch clamp. Cell Calcium 56(2):96–107, DOI: 10.1016/j.ceca.2014.05.004 89, 135

Walker MA, Williams GSB, Kohl T, Lehnart SE, Jafri MS, Greenstein JL, Lederer WJ, Winslow RL (2014) Superresolution modeling of calcium release in the heart. Biophys J 107(12):3009–20, DOI: 10.1016/j.bpj.2014.11.003 302, 302, 302, 303, 303

Wallach G, Lallouette J, Herzog N, De Pittà M, Ben Jacob E, Berry H, Hanein Y (2014) Glutamate mediated astrocytic filtering of neuronal activity. PLoS Comput Biol 10(12):e1003,964, DOI: 10.1371/journal.pcbi.1003964 294

Walter L, Hajnóczky G (2005) Mitochondria and endoplasmic reticulum: the lethal interorganelle cross-talk. J Bioenerg Biomembr 37(3):191–206, DOI: 10.1007/s10863-005-6600-x 52

Wang I, Politi AZ, Tania N, Bai Y, Sanderson MJ, Sneyd J (2008) A mathematical model of airway and pulmonary arteriole smooth muscle. Biophys J 94(6):2053–64, DOI: 10.1529/biophysj.107.113977 334, 336

Wang IY, Bai Y, Sanderson MJ, Sneyd J (2010) A mathematical analysis of agonist- and KCl-induced Ca^{2+} oscillations in mouse airway smooth muscle cells. Biophys J 98(7):1170–81, DOI: 10.1016/j.bpj.2009.12.4273 317, 319, 319, 321, 323, 324, 331, 334, 335, 336

Wang YX (ed) (2014) Calcium Signaling In Airway Smooth Muscle Cells. Springer, New York, DOI: 10.1007/978-3-319-01312-1 315

Watts M, Sherman A (2014) Modeling the pancreatic α-cell: dual mechanisms of glucose suppression of glucagon secretion. Biophys J 106(3):741–51, DOI: 10.1016/j.bpj.2013.11.4504 365, 368, 368, 371

Wechselberger M (2005) Existence and bifurcation of canards in R^3 in the case of a folded node. SIAM J Appl Dyn Syst 4(1):101–39, DOI: 10.1137/030601995 222, 225

Wei F, Shuai J (2011) Intercellular calcium waves in glial cells with bistable dynamics. Phys Biol 8(2):026,009, DOI: 10.1088/1478-3975/8/2/026009 149

Weiss J, Morgan P, Lutz M, Kenakin T (1996) The cubic ternary complex receptor-occupancy model I. Model description. J Theor Biol 178(2):151–67, DOI: 10.1006/jtbi.1996.0014 33, 33

Wier WG, Egan TM, López-López JR, Balke CW (1994) Local control of excitation-contraction coupling in rat heart cells. J Physiol 474(3):463–71, DOI: 10.1113/jphysiol.1994.sp020037 297, 298

Wierschem K, Bertram R (2004) Complex bursting in pancreatic islets: a potential glycolytic mechanism. J Theor Biol 228(4):513–21, DOI: 10.1016/j.jtbi.2004.02.022 358, 364

Williams GSB, Huertas MA, Sobie EA, Jafri MS, Smith GD (2007) A probability density approach to modeling local control of calcium-induced calcium release in cardiac myocytes. Biophys J 92(7):2311–28, DOI: 10.1529/biophysj.106.099861 306, 309, 310

Williams GSB, Molinelli EJ, Smith GD (2008) Modeling local and global intracellular calcium responses mediated by diffusely distributed inositol 1,4,5-trisphosphate receptors. J Theor Biol 253(1):170–88, DOI: 10.1016/j.jtbi.2008.02.040 310

Williams GSB, Smith GD, Sobie EA, Jafri MS (2010) Models of cardiac excitation-contraction coupling in ventricular myocytes. Math Biosci 226(1):1–15, DOI: 10.1016/j.mbs.2010.03.005 297, 297

Wiltgen SM, Smith IF, Parker I (2010) Superresolution localization of single functional IP_3R channels utilizing Ca^{2+} flux as a readout. Biophys J 99(2):437–46, DOI: 10.1016/j.bpj.2010.04.037 166

Wiltgen SM, Dickinson GD, Swaminathan D, Parker I (2014) Termination of calcium puffs and coupled closings of inositol trisphosphate receptor channels. Cell Calcium 56(3):157–68, DOI: http://dx.doi.org/10.1016/j.ceca.2014.06.005 169

Winfree A (1980) The Geometry of Biological Time. Springer-Verlag, Berlin, Heidelberg, New York 118

Wingrove DE, Gunter TE (1986) Kinetics of mitochondrial calcium transport. II. a kinetic description of the sodium-dependent calcium efflux mechanism of liver mitochondria and inhibition by ruthenium red and by tetraphenylphosphonium. J Biol Chem 261(32):15,166–71 56

Winslow RL, Rice J, Jafri S, Marbán E, O'Rourke B (1999) Mechanisms of altered excitation-contraction coupling in canine tachycardia-induced heart failure, II: model studies. Circ Res 84(5):571–86, DOI: 10.1161/01.res.84.5.571 300

Wong AY, Klassen GA (1993) A model of calcium regulation in smooth muscle cell. Cell Calcium 14(3):227–43, DOI: 10.1016/0143-4160(93)90070-m 327

Woods NM, Cuthbertson KS, Cobbold PH (1986) Repetitive transient rises in cytoplasmic free calcium in hormone-stimulated hepatocytes. Nature 319(6054):600–2, DOI: 10.1038/319600a0 258

Woods NM, Cuthbertson KS, Cobbold PH (1987) Agonist-induced oscillations in cytoplasmic free calcium concentration in single rat hepatocytes. Cell Calcium 8(1):79–100, DOI: 10.1016/0143-4160(87)90038-8 259

Worley JF 3rd, McIntyre MS, Spencer B, Dukes ID (1994) Depletion of intracellular Ca^{2+} stores activates a maitotoxin-sensitive nonselective cationic current in beta-cells. J Biol Chem 269(51):32,055–8 369

Yang J, Clark Jr JW, Bryan RM, Robertson C (2003) The myogenic response in isolated rat cerebrovascular arteries: smooth muscle cell model. Med Eng Phys 25(8):691–709, DOI: 10.1016/S1350-4533(03)00100-0 329

Yao Y, Parker I (1994) Ca^{2+} influx modulation of temporal and spatial patterns of inositol trisphosphate-mediated Ca^{2+} liberation in *Xenopus* oocytes. J Physiol 476(1):17–28 113, 246

Young KW, Nash MS, Challiss RAJ, Nahorski SR (2003) Role of Ca^{2+} feedback on single cell inositol 1,4,5-trisphosphate oscillations mediated by G-protein-coupled receptors. J Biol Chem 278(23):20,753–60, DOI: 10.1074/jbc.M211555200 287

Yu X, Carroll S, Rigaud JL, Inesi G (1993) H^+ countertransport and electrogenicity of the sarcoplasmic reticulum Ca^{2+} pump in reconstituted proteoliposomes. Biophys J 64(4):1232–42, DOI: 10.1016/S0006-3495(93)81489-9 43

Yule D, Lawrie A, Gallacher D (1991) Acetylcholine and cholecystokinin induce different patterns of oscillating calcium signals in pancreatic acinar cells. Cell Calcium 12:145–51, DOI: 10.1016/0143-4160(91)90016-8 25

Yule DI, Stuenkel E, Williams JA (1996) Intercellular calcium waves in rat pancreatic acini: mechanism of transmission. Am J Physiol 271(4):C1285–94 27

Yule DI, Straub SV, Bruce JIE (2003) Modulation of Ca^{2+} oscillations by phosphorylation of Ins(1,4,5)P_3 receptors. Biochem Soc Trans 31(5):954–7, DOI: 10.1042/ 277

Yule DI, Betzenhauser MJ, Joseph SK (2010) Linking structure to function: Recent lessons from inositol 1,4,5-trisphosphate receptor mutagenesis. Cell Calcium 47(6):469–79, DOI: 10.1016/j.ceca.2010.04.005 11

Zahradníková A, Zahradník I (1995) Description of modal gating of the cardiac calcium release channel in planar lipid membranes. Biophys J 69(5):1780–8, DOI: 10.1016/s0006-3495(95)80048-2 89

Zeller S, Rüdiger S, Engel H, Sneyd J, Warnecke G, Parker I, Falcke M (2009) Modeling of the modulation by buffers of Ca^{2+} release through clusters of IP_3 receptors. Biophys J 97(4):992–1002, DOI: 10.1016/j.bpj.2009.05.050 160, 161, 161

Zhabotinsky AM (2000) Bistability in the Ca^{2+}/calmodulin-dependent protein kinase-phosphatase system. Biophys J 79(5):2211–21, DOI: 10.1016/S0006-3495(00)76469-1 348, 352

Zhang J, Désilets M, Moon TW (1992) Evidence for the modulation of cell calcium by epinephrine in fish hepatocytes. Am J Physiol 263(3):E512–9 273

Zhang M, Goforth P, Bertram R, Sherman A, Satin L (2003) The Ca^{2+} dynamics of isolated mouse beta-cells and islets: implications for mathematical models. Biophys J 84(5):2852–70, DOI: 10.1016/S0006-3495(03)70014-9 232, 358

Zhang W, Kirk V, Sneyd J, Wechselberger M (2011) Changes in the criticality of Hopf bifurcations due to certain model reduction techniques in systems with multiple timescales. J Math Neurosci 1(1):9, DOI: 10.1186/2190-8567-1-9 216, 227

Zhang W, Krauskopf B, Kirk V (2012) How to find a codimension-one heteroclinic cycle between two periodic orbits. Discrete Cont Dyn S 32(8):2825–51, DOI: 10.3934/dcds.2012.32.2825 236

Zhu MX, Ma J, Parrington J, Calcraft PJ, Galione A, Evans AM (2010) Calcium signaling via two-pore channels: local or global, that is the question. Am J Physiol 298(3):C430–41, DOI: 10.1152/ajpcell.00475.2009 9

ZhuGe R, Bao R, Fogarty KE, Lifshitz LM (2010) Ca^{2+} sparks act as potent regulators of excitation-contraction coupling in airway smooth muscle. J Biol Chem 285(3):2203–10, DOI: 10.1074/jbc.M109.067546 317, 326
Zucker RS, Regehr WG (2002) Short-term synaptic plasticity. Ann Rev Physiol 64(1):355–405, DOI: 10.1146/annurev.physiol.64.092501.114547
339

Index

G. Dupont et al., *Models of Calcium Signalling*, Interdisciplinary Applied Mathematics 43, DOI 10.1007/978-3-319-29647-0

Printed in the United States
By Bookmasters